D0122276

Environmental Nanotechnology

Applications and Impacts of Nanomaterials

Editors

Mark R. Wiesner, Ph.D., P.E.

Jean-Yves Bottero, Ph.D.

New York Chicago San Francisco Lisbon London Madrid
Mexico City Milan New Delhi San Juan
Seoul Singapore Sydney Toronto

The McGraw·Hill Companies

Cataloging-in-Publication Data is on file with the Library of
Congress

McGraw-Hill books are available at special quantity discounts to use
as premiums and sales promotions, or for use in corporate training
programs. For more information, please write to the Director of Special
Sales, Professional Publishing, McGraw-Hill, Two Penn Plaza, New York,
NY 10121-2298. Or contact your local bookstore.

**Environmental Nanotechnology: Applications and Impacts
of Nanomaterials**

1 2 3 4 5 6 7 8 9 0 DOC/DOC 0 1 9 8 7

ISBN-13: 978-0-07-147750-5
ISBN-10: 0-07-147750-0

Sponsoring Editors
Kenneth P. McCoombs
Steve Chapman

Editorial Supervisor
Jody McKenzie

Project Manager
Vastavikta Sharma, International
Typesetting and Composition

Acquisitions Coordinator
Alexis Richard

Copy Editor
Lunaea Weatherstone

Indexer
Kevin Broccoli

Production Supervisor
George Anderson

Composition
International Typesetting and
Composition

Illustration
International Typesetting and
Composition

Art Director, Cover
Jett Weeks

Cover Design and Illustration
Stanislav Jourin

Contents

Part III Environmental Applications of Nanomaterials

Part IV Potential Impacts of Nanomaterials

Acknowledgments

Portions of the work presented in this book were supported by grants from the US National Science Foundation, the US Environmental Protection Agency, and the ECCO-Dyn program of France's CNRS-FNS. Support from the Office of Science and Technology of the French Consulate (Houston), and Rice's Environmental and Energy Systems Institute in organizing the symposia that led to this effort are also gratefully acknowledged.

About the Contributors

Mark R. Wiesner, Ph.D., is a professor of Environmental Engineering at Duke University where he holds the James L. Meriam Chair in Civil and Environmental Engineering. His work has focused on applications of emerging nanomaterials to membrane science and water treatment, and an examination of the fate, transport, and effects of nanomaterials in the environment. Before joining the Duke University faculty in 2006, he served on the Rice University faculty for 18 years in the Departments of Civil and Environmental Engineering and Chemical Engineering, and as director of the Environmental and Energy Systems Institute. He is a co-founder of the Houston-based nanomaterials company Oxane Materials. Dr. Wiesner holds a B.A. in Mathematics and Biology from Coe College, an M.S. in Civil and Environmental Engineering from the University of Iowa, a Ph.D. in Environmental Engineering from the Johns Hopkins University, and has completed postdoctoral training at the École Nationale Supérieure des Industries Chimiques (ENSIC). In 2004, Dr. Wiesner was named a "de Fermat Laureate" and was awarded an International Chair of Excellence at the French Polytechnic Institute in Toulouse.

Dr. Jean-Yves Bottero is a senior research director with France's Centre National de la Recherche Scientifique (CNRS), and director of France's Geoscience and Environment Lab (CEREGE) associated with the University of Paul Cézanne, Aix-Marseille. He also holds an appointment as Adjunct Professor at Duke University. His research addresses physico-chemical phenomena of surfaces and particles. His early work addressed the structure of materials used in water treatment at the nanometric scale, and most notably demonstrated for the first time the existence of the Al_{13} species that controls the chemistry of the now widely used "polyaluminum" coagulants. He has worked extensively on topics ranging from particle aggregation and membrane filtration to solid waste disposal and reuse. More recently, he has been a senior spokesman in Europe in advancing the agenda for research on possible environmental and health impacts of nanomaterials.

Wade Adams, Ph.D., is the director of the Richard E. Smalley Institute for Nanoscale Science and Technology at Rice University. Before heading the Smalley Institute, Dr. Adams was Chief Scientist of the Materials and Manufacturing Directorate, Air Force Research Laboratory, Wright-Patterson Air Force Base, Dayton, Ohio. Dr. Adams was educated at the US Air Force Academy, Vanderbilt University, and the University of Massachusetts. For the past 33 years he has conducted research in polymer physics, concentrating on structure-property relations in high-performance organic materials. He is a Fellow of the American Physical Society and the Air Force Research Laboratory. Dr. Adams retired from the Air Force Reserve in the rank of colonel in 1998.

Pedro J.J. Alvarez, Ph.D., is the George R. Brown Professor and Chair of Civil and Environmental Engineering at Rice University. He received a degree in Civil Engineering from McGill University and M.S. and Ph.D. degrees in Environmental Engineering from the University of Michigan. He is a diplomate of the American Academy of Environmental Engineers and a Fellow of the American Society of Civil Engineers. Honors include being elected president of the Association of Environmental Engineering and Science Professors (2005–2006), the cleanup project of the year award from SERDP (2002), and the Button of the City of Valencia (2000).

Dr. Mélanie Auffan holds a doctoral degree from the University of Paul Cézanne in Aix-Marseille where she performed research at the Geoscience and Environment Lab (CEREGE) on the transport and transformation of manufactured nanoparticles in the environment. Her work addresses cellular interactions of mineral nanoparticle and the use of iron nanoparticles for the adsorption of arsenic.

Andrew R. Barron, Ph.D., is the Charles W. Duncan, Jr.–Welch Chair of Chemistry, Professor of Materials Science, and Associate Dean of Industry Interactions and Technology Transfer at Rice University. Prior to moving to Rice University in 1995 he spent eight years on the faculty at Harvard University. He received his Ph.D. from the Imperial College of Science and Technology, University of London, and served as a post-doctoral research associate at the University of Texas. Dr. Barron currently sits on the editorial boards of three chemistry and materials science journals, is a Fellow of the Royal Society of Chemistry, and is the 1997 recipient of the Humboldt Senior Service Award.

Jonathan Brant, Ph.D., is a research associate at Duke University in the Department of Civil and Environmental Engineering. He obtained a Ph.D. in environmental engineering at the University of Nevada, Reno, where his research focused on the characterization of interfacial interactions between water-treatment membranes and organic and inorganic materials with the purpose of reducing fouling. Upon completion of his Ph.D. he completed a two-year postdoctoral research assignment at Rice University where he studied the behavior of fullerene nanomaterials in environmental systems. His areas of research focus on characterizing surfaces to predict and understand the impact of materials in environmental processes.

Michael Hoffmann, Ph.D., received a B.A. in chemistry from Northwestern University and a Ph.D. in chemical kinetics from Brown University. In 1973, he was awarded an NIH postdoctoral training fellowship in Environmental Engineering Science at the California Institute of Technology. From 1975 to 1980, he was member of the faculty at the University of Minnesota and since 1980 a member of the faculty at Caltech (Engineering and Applied Science). Dr. Hoffmann has published more than 220 peer-reviewed professional papers and is the holder of seven patents. In 2001, Dr. Hoffmann was presented with the American Chemical Society Award for Creative Advances in Environmental Science and Technology and received the Jack E. McKee Medal for Groundwater Protection by the Water Environment Federation in October 2003.

Ernest (Matt) Hotze is a doctoral candidate at Duke University where he is performing research on Reactive Oxygen Production by nanoparticles. He holds an M.S. in Environmental Engineering from Rice University and a B.S. in Chemistry from Notre Dame.

Amy Myers Jaffe is the Wallace S. Wilson Fellow in Energy Studies at the James A. Baker III Institute for Public Policy and associate director of the Rice University energy program. Her research focuses on the subject of oil geopolitics, strategic energy policy including energy science policy and energy economics. Ms. Jaffe is widely published in academic journals and numerous book volumes and served as coeditor of *Energy in the Caspian Region: Present and Future* (Palgrave, 2002) and *Natural Gas and Geopolitics: From 1970 to 2040* (Cambridge University Press, 2006). She served as a member of the reconstruction and economy working group of the Baker/Hamilton Iraq Study Group, and as project director for the Baker Institute/Council on Foreign Relations task force on Strategic Energy Policy.

Dr. Jean-Pierre Jolivet is a professor at the Université Pierre et Marie Curie (Paris 6), in Paris, France, where he teaches inorganic chemistry. His research activities are focused on the synthesis of metal and metal oxide nanoparticles with controlled characteristics (crystalline structure, morphology, size, and dispersion state in various solid or liquid media) for various application areas such as optics, electrochemistry, catalysis, and nanomagnetism. He is equally involved in research addressing the environmental impact of nanomaterials. He is the author of a widely cited book, *Metal Oxide Chemistry and Synthesis: From Solution to Solid State.*

Dr. Jérôme Labille is a researcher with the French National Research Center (CNRS) at the Geosciences and Environment Lab (CEREGE). He obtained a Ph.D. in physical chemistry of Geosciences in the French Institute INPL and did subsequent postdoctoral work at the Analytical Center for Biophysicochemistry of the Environment (CABE) in Geneva. He has been working at CEREGE for three years on the environmental fate of manufactured nanoparticles, considering and characterizing the numerous conditions that control their bioavailability and toxicity, such as surface reactivity, colloidal dispersion, and interaction with organics or pollutants.

Gregory V. Lowry, Ph.D., is an associate professor of Environmental Engineering at Carnegie Mellon University in Pittsburgh, Pennsylvania. His research interests are broadly defined as transport and reaction in porous media, with a focus on the fundamental physical/geochemical processes affecting the fate of inorganic and synthetic organic contaminants and engineered nanomaterials in the environment. He is an experimentalist and works on a variety of application-oriented research projects developing novel environmental technologies for restoring contaminated sediments and groundwater, including reactive nanoparticles for efficient *in situ* remediation of entrapped NAPL and innovative sediment caps for *in situ* treatment and management of PCB-contaminated sediments.

Delina Y. Lyon is a doctoral student in the Civil and Environmental Engineering Department at Rice University. She received a B.A. from St. Mary's College of Maryland and an M.S. in Microbiology from the University of Georgia.

Nancy Ann Monteiro-Riviere, Ph.D., is a Professor of Investigative Dermatology and Toxicology at the Center for Chemical Toxicology Research and Pharmacokinetics, North Carolina State University (NCSU), a professor in the Joint NCSU/UNC-Chapel Hill Biomedical Engineering Faculty, and Research Adjunct Professor of Dermatology at the School of Medicine at UNC Chapel Hill. Dr. Monteiro-Riviere received an M.S. and Ph.D. from Purdue University. She completed postdoctoral training at Chemical Industry Institute of Toxicology in Research Triangle Park, North Carolina. Dr. Monteiro-Riviere was president of the Dermal Toxicology and In Vitro Toxicology Specialty Sections of the National Society of Toxicology and currently serves as chairperson of the Board of Publications. She is a Fellow in the Academy of Toxicological Sciences and in the American College of Toxicology.

André E. Nel, M.D., Ph.D. is a Professor of Medicine and Chief of the Division of NanoMedicine at UCLA. He runs the Cellular Immunology Activation Laboratory in the Johnson Cancer Center at UCLA. Dr. Nel obtained his M.B., Ch.B. (M.D.), and Doctorate of Medicine (Ph.D. equivalent) degrees from the University of Stellenbosch in Cape Town, South Africa, and subsequently did Clinical Immunology and Allergy training at UCLA. Dr. Nel is the principal investigator of the UCLA Asthma and Immunology Disease Center, codirector of the Southern California Particle Center, and director of the University of California Nanotoxicology Research and Training program. He served as chair of the Allergy Immunology Transplant Research Committee at the NIAID and is chair of the Air Pollution Committee in the AAAAI. Dr. Nel is a member of the ASCI, AAAAI, AAI, and the Western Association of Physicians.

Dr. Christine Ogilvie Robichaud is a doctoral candidate at Duke University where she is engaged in research on assessing life-cycle risks of nanomaterials, targeting use in energy technologies. Ms. Robichaud holds an M.S. in Environmental Analysis and Decision Making from Rice University, and a B.S. in Industrial Engineering from Texas A&M University. Prior to graduate school she worked in energy supply chain consulting and in the biofuels industry.

Dr. Thierry Orsiére has been a Research Scientist at the Université de la Méditerranée (Faculty of Medicine) since 1996. He obtained a Master's degree in Biochemistry from the Université de Provence (Aix-Marseille) and a doctorate of Pharmacology from the Université de la Méditerranée (Aix-Marseille). His research has focused on the effects of DNA damage and changes in processes governing cell division on human cells. His work has included studies of worker exposure to mutagens, determination of the ability of contaminants to induce chromosome aberrations, and the genotoxic properties of mineral nanoparticles.

Dr. Jérôme Rose is a senior scientist at the CEREGE (CNRS) since 1997 and serves as adjunct faculty at Rice University and Columbia University. He obtained an Engineering degree in geosciences and a doctorate from the Lorraine National Polytechnic Institute (France). He has supervised twelve Ph.D. students and two postdoctoral researchers. Dr. Rose was the 2006 recipient of the bronze medal from the CNRS. His research focuses on the behavior and toxicity of colloids and contaminants from laboratory to field scale. He is employing intensively synchrotron-based techniques to study mechanisms at a molecular level.

Dr. Rose has been involved in research on the environmental impact of nanotechnology since at least 2001 and is one of the inventors of the ferroxane nanoparticles.

Heather J. Shipley is a Ph.D. candidate at Rice University working on arsenic adsorption with iron oxide nanoparticles. She also has done research with the Brine Chemistry Consortium at Rice University on iron sulfides and inhibitor adsorption. Previously, she conducted research with the Hazardous Substance Research Center South/Southwest on the resuspension of sediments to predict the amount of metals that can become available. Ms. Shipley holds a B.S. degree in chemistry from Baylor University, Waco, Texas. and an M.S. in Environmental Engineering from Rice University.

Dicksen Tanzil, Ph.D., is a sustainability specialist at Golder Associates Inc. He received his doctorate from Rice University and B.S. from Purdue University, both in chemical engineering. His work focuses on the assessment of environmental and social impacts of industrial operations, and the incorporation of sustainability considerations in engineering design and business decision-making. He is co-editor of the book *Transforming Sustainability Strategy into Action: The Chemical Industry*, published in 2005.

Dr. Antoine Thill is a researcher at the Commissariat de l'Energy Atomique in Saclay (Paris) where he is in charge of the Ultra Small Angle X-ray Scattering facility at the Laboratoire Interdisciplinaire sur l'Organisation Nanométrique et Supramoléculaire in the Department of Material Science. He holds a doctorate in Geosciences and Engineering from the University of Aix-Marseille III where he studied the aggregation of natural suspended matter in estuaries and developed methods to characterize particle size and agglomeration states by light scattering and confocal microscopy. His work addresses the structure and dynamics of nanoparticles in complex samples through an intensive use of scattering techniques as well as a consideration of properties of these particles as they affect bacterial toxicity of CeO_2 and other nanoparticles.

Joanne I. Yeh, Ph.D., obtained her Ph.D. in 1994 from the Chemistry Department at the University of California, Berkeley, where she studied macromolecular X-ray crystallography as a NSF predoctoral fellow under the supervision of Professor Sung-Hou Kim. Dr. Yeh was a NIH post doctoral fellow with Professor Wim G.J. Hol at the University of Washington/Howard Hughes Medical Institute, studying through structural characterization soluble and membrane proteins involved in oxidative and glycerol metabolism pathways.

Nanotechnology as a Tool for Sustainability

Nanotechnology and the Environment

Mark R. Wiesner   *Duke University, Durham, NC*
Jean-Yves Bottero   *CNRS-University of Aix-Marseille, Aix-en-Provence, France*

Advances in information technologies, materials science, biotechnology, energy engineering, and many other disciplines—including environmental engineering—are converging at the quantum and molecular scales. This molecular terrain is common ground for interdisciplinary research and education that will be an essential component of science and engineering in the future. Much like the digital computer and its impact on science and technology in the 20th century, the tools that serve as portals to the molecular realm will act as both instruments of discovery and rallying points for social interaction between researchers from many disciplines. In this setting, environmental engineers and scientists will take on new roles in collaborating with materials scientists, molecular biologists, chemists, and others to address the challenges of meeting society's needs for energy and materials in an environmentally responsible fashion.

Nanotechnology is defined as a branch of engineering that deals with creating objects smaller than 100 nm in dimension. Behind this definition is a vision of building objects atom by atom, molecule by molecule [1] by self-assembly or molecular assemblers [2]. Activities spawned by a "nanomotivated" interdisciplinarity will affect the social, economic, and environmental dimensions of our world, often in ways that are entirely unanticipated. We focus here on the potential impacts of nanomaterials on human health and environment. Many of these impacts will be beneficial. In addition to a myriad of developments in medical science, there is considerable effort underway to explore uses of nanomaterials in applications such as membrane separations, catalysis, adsorption, and analysis with the goal of better protecting environmental quality.

However, along with these innovations and the growth of a supporting nanomaterials industry, there is also the need to consider impacts of nanomaterials on environment and human health. Past technological accomplishments such as the development of nuclear power, genetically modified organisms, information technologies and synthetic organic chemistry have generated public cynicism as some of the consequences of these technologies, often environmental, become apparent. Even potable water disinfection, the single most important technological advance with regard to prolonging human life expectancy, has been found to produce carcinogenic by-products. Some groups have called on industry and governments to employ the precautionary principle while conducting more research in toxicology and transport behaviors [3, 4]. The precautionary principle, often associated with the Western European approach to regulation, might be summarized as "no data, no market." In contrast, the risk-based approach that has come to typify regulatory development in the United States, might be reduced to the philosophy of "no data, no regulation." Both approaches require reliable data. Although studies are beginning to appear in the literature addressing the toxicity of various nanomaterials [5–10] and their potential for exposure [11, 12], at this stage definitive statements regarding the impacts of nanomaterials on human health and the environment remain sketchy.

In this book, we consider the topic of nanomaterials through the lens of environmental engineering. A key premise of our approach is that the nanomaterials industry is an emerging case study on the design of an industry as an environmentally beneficial system throughout the life cycle of materials production, use, disposal, and reuse. One element of this socio-industrial design process is an expansion of the training and practice of environmental engineering to include concepts of energy and materials production and use into environmental engineering education and research.

Nanoconvergence and Environmental Engineering

Environmental engineering evolved from an interdisciplinary approach to solving water quality problems that traces its origins to the latter part of the 19th century. The concept of bringing microbiologists, stream ecologists, traditional civil engineers, and aquatic chemists together to resolve problems with dissolved oxygen in surface water originating from waste discharges was revolutionary in its time. This interdisciplinary model for what was first called sanitary engineering, was expanded during the last century to include the study of impacts of human activities on air, soils, groundwater, and other environmental media. Today, the standard model for teaching and research in an environmental engineering program covers bases in the areas of air, water, and soil with crosscutting areas of expertise such as fluid mechanics, chemistry, microbiology, physical chemical processes, and, occasionally, policy. Some programs additionally also include a consideration of impacts on human health and ecosystems.

It is increasingly clear that long-range solutions to environmental problems must also address "upstream" issues associated with the inputs of energy and materials to our society. The recent concern with potential environmental impacts of nanomaterials is one manifestation of the need for such an approach. Students of environmental engineering must be able to address the impacts of energy and materials as they are used by our society on downstream receptors such as air, water, soil, human health, and ecosystems. Our energy systems and systems for producing, using, and disposing of materials must be designed with a view toward the full life-cycle effects of these systems. Environmental engineers should be literate in processes for energy generation, raw materials acquisition, and materials processing that will allow them to work with other engineers and scientists to design systems in this life-cycle context.

The urgent demands for interdisciplinary solutions to environmental problems that stem from materials and energy use coincide with a convergence of disciplines at the nanometric scale. In contrast with a view of interdisciplinarity that entails an ever-expanding grouping of overlapping disciplines from the top down, convergence at the nanoscale is a vision of interdisciplinarity from the bottom up that can be particularly powerful as a basis for rigorous science and engineering education. Nowhere is such an approach more appropriate than in the environmental arena, where problem solving draws on principles from multiple disciplines, including biology, chemistry, physics, and information science.

Science is social. From the office water cooler and coffee machine to the conferences we organize, great effort is placed in the scientific community on creating conditions that nurture social interactions between researchers, facilitating the improvisation and collective creativity that are built on an exchange of ideas. For the time being, advanced instrumentation is the coffee machine of this nanoscale interdisciplinarity as well as the portal to the molecular realm. This will change with time as these techniques become more commonplace (just as the use and availability of digital computers have become widespread). However, the current need for researchers to pass through

the same analytical doors of perception on their way to studying phenomena at the scale of atoms and molecules inevitably produces interactions between individuals who are motivated by widely different
problems and applications. To participate in this dialogue, environmental engineers must be conversant in the language of these enabling
analytical technologies.

Origin and Organization of This Book

The origin of this book was a dialogue between the Wiesner and Bottero
groups that began when Wiesner was on sabbatical from Rice University
in 1998 at the Centre Européen de Recherche et d'Enseignement des
Géosciences de l'Environnement (CEREGE) in Aix-en-Provence. There
was great interest in nanotechnology on the Rice University campus,
where the first nanotechnology institute in the world—today the
Smalley Institute—was established in 1993. The physical-chemical
processes group at the CEREGE had for many years taken an atomic-
scale approach to addressing water treatment and solid waste issues,
employing many of the tools that enable nanochemistry research. We
perceived a need for introducing a materials science dimension to environmental engineering curricula, and this need resonated with a growing interest in nanotechnology. Although efforts were being made at
that time to import newly developed nanomaterials to environmental
applications, apart from futuristic scenarios of "gray-goo"–producing
nanobots [13], the environmental impacts of nanomaterials produced in
the near term had not yet been addressed.

With financial support from the Office of Science and Technology of
the French Consulate, in December 2001 we organized the first-ever
public forum addressing the environmental implications of nanotechnologies. This event, sponsored by the Environmental and Energy
Systems Institute at Rice, brought together nanochemistry and environmental researchers from various laboratories in France and from
Rice to speculate on health and environmental impacts of nanomaterials. The repercussions from this event have been widespread and
included the launch of a newly minted NSF center (CBEN) and the
creation of an international consortium of researchers addressing the
impacts of nanomaterials (I-CENTR). Four years of collaboration later,
and with a substantially enlarged community of researchers around the
world engaged in research in this area, we reconvened in December
2005 to consider the progress made in applying nanomaterials to environmental technologies and in understanding the possible impacts of
these materials on health and the environment. The outline of this
book follows the agenda of this second symposium. This agenda can be
presented in the context of the information needed for risk assessment

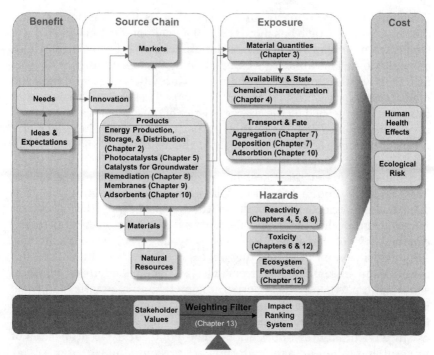

Figure 1.1   Elements of environmental risk assessment focusing on the use of nanomaterials in environmental technologies.

and risk management (Figure 1.1). Considerations in integrating this information to arrive at a risk assessment for a given nanomaterial are presented in Chapter 13.

Properties and principles of nanomaterials

Nanoscale particles have long been present in environmental systems and play a significant role in the function of many natural processes. Examples of naturally occurring nanoscale particles include those resulting from mineral weathering (e.g., iron oxides and silicates) and emissions from combustion processes (i.e., carbon soot). Recent advances in the area of nanoscience have provided a means to characterize and manipulate naturally occurring nanomaterials as well as to manufacture engineered nanoparticles (e.g., metal oxides, carbon nanotubes, and buckminsterfullerene). It is now possible to control the chemical and physical properties of nanoparticles and to tailor them for specific applications. In Chapter 3, methods for fabricating nanomaterials and some of the unique properties of these materials are presented. Methods for structural and chemical characterization are presented in Chapter 4.

Nanoparticles are nearly all surface. As an approximation, a 4 nm diameter solid particle has more than 50 percent of its atoms at surface. One gram of single-walled carbon nanotubes (SWNTs) has over 10 m^2 of available surface area. This results in nanoparticles being highly surface reactive and implies that their behavior will, to a great degree, be mediated by interfacial chemical interactions. These interactions should therefore be governed by the characteristics of these surfaces as reflected in adsorption energies, the change in the surface energy heterogeneity due to the change of lattice parameters, the distortion of the bonds, and the increase of the surface pressure.

Because atoms at interfaces behave differently, nanomaterials are likely to have unique properties compared to larger bulk materials of the same general composition. Also, as the size of particles of a given material approaches the nanoscale, material properties such as electrical conductivity, color, strength, and reactivity may change. These changes are in turn related to underlying effects of size that include quantum confinement in semiconductor particles, surface plasmon resonance in some metal particles, and superparamagnetism in magnetic materials. Greater reactivity, and the ability of some nanoparticles to act as electron shuttles or, in other cases, as photocatalysts, holds particular interest in environmental applications. The photocatalytic properties of mineral and fullerene nanomaterials are presented in Chapter 5.

The ability of some nanomaterials to photocatalytically produce reactive oxygen species (ROS) that may be used to oxidize contaminants or inactivate microorganisms may also present a risk of toxicity to organisms. A methodology for assessing nanoparticle toxicity is presented in Chapter 6. While the toxicity of some nanomaterials may be related to ROS production, there may be other possible mechanisms as discussed in Chapters 11 and 12.

For nanomaterials to present a risk, there must be both a hazard, such as toxicity, and potential for exposure to these materials. The interfacial chemical properties of nanoparticles in aqueous media affect particle aggregation and deposition processes that in turn affect exposure. Nanoparticle stability and transport are important in determining whether these materials can be removed by water treatment technologies, or whether nanoparticles have a high potential for mobility in the environment. The mobility of submicron particles and the factors that control particle transport, aggregation, and deposition in aqueous systems have been explored extensively, particularly for cases such as silica and latex suspensions [14–16]. However, despite a large number of publications describing procedures for producing nanoparticles of specific size, shape, and composition displaying unique properties of reactivity and mobility compared with better-known bulk phases, there has been little theoretical consideration of the special properties of nanoparticles that might affect their potential for aggregation [17] and little evaluation of the transport properties of these new materials in aqueous systems. In Chapter 7, the chemistry of nanoparticles is examined from a colloid science perspective, considering factors that may be different at the nanoscale in determining the nanoparticle stability and transport.

Environmental applications of nanomaterials

The products of nanochemistry are being been used to create new generations of technologies for curing environmental maladies and protecting public health. Water pollution control, groundwater remediation, potable water treatment, and air quality control are being advanced through nanomaterial-based membrane technologies, adsorbents, and catalysts. The use of nanomaterials as photocatalysts is discussed in Chapter 5. Nanomaterials are also inspiring new solutions for providing and using energy that are more environmentally neutral than conventional approaches, as introduced in Chapter 2. Advances in fuel cells, photovoltaics, and electrical transmission, as well as solutions for managing air and water pollution generated by fossil fuels, have been enabled through developments in new materials.

Groundwater remediation, water treatment, and fuel cell development are among the applications in which nanomaterials are finding their way into environmental engineering practice. The remediation of contaminated groundwater is a costly problem that has been approached in environmental engineering using both pump-and-treat and *in situ* technologies. In most cases, pumping contaminated groundwater to the surface to remove contaminants and reinjecting the treated water has proven to be both cost-prohibitive and incapable of meeting cleanup goals. As a result, *in situ* treatments such as biodegradation have been explored extensively. Physical chemical approaches to *in situ* treatment have included the use of zero valent iron and catalysts to promote redox reactions that degrade contaminants. Nanomaterials have been developed to promote such reactions at high rates; however, successful application of this technology will require a high degree of control of nanoparticle mobility, reactivity, and ideally, specificity for the contaminant of interest. Background on groundwater remediation and the development of nanomaterials for *in situ* treatment is presented in Chapter 8.

Membrane technologies are playing an increasingly important role as unit operations for environmental quality control, resource recovery, pollution prevention, energy production, and environmental monitoring. In water treatment they can be used for a wide spectrum of applications, ranging from particle removal to organic removal and desalination. Membranes are also at the heart of fuel cell technologies. Fuel cells can be viewed as vehicles for both electricity generation and water fabrication. Nanoscale control of membrane architecture may yield membranes of greater selectivity and lower cost in both water treatment and water fabrication. Principles of membrane processes are summarized in Chapter 9, and examples of the use of nanomaterials to create new membrane systems are described.

The ability to tailor surfaces for reactivity suggests that nanomaterials may find considerable use as adsorbents. Polymeric nanomaterials can be adapted to molecular templating methods to create adsorption sites that recognize a specific contaminant. Nanobio conjugates such as antibodies attached to mineral nanoparticles have been made to complex specific contaminants and also allow for quantitative measurements. While nanobio conjugates have been focused largely on analytical techniques and medical applications, it is possible to imagine adsorbent/measurement systems that allow adsorbency for a targeted contaminant to be optimally dosed as a function of the level of contaminant in the water.

Stimulated in part by research on health effects from low levels of arsenic exposure and changes in drinking water standards, there is an extensive body of literature growing on the use of various iron nanoparticles for arsenic adsorption. Nanostructured ferric oxides (derived in some cases from ferroxanes), magnetite (for subsequent magnetic separation), maghemite, mackinawite, and other forms of iron have been investigated. These materials can be easily integrated into existing treatment trains and the most promising materials are likely to see widespread use in potable water treatment. Chapter 10 focuses on the development of iron-based nanomaterials as adsorbents for contaminants such as arsenic.

Potential impacts of nanomaterials on organisms and ecosystems

Possible risks associated with nanomaterial exposure may arise during nanomaterial fabrication, handling of nanomaterials in subsequent processing to create derivative products, product usage, and as the result of postusage or waste disposal practices. The quantities of nanomaterials produced per year are large and increasing rapidly. Current production capacities for C_{60} fullerene are in excess of 2,000 tons per year, while the carbon nanotube production capacity in 2006 is in the hundreds of tons per year. These volumes are small compared with the production of more conventional TiO_2 nanoparticles, silica nanoparticles and other materials with a longer history of commercialization. This level of production, fueled by growing markets for products that incorporate these materials, will inevitably lead to the appearance of nanomaterials in air, water, soils, and organisms.

Ironically, the properties of nanomaterials that may create concern in terms of environmental impact (such as nanoparticle uptake by cells) are often precisely the properties desired for beneficial uses, such as in medical applications. While it may be desirable from the standpoint of medical therapeutics to have nanoparticles readily taken up by cells, this same trait may also have negative implications in an environmental context. Early efforts addressing the toxicity of nanomaterials have sometimes yielded results that may be difficult to interpret or even contradictory, due to evolving methodologies and limited resources. Studies surrounding the possible health effects of a class of carbon-based nanomaterials referred to as fullerenes serve as one example. Medical studies dating as far back as ten years report that the soccer ball–shaped fullerene molecules known as "buckyballs" are powerful antioxidants, comparable in strength to vitamin E, while other studies report that some types of buckyballs can be toxic to tumor cells, cleave DNA, and inhibit bacterial growth [18–23].

Exposure via inhalation has been an important concern, particularly in the context of nanomaterials fabrication. Some studies have attempted to simulate the inhalation of carbon nanotubes, exploring the possibility that these nanomaterials might damage lungs, as has been observed with other particles such as silica and asbestos. Although carbon nanotubes tend to form large aggregates that may have little opportunity for transport into the lungs, smaller concentrations of material might be present in air during fabrication that might not encounter each other enough to cluster together. Experiments have been conducted wherein nanotubes were introduced by washing the lungs of laboratory animals with solutions containing the carbon nanotubes. It is not surprising that aggregation within the lungs and subsequent suffocation appear to be the major risks presented in this mode of exposure [7]. Dermal exposure is also likely to be important for assessing the risks of both nanomaterial fabrication and product use. Work on the toxicity of nanomaterials focused on possible human exposure is reviewed in Chapter 11.

Ecotoxicological studies of nanomaterials are evolving in parallel with work on toxicity to human cells. Much of the earliest work focused on fullerene-based nanomaterials. However, studies concluding that buckyballs could impair brain functions in fish [8] and were highly toxic to human tissue cultures [24] are difficult to interpret, in part because reproducibility in characteristics of the materials used appears to be elusive, and because the nanomaterials used in these studies were contaminated with an organic solvent added as part of a process to mobilize the fullerenes in water. A subsequent study of fullerene toxicity conducted by another group of researchers [25] found no significant toxicity for buckyballs, but did observe a toxic response of cell cultures to a second group of fullerenes, single-wall nanotubes. Very little is known about interactions between fullerenes and microbial populations. This is a critical knowledge gap not only because of the potential to impact microorganisms themselves, but also because these organisms serve as the basis of the food chain and as primary agents for global biogeochemical cycles. Ecotoxicological aspects of nanomaterials are discussed in Chapter 12.

The matter of determining whether a substance is "dangerous" involves not only determining the material's toxicity or hazard, but also its exposure or probability of coming in contact with an organism. When materials are persistent and resist degradation, they may remain in the environment for long periods of time, contributing to increased chances of exposure. Persistent materials have a greater chance of interacting with the living environment. The roles of particle deposition and aggregation in determining nanomaterials' persistence in the environment are discussed in Chapter 7. The extent to which nanomaterials may be degraded and the conditions that may lead to their breakdown, including metabolism and degradation by bacteria, remain virtually unknown. While it is possible that bacteria may accelerate the dissolution of mineral nanoparticles through, for example, their impact on redox conditions, little work has been done examining the interactions between bacteria and these materials. Early efforts exploring biodegradation of the carbon-based fullerenes revealed that, at best, these materials will be difficult to degrade and therefore should be scrutinized for their persistence in the environment.

Chapter 13 addresses the framework for assessing the risks of nanomaterials, and why such a framework may diverge from the methods of assessing conventional materials. There is some temptation to use the information known about bulk materials of similar composition to describe nanomaterials in assessing risk. For instance, the material safety data sheet (MSDS) for carbon black has been applied for C_{60} since they both consist of only carbon atoms. Using such data for the bulk-phase format of a material may be a useful exercise as a first step if the results are interpreted with the appropriate degree of skepticism. However, the novel properties of these materials must be incorporated into risk models if, as we suspect, they are found to change their interactions and effects.

While much more study will be needed to assess the true risks to health and the environment posed by nanomaterials, there is an immediate and urgent need for information assessing the risks of producing these materials.

Indeed, many of the materials used to make nanomaterials are currently known to present a risk to human health in their own right. In one recent study [26], the procedures for making nanomaterials were assessed using methods employed by the insurance industry to quantify risk and estimate premiums for chemical manufacturers. An encouraging trend we have observed is that the methods for producing nanomaterials often appear to become "greener" as they move from laboratory to industrial production. Moreover, setting aside the highly speculative issue of nanomaterials toxicity, these preliminary results suggest that the fabrication of nanomaterials entails risks that are less than or comparable to those associated with many current industrial activities, such as computer manufacturing.

It would be naïve to imagine that the emerging nanomaterials industry will not leave unforeseen and undesirable traces on our environment. To minimize these impacts, environmental engineers and scientists should be actively engaged in ensuring that this industry is conceived from the outset as a system with global consequences that begin at the molecular scale. However, they must also be positioned to apply convergent knowledge at the molecular scale to respond to some of the world's most pressing problems. Dysentery resulting from a lack of access to sanitation facilities and potable water remains the single largest killer on our planet. Agricultural productivity will continue to be limited in many regions by water availability.

Water and sanitation are closely linked with access to a sustainable supply of energy, while energy production and use have proven to have major consequences for human health and the environment. Today, those of us living in highly industrialized nations account for less than 13 percent of the world's population and half of the world's energy consumption. More than a quarter of the world's population has no access to electricity today, and even more must rely on firewood, animal waste, and other sources of combustible biomass for basic cooking and heating needs. In both the developed and developing nations, the pollution generated in the combustion of oil, gas, and biomass pose numerous threats to human health and the environment in the form of indoor air pollution, acidification of water and soil, and global warming, to name a few. The convergence of disciplines at the nanoscale creates opportunities for discovering new solutions to these problems as well as the challenge of ensuring the cure is not worse than the disease.

References

1. Feynman, R., *There's plenty of room at the bottom, speech presented at the annual meeting of the American Physical Society, California Institute of Technology, December 29, 1959.*
2. Drexler, K.E., *Molecular engineering: An approach to the development of general capabilities for molecular manipulation.* Proceedings of the National Academy of Sciences USA, 1981. 78:5275–5278.
3. Commission, E., *Nanotechnologies: A preliminary risk analysis on the basis of a workshop March 2004* in *Nanotechnologies: A Preliminary Risk Analysis*, R.A.U.P.H.a.R.A. Directorate, Editor. 2004, European Commission Community Health and Consumer Protection Directorate General of the European Commission: Brussels, pp. 11–29.

4. Commission, E., *Opinion on the appropriateness of existing methodologies to assess the potential risks associated with engineered and adventitious products of nanotechnologies.* 2005, Scientific Committee on Emerging and Newly Identified Health Risks, pp. 41–58.

5. Limbach, L.K., et al., *Oxide nanoparticle uptake in human lung fibroblasts: Effects of particle size, agglomeration, and diffusion at low concentrations.* Environmental Science and Technology, 2005. 39(23):9370–9376.

6. Lam, C.W., et al., *Pulmonary toxicity of single-wall carbon nanotubes in mice 7 and 90 days after intratracheal installation.* Toxicological Sciences, 2004. 77:126–134.

7. Warheit, D.B., et al., *Comparative pulmonary toxicity assessment of single-wall carbon nanotubes in rates.* Toxicological Sciences, 2004. 77:117–125.

8. Oberdörster, E., *Manufactured nanomaterials (fullerenes, C_{60}) induce oxidative stress in the brain of juvenile largemouth bass.* Environmental Health Perspectives, 2004. 112(10):1058–1062.

9. Xia, T., et al., *Comparison of the abilities of ambient and manufactured nanoparticles to induce cellular toxicity according to an oxidative stress paradigm.* Nano Letters, 2006. 6(8):1794–1807.

10. Auffan, M., et al., *In vitro interactions between DMSA-coated maghemite nanoparticles and human fibroblasts: A physicochemical and cyto-genotoxical study.* Environmental Science and Technology, 2006. 40(14):4367–4373.

11. Lecoanet, H., J. Bottero, and M. Wiesner, *Laboratory assessment of the mobility of nanomaterials in porous media.* Environmental Science and Technology, 2004. 38(16):4377–4382.

12. Lecoanet, H., and M. Wiesner, *Velocity effects on fullerene and oxide nanoparticle deposition in porous media.* Environmental Science and Technology, 2004. 38(16):4377–4382.

13. Drexler, K.E., *Engines of destruction,* in *engines of creation: The Coming Era of Nanotechnology.* 1986, Anchor Books: New York, NY.

14. Elimelech, M., and C.R. O'Melia, *Effect of particle size on collision efficiency in the deposition of brownian particles with electrostatic energy barriers.* Langmuir, 1990. 6:1153–1163.

15. Tobiason, J.E., and C.R. O'Melia, *Physicochemical aspects of particle removal in depth filtration.* Journal of American Water Works Association, 1988. 80(12):54–64.

16. Veerapaneni, S., and M.R. Wiesner, *Role of suspension polydispersivity in granular media filtration.* Journal of Environmental Engineers, ASCE, 1993. 119(1):172–190.

17. Kallay, N., and S. Zalac, *Stability of nanodispersions: A model for kinetics of aggregation of nanoparticles.* Journal of Colloid and Interface Science, 2002. 253:70–76.

18. Kasermann, F., and C. Kempf, *Buckminsterfullerene and photodynamic inactivation of viruses.* Reviews in Medical Virology, 1998. 8(3):143–151.

19. Nakamura, E., et al., *Biological activity of water-soluble fullerenes. Structural dependence of DNA cleavage, cytotoxicity, and enzyme inhibitory activities including HIV-protease inhibition.* Bulletin of the Chemical Society of Japan, 1996. 69(8):2143–2151.

20. Tabata, Y., Y. Murakami, and Y. Ikada, *Antitumor effect of poly(ethylene glycol) modified fullerene.* Fullerene Science and Technology, 1997. 5(5):989–1007.

21. Tabata, Y., Y. Murakami, and Y. Ikada, *Photodynamic effect of polyethylene glycol-modified fullerene on tumor.* Japanese Journal of Cancer Research, 1997. 88:1108–1116.

22. Tokuyama, H., S. Yamago, and E. Nakamura, *Photoinduced biochemical activity of fullerene carboxylic acid.* Journal of the American Chemical Society, 1993. 115:7918–7919.

23. Tsao, N., et al., *Inhibition of Escherichia coli-induced meningitis by carboxyfullerence.* Antimicrobial Agents and Chemotherapy, 1999. 43:2273–2277.

24. Sayes, C.M., et al., *The differential cytotoxicity of water-soluble fullerenes.* Nano Letters, 2004. 4(10):1881–1887.

25. Jia, G., et al., *Cytotoxicity of carbon nanomaterials: Single-wall nanotube, multi-wall nanotube, and fullerene,* Environmental Science and Technology, 2005. 39(5):1378–1383.

26. Robichaud, C.O., et al., *Relative risk analysis of several manufactured nanomaterials: An insurance industry context.* Environmental Science and Technology, 2005. 39(22):8985–8994.

2

Nanotechnology and Our Energy Challenge

Wade Adams   *Richard E. Smalley Institute for Nanoscale Science and Technology, Rice University*

Amy Myers Jaffe   *James A. Baker III Institute for Public Policy, Rice University*

Oil was, unquestionably, the basis for prosperity for the United States and the planet in the last half of the past century. But continuing on an oil-based path into the 21st century is not a sustainable path for humanity. As the late Nobel laureate Richard E. Smalley points out, "It is very clear to many of us, including leading scientists and policy makers, that if oil remains the basis for prosperity for the world throughout this century, it cannot be a very prosperous or happy century."

Among the most important technical challenges facing the world in the 21st century will be sustainable energy supply. Lack of access by the poor to modern energy services constitutes one of the most critical links in the poverty cycle in Africa, Asia, and Latin America. Despite great advances in oil and gas drilling techniques and progress in renewable fuels, more than a quarter of the world's population has no access to electricity today, and two-fifths are forced to rely mainly on traditional biomass—firewood and animal waste—for their basic cooking and heating needs. Indoor air pollution from this traditional energy source is responsible for the premature death of over 2 million women and children a year worldwide from respiratory infections, according to the World Health Organization. Without a major technological breakthrough, well over 1 billion people will still be without modern electricity in 2030, according to calculations by the International Energy Agency.

The most developed countries, with 13 percent of the world's population, account for half of the world's annual energy use. The rate of use of energy in the wealthiest countries, per person, is about eight kilowatts, compared to one kilowatt in the less developed world. Developing countries, such as China and India, are rapidly increasing their energy consumption as they improve standards of living. The consequences of this rise in demand for energy in the developing world, coupled with consistent rises in energy use in the United States, will pose serious risks to the global system if new technologies are not developed. Indeed, we are already seeing serious global conflicts that are, in part, due to increasing concerns about meeting the world's energy needs.

The need for breakthrough energy solutions is all the more important because scientists have become increasingly convinced that the consequences of continuing to burn fossil fuels at current or expanding rates will have deleterious impacts on the global climate.

Martin Hoffert, professor of physics at New York University and author of the widely quoted *Science* article "Advanced Technology Paths to Global Climate Stability: Energy for a Greenhouse Planet," has cost the issue of stabilizing the carbon dioxide–induced component of climate change as an energy problem. He notes that stabilization will not only require an effort to reduce end-use energy demand, but also the development of primary energy sources that do not emit carbon dioxide into the atmosphere. Hoffert argues that a broad range of intensive research and development is urgently needed to produce energy technological options that can allow both climate stabilization and economic development [1].

Under a business-as-usual energy supply scenario, carbon concentrations in the atmosphere would rise to 750 ppm by the end of the century, a concentration level that would melt the West Antarctic ice sheets and erode coastlines around the globe, Dr. Hoffert told the conference. In order to hold atmospheric CO_2 concentrations to 350 ppm by midcentury—the level targeted by environmental scientists as preventing catastrophic changes—at least 15 terawatts of nonfossil fuel energy will be needed to reduce CO_2 levels to modest targets of 550 ppm by 2050. To reach the goal of 350 ppm, at least 30 terawatts would need to be derived from nonfossil sources.

The very large projected growth in world demand for carbon-free energy in the coming decades, even under the most conservative assumptions, cannot be met with existing technologies. New technologies will require a much larger energy R&D effort—combining government and industry—than in recent years. That will require significant multiyear increases in the federal budgets for energy-related research in several agencies; improved coordination across government agencies and national laboratories; enhanced partnerships among national laboratories, universities, and industry; and increased international cooperation.

Looking at these looming challenges, Dr. Richard Smalley began a campaign to promote the utilization of nanotechnology to find a solution to the energy problem. Dr. Smalley noted in a speech at Rice University in 2004.

> We need to find an economic alternative to oil. We need a new basis for energy prosperity. Ten billion people on the planet—that is our challenge. I believe this challenge is vastly greater than we admit. Between where we are right now and where we need to get to, we really have to find a new oil. I do not mean a liquid; I mean a technology that makes us energy rich again in an environmentally acceptable fashion for 10 billion people. Between here and where we need to be, there is something like ten miracles. The good news is that miracles do happen. I have been involved in physical sciences long enough to see many of them happen: lasers, high temperature super conductors, and so forth. But at the rate that they have been happening, over my lifetime, I am beginning to appreciate the magnitude of the breakthroughs that need to happen. I am not by any means convinced that we will get there soon enough. That is the reason I feel strongly that we ought to get much more intense about this issue than we have in the past and launch a major new energy research program to get this problem solved. [2]

Energy is a "quantitative business." Worldwide energy use is at the rate of 13–14 terawatts, the equivalent of 200–210 million barrels of oil per day. Analysts project that we will need at least twice as much energy in the next fifty years, but even doubling current resources and finding a way to satisfy twice our current levels of consumption for the next half-century would not be enough to give each individual on the planet a life comparable to that of citizens in the developed world.

Nanotechnology is the art and science of building materials that act at the nanometer scale. The ultimate nanotechnology builds at the ultimate level of finesse, one atom at a time. The "wet side" of nanotechnology includes all the nanomachinery of cellular life and viruses and manifests itself as biotechnology. The "dry side" of nanotechnology, which relates to energy, includes electrical and thermal conduction and provides great strength, toughness, and high temperature resistance, properties not found in biotechnology. Applied nanotechnology holds great promise in the energy area.

There are many ways nanotechnology may play a role in finding solutions to the energy problem. Funding committed to nanoscience and energy has great distributive benefits as it is a *crosscutting* research area. Incremental discoveries, as well as disruptive discoveries, could have implications for many fuel and energy sources as well as storage and delivery systems.

Nanotechnology could also play a pivotal role in providing stronger, lighter materials to build lighter-weight vehicles and to provide safer, more cost-effective storage for hydrogen fuels. Nanotechnology can play a key role in the development of sturdier fuel cells and improved membrane technology by providing new, light materials that can withstand the large changes in temperatures required in automotive operations.

At present, polymer electrolyte membranes are the most common membranes commercially available. But scientists are working to develop ceramic electrolyte membranes that will be more durable under extreme conditions. Nanostructured ceramic membranes, derived from metal-oxane nanoparticles, could present an improvement in the efficiency of fuel cells.

The Materials Nanotechnology Research Group at the University of Nevada, Reno, under the direction of Dr. Manoranjan Misra, professor of materials science in the Department of Chemical and Metallurgical Engineering, has developed titanium dioxide nanotube arrays for generating hydrogen by splitting water using sunlight. Once the process is scaled up to generate a lot of hydrogen from water, it will have great potential as a clean energy resource.

This new method splits water molecules, creating hydrogen energy more efficiently than currently available. The fabrication and production of nanotubes is done by a simple electrochemical method. University scientists add different tubular materials to increase the water-splitting efficiency and using sunlight.

"We can put one trillion nanotube-holes in solid titanium oxide substrate, which is approximately the size of thumbnails," said Misra. Each of these holes, a thousand times smaller than a human hair, acts as nanoelectrodes.

The hydrogen project also stores hydrogen in nanoporous titanium and carbon nanotube assemblies. These nanomaterials are powerful enough to maintain hydrogen for use in vehicles.

Among the major energy nanotechnology grand challenges are the following, according to a Smalley Institute for Nanoscale Science and Technology/James A. Baker III Institute for Public Policy study on the subject:

- Lower costs of photovoltaic solar energy by tenfold
- Achieve commercial photocatalytic reduction of CO_2 to methanol
- Create a commercial process for direct photoconversion of light and water to produce hydrogen
- Lower the costs of fuel cells by ten- to a hundred-fold and create new, sturdier materials

- Improve the efficiency/storage capacity of batteries and supercapacitors by ten- to a hundred-fold for automotive and distributed generation applications

- Create new lightweight materials for hydrogen storage for pressure tanks, LH2 vessels, and an easily reversible hydrogen chemisorption system

- Develop power cables, superconductors, or quantum conductors made of new nanomaterials to rewire the electricity grid and enable long distance, continental, and even international electrical energy transport, and reducing or eliminating thermal sag failures, eddy current losses, and resistive losses by replacing copper and aluminum wires

- Enable nanoelectronics to revolutionize computers, sensors, and devices for the electricity grid and other applications

- Develop thermochemical processes with catalysts to generate hydrogen from water at temperatures lower than 900 C and at commercially viable costs

- Create super-strong, lightweight materials that can be used to improve efficiency in cars, planes, and space travel; the latter, if combined with nanoelectronics-based robotics, possibly enabling solar structures on the moon or in space

- Create efficient lighting to replace incandescent and fluorescent lights

- Develop nanomaterials and coatings that will enable deep drilling at lower costs to tap energy resources, including geothermal heat, in deep strata

- Create CO_2 mineralization methods that can work on a vast scale without waste streams (possibly basalt-based)

Nanotechnology and Renewable Energy

Use of renewable energy is an extremely promising option for both reducing greenhouse gas emissions and enhancing diversity of energy supplies. Unlike nuclear energy or coal-derived fuel, solar-derived energy has no massive scale waste product requiring expensive and environmentally challenging disposal. Environmentally driven carbon taxes that favored renewable energy might be one policy route that would propel the use of solar technologies. But so far, many countries have favored direct subsidies to investors in renewable energy and imposition of renewable energy target standards. China, with the highest energy-use growth rate in the world, has set a target of 10 percent renewable energy by 2010. The European Union (EU) directive on renewable energy sources sets a target of 12 percent of energy and 22 percent of electricity from renewable sources by 2010, including hydroelectric power.

In the United States, state governments are leading the way for the promotion of solar energy. More than twenty states have now passed Renewable Portfolio Standards, while fourteen states have set up Renewable Energy Funds to subsidize or promote development of new renewable technologies, such as solar and wind power. Clean Edge, a research firm in Oakland, California, predicts that spending in renewable energy will jump to $89 billion by 2012, from $10 billion today.

With a production cost of around 20 to 30 cents per KWh for solar energy, solar energy is not yet positioned to be a major competitor to fossil fuels, whose electricity generation costs are as low as 2 to 3 cents per KWh. However, distributed customer-sited photovoltaics (PVs), where transmission and most distribution costs are avoided, is currently competitive as a peaking technology with small subsidies in regions with high levels of solar radiation. In dense urban areas with constrained underground transmission and distribution networks, such as San Diego, California, PVs can be competitive if the retail pricing fairly reflects the full value of generation at peak. Solar energy has also made inroads in Germany and Japan where overall retail electricity prices are higher than in the United States.

Renewable Resource	Approximate Price per kilowatt hour (1980)	Approximate Price per kilowatt hour (2003)	R &D Goal Approximate Price Target
Wind	$0.80	$0.05	$0.03 (2012)
Solar (PV)	$2.00	$0.20 – 0.30	$0.06 (2020)
Biomass	$0.20	$0.10	$0.06 (2020)
Geothermal	$0.15	$0.05 – 0.08	$0.04 (2010)

SOURCE: U.S. Department of Energy

Numerous challenges must be overcome to propel renewable energy to replace fossil fuels. Solar energy can be generated through the use of plants, through photovoltaic semiconductor junctions, and through catalysis (in which water is split using sunlight, producing relatively cheap hydrogen to produce electricity). Researchers will need to be able to offer disruptive solar technology with inexpensive conversion systems and effective storage systems. One option is to reduce the costs by improving the efficiency of photovoltaic cells. Another is to lower costs by enhancing systems to generate thermal solar energy on a larger, more cost-effective scale. New catalysts and new integrated systems need to be developed to help convert intermittent power into base-load power, including new materials to convert sunlight to hydrogen and oxygen.

In the May 2004 issue of *Physical Review Letters*, a team from Los Alamos National Laboratory found that quantum dots produce as many as three electrons from one high energy photon of sunlight. When today's photovoltaic solar cells absorb a photon of sunlight, the energy gets converted to one electron, and the rest is lost as heat. This nanotechnology method could boost the efficiency from today's solar cells of 20–30 percent to 65 percent.

Global electricity demand has been expanding at a rate of 3.0 percent per year since 1980, resulting in an overall increase of 88 percent to 13,934 bkwh, up from 7,417 bkwh. World electricity demand is expected to double by 2030, growing at an annual rate of about 2.4 percent, as economic activity is enhanced in developing nations such as China and India. U.S. electricity demand grew from 2,094 bkwh in 1980 to 3,602 bkwh in 2005, or an average annual rate of 2.6 percent. U.S. electricity demand is projected to increase by 1.9 percent per annum by 2020.

Still, much of the world's population will remain without modern energy services unless new, aggressive policies and emerging technologies are launched in the coming years. The global electricity sector will require as much as $10 trillion in new investments over the next three decades, according to the International Energy Agency (IEA). This is close to three times higher in real terms than the investment made in the sector over the past three decades. Substantial investment will go into transmission and distribution networks. In the developing world alone, $5 trillion in spending in new electricity infrastructure will be needed to meet projected targets for economic growth and social development.

The advantages of developing a new, improved, and more efficient grid system are tremendous. But there are clear technological and political hurdles that must be overcome to achieve this target. New materials and new technical approaches will need to be developed and an elaborative plan must be sculpted to map a smooth transition into an electrically digital society. Nanotechnology holds great promise for the electricity sector through its ability to enhance the new grid by introducing post-silicon power electronics and complex, iterative, adaptive controls.

By supplying electrical systems with nanosensors and nanosources as well as nanochips able to apply concepts of distributed business, adaptive learning, simulation, micro-real options, and work-flows while performing peer-to-peer assessment, major changes can occur in terms of energy efficiency and energy supply.

A new national initiative is being pursued to produce solid-state lighting—light emitting diodes (LEDs) and lasers—that promises to be ten times more efficient and two times brighter than incandescent and fluorescent lights, respectively. General lighting is responsible for 20 percent of global energy consumption, and conventional light sources offer very low energy efficiencies of 5 percent for incandescent and 25 percent for fluorescent bulbs. Department of Energy (DOE) road mapping studies predict that by 2025, government investments in nano-layered solid-state lighting (SSL) will result in a 50 percent decrease in the amount of U.S. electricity used for lighting and a 10 percent decrease in the total U.S. electricity consumption overall. This will translate into a 17 gigawatt reduction in U.S. demand for electrical generating capacity and the equivalent of more than 28 megatons per year reduction in U.S. carbon emissions.

Nanomaterial applications are expected to play a major economic role in increasing the efficiency of light sources, motors, electrodes, and efficient wear-resistant material. In addition, nanoclusters, able to increase the efficiency of catalytic processes, are thought to hold the answer for reducing the emission of nitrogen oxides. Beyond applications in materials science and catalysis, there is also great potential for first order interaction between nanoscience and energy.

According to Dr. Timothy Fisher, associate professor of mechanical engineering at Purdue University, nanoscience could have profound implications for energy conversions and efficiency. "When materials are being spatially confined, the energy states of the energy carriers change," he explains. This change in behavior can be particularly useful in direct energy conversion technologies and energy transport in electron emission processes. Direct thermal-electrical conversion is particularly appealing from an engineering point of view due to its ability to eliminate moving parts. To reach the phase where nanoenergy conversion devices could be produced, advancements in the field of nanoscale thermoelectrics are needed. According to research conducted at the Research Triangle Institute, the use of nanoscale structures is thought to significantly improve thermoelectric performance.

Nanomaterials (e.g., carbon nanotubes) also offer a solid path to updating energy transmission. Since they are strong and conduct energy six times better than copper, they make sense as a new option. Their size is another advantage, especially in places like New York City that have run out of underground real estate. In NYC, the underground utility corridors are so full of copper wires, it is joked that they have the greatest supply of copper in the world.

Carbon nanotube cables would be small enough to be added to existing corridors. When demand outpaced supply, metropolitan utility companies would have another alternative than buying expensive real estate.

Smalley Electricity Vision

One attractive candidate for the "new oil" fuel of the coming century is electricity, with local storage technology and long distance transmission holding the key to a new energy world. The single biggest problem of electricity is storing it. Approaches that entail production and storage of electricity on a vast scale are daunting, but technologies could be developed to attack the energy storage problem locally, at the scale of a house or small business. A local storage based system would allow users to buy energy supplies off the grid when supplies are cheapest, unlike the current centralized plant system where almost twice as much generation capacity is needed to fulfill peak time demand.

One vision of such a distributed storage/generation grid for 2050 includes a vast electrical continental power grid with over 100 million asynchronous local storage units and generation sites, including private households and businesses. This system would be continually innovated by free enterprise, with local generation buying low and selling high to the grid network. Optimized local storage systems would be based on improved batteries, hydrogen conversion systems, and flywheels, while mass primary power input to the grid could come from remote locations with large-scale access to cleaner energy resources (solar farms, stranded natural gas, closed-system clean coal plants, and wave power) to the common grid via carbon nanotubes, high-voltage wires that minimize loss. Excess hydrogen produced in the system could be used in the transportation sector, and excess residential electricity could be used to recharge plug-in hybrid electric vehicles. Innovative technological improvements in long distance, continental power grids that could transport hundreds of gigawatts over a thousand miles instead of a hundred megawatts over the same distance would permit access to very remote sources, including large solar farms in the deserts, where local storage can be used as a buffer. Remote nuclear power sources could be located far from populated areas and behind military fences, to address proliferation concerns. Clean coal plants could be located wherever it is convenient and economical to strip out and sequester the CO_2.

To Richard Smalley, breaking down the electricity issue to the localized level was a critical part of creating new energy solutions. He noted in his energy lecture.

> When we are trying to find a way to store electrical energy on a vast scale, as we generally need energy in gigawatt power plants, there are very few options that one can imagine on that large scale for energy storage. But if you imagine attacking the energy storage problem locally, at the scale of a house or a small business, the problem becomes vastly more solvable because there must be many more technologies that are accessible at the smaller scale. As a scientist, I would rather fight the battle of energy storage locally than on huge centralized scales.

Under a change of approach toward distributed energy and localized residential storage, reliability of the electrical grid becomes less important. The local residential and business sites can determine what period of time they want to be buffered by the grid and when to rely on storage.

Increasingly, the primary energy producers that put electrical power in the world can simply dump the power onto the grid in the cheapest possible way, and locally, the local storage buys it off the grid when it is cheapest, when it is, of course, most abundant. Such a system would relieve the pressures of having to provide almost twice the generating capacity than is used on average to account for the peaks and the lows of electricity demand. Competition at the residential storage appliance level allows the electrical energy grid to transform itself, with a time period of a couple of years rather than decades. By allowing the possibility to mix locally produced electricity with grid-delivered centralized sources, the energy system becomes more robust.

The shift to a nanosupported, localized distributed energy system has the advantage that it does not require tearing down existing centralized facilities and building completely new and separate infrastructure. New distributed facilities can be initiated now and connected to the existing plants providing power to the grid. The value is that it would facilitate not only residential solar, but also other very remote energy sources to this same grid, including vast solar farms in the deserts, where local storage serves as a buffer supply for when the sun is down, as well as storing energy from wind when it is blowing. Vast amounts of electrical power can also be imported from remote nuclear power sources way out in somebody else's backyard, behind a military fence, where one can be absolutely sure there is no nuclear weapons risk or accident risk associated with plant operations. Links are also possible to electricity plants generating power from clean coal, wherever there is a site where CO_2 can be sequestered cheaply and not have it come back at greater than 0.1 percent leakage per year. Distributed electricity systems can be used to fuel plug-in electric hybrid vehicles.

Conclusion

Our current energy predicament requires a bold new energy science and technology program, as well as an enlightened federal policy to map out the path to development of new sources for a better energy and environmental future for the 21st century. Such a path will have to be guided by an enlightened federal energy policy that goes well beyond anything we have had or have today. Elements of a new energy policy must include the means and incentives to rapidly develop, demonstrate, and deploy cheaper, more efficient, and environmentally sound energy supplies to protect the global environment while improving the quality of life in developing countries. With visionary leadership at the highest levels of government—and sound national science, technology, and energy policies to match—larger numbers of talented and motivated young people might well find the world's energy challenge sufficiently compelling to attract them into careers in science and engineering.

In the case of energy policy, scientists should focus on guiding federal decision making so that limited resources are well spent, and solutions to critical challenges such as energy and the environment can be developed and implemented in an efficient and cost-effective manner. National strategies should reflect the best range of alternatives so that markets are able to select technological solutions that will meet national goals. The United States has a leading role to play, working in partnership with other nations of the world, including the least developed countries, in dealing with this truly global energy, environmental, and security challenge.

References

1. Hoffert, Martin, et al., *Advanced technology paths to global climate stability: Energy for a greenhouse planet*, Science, 298(5595):981–987.
2. Lecture at Rice University, May 14, 2004, available at www.rice.edu/energy

Part

Principles and Methods

Nanomaterials Fabrication

Jean Pierre Jolivet   *Université Pierre et Marie Curie-Paris, France*
Andrew R. Barron   *Rice University, Houston, Texas, USA*

The ability to fabricate nanomaterials (often in the form of nanoparticles) with strictly controlled size, shape, and crystalline structure, has inspired the application of nanochemistry to numerous fields, including catalysis, optics, and electronics. The use of nanomaterials in such applications also requires the development of methods for nanoparticle assembly or dispersion in various media. Although much progress has been realized during the last decades in the development of highly advanced analytical tools enabling the characterization of nanostructures and an understanding of their physical properties, the synthesis of well-defined nanoparticles has resulted in several prominent milestones in the progress of nanoscience, including the discovery of fullerenes [1], carbon nanotubes [2, 3], the synthesis of well-defined quantum dots [4–6] and the shape control of semiconductor CdSe nanocrystals [7]. However, despite a vigorous expansion in the methods of nanoparticles synthesis, it is still difficult to generalize underlying physical or chemical principles behind existing synthesis strategies to any arbitrary nanomaterial. A general, mechanistic understanding of nanoparticle formation that might guide the development of new materials remains lacking [8]. Though the synthesis of nanoparticles with control over size, shape, and size distribution has been a major part of colloid chemistry for decades, it remains an intensely studied topic as is evident by a substantial body of literature. In this chapter, we provide an overview of the main methods that have proved to be successful for the fabrication of several classes of nanomaterials: specifically, oxides, chalcogenides, metals, and fullerenes.

Specificity and Requirements
in the Fabrication Methods of Nanoparticles

Ultra-dispersed systems, such as dispersions of nanoparticles, are intrinsically thermodynamically metastable, in large part due to the very high interfacial areas. Nanoparticle surface area represents a positive contribution to the free enthalpy of the system. If the activation energies are not too high, spontaneous evolution of a nanoparticle dispersion can occur causing an increase in nanoparticle size or the formation of nanostructured domains and leading to the decrease of the surface area. Consequently, it follows that:

- An ultra-dispersed system with a high surface energy can be only kinetically stabilized.

- Ultrafine powders cannot be synthesized by methods involving energies that exceed a threshold, but rather through methods of "soft chemistry" that maintain the forming particles in a metastable state.

- Additives and/or synthesis conditions that reduce the surface energy are needed to form nanoparticles stabilized against sintering, recrystallization, and aggregation.

Under these conditions, any solid matter such as metal oxides, chalcogenides, metals, or carbon can be obtained at the nanometric scale.

Synthesis methods for nanoparticles are typically grouped into two categories:

- The first involves division of a massive solid into smaller portions. This "top-down" approach may involve milling or attrition (mecanosynthesis), chemical methods for breaking specific bonds (e.g., hydrogen bonds) that hold together larger repeating elements of the bulk solid, and volatilization of a solid by laser ablation, solar furnace, or some other method, followed by condensation of the volatilized components.

- The second category of nanoparticle fabrication methods involves condensation of atoms or molecular entities in a gas phase or in solution. This is the "bottom-up" approach in which the chemistry of metal complexes in solution holds an important place. This approach is far more popular in the synthesis of nanoparticles, and many methods have been developed to obtain oxides, chalcogenides, and metals.

The liquid-phase colloidal synthetic approach is an especially powerful tool for convenient and reproducible shape-controlled synthesis of nanocrystals—not only because this method allows for the resulting nanocrystals to be precisely tuned in terms of their size, shape, crystalline structure, and composition on the nanometer scale, but also because it allows them to be dispersed in either an aqueous or a non-aqueous medium. Moreover, these nanoparticles can be modified in liquid suspension by treatment with various chemical species for application and use in a diverse range of technical or biological systems.

Oxides

The most widespread route to fabrication of metal oxide nanoparticles involves the bottom-up approach by the precipitation in aqueous solution from metal salts. Organometallic species can also be used in hydrolytic or nonhydrolytic pathways, but due to their cost and the difficulty in manipulating these compounds, they are used less frequently and primarily for high-tech applications. An alternative top-down approach has been demonstrated for aluminum and iron oxide nanoparticles; however, it is possible that this methodology could be extended to other oxides.

From molecular species to nanoparticles

One approach to the creation of oxide nanoparticles is to build from the "bottom-up," beginning with individual ions or molecular complexes of metals. Variations on this approach include the hydroxylation of metal cations in aqueous solutions, the use of metal alkoxides, nonhydrolytic routes such as those employing metal halides.

Hydroxylation of metal cations in aqueous solution and condensation: Inorganic polymerization.   The metal cations issued for the dissolution of salts in aqueous solution form true coordination complexes in which water molecules form the coordination sphere. The chemistry of such complexes, and especially their acid behavior, provides a framework for understanding how the solid (oxide) forms via inorganic polycondensation [9, 10].

The binding of water molecules to a cation involves an orbital interaction allowing an electron transfer from a water molecule to a cation following Lewis's acid-base concept of the coordination bond. Such a transfer drives the electronic density of water molecules toward the cation and weakens the O-H bond of the coordinated water molecules. They are consequently stronger Brønsted acids than the water molecules in the solvent itself, and they tend to be deprotoned spontaneously according to the hydrolysis equilibrium:

$$[M(H_2O)_n]^{z+} + h\ H_2O \Leftrightarrow [M(OH)_h(H_2O)_{n-h}]^{(z-h)+} + h\ H_3O^+$$

or by neutralization with a base:

$$[M(H_2O)_n]^{z+} + h\ HO^- \Leftrightarrow [M(OH)_h(H_2O)_{n-h}]^{(z-h)+} + h\ H_2O$$

In these equilibria, h is the hydroxylation ratio of the cation. It represents the number of hydroxo ligands (OH) present within the coordination sphere, or the number of protons eliminated from the coordination sphere of the aqua cation.

The acidity of the aqua cation strongly depends on the strength of the M-O bond—that is, the magnitude of the electron transfer from oxygen toward the metal. The acidity can be related to the polarizing character of the cation—that is, the ratio of formal charge (oxidation state) to its size. The equilibrium constant of the first step of hydroxylation (h = 1) for many cations where d stands for the M-O distance can be empirically expressed by:[9]

$$\log K \approx -20 + 11(z/d) = -\Delta G^\circ \cdot RT$$

The hydrolysis rate of cations in aqueous solution also depends strongly on the pH of the medium because the equilibrium involves the transfer of protons. From the acidity constants of medium-sized cations [9, 11], a charge-pH diagram was established [12] in which three domains are plotted (Figure 3.1). The lower domain corresponds to the existence of aqua-cations $[M(H_2O)_n]^{z+}$, the upper to oxo-anions $[MO_n]^{(2n-z)-}$, and the intermediate domain corresponds to hydroxylated complexes containing at least one hydroxo ligand. These domains are separated by two lines corresponding to h = 1 and h = 2n−1, respectively. This diagram is a useful guide to understand the condensation and precipitation phenomena involved in the synthesis of particles. Condensation between species in solution becomes possible only when they are hydroxylated.

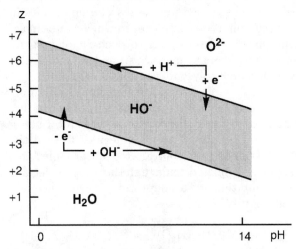

Figure 3.1 Nature of the ligand in the coordination sphere of a cation as function of its formal charge, z, and the pH of the medium [12]. Possible initiation methods of condensation reactions are depicted.

Therefore, condensation can be initiated by varying the pH of the solution by addition of a base on an acid (element M with $z \leq 4$) or by addition of an acid on a base (element M with $z > 4$). Condensation can also be initiated via redox reactions with elements having several stable oxidation states (Figure 3.1).

Hydrolysis is, strictly speaking, a neutralization reaction carried out by the water molecule:

$$[M(H_2O)_n]^{z+} + h\ H_2O \rightarrow [M(OH)_h(H_2O)_{n-h}]^{(z-h)+} + h\ H^+_{solvated}$$

For this reaction, it has been shown [9]:

$$\Delta H^\circ = (75.2 - 9.6\ z)\ kJ\ mol^{-1},\ \Delta S^\circ = (-148.4 + 73.1\ z)\ J\ mol^{-1}$$

and:

$$\Delta G^\circ_{298} = (119.5 - 31.35\ z)\ kJ\ mol^{-1}$$

The reaction is spontaneous ($\Delta G^\circ < 0$) for elements of charge equal to or greater than 4. Therefore, at room temperature, tetravalent elements do not exist as purely aquo complexes, even in strongly acidic medium. For elements with a charge, z, smaller than 4, ΔG° becomes negative only if the temperature is higher than 298 K. Therefore, it is necessary to heat the solution in order to carry out hydrolysis of the cation (forced hydrolysis or thermohydrolysis).

The monomeric, electrostatically charged, hydroxylated species $[M(OH)_h(H_2O)_{n-h}]^{(z-h)+}$ or $[MO_{n-h}(OH)_h]^{(2n-z-h)-}$ are generally observed in solution only at very low concentrations. More often, they condense and form soluble polynuclear species, polycations ($h < z$), and polyanions ($h > z$) respectively, in which the cations are bounded by hydroxo or oxo bridges [10]. These entities are generally of molecular size, although giant polyanions of molybdenum containing up to 368 Mo ions have been recently synthesized! [13] The condensation of neutral complexes with $h = z$ is, in general, not limited and continues to the formation of a solid.

Hydroxylated complexes condense via two basic mechanisms of nucleophilic substitution, depending on the nature of the coordination sphere of the cations. In all cases, the driving force to the condensation is the nucleophilicity of the hydroxo ligand. The cation must also have an electrophilic character high enough to be subjected to the nucleophilic attack. Condensation of aquohydroxo complexes proceeds by elimination of water and formation of hydroxo bridges (olation):

For oxohydroxo complexes, there is no water molecule in the coordination sphere of the complexes and therefore no leaving group. Condensation has to proceed in this case via a two-step associative mechanism leading to the formation of oxo bridges (oxolation):

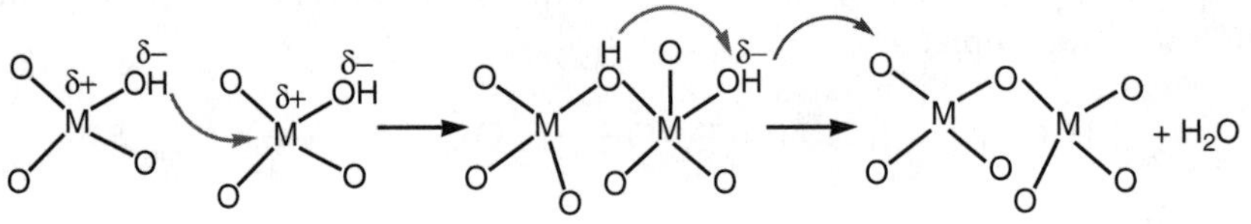

The hydroxylation rate, h, of the complexes represents their functionality toward condensation, and it controls the type and the structure of condensed species. It is obvious that h is a function of the pH of the medium. Its also depends on the characteristics of the cation such as size, formal charge, and electronegativity.

Condensation of hydroxylated and electrically charged complexes (h < z) *always* ends at a more or less advanced stage, leaving discrete species in solution, either polycations or polyanions, depending on whether the monomeric complex is a cation or an anion. Indeed, electrical charges cannot indefinitely accumulate on a metal-oxo-polymer, and condensation stops as soon as conditions allowing nucleophilic substitution are no longer present. As condensation causes water elimination, there is a change in composition of the reaction product that produces a variation of its average electronegativity, causing charge redistribution within its structure and, therefore, a change in the reactivity of the functional groups [10]. Hence, OH ligands in the growing species may lose their nucleophilic character, and cations may lose their electrophilic character. Usually, during condensation, the nucleophilic character of hydroxo ligands cancels in polycations, and the electrophilic character of the cation cancels in polyanions. Condensation of electrically neutral ions (h = z) continues always indefinitely until there is precipitation of a solid (hydroxide, oxyhydroxide, or more or less hydrated oxide) or of a basic salt in the presence of complexing ligands. Elimination of water from noncharged complexes never leads to a sufficient change in the average electronegativity to cancel the reactivity of functional groups.

In theory, an hydroxide $M(OH)_z$ is formed via endless condensation of aquo-hydroxo complexes. However, the hydroxide may not be stable. Its spontaneous dehydration, more or less rapid and extensive, generates an oxyhydroxide $MO_x(OH)_{z-2x}$ or a hydrated oxide $MO_{z/2} \cdot (H_2O)_x$. The reaction takes place via oxolation in the solid phase with elimination of water from hydroxo ligands. The reaction is associated with structural changes in order to preserve the coordination of the cation. Usually, elements with a $+2$ charge precipitate as hydroxides, and those with a $+3$ charge as oxyhydroxides (the final stage of evolution is the oxide). Those of higher charge form oxides of various level of hydration [14]. This sequence is a clear illustration of the increasing polarization of the hydroxo ligands by the cation, which is associated with the covalent nature of the metal-oxygen bond.

In summary, condensation of cations in solution is initiated when acidity allows the presence of the hydroxo ligand in the coordination sphere of the cation. This occurs through addition of a base to aquo complexes of elements of formal charge equal to or smaller than 4, or through addition of an acid to oxo complexes of elements of charge equal to or greater than 4 (Figure 3.1). Two reactions, olation and oxolation, respectively, ensure the development of condensation. The condensation of cationic and anionic hydroxylated complexes is always limited. It leads to polycations and polyanions, respectively. Formation of a solid requires the presence of zero-charge complexes. It is also possible (although less common) to involve redox phenomena in order to decrease the formal charge on the metal and force the appearance, under given acidic conditions, of the hydroxo ligand in the coordination sphere of the cation.

Hydroxide, oxyhydroxide, or hydrated oxide solid phases obtained via precipitation are made of particles whose average size may range from a few nanometers to a few microns. Particle morphology may vary depending on synthesis conditions. Moreover, aging in aqueous solution may bring about significant dimensional, morphological, and structural changes. In order to understand how small particles form and what role the experimental parameters play on their characteristics and on evolution, it is useful to review the kinetic aspects of condensation reactions.

The precipitation of a solid involves four kinetic steps [15–17]:

1. Formation of the zero-charge precursor $[M(OH)_z(H_2O)_{n-z}]^0$, which is able to condense and form a solid phase. Hydroxylation of the cation is a very fast acid/base reaction, but the rate of formation of the zero-charge precursor in solution can largely vary depending on whether the reaction starting from cationic complexes for example, takes place through addition of a base, thermohydrolysis, or slow thermal decomposition of a base such as urea.

2. Creation of nuclei, through condensation (olation or oxolation) of zero-charge precursors. The condensation rate is a function of precursor concentration, and as long as it is small at the onset of cation hydroxylation, the rate is almost zero (zone I, Figure 3.2a). Beyond a critical concentration C_{min}, the condensation rate increases abruptly

and polynuclear entities—the nuclei—are formed in an "explosive" manner throughout the solution (zone II, Figure 3.2c). Indeed, nucleation is an abrupt kinetic phenomenon because, since its order is high compared to the precursor concentration, it is either extremely fast or nonexistent within a narrow concentration range (Figure 3.2b and 3.2c). If the rate of generation of the precursor is significantly smaller than the condensation rate, nucleation sharply reduces the precursor concentration and the condensation rate decreases equally rapidly. When the precursor condensation is again close to C_{min}, formation of new nuclei is no longer possible.

3. Growth of the nuclei through addition of matter, until the primary particle stage is reached. This step follows the same chemical mechanisms as nucleation: olation or oxolation. However, for a concentration close to C_{min}, the nucleation rate is very small and precursors condense preferentially on existing nuclei. Nuclei grow until the precursor concentration reaches the solution saturation (in other words, the solubility limit) of the solid phase (zone III, Figure 3.2b,c). Growth, having kinetics of first or second order, is a somewhat faster process. Precursor condensation during precipitation is a function of the respective rates of precursor generation and nucleation. Nucleation and growth phases may therefore be consecutive or overlap and occur simultaneously if the precursor concentration stays higher than C_{min}.

The number, and therefore the size, of the primary particles that form from a given quantity of matter is linked to the relative nucleation

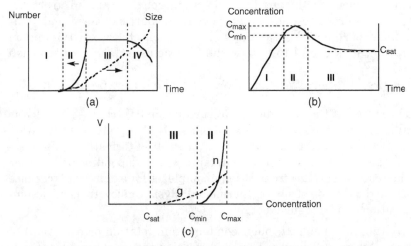

Figure 3.2   Change (a) in the number and sizes of particles formed in solution, and (b) in concentration C of the soluble precursor of the solid phase during precipitation [18]. Condensation rate, which is zero for $C < C_{min}$, becomes infinite for $C \geq C_{max}$. C_S is the solubility of the solid phase. (c) Nucleation (n) and growth (g) rates as a function of precursor concentration in solution.

and growth rates (Figure 3.2a). In order to obtain particles of homogeneous size, it is necessary that the nucleation and growth steps be separated to ensure that a single nucleation stage takes place, and that growth, via accumulation of all remaining matter, be controlled. This implies that the nucleation rate should be much greater than the rate at which the precursor is generated. Under these conditions, nucleation is very brief and clearly decoupled from the growth phase. If the nucleation rate is not high enough compared to the rate of generation of the precursor, precursor concentration remains higher than C_{min} throughout the reaction, and nucleation and growth are simultaneous. The growth of the first nuclei is much larger than that of the younger ones, which leads to a large particle size distribution.

4. Nucleation and growth steps form particles under kinetic control following a reaction path of minimum activation energy under conditions imposed to the system (acidity, concentration, temperature), but the products are not necessarily thermodynamically stable. Aging of the suspensions, which may take place over a very large time scale (hours, days, or months), allows the system to tend toward or reach stability, and it is often associated with modifications of some physical or chemical characteristics of the particles. "Ostwald ripening" leads to an increase in the average particle size and possible aggregation (zone IV, Figure 3.2a). Aging may also trigger a change in morphology and crystalline structure or even cause crystallization of amorphous particles. In fact, aging is one of the most important phenomena that must be considered, because it determines the characteristics of the particles after precipitation.

Control of particle size, crystalline structure, and morphology. There are different techniques to form the complex of zero charge and to obtain a solid. The most common method consists of introducing a base into the acid solution of a metal salt at room temperature. When solutions such as these are mixed, a high concentration of hydroxylated complexes rapidly forms along with induced local pH gradients. Inhomogeneities in the hydrolysis products often present during such a missing procedure may result in random condensation and the formation of an amorphous solid with an ill-defined chemical composition. Such a result is exemplified by the case of ferric ions. They precipitate quasi-instantaneously at pH $\geq$ 3 into a poorly defined, highly hydrated phase, called 2-line ferrihydrite [19]. (This phase takes its name from its X-ray diffraction pattern, which exhibits only two broad bands.) In similar conditions, Al^{3+} ions form a transparent amorphous gel [20]. At pH $\geq$ 2, Ti^{4+} ions form an amorphous oxyhydroxide with a composition near to $TiO_{0.3}(OH)_{3.4}$ [21]. These solids are formed of very small size particles, around 2–3 nm in diameter, and are strongly metastable. They evolve spontaneously in suspension more or less quickly to form crystalline nanoparticles, with possibly an increase in particle size, releasing simultaneously the lattice energy (and decreasing the surface energy) to decrease the free enthalpy of the system. The acidity of the suspension during evolution is the most important parameter to control crystalline structure and the size of the final particles. Two distinct mechanisms are involved in the transformation.

When the suspensions are aged at a pH where the solid is partially soluble, the concentration in solution may be enough to feed nuclei of a more stable crystalline phase. A transfer of matter occurs via the solution from the soluble amorphous phase toward a less soluble crystalline phase during a slow dissolution-crystallization process allowing formation of well-crystallized particles. Such a process is involved in the formation of goethite, α-Fe(O)(OH), during aging of ferrihydrite in suspension at pH < 5 or pH > 10. Because of structural anisotropy of goethite, rod-like particles of mean dimensions $150 \times 25 \times 15$ nm are obtained (Figure 3.3). These particles, anisotropic in shape, form very stable concentrated suspensions, which behave as nematic lyotropic liquid crystals exhibiting very interesting magnetic properties [22]. The nematic phase aligns in a very low magnetic field (20 mT for samples 20 mm thick). The particles orient along the field direction at intensities smaller than 350 mT but reorient perpendicular to the field beyond 350 mT. This behavior could have interesting applications.

In similar ranges of acidity, the aluminate gel is transformed into platelets of hydroxide $Al(OH)_3$, gibbsite at pH < 5, and bayerite at pH > 8 [20]. In a rather acidic medium (pH < 1), the same dissolution-crystallization mechanism transforms the amorphous titanium oxyhydroxide into elongated TiO_2 rutile nanoparticles. In these examples, the final size of particles depends on the acidity of the medium: the particle size increases when the acidity is strong.

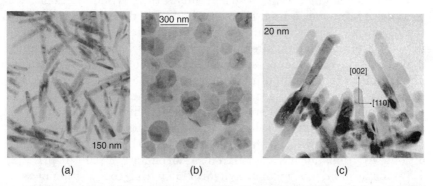

Figure 3.3 Particles of (a) goethite α-Fe(O)(OH), (b) gibbsite $Al(OH)_3$, and (c) rutile TiO_2 synthesized in aqueous medium.

If the suspensions are aged at an acidity where the solubility of the solid is very low or at a minimum, the concentration of soluble species in equilibrium with the solid phase does not allow an efficient transport of matter, and crystallization of the early amorphous material will occur more easily by a transformation *in situ*, in the solid state. The transformation involves the diffusion of ions within the solid with partial dehydration, and the formation of crystalline domains of very small size. Nanoparticles of hematite, α-Fe_2O_3, are so obtained from ferrihydrite at $6 \leq pH \leq 8$ [23]. Very small nanoparticles of boehmite, γ-$Al(O)(OH)$, (around 300 $m^2 \cdot g^{-1}$) are similarly obtained by aging of aluminate gels at the same pH range (6 to 8) [20]. Although boehmite is not the most thermodynamically stable phase at room temperature, it is probably kinetically stabilized because the system is constrained to evolve without heating and transforms on the lowest activation energy path. Between pH 2 and pH 7, where the solubility of titania is very low, the amorphous solid is transformed into TiO_2 anatase nanoparticles [21]. Over this acidity range, the particle size of anatase depends on the pH of precipitation and aging. This effect of acidity on particle size will be discussed later.

Precipitation by addition of a base at room temperature may also lead to stable crystalline nanoparticles without involving any transformation by the above mechanisms. For instance, magnetite Fe_3O_4 is easily obtained by coprecipitating aqueous Fe^{3+} and Fe^{2+} ions with $x = 0.66$ [24]. Iron ions are distributed into the octahedral (Oh) and tetrahedral (Td) sites of the face centered cubic (fcc) stacking of oxygen according to $[(Fe^{3+})_{Td}(Fe^{3+}Fe^{2+})_{Oh}O_4]$. Magnetite is characterized by a fast electron hopping between the iron cations on the octahedral sublattice. Crystallization of spinel is quasi-immediate at room temperature, and electron transfer between Fe^{2+} and Fe^{3+} ions plays a fundamental role in the process [25, 26]. In effect, maghemite, γ-Fe_2O_3, $[(Fe^{3+})_{Td}(Fe^{3+}_{5/3}V_{1/3})_{Oh}O_4]$ (where V stands for a cationic vacancy) does not form directly in solution by precipitation of ferric ions, but a small proportion of Fe^{2+} (≤ 10 mol %) induces the crystallization of all the iron into spinel. Studies of the early precipitate revealed that all Fe^{2+} ions were incorporated into a Fe^{2+}-ferrihydrite, forming a short-range ordered, mixed valence material exhibiting fast electron hopping, as evidenced by Mössbauer spectroscopy [26]. Electron mobility brings about local structural rearrangements and drives spinel ordering. Besides this topotactic process, crystallization of spinel can also proceed by dissolution crystallization, resulting in two families of non-stoichiometric spinel particles $[(Fe^{III})_{Td}(Fe^{III}_{1+2z/3}Fe^{II}_{1-z}V_{z/3})_{Oh}O_4]$ with very different mean size [25]. The relative importance of these two pathways depends on the Fe^{2+} level in the system, and the end products of the coprecipitation are single phase only for $0.60 \leq x \leq 0.66$. The comparison with the cases where M^{2+} is different from Fe^{2+} emphasizes the role of electron mobility between Fe^{2+} and Fe^{3+} ions in the crystallization process. With other divalent cations, intervalence transfers are negligible and a spinel ferrite forms only by dissolution-crystallization [24]. With $x = 0.66$, corresponding to stoichiometric magnetite, the mean particle size is controlled on the range 2–12 nm by the conditions of the medium, pH and ionic strength (I), imposed by a salt ($8.5 \leq pH \leq 12$ and $0.5 \leq I \leq 3$ mol $\cdot L^{-1}$) (Figure 3.4) [27]. Such an influence of acidity on the particle size is relevant to thermodynamics rather than kinetics (nucleation and growth processes). Acidity and ionic strength act on protonation–deprotonation equilibria of surface hydroxylated groups and, hence, on the electrostatic surface charge. This leads to a change in the chemical composition of the interface, inducing a decrease of the interfacial tension, γ, as stated by Gibbs's law, $d\gamma = -\Gamma_i d\mu_i$, where Γ_i is the density of adsorbed species i with chemical potential μ_i. Finally, the surface contribution, $dG = \gamma dA$ (A is the surface area of the system), to the free enthalpy of the formation of particles is lowered, allowing the increase in the system surface area [28].

Due to the high electron mobility in the bulk, magnetite nanoparticles give rise to an interesting surface chemistry involving interfacial transfer of ions and/or electrons and allowing us to consider spinel iron oxide nanoparticles as refillable nanobatteries. Nanoparticles of magnetite are very sensitive to oxidation and transform into maghemite $[(Fe^{3+})_{Td}(Fe^{3+}_{5/3}V_{1/3})_{Oh}O_4]$. The high reactivity is obviously due to the high surface-to-volume ratio, and a controlled synthesis of particles requires strictly anaerobic conditions. However, aerobic oxidation is not the only way to go to maghemite. Different interfacial ionic and/or electron transfers that depend on the pH of the suspension can be involved in the transformation. In basic media, the oxidation of magnetite proceeds by oxygen reduction at the surface of the particles (electron transfer only) and coordination of oxide ions, while in acidic

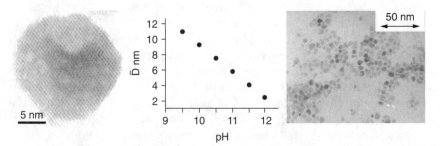

Figure 3.4 Electron micrographs of magnetite particles synthesized by precipitation in water and particle size variation against pH of precipitation [27].

medium and anaerobic conditions [29], surface Fe^{2+} ions are desorbed as hexa-aquo complexes in solution (electron and ion transfer) according to:

$$[Fe^{3+}]_{Td}[Fe_2^{2.5+}]_{Oh}O_4 + 2H^+ \rightarrow 0.75\ [Fe^{3+}]_{Td}[Fe^3{}_{5/3}^{+}V_{1/3}]_{Oh}O_4$$
$$+ Fe^{2+}_{aq} + H_2O$$

In both cases, the oxidation of Fe^{2+} ions is correlated with the migration of cations through the lattice framework, creating cationic vacancies in order to maintain the charge balance (Figure 3.5). The mobility of electrons on the octahedral sublattice renews the surface ferrous ions allowing the reaction to go to completion. The oxidation in acidic medium (pH $\approx$ 2) does not lead to noticeable size variation.

A very interesting technique for obtaining oxide nanoparticles is the thermolysis (or forced hydrolysis) of acidic solutions. Heating of a solution to approximately 50–100°C enables, particularly with trivalent and tetravalent elements (Al, Fe, Cr, Ti, Zr, etc.), a homogeneous hydrolysis in conditions close to thermodynamic equilibrium [30]. Under such conditions, the slow speed of formation of the hydrolyzed precursors allows decoupling of the nucleation and growth steps, from a kinetic standpoint. As a result, narrow particle size distributions can be obtained.

Thermolysis at 90–100°C of acidic ferric solutions (pH $\leq$ 3) forms hematite [10, 31]. In these conditions, olation and oxolation compete and acidity facilitates oxolation leading to oxide. The acidity and the nature of the anions are, however, crucial for the control of the size of particles. At low concentration of chloride ($C < 10^{-3}$ mol $\cdot$ L^{-1}), 6-line ferrihydrite forms initially [31, 32]. It transforms into hematite during thermolysis, but the particle size depends strongly on the acidity of the medium

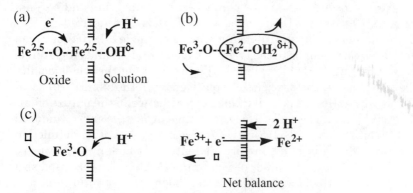

Figure 3.5 Oxidation mechanism of magnetite to maghemite in acidic medium [29].

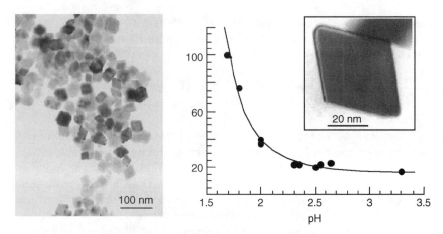

Figure 3.6 Particles of hematite obtained by thermolysis at 95°C of ferric nitrate solutions. Influence of the pH of the medium on the mean size of particles [19].

(Figure 3.6). At high concentration of chloride, akaganeite, β–Fe(O)(OH), is first formed [33]. This metastable phase is slowly transformed into hematite during thermolysis, and large (μm-sized) polycrystalline particles with various morphologies are obtained depending on the nature of anions in the medium [34–36].

Thermolysis at 95°C of aluminum nitrate solutions for one week produces exclusively boehmite, γ-Al(O)(OH), in avoiding the formation of hydroxide Al(OH)$_3$, which is thermodynamically less stable at this temperature. The change in the acidity over a large range allows modification of the shape of the nanoparticles. At pH 4–5, heating produces fibers or rods around 100 nm in length. The fibers are formed by aggregation of very small platelets 3 nm thick and 6 nm wide, exhibiting (100) lateral faces and (010) basal planes. The particles synthesized at pH = 6.5 are pseudohexagonal platelets 10–15 nm wide and 4–5 nm thick with (100) and (101) lateral faces, while those synthesized at pH = 11.5 are diamond-shaped, 10–25 nm wide. The angle of ~104° between lateral faces corresponds to the angle between the (101) and (10-1) directions, suggesting (101) lateral faces (Figure 3.7). On the whole acidity range of synthesis, the particles are platelets with the same (010) basal faces but with different lateral faces. Such a change in the nature of lateral faces of particles results from the change in surface energy induced by the variation in the electrostatic surface charge density as a function of the pH [28]. This is an important feature of boehmite particles, because they are the precursor of γ-alumina, γ-Al$_2$O$_3$, largely used as a catalyst. As the thermal transformation boehmite → γ-alumina is a topotactic transformation, which maintains the morphology of particle, the control of the shape of boehmite particles enables the development

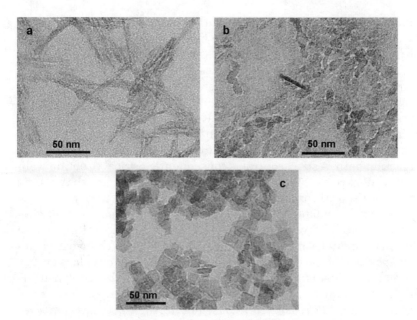

Figure 3.7 TEM micrographs of boehmite nanoparticles synthesized at (a) pH = 4.5, (b) pH = 6.5, (c) pH = 12.

of corresponding faces on γ-alumina nanoparticles and consequently adjustment of their catalytic activity toward a given reaction [20].

Thermolysis of strongly acidic $TiCl_4$ solutions enables a very efficient structural and morphological control of titanium oxide nanoparticles. After heating for one day at 90°C, $TiCl_4$ in concentrated perchloric acid solutions ($HClO_4$, 1–5 M) forms mixtures containing various proportions of the different TiO_2 polymorphs (anatase, brookite, and rutile). After heating for one week, the metastable phases, anatase and brookite, disappear through transformation into rutile with very different shapes depending on the acidity. This can be explained by the amount of metastable material transformed into rutile by a dissolution-crystallization process: when the amount of metastable phases is high, the initial rutile particles are strongly fed and their growth leads to elongated rods. It is thus possible to adjust the aspect ratio of rutile nanoparticles from around 5:1 to 15:1.

When $TiCl_4$ is thermolyzed in concentrated hydrochloric acid (HCl 1–5M), brookite nanoplatelets are stabilized and it is possible to obtain them as the main product when the stoichiometries of Cl/Ti and H^+/Ti are optimized (Figure 3.8) [37]. Brookite is currently obtained in hydrothermal conditions at elevated temperature in the form of large particles [38]. Nanoparticles of brookite apparently are never obtained except as byproducts of various reactions [39]. Quasi-quantitative synthesis of brookite nanoparticles seems to result from a specific precursor, $Ti(OH)_2Cl_2(H_2O)_2$, containing chloride as ligands in the early complexes formed in solution. It has been proposed [37] that chloride ligands orient the early stages of condensation in the formation of brookite. As long as chloride ions are present in suspension, brookite nanoparticles remain stable, while if chloride ions are replaced by perchloric anions, brookite transformation into rutile is complete after several hours at 90°C.

These examples underscore the versatility of oxide nanoparticle chemistry in an aqueous medium. The main parameter allowing the control of nanoparticle morphology (size, crystalline structure) is the acidity of the reaction bath. A strict control is consequently critical to obtain well-defined nanoparticles. It is however interesting to distinguish two sorts of physico-chemical conditions in these syntheses.

In moderately acidic or basic media, the sign and the density of the electrostatic surface charge of particles varies as a function of pH due to proton adsorption-desorption equilibria. This involves a change in chemical composition of the surface and therefore a change in surface energy of the particle during formation. When the surface charge density is high (the pH is far from the surface point of zero charge), the surface energy is strongly decreased. As a consequence, the size of nanoparticles decreases because the energetic penalty to develop surface is notably reduced. A semiquantitative model [28] works well to account for this size effect for anatase and magnetite and to explain the change in shape of

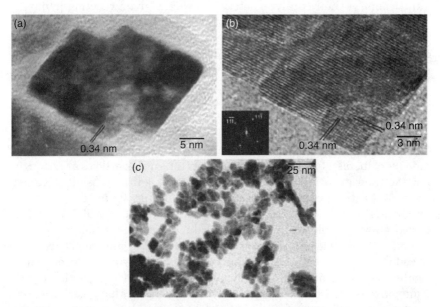

Figure 3.8 HRTEM micrographs of particles of brookite after one month of thermolysis at 100°C ($TiCl_4$ 0.15 mol dm^{-3}, HCl 3 mol · dm^{-3}); inset, the local electron diffraction pattern [37].

boehmite nanoparticles. In strongly acidic media used for thermolysis, the solubility of the solid is high because the surface is totally protonated and the ionic atmosphere near the surface of particles during formation is very likely high and constant, so that the surface energy is always low. Depending on their structure, some faces can be energetically favored, but dissolution-crystallization processes very likely play a role in the growth of particles. Other parameters such as thermolysis temperature, concentration, and presence of specific ligands have to be taken into account in the describing nanoparticle formation.

Hydrolysis of metallo-organic compounds.   Metallo-organic compounds, and especially metal alkoxides [40, 41], are largely involved in so-called sol-gel chemistry of oxide nanomaterials [42]. Metal alkoxides are also precursors of hybrid organic-inorganic materials, because such compounds can be used to introduce an organic part inside the mineral component [43–45].

Sol-gel chemistry mainly involves hydrolysis and condensation reactions of alkoxides $M(OR)_z$ in solution in an alcohol ROH, schematically represented as:

$$M(OR)_z + z \; H_2O \rightarrow M(OH)_z + z \; ROH \rightarrow MO_{z/2} + z/2 \; H_2O$$

These two reactions, hydroxylation and condensation, proceed by nucleophilic substitution of alkoxy or hydroxy ligands by hydroxylated species according to:

$$M(OR)_z + x \; HOX \rightleftharpoons [M(OR)_{z-x}(OX)_x] + x \; ROH$$

If X = H, the reaction is a hydroxylation. For X = M, it is a condensation (oxolation) and if X represents an organic or inorganic ligand, the reaction is a complexation. There is a deep difference with the processes in aqueous medium where condensation and complexation are nucleophilic substitutions while hydroxylation is an acid-base reaction. In organic medium, both hydrolysis and condensation follow an associative SN_2 mechanism in forming intermediate species in transition states in which the coordination number of the metal atom is increased. That explains why the reactivity of metal alkoxides toward hydrolysis and condensation is governed by three main parameters: the electrophilic character of the metal (its polarizing power), the steric effect of the alkoxy ligands, and the molecular structure of the metal alkoxide.

Generally, the reactivity of alkoxides toward substitutions increases when the electronegativity of the metal is low and its size is high. That lowers the covalence of the M-O bond and enhances the reaction rates. Silicon alkoxides are weakly reactive in the presence of water ($\chi_{Si} = 1.74$) while titanium alkoxides ($\chi_{Ti} = 1.32$) are very sensitive to moisture.

TABLE 3.1   Gelation Time of Silicon Alkoxides as a Function of Alkoxy Groups
Si(OR)$_4$ at Several Values of pH in Water and in 4-(Dimethylamino)Pyridine (DMAP)

	Gel Time (h) at RT			
Alkoxide	pH 7	pH 1 (HCl, HNO$_3$)	pH 9 (NH$_3$)	DMAP
Si(OMe)$_4$	44			
Si(OEt)$_4$	242	10	10	5 min
Si(OBu)$_4$	550			

Alkoxides of low electronegative elements have to be handled with care, under dry atmosphere, because traces of water can be enough to provoke precipitation. By comparison, alkoxides of very electronegative elements such as O = P(OEt)$_3$ (χ_P = 2.11) are quite inert and do not react with water in normal conditions.

The reactivity of metal alkoxides is also very sensitive to the steric hindrance of the alkoxy groups. It strongly decreases when the size of the OR group increases. For instance, the rate constant, k, for hydrolysis of Si(OR)$_4$ at 20°C decreases from 5.1×10^{-2} L · mol^{-1}s^{-1} for Si(OMe)$_4$ to 0.8×10^{-2} L · mol^{-1}s^{-1} for Si(OBu)$_4$ and the gelation time is increased by a factor of 10 (Table 3.1).

The acidity of the medium also influences the rate of hydrolysis and condensation reaction to a great extent as well as the morphology of the products. In an excess of water and in acidic medium (pH $\leq$ 4), the silicon alkoxides form transparent polymeric gels while in basic medium (pH $\geq$ 8); the condensation is also accelerated relatively to the reaction in neutral medium (Table 3.1) and leads to perfectly spherical and monodispersed particles of hydrated silica, as exemplified by Stöber's method (Figure 3.9) [46]. These variations reflect the acid or basic catalysis of the involved reactions.

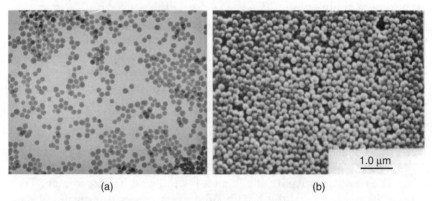

(a)   (b)

Figure 3.9   SEM of silica nanoparticles (23 nm in mean size) synthesized following Stöber's method from hydrolysis of Si(OEt)$_4$ in water/ethanol with 5% NH$_3$. (b) SEM of TiO$_2$ nanoparticles resulting from hydrolysis of Ti(OEt)$_4$ with water (pH 7).

It is possible to explain the overall structure of the silica polymer by considering at an early stage of condensation a branched chain of silicic tetrahedra containing several types of groups:

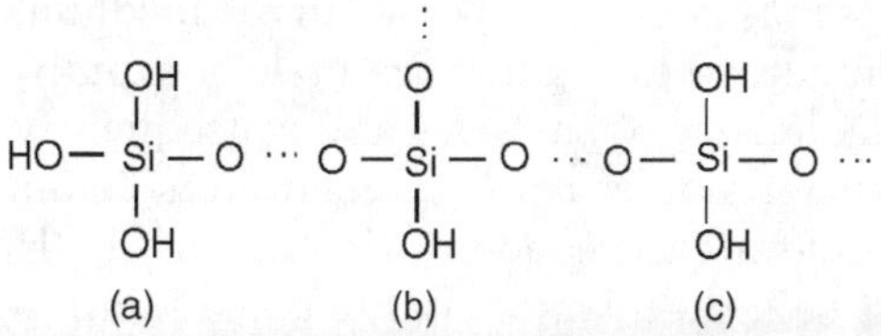

such a chain being formed by the oxolation reaction:

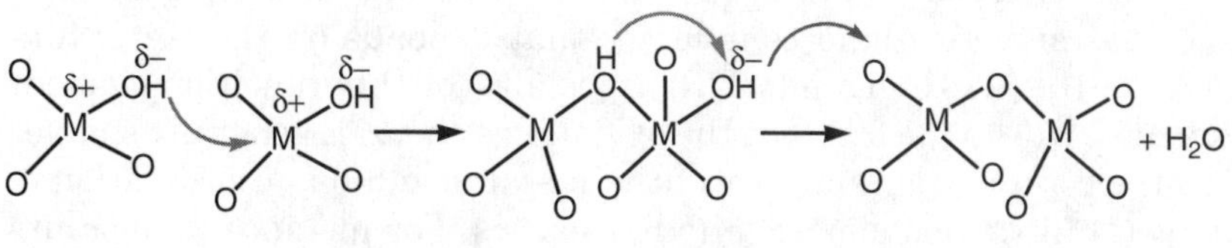

One may consider three different reactive groups: terminal or monocoordinated [A], branched or tricoordinated [B], and middle or dicoordinated [C]. Using the Partial Charges Model [10], it is possible to estimate the relative partial charges on the sites A, B, and C (Table 3.2).

In an alkaline medium, catalysis involves the first step of the condensation mechanism—that is, nucleophilic attack by the anionic forms (or OH^-). It must take place preferentially on sites with the highest partial charge—in the middle of the chain (sites B and C), leading to cross-linked polymers forming dense particles, in agreement with experimental observation. In an acidic medium, catalysis impacts the second step of condensation. Elimination of the proton from the alcohol bridge in the transition state is eased by the protonation of an OH ligand, which favors formation of the leaving group (aquo ligand). The OH groups concerned are those located at the ends of chains, which bear the highest negative partial charge, or even those of the $Si(OH)_4$ monomer. As a result, poorly cross-linked and poorly condensed chains

TABLE 3.2   Partial Charge Calculated of the Various Sites into a Chain of Silica Tetrahedral

Site	$\delta(Si)$	$\delta(OH)$
A	+0.50	−0.06
B	+0.58	+0.06
C	+0.54	0

are formed. Therefore, the morphology of the particles is heavily depend-ent upon the conditions of acidity in which condensation takes place. The catalysis of silica condensation may also be affected by nucleophilic acti-vation using additives such as 4-dimethylaminopyridine (DMAP, see Table 3.1). Particles and polymers may remain dispersed in the medium, forming sols, or they can agglomerate and gel more or less rapidly, depending on the surface charge density of particles and consequently on the pH of the medium. On either side of pH = 2, gelation is faster because acid or base catalysis accelerate the condensation rate of Si-OH groups between particles. At pH < 2, the surface charge is too small to provide efficient repulsion between particles. At pH > 2, base catalysis of oxolation has the same effect, which is maximum for pH = 6. For pH > 6, the surface charge is high enough for the sol to remain stable.

The reactivity of metal alkoxides is also deeply influenced by their molecular structure and complexity that depends on the steric hin-drance of the alkoxo ligands, OR, especially for the transition element alkoxides. Due to the fact that the oxidation state, z, is generally smaller than the coordination number of the metal, it inhibits coordination of the metal in the monomeric $M(OR)_z$ species. For instance, this occurs in the case of titanium alkoxide $Ti(O^iPr)_4$, which is a monomer in iso-propanol. The coordination of titanium is only four and the reaction with water leads to instantaneous precipitation of heterogeneous and amor-phous titania particles. With ethoxy ligands, titanium forms oligomeric species $[Ti(OEt)_4]_n$ (n = 3 in benzene, n = 2 in EtOH) in which the tita-nium coordination is higher, n = 5 in the trimer, n = 6 in the dimer because of the formation of a solvate $[Ti(OEt)_4]_2 \cdot (EtOH)_2$. Monodispersed spherical particles have been synthesized by controlled hydrolysis of a diluted solution of $Ti(OEt)_4$ in EtOH [47]. The monodispersity clearly results from slower hydrolysis and condensation reactions with less reactive precursors allowing decoupling of the nucleation and growth steps. It is however possible to control the reactivity of low coordinated titanium in the presence of specific ligands. For instance, hydrolysis at 60°C of titanium butoxide $Ti(OBu)_4$ in the presence of acetylacetone forms monodispersed 1–5 nm TiO_2 anatase nanoparticles [48]. A very ele-gant design of the shape of anatase nanospheres and nanorods is obtained by controlling the rate of hydrolysis of $Ti(O^iPr)_4$ at 80°C in the presence of oleic acid.

In a general way, the rate of reactions and the nature of condensed species obtained depend also on the hydrolysis ratio defined as h = H_2O/M.

- Molecular clusters are formed with very low hydrolysis ratios (h < 1). The condensation reactions are relatively limited. Hydrolysis of $[Ti(OEt)_4]_2 \cdot 2(EtOH)$ forms soluble species such as $Ti_7O_4(OEt)_{20}$ (h = 0,6), $Ti_{10}O_8(OEt)_{24}$ (h = 0,8) or $Ti_{16}O_{16}(OEt)_{32}$ (h = 1). A variety of such clusters have been isolated and characterized by X-ray diffraction. They can assemble themselves into nanostructures enabling the formation of hybrid organic-inorganic materials [49].

- Addition of water in substoichiometric amounts does not allow the substitution of all alkoxo ligands that otherwise leads to oxopolymers. Such precursors are well designed for obtaining coatings or thin films. The residual OR groups can react with surface hydroxyl groups of the substrate forming covalent bonds. The films are strongly adhesive and the organic residues can be then eliminated by thermal treatment.

- All alkoxo groups are eliminated in the presence of a large excess of water (h >>10), leading to oxide nanoparticles in suspension. Because of the high dielectric constant of the medium, the surface hydroxylated groups are mainly ionized allowing formation of sols or gels similar to those obtained in aqueous solution.

Nonhydrolytic routes to oxide nanoparticles.   Nonhydrolytic sol-gel chemistry has proved to be a promising route to metal oxides, as demonstrated by the work of Corriu and Vioux on silica, titania, and alumina [50]. It has become a widely explored approach to synthesize metal oxide nanoparticles under various conditions [8].

In nonaqueous media in the absence of surfactant, one possibility is the use of metal halides and alcohols (Nierderberger). This approach is based on the general reaction scheme:

$$\equiv M\text{-}X + ROH \rightarrow\ \equiv M\text{-}OH + RX$$

$$\equiv M\text{-}OH + \equiv M\text{-}X \rightarrow\ \equiv M\text{-}O\text{-}M \equiv\ + HX$$

It is widely observed that complexation of water to a transition metal results in an increase in its Brønsted acidity [9]. Similarly, an increased acidity of water upon complexation to main group compounds has been inferred from NMR data. The recent isolation of a series of amine substituted alcohol complexes [51] has allowed for an estimation of the change in the acidity of alcohols upon coordination to a metal. Complexation of a protic Lewis base (e.g., ROH, R_2NH, etc.) results in the increase in Brønsted acidity discerned by a decrease in pK_a of about 7 for the α-proton. This activation of the coordinated ligand by increasing the formal positive charge on the α-substituent is analogous to the activation of organic carbonyls toward alkylation and/or reduction by aluminum alkyls [52–54].

While reaction of primary and secondary alcohols with tetrachlorosilane is the usual method for preparing tetraalkoxysilanes [40], the same

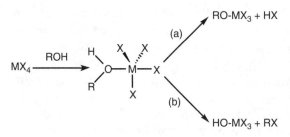

Figure 3.10 Possible reactions for the alcoholysis of MX_4.

reaction with tertiary alcohols and benzylic alcohols form silica and the corresponding alkyl halide, RCl. The two modes of reaction involve initially the coordination of a lone pair of electrons of an alcoholic oxygen atom to the silicon center, followed by the cleavage of either the hydroxyl or alkoxyl group (Figure 3.10). Electron-donor substituent groups in the alkyl radical direct the process to hydroxylation (pathway b) with the liberation of RCl) by favoring the nucleophilic attack of chloride on the carbon group, due to its increased cationic character.

Hydroxylated species so formed react with unsolvolyzed compound according to:

$$\equiv M\text{-}OH + X\text{-}M \equiv \rightarrow \equiv M\text{-}O\text{-}M \equiv + HX$$

Benzyl alcohol seems to be well-designed for synthesis of various oxide nanoparticles. Typically, anhydrous metal chloride is introduced in benzyl alcohol under vigorous stirring in order to avoid precipitation and then the mixture is heated under stirring for days (2 to 20), depending on the metal chloride. Nanoparticles of titania (anatase) with size varying from 4 to 8 nm are obtained at temperatures from 40 to 150°C with different concentrations of $TiCl_4$ [55]. In similar conditions, $VOCl_3$ forms nanorods (approximately 200×35 nm) of vanadium oxide and WCl_6 forms platelets (approximately 30 to 100 nm, thickness 5 to 10 nm) of tungsten oxide [56].

Alkyl halide elimination also occurs between metal chloride and metal alkoxide following the reaction:

$$\equiv M\text{-}Cl + RO\text{-}M \equiv \rightarrow \equiv M\text{-}O\text{-}M \equiv + RCl$$

Such a reaction between TiX_4 and $Ti(OR)_4$ in heptadecane in the presence or trioctylphosphine oxide (TOPO) at 300°C produces spherical nanoparticles of TiO_2 anatase, around 10 nm in diameter [57]. Here, TOPO acts as a nonselectively adsorbed surfactant, which slows down the rate of reaction, allows the control of particle size, and avoids the formation of other TiO_2 polymorphs (brookite or rutile). In the presence of the mixed

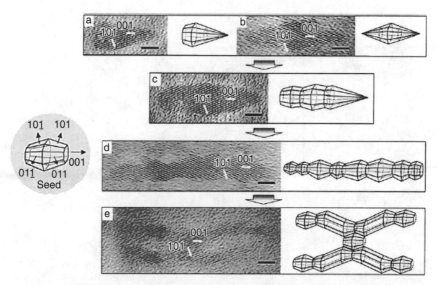

Figure 3.11 HRTEM analyses of TiO$_2$ anatase nanoparticles and simulated three-dimensional shape of (a) a bullet, (b) a diamond, (c) a short rod, (d) a long rod, and (e) a branched rod. The long axes of the nanocrystals are parallel to the c-axis of the anatase structure, while the nanocrystals are faceted with (101) faces along the short axes. Hexagon shapes (the [010] projection of a truncated octagonal bipyramid) truncated with two (001) and four (101) faces are observed either at the one end or at the center of the nanocrystals. The branched shape is a result of the growth along (101) directions starting from the hexagon shape. Scale bar = 3 nm [58].

surfactant system, TOPO and lauric acid (LA), with increasing ratios LA/TOPO, a spectacular control of the shape of TiO$_2$ anatase nanorods is obtained (Figure 3.11). The specifically strong adsorption of LA onto (001) faces slows down the growth along [001] directions, thereby inducing growth along [101] directions that results in the formation of rods.

Another nonhydrolytic synthesis of oxide nanoparticles involves thermal decomposition of metal organic complexes in solution in the presence of surfactant. In fact, since water may be produced by the thermolysis of the organic derivatives, a hydrolytic pathway cannot be excluded. One of the most studied approaches involves the thermolytic decomposition of an inorganic complex at high temperatures. Two approaches include: the decomposition of Fe(acac)$_3$ or FeCl$_3$ and M(acac)$_2$ salts [59–61], and the decomposition of Fe(CO)$_5$ and M(acac)$_2$ salts [62, 63].

For simple oxides (e.g., Fe$_3$O$_4$) the precursor (e.g., Fe(acac)$_3$) is added to a suitable solvent heated to a temperature that allows for the rapid decomposition of the precursor. The choice of temperature and the temperature control (i.e., variation of the temperature during the reaction) are important in defining the resulting nanoparticle size and size

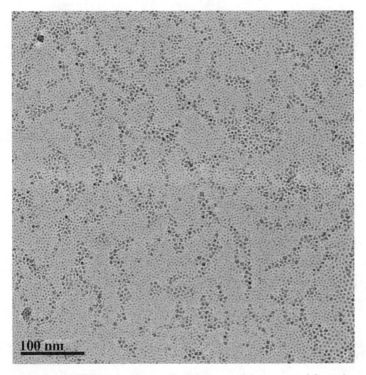

Figure 3.12 TEM image of 4 nm Fe_3O_4 nanoparticles prepared from the thermal decomposition of $Fe(acac)_3$.

distribution. By this method, highly uniform nanoparticles can be obtained (see Figure 3.12).

In addition to simple metal oxides (M_xO_y) a range of mixed metal oxides can also be prepared. For example, nanospheres and nanocubes of cobalt ferrite can be obtained from cobalt and iron acetylacetonates, $Co(acac)_2$ and $Fe(acac)_3$ in solution in phenylether and hexadecanediol in the presence of oleic acid and oleylamine [64]. Heating at 260°C forms $CoFe_2O_4$ spherical nanocrystals with a diameter of 5 nm. These nanocrystals serve as seeds for a new growth as the second step of the synthesis, giving perfect nanocubes from 8 to 12 nm, depending on the conditions. Nanocubes in the 8 nm range can also be used as seeds to obtain spheres (Figure 3.13). The tuning of the shape of ferrite nanocrystals is managed by the parameters of growth, such as heating rate, temperature, reaction time, ratio of seed to precursors, and ratio of oleic acid, acting as surfactant stabilizing the nanocrystal, to oleylamine providing basic conditions needed for the formation of spinel oxide. Variations in the morphology of numerous oxide nanocrystals, including nanocrystals of Fe, Co, Mn ferrites, Co_3O_4, Cr_2O_3, MnO, NiO, ZnO, and others, have been obtained by pyrolysis of metal carboxylates in the presence of different fatty acids (oleic, myristic) [65–67].

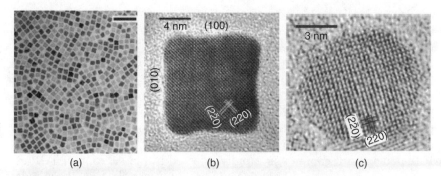

(a)   (b)   (c)

Figure 3.13 (a) TEM micrographs with the scale bar as 50 nm of cubic $CoFe_2O_4$ nanocrystals. (b) HRTEM micrographs showing a 12-nm cubic $CoFe_2O_4$ nanocrystal and (c) an 8-nm spherical $CoFe_2O_4$ nanocrystal [64].

Metal carbonyl complexes are also interesting precursors to synthesize uniform metal oxide nanoparticles. Thermal decomposition at 100°C of iron pentacarbonyl, $Fe(CO)_5$, in octyl ether in the presence of oleic acid forms iron nanoparticles which are then transformed to monodisperse spherical γ-Fe_2O_3 nanoparticles by trimethylamine oxide acting as a mild oxidant (Figure 3.14) [68]. Particle size can be varied from 4 to 16 nm by

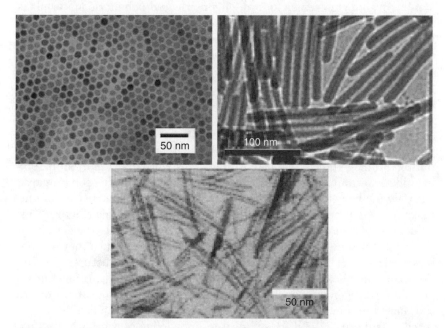

Figure 3.14 TEM image of (a) a two-dimensional hexagonal assembly of 11 nm γ-Fe_2O_3 nanocrystallites (from Hyeon 2001) and (b) 8 × 140 nm sized MnO nanorods (from Park 2004a), (c) 75 ± 20 nm tungsten oxide nanorods [71].

controlling the molar ratio $Fe(CO)_5$ to oleic acid. Thermal decomposition of $Fe(CO)_5$ in solution containing dodecylamine as a capping ligand and under aerobic conditions forms also γ-Fe_2O_3 nanoparticles with diamond, sphere, and triangle shapes with similar 12-nm size [69]. Uniform-sized MnO nanospheres and nanorods are obtained by heating at 300°C the mixture of $Mn_2(CO)_{10}$ with oleylamine in trioctylphosphine (TOP) [70]. The size of nanospheres can be varied from 5 to 40 nm depending on the duration of heating, using phosphines both as solvent and stabilizing agent (Figure 3.14). With TOP, 10 nm MnO particles can be obtained. If the surfactant complex is rapidly injected into a solution of TOP at 330°C, nanorods 8 × 140 nm of MnO are produced. In fact, these rods are polycrystalline. They are formed by an aggregation of spheres with oriented attachment and having a core shell structure with a thin Mn_3O_4 shell. Heating of $W(CO)_6$ at 270°C for 2 hours in trimethylamine oxide in the presence of oleylamine forms uniform nanorods of tungsten oxide with an X-ray diffraction pattern matching the $W_{18}O_{49}$ reflections [71]. The lengths of the nanorods are controlled by the temperature and the amount of oleylamine.

From minerals to materials

As discussed earlier, precursor sol-gels are traditionally prepared via the hydrolysis of metal compounds. This "bottom-up" approach of reacting small inorganic molecules to form oligomeric and polymeric materials is a common approach for a wide range of metal and nonmetal oxides. However, in the case of aluminum oxide nanoparticles, the relative rate of the hydrolysis and condensation reactions often makes particle size control difficult. The aluminum-based sol-gels formed during the hydrolysis of aluminum compounds belong to a general class of compounds: alumoxanes. The term alumoxane is often given to aluminum oxide macromolecules formed by the hydrolysis of aluminum compounds or salts, AlX_3 where X = R, OR, $OSiR_3$, or O_2CR; however, it may also be used for any species containing an oxo (O^{2-}) bridge binding (at least) two aluminum atoms—that is, Al-O-Al. Alumoxanes were first reported in 1958 by Andrianov and Zhadanov [72], however, they have since been prepared with a wide variety of substituents on aluminum. The structure of alumoxanes was proposed to consist of linear or cyclic chains (Figure 3.15) analogous to that of poly-siloxanes [73]. Strictly speaking, the classification of alumoxanes as polymers is slightly misleading since they are not polymeric per se, but exist as three-dimensional cage structures [74–76]. For example, siloxy-alumoxanes, $[Al(O)(OH)_x(OSiR_3)_{1-x}]_n$, consist of an aluminum-oxygen nanoparticle ore structure (Figure 3.15c) analogous to that found in the mineral boehmite, $[Al(O)(OH)]_n$, with a siloxide substituted periphery [77–79]. Based on the knowledge of the boehmite-like nanoparticle core structure of hydrolytically stable alumoxanes, it was

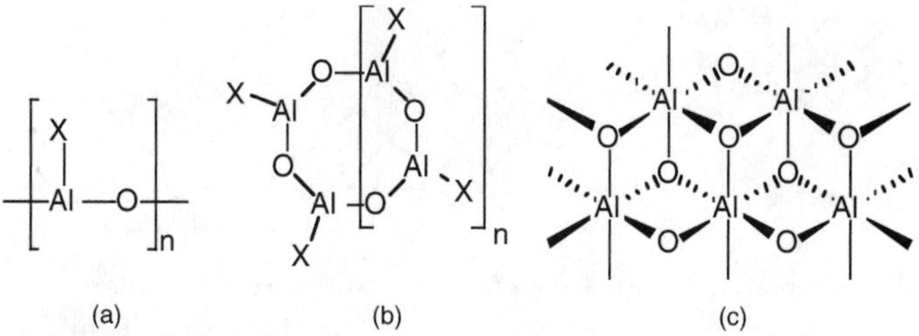

Figure 3.15 Structural models proposed (a and b) and observed (c) for aluminum oxide nanoparticles formed from the hydrolysis of aluminum compounds.

proposed that alumoxanes could be prepared directly from the mineral boehmite. Such a "top-down" approach represented a departure from the traditional synthetic methodologies.

Assuming that hydrolytically stable alumoxanes have the boehmite-like core structure (Figure 3.15c), it would seem logical that they could be prepared directly from the mineral boehmite. The type of capping ligand used in such a process must be able to abstract and stabilize a small fragment of the solid-state material. In the siloxy-alumoxanes it was demonstrated that the "organic" unit itself contains aluminum, as shown in Figure 3.16a. Thus, in order to prepare the siloxy-alumoxane the "ligand" $[Al(OH)_2(OSiR_3)_2]^-$, would be required as a bridging group; adding this unit clearly presents a significant synthetic challenge. However, the carboxylate anion, $[RCO_2]^-$, is an isoelectronic and structural analog of the organic periphery found in our siloxy-alumoxanes (Figure 3.16).

Thus, it has been shown that carboxylic acids (RCO_2H) react with boehmite, $[Al(O)(OH)]_n$, to yield the appropriate carboxy-alumoxane:

$$[Al(O)(OH)]_n \xrightarrow{\text{HO}_2\text{CR}} [Al(O)_x(OH)_y(O_2CR)_z]_n$$

Initial syntheses were carried out using the acid as the solvent or xylene [80, 81], however, subsequent research demonstrated the use of water as a solvent and acetic acid as the most convenient capping agent [82]. A solventless synthesis has also been developed [83]. Thus, the synthesis

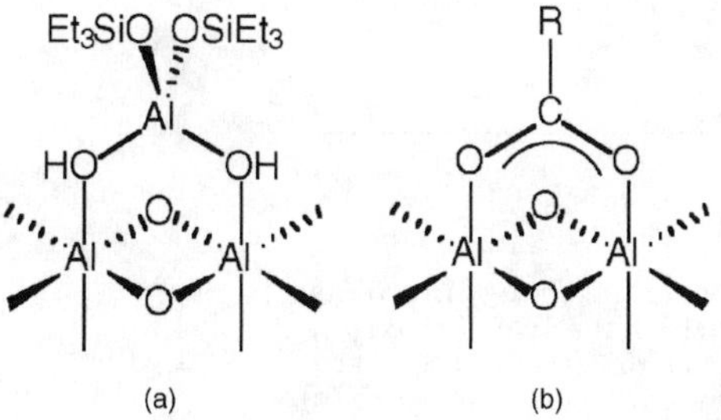

Figure 3.16 Structural relationship of the capping ligand for (a) siloxy and (b) carboxylate alumoxane nanoparticles.

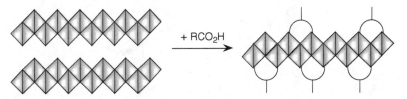

Figure 3.17 Pictorial representation of the reaction of boehmite with carboxylic acids.

of alumoxane nanoparticles may be summarized as involving the reaction between dirt (boehmite), vinegar (acetic acid), and water. The function of the acid is twofold. First, to cleave the mineral lattice and "carve out" nanoscale fragment, and second, to provide a chemical cap to the fragment (Figure 3.17).

The carboxylate-alumoxane nanoparticles prepared from the reaction of boehmite and carboxylic acids are air and water stable. The soluble carboxylate-alumoxanes can be dip-coated, spin-coated, and spray-coated onto various substrates. The physical properties of the alumoxanes are highly dependent on the identity of the substituents. The size of the alumoxane nanoparticles is dependant on the substituents, the reaction conditions (concentration, temperature, time, etc.), and the pH of the reaction solution (Figure 3.18) [84]. Unlike other forms of oxide nanoparticle, the alumoxanes are not monodispersed but have a range of particle sizes.

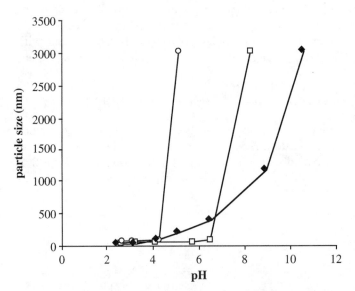

Figure 3.18 Unimodal analysis plot showing the change in average particle size with increasing pH for each of the five carboxylate-alumoxanes: acetic acid-alumoxane (□), methoxy(ethoxy)acetic acid-alumoxane (o), and methoxy(ethoxyethoxy)acetic acid-alumoxane (■).

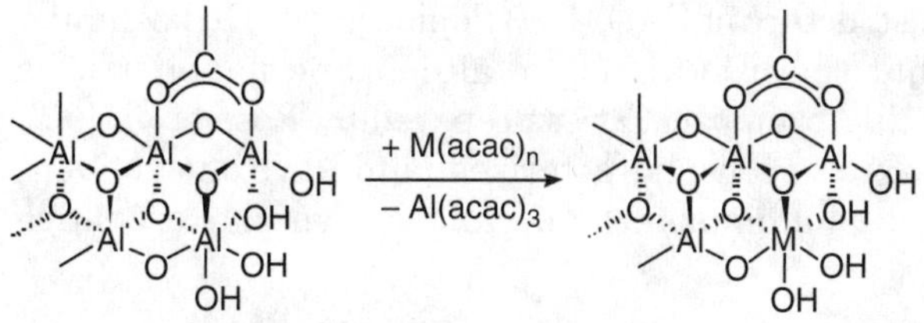

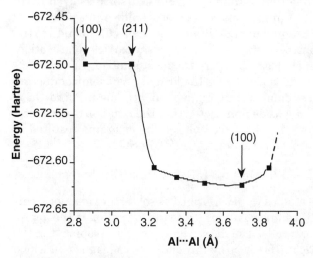

Figure 3.19 Schematic representation of the exchange reaction that occurs between a metal complex and the core of the alumoxane nanoparticle.

Also, unlike other metal oxide nanoparticles, the core of the alumoxane can undergo a low temperature reaction that allows for the incorporation of other metals (e.g., Ti, La, Mo, V, Ca). This occurs by reaction of metal acetylacetenoates [$M(acac)_n$] with the carboxylate alumoxane (Figure 3.19) [85–87].

The reason for the particular stability and usefulness of carboxylic acids in the cleavage of the boehmite structure is due to the particular bridging binding mode the carboxylate ligand adopts in aluminum-oxide systems [88]. Using a combination of X-ray crystallography and *ab initio* calculations it has been shown that the carboxylate ligand is therefore near perfectly suited to bind to the (100) surface of boehmite (Al $\cdots$ Al = 3.70 Å), and hence stabilize the boehmite-like core in carboxylate alumoxanes (Figure 3.20) [89].

Figure 3.20 Total energy calculation of a carboxylic acid interacting with an Al_2 unit as a function of the Al $\cdots$ Al distance. The Al $\cdots$ Al distances present on the [100] and [211] crystallographic planes of boehmite are marked.

Given the analogous structure of Fe(O)(OH) (lepidocrocite) to boehmite, it is not surprising that the iron analog of alumoxane nanoparticles (i.e., ferroxanes) is readily prepared. First prepared by Rose et al. [90], ferroxanes have been extensively characterized, and have shown identical structural features to alumoxanes and undergo similar exchange reactions [91].

Semiconductor Nanoparticles
(Quantum Dots and Quantum Rods)

The synthesis of semiconductors as nanoscale particles yields materials with properties of absorbance and fluorescence that differ considerably from those of the larger, bulk-scale material. Highly specific bands of absorbance or fluorescence arise from the quantum confinement of electrons that are excited in these materials when exposed to light. These materials are therefore of great interest in applications ranging from medical imaging to tagging and sensing.

Solution processes

The most studied nonoxide semiconductors are cadmium chalcogenides (CdE, with E = sulfide, selenide, and telluride). CdE nanocrystals were probably the first material used to demonstrate quantum-size effects corresponding to a change in the electronic structure with size—that is, the increase of the band gap energy with the decrease in size of particles [4–6, 92, 93]. These semiconductor nanocrystals are commonly synthesized by thermal decomposition of an organometallic precursor dissolved in an anhydrous solvent containing a source of chalcogenide and a stabilizing material (polymer or capping ligand). Stabilizing molecules bound to the surface of particles control their growth and prevent particle aggregation.

Although cadmium chalcogenides are the most studied semiconducting nanoparticles, the methodology for the formation of semiconducting nanoparticles was first demonstrated independently for InP and GaAs [94, 95]. In both cases it was demonstrated that the reaction of the metal halide with the trimethylsilyl–derived phosphine or arsine resulted in the formation of the appropriate pnictide and Me_3SiCl:

$$InCl_3 + P(SiMe_3)_3 \rightarrow InP + 3\ Me_3SiCl$$

Although these initial studies were performed as solid-state reactions, carrying them out in high boiling solutions led to the formation of the appropriate nanoparticle materials. The most widely used development of the Barron/Wells synthetic method was by Bawendi and colleagues[7] in which an alkyl derivative was used in place of the halide. Dimethylcadmium $Cd(CH_3)_2$ is used as a cadmium source and bis(trimethylsilyl)sulfide, $(Me_3Si)_2S$, trioctylphosphine selenide or telluride (TOPSe, TOPTe) serve as sources of selenide in trioctylphosphine oxide (TOPO) used as solvent and capping molecule. The mixture is heated at 230–260°C. It is best to prepare samples over a period of a few hours of steady growth by modulating the temperature in response to changes in the size distribution as estimated from the absorption spectra of aliquots removed at regular intervals. Temperature is lowered in response to a spreading of size distribution and increased when growth appears to stop. Using this method, a series of sizes of CdSe nanocrystals ranging from 1.5 to 11.5 nm in diameter can be obtained (Figure 3.21). These particles, capped with

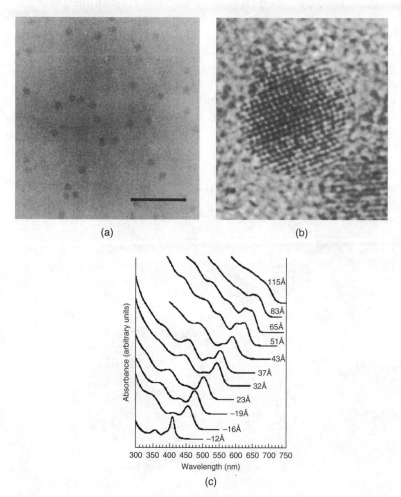

Figure 3.21 TEM image of (a) well-dispersed CdSe nanoparticles of around 3.5 nm in mean diameter (bar: 20 nm); (b) CdSe crystallite 8 nm in size showing stacking faults in the (002) direction (bar 5 nm); (c) room temperature optical absorption spectra of CdSe nanocrystallites dispersed in hexane and ranging in size from 1.2 to 11.5 nm [7].

TOP/TOPO molecules, are nonaggregated and easily dispersible in organic solvents forming optically clear dispersions.

When similar syntheses are performed in the presence of surfactant, strongly anisotropic nanoparticles are obtained. From $Cd(CH_3)_2$ and TOPSe in hot mixture of TOPO and hexylphosphonic acid (HPA) rod-shaped CdSe nanoparticles can be obtained [96]. The role of adsorption of HPA is clearly demonstrated by the change in shape of nanoparticles with the increase in HPA concentration. An increase in the concentration of Cd precursor into the reactor also induces a higher aspect ratio of nanoparticles exhibiting quasi-perfect wurtzite crystalline structure (Figure 3.22). A further slow addition of monomer sustains growth of (00-1) faces without additional nucleation giving rods exceeding 100 nm in length with an aspect ratio over 30:1.

The role of HPA seems to be to increase the growth rate of the (00-1) faces relative to all other faces. Further support for this argument comes from the formation of multipods, especially tetrapods. These are remarkable single-crystal particles consisting of a tetrahedral zinc blende core

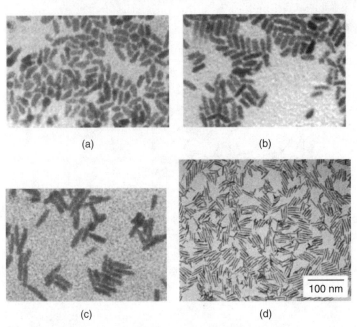

(a)   (b)

(c)   (d)

Figure 3.22 TEMs of CdSe nanoparticles obtained with surfactant to Cd ratio of 20% HPA in TOPO. The single-injection volumes increase from (a) to (c). Greater injection volume favors rod growth. (d) TEM of a typical multiple injection extended rod synthesis. The average length is 34.5 ± 4.4 nm with an aspect ratio of 10:1 [96].

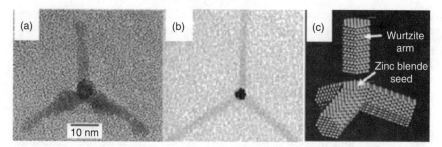

Figure 3.23 Tetrapod-shaped nanocrystals of (a) CdSe and (b) MnS formed by epitaxial growth of four wurtzite arms on a tetrahedral zinc blende seed results in tetrapod-shaped nanocrystals (c) [96, 97].

with four wurtzite arms (Figure 3.23). They are formed when a CdSe or CdTe nanocrystal nucleates in the zinc blende structure instead of the wurtzite structure. The wurtzite arms grow out of the four (111) equivalent faces of the tetrahedral zinc blende core. A key parameter for achieving tetrapod growth is the energy difference between the two structural types, which determines the temperature range in which one structure is preferred during nucleation and the other during growth. For CdS and CdSe, the energy difference is very small and it is difficult to isolate controllably the growth of one phase at a time. With CdTe, the energy difference is large enough that, even at the elevated temperatures preferred for wurtzite growth, nucleation can occur selectively in the zinc blende structure, the wurtzite growth being favored by using higher temperatures in the presence of surfactant.

Because $Cd(CH_3)_2$ is extremely toxic, pyrophoric, and explosive at elevated temperature, other Cd sources have been used. CdO appears to be an interesting precursor [98, 99]. The reddish CdO powder dissolves in TOPO and HPA or TDPA (tetradecylphosphonic acid) at about 300°C giving a colorless homogeneous solution. By introducing selenium or tellurium dissolved in TOP, nanocrystals grow at 250°C to the desired size. Nanorods of CdSe or CdTe can also be produced by using a greater initial concentration of cadmium as compared to reactions for nanoparticles. The evolution of the particle shape results from a diffusion-controlled growth mechanism. The unique structural feature of (00-1) facets of the wurtzite structure and the high chemical potential on both unique facets makes the growth reaction rate along the c-axis much faster than that along any other axis. The limited amount of monomers maintained by diffusion is mainly consumed by the quick growth of these unique facets. As a result, the diffusion flux goes to the c-axis exclusively, which is the long axis of the quantum rods [100]. This approach further enables large-scale production of Cd chalcogenide quantum dots and quantum rods with controllability of their size, monodispersity, and aspect ratio. This approach has been successfully applied for synthesis of numerous other metal chalcogenides, including ZnS, ZnSe, and $Zn_{1-x}Cd_xS$ [101]. CdS nanorods have also been obtained from $Cd(S_2CNCH_2CH_3)_2$, an air-stable compound, thermally decomposed in hexadecylamine HDA at around 300°C [102]. Various shapes of CdS nanocrystals are obtained in changing the growth temperature. Rods are formed at elevated temperatures (300°C) and armed rods (bipods, tripods) are obtained as the growth temperature is decreased to 180°C. Around 120°C, tetrapods of four armed rods are dominant.

A similar procedure, using $Mn(S_2CNCH_2CH_3)_2$, enables formation of MnS nanocrystals with various shapes including cubes, spheres, monowires, and branched wires (bi-, tri- and tetrapods) [103]. Nanorods of diluted magnetic semiconductors, $Cd_{1-x}Mn_xS$, have also been obtained by this procedure. Various shapes of PbS nanocrystals have similarly been produced from $Pb(S_2CNCH_2CH_3)_2$ [104]. NiS nanocrystals, elongated along the 110 direction, were prepared by solventless thermal decomposition of a mixture of nickel alkylthiolate and octadecanoate. Similarly, Cu_2S nanorods or nanodisks are obtained by solventless thermal decomposition of a copper alkylthiolate precursor [105]. Finally, a very interesting design of nano-objects with advanced shapes results from oriented attachment of nanoparticles. PbSe nanowires of 3.5 to 18 nm in diameter and 10 to 30 mm in length (Figure 3.17) are obtained from the reaction between lead oleate with TOPSe at 250°C in solution in diphenylether in the presence of TDPA [106]. In the presence of hot (250°C) hexadecylamine in diphenylether, lead oleate and TOPSe form PbSe nanorings resulting very likely from a similarly oriented attachment of nanoparticles (Figure 3.24).

Nanoparticles from the vapor phase

The chemical vapor deposition (CVD) of semiconductors from molecular precursors has been extensively studied. One class of precursors is the so-called single source precursors, those in which all of the desired elements are in the same molecule. The use of single-source precursors allows for the structure of films grown by CVD to be controlled by the structure of the precursor molecule employed [107–109]. Such a process requires the precursor structure to remain intact during deposition [110]. However, vapor phase molecular cleavage can alter film morphology as well as influence phase formation. It has been observed that a major consequence of precursor decomposition in the vapor phase is cluster formation, leading to a rough surface morphology [111]. While particulate growth during CVD is often an undesirable component of the film deposition process, it is possible to prepare highly uniform

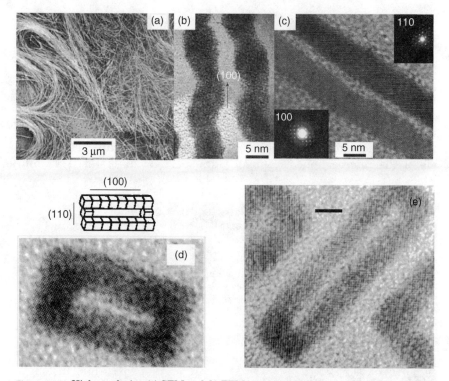

Figure 3.24 High-resolution (a) SEM and (b) TEM images of PbSe nanowires grown in solution in the presence of oleic acid. (c) high-resolution TEM images of PbSe nanowires formed in the presence of oleic acid and n-tetradecylphosphonic acid. Selected area electron diffraction from single nanowires imaged along the (100) and (110) zone axes (insets to c). (d, e) TEM image of PbSe nanorings (bar 5 nm) [106].

well-defined nanoparticles of InSe and GaSe from the vapor phase thermolysis of precursor molecules under CVD conditions.

CVD growth using the cubane precursors [(^{t}Bu)GaSe]$_4$ (Figure 3.25a) and [(EtMe$_2$C)InSe]$_4$ (Figure 3.25b) carried out on single crystal KBr

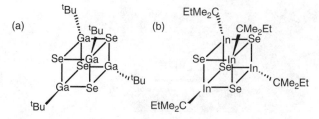

Figure 3.25 The structure of the cubane single source precursors to GaSe and InSe nanoparticles [113, 114].

substrates in a hot-walled CVD system allows for the formation of spherical nanoparticles [112].

The InSe grown at 290°C from $[(EtMe_2C)InSe]_4$ consist of spheres (Figure 3.26) with a mean diameter of 88 nm. The electron diffraction pattern exhibits well-defined rings, consistent with a polycrystalline hexagonal InSe. As with InSe, those grown from $[(^tBu)GaSe]_4$ at 335°C consist of pseudospherical nanoparticles; however, unlike the InSe films, these appear as "strings of pearls" which retain their connectivity and remain intact after being floated from the growth substrate. From analysis of the micrograph, the mean GaSe particle diameter is 42 nm.

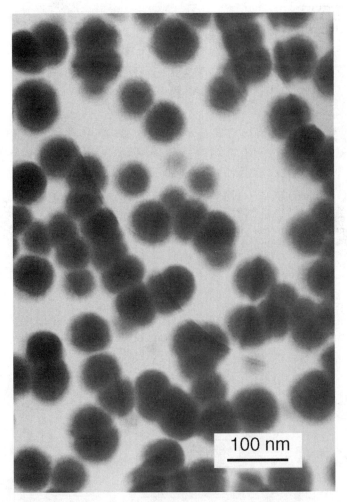

Figure 3.26   TEM micrograph of InSe nanoparticles grown from $[(EtMe_2C)InSe]_4$ at 290°C under vacuum.

A similar "strings of pearls" morphology has been observed for the solution decomposition of $Co_2(CO)_8$ in the presence of soluble organic polymers [115].

The small sizes and pseudospherical geometry of the InSe and GaSe particles suggests a vapor phase component is operative during the deposition of these metal selenide films. The fragmentation of the precursors in the vapor phase may combine to form particles that will remain in the vapor phase until their size is such as to "precipitate" out of the gas stream. The deposition of particles from the vapor phase due to gravitational or thermophoretic forces has been studied for SiC [116]. Factors that may control the particle size include: precursor concentration, vacuum versus atmospheric growth, growth temperature, and the thermal stability of the precursor.

Metallics, Bimetallics, and Alloys

The most currently synthesized metallic nanoparticles include two categories of metals: noble or precious metals (Ag, Au, Pd, Pt), less extensively Ru and Cu, and ferromagnetic metals (Co, Fe, Ni). Silver, gold, and copper are essentially used for their color, yellow to red, due to their plasmon resonance located within the visible domain of the electromagnetic spectrum. Palladium, platinum, and ruthenium are largely used for heterogeneous catalysis. Cobalt, iron, and nickel are interesting as magnetic nanomaterials for various applications such as information storage in recording devices, ferrofluids, and microwave composite materials.

The two main routes for the synthesis of metallic nanoparticles are the reduction of metallic salts in solution, involving a large variety of salts and reducing agents, and the decomposition of zerovalent metal compounds. Whatever the reaction involved, the formation of monosized nanoparticles is achieved by a combination of a low concentration of solute and a protective layer (polymer, surfactant, or functional groups) adsorbed or grafted onto the surfaces. Low concentrations are needed to control the nucleation/growth steps, and polymeric layers reduce diffusion causing diffusion to be the rate-limiting step of nuclei growth, resulting in uniformly sized nanoparticles. The protective layer also limits or avoids irreversible aggregation of nanoparticles.

Reduction mechanism

The basic reaction involved in the production of metallic nanoparticles is the reduction of metal cations in solution:

$$M^{z+} + Red \rightarrow M^0 + Ox$$

TABLE 3.3   Guidelines for the Choice of Reducing Agents and Reaction Conditions in the Precipitation of Metal Particles [117]

Metal species	E^0 (V)	Reducing Agent	Conditions	Rate
Au^{3+}, Au^+, Pt^{4+}, Pt^{2+}, Pd^{2+}	≥ 0.7	organic acids, alcohols, polyols	$\geq 70°C$	Slow
Ag^+, Rh^{3+}, Hg^{2+}, Ir^{3+}		aldehydes, sugars	$<50°C$	Moderate
		hydrazine, H_2SO_3	Ambient	Fast
		$NaBH_4$, boranes,	Ambient	Very fast
Cu^{2+}, Re^{3+}, Ru^{3+}	0.7 and ≥ 0	polyols	$> 120°C$	Slow
		aldehydes, sugars	70–100°C	Slow
		hydrazine, hydrogen	$< 70°C$	Moderate
		$NaBH_4$	Ambient	Fast
Cd^{2+}, Co^{2+}, Ni^{2+}, Fe^{2+}	< 0 and ≥ -0.5	polyols	$>180°C$	Slow
In^{3+}, Sn^{2+}, Mo^{3+}, W^{6+}		hydrazine	70–100°C	Slow
		$NaBH_4$, boranes	Ambient	Fast
		hydrated e^-, radicals	Ambient	Very fast
Cr^{3+}, Mn^{2+}, Ta^{5+}, V^{2+}	< 0.6	$NaBH_4$, boranes	T, P > ambient	Slow
		hydrated e^-, radicals	Ambient	Fast

The driving force of the reaction is the difference, $\Delta E°$, between the standard redox potentials of the two redox couples implicated, $E(M^{z+}/M^0)$ and $E(Ox/Red)$. The value of $\Delta E°$ determines the composition of the system through the equilibrium constant K given by:

$$\ln K = nF \, \Delta E°/RT$$

The reaction is thermodynamically possible if $\Delta E°$ is positive, but practically, its value must be at least 0.3 to 0.4, otherwise the reaction proceeds too slowly to be useful. Thus, highly electropositive metals (Ag, Au, Pt, Pd, Ru, Rh) with standard potentials $E^0(Mn^+/M^0) > 0.7$ V react with mild reducing agents while more electronegative metals (Co, Fe, Ni) with $E^0(Mn^+/M^0) < -0.2$ V need strongly reducing agents and have to be manipulated with care because the metallic nanoparticles are very sensitive to oxidation. Some widely used reactions are listed in Table 3.3. Complexation of cations in solution plays an important role on their reducibility and, as the stability of complexes increases, reduction is more difficult (Table 3.4).

TABLE 3.4   Influence of Silver Ion Complexation on the Redox Potential

Couple	logK	E°(V)
$Ag^+ + e \rightarrow Ag^0$	–	+0.80
$Ag(NH_3)_2^+ + e \rightarrow Ag^0 + 2\,NH_3$	−7.2	+0.38
$Ag(S_2O_3)_2^{3-} + e \rightarrow Ag^0 + 2\,S_2O_3^-$	−13.4	+0.01
$Ag(CN)_3^{2-} + e \rightarrow Ag^0 + 3\,CN_3^-$	−22.2	−0.51

A very large majority of redox systems in aqueous media is also strongly dependent on pH because the nature of the metal complex in solution is highly dependent on the acidity of the medium. Consequently, the control of metal complex chemistry in solution by complexation and/or by adjusting pH allows control of the reactivity of the species in a given metal-reducing agent system. It is also important to note that reduction can be performed in aqueous or nonaqueous media. However, synthesis in a nonaqueous medium presents serious difficulties because the metallic salts are generally very weakly soluble. Organometallic compounds can be used with suitable reducing agents, which are soluble in a specific medium. In some cases, the solvent can play a multiple role—for instance, primary alcohols (methanol, ethanol), which act as both solvent and reducing agent in the synthesis of Au, Pd, Pt nanoparticles, and polyols (e.g., glycerol, diethyleneglycol) can play a triple role as solvent, reducing agent, and stabilizer in the preparation of a lot of metallic nanoparticles, including noble metals and transition metals.

Whatever the reaction involved, the formation of particles occurs, as for other systems (oxides, chalcogenides), in two steps: nucleation and growth. Since the metal atoms are highly insoluble, as soon as they are generated they aggregate by a stepwise addition, forming clusters and then nuclei when they reach a critical size. This step is favored by a high supersaturation of metal atoms, and consequently by a high rate of reduction or decomposition of precursor species in solution. The growth of nuclei proceeds at first by stepwise addition of new metal atoms formed in solution, leading to primary particles. Particle growth may continue by further addition of metal atoms controlled by diffusion toward the surface, by incorporation onto the surface, or by coalescence of primary particles and formation of larger secondary ones. In order to avoid coalescence and limit the growth and also to form stable dispersions of nanoparticles, protective agents are introduced into the reaction medium, allowing, as already indicated before, formation of a surface layer limiting the diffusion and growth, and also the nanoparticle aggregation.

Nanoparticles of noble or precious metals

Gold and silver nanoparticles have been synthesized and studied extensively for a long time. Colloidal gold, as described in 1857 by M. Faraday [118], is probably the first monodispersed system ever reported in the literature (apart from carbon black used for Egyptian ink, already known in the time of the Pharaohs!). Reduction of chloroauric acid, $HAuCl_4$, in aqueous solution by citrate at 100°C forms spherical nanoparticles. Their mean diameter increases from around 10 to

20 nm when the concentration of chloroauric acid decreases [119]. Here, citrate acts as a reducing agent and also as a stabilizer. Organic ligands, with a group such as phosphine, having a strong affinity for metal are often used to stabilize metallic nanoparticles. For gold, thiol derivatives are strong stabilizers and the reduction of Au(III) ions by citrate or borohydride in the presence of a thiol ligand gives uniform Au nanoparticles, the Au/thiol ratio controlling the mean size of the nanoparticles [120]. Silver nanoparticles are similarly obtained by reduction of silver nitrate by ferrous citrate in an aqueous medium, their stabilization in solution resulting from silver citrate adsorption [121]. Very uniform silver nanoparticles have also been obtained by reduction in nonaqueous media [122]. Aqueous $AgNO_3$ was vigorously mixed with chloroform containing tetra n-octylammonium bromide, $[(C_8H_{17})_4N]Br$, acting as a catalyst for phase transfer. 1-nonanethiol was first added to the gray organic phase collected followed by an aqueous solution of sodium borohydride ($NaBH_4$). A stable dispersion of nearly spherical 1-nonanethiol–capped silver nanoparticles in chloroform was obtained. The evaporation of the solvent onto a carbon-coated microscope grid forms a 2D hexagonal superlattice of nanoparticles (Figure 3.27).

Gold and silver nanoparticles are also obtained by various methods such as γ-radiolytic reduction [123] or photochemical reduction [124], always in the presence of various protective agents. Gold and silver nanoparticles obtained in this fashion are generally spherical. However, the shape of nanoparticles can be varied and controlled using different methods. Gold truncated tetrahedra, octahedra, icosahedra, and cubes

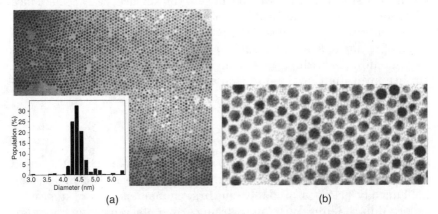

(a)   (b)

Figure 3.27 (a) TEM image of a two-dimensional silver nanoparticle superlattice and (inset) the histogram of the nanoparticles. (b) TEM image of the selected area in (a) [122].

Figure 3.28 (a) TEM image of truncated tetrahedral gold nanocrystals. (b) SEM images of several partially developed gold tetrahedra. (d) SEM image of icosahedral gold nanoparticles. (d and e) Gold nanocubes dispersed on a TEM grid and a silicon substrate [125].

have been obtained by the polyol process (Figure 3.28) [125]. Chloroauric acid and polyvinyl pyrolidone (PVP) are introduced in boiling ethylene glycol (EG). EG serves as both solvent and reducing agent, PVP stabilizes the particles and also, in conjunction with the concentration of the gold precursor, controls their shape. A low concentration of silver ions in the medium orients the process toward the formation of gold nanocubes.

Another interesting seeding growth method has been used to produce nanorods [126]. The basic principle for the shape-controlled synthesis involves two steps: first, the preparation of spherical gold nanoparticles of around 3,5 nm in diameter (seeds), by reduction at RT of $HAuCl_4$ by $NaBH_4$ in the presence of citrate [citrate serves only as a capping or protective agent because it cannot, at room temperature, reduce Au(III)]; second, growth of the seeds in rod-like micellar environments acting as templates [127]. The growth solution contains $HAuCl_4$, CTAB (cetyltrimethylammonium bromide), and ascorbic acid to which the seed solution is added. Ascorbic acid is a mild reducing agent, which cannot reduce the gold salt in the presence of micelles without the presence of seeds. Gold nanorods are so obtained with various aspect ratios (Figure 3.29). The aspect ratio is controlled by varying the ratio of metal salt to seed, if some Ag^+ ions are present [128]. Silver ions are not reduced under these experimental conditions, and their role in controlling the shape of gold nanorods

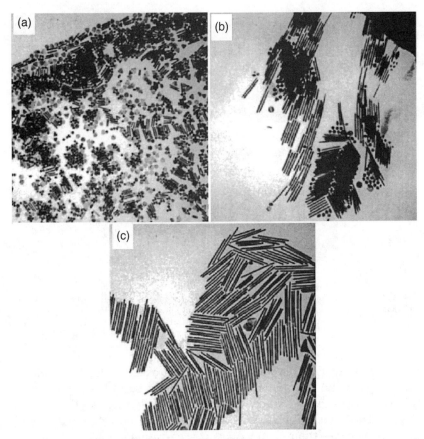

Figure 3.29 (a) TEM images of 4.6 aspect ratio gold nanorods, (b) shape-separated 13 aspect ratio gold nanorods, and (c) shape-separated 18 aspect ratio gold nanorods [126].

is not yet completely understood. Replacement of citrate by CTAB in the seed formation step and the use of a binary surfactant mixture improve the procedure in terms of selectivity in particle shape and amount of gold nanorods [129]. A similar procedure with silver nitrate allows the formation of silver nanorods and nanowires [130]. More generally, templating syntheses involving surfactants have been used to control the morphology of various metallic materials [127].

Silver nanowires were also obtained by an intriguing sono-assisted self-reduction template process [131]. A solution of a suitable precursor, the adduct Ag(hexafluoro-acetylacetone)-tetraglyme (tetraglyme = 2,5,8,11,14-pentaoxadecane) in ethanol/water 1:1 is loaded into the pores of an anodic aluminum oxide membrane consisting of ordered hole arrays of 200 nm in diameter and 60 μm in thickness, which is

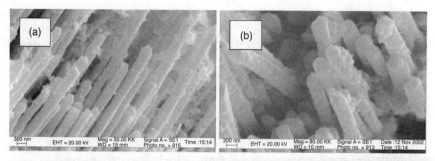

Figure 3.30 SEM images of free-standing Ag nanowires at low (a) and high (b) magnification [131].

used as a noninteracting template (Figure 3.30). The Ag self-reduction process is activated by sonication for 2 hours at 45°C. This process does not require any reducing agent and directly forms crystalline silver nanowires. They are recovered by dissolving the templating membrane in NaOH solution at RT. The nanowire dimensions are determined by the pore dimensions and thickness of the membrane.

Nanoparticles of ferromagnetic 3D transition metals (Co, Fe)

The strategy used for the synthesis of 3D transition metals nanoparticles is analogous to that involved for precious metals, but with harder conditions because the redox potential of 3D transition metal elements is much lower. The strategy involves either injecting a strong reducing agent into a hot nonaqueous solution of a metal precursor containing surfactants, or injecting a thermally unstable zerovalent metal precursor into a hot solution containing stabilizers. Adjusting the temperature and the metal precursor to surfactant ratio controls the nanoparticle size. Higher temperatures and larger metal precursor to surfactant ratios produce bigger nanoparticles. Surfactants not only stabilize the nanoparticles in dispersion but also prevent or limit their oxidation [132]. The following illustrative examples show the variety of methods developed.

Reduction of cobalt salts by polyalcohols (polyols) at temperatures between 100°C and 300°C in the presence of stabilizers produce Co nanoparticles with diameters of 2–20 nm [132, 133]. In a typical synthesis, cobalt acetate and oleic acid are heated at 200°C in diphenylether in the presence of trioctylphosphine (TOP). Reduction is started by addition of 1,2-dodecanediol solubilized in hot and dry diphenylether and heating at 250°C for 15–20 minutes. With a Co to oleic acid molar ratio of 1 and Co to TOP molar ratio of 2, 6–8 nm crystalline (hcp) Co particles are obtained.

Increasing the concentration of oleic acid and TOP by a factor of 2 yields smaller, 3–6 nm nanoparticles [132].

Reduction by superhydride LiBEt$_3$H of anhydrous cobalt chloride at 200°C in solution in dioctylether in the presence of oleic acid and alkylphosphine yields ε Co nanoparticles with a complex ε crystalline structure (β-Mn). The average particle size is coarsely controlled by the type of phosphine. Bulky P(C$_8$H$_{17}$)$_3$ limits the growth and produces 2–6 nm particles, while less bulky P(C$_4$H$_9$)$_3$ leads to larger (7–11 nm) particles [134]. The combination of oleic acid and trialkylphosphine produce a tight ligand shell, which allows the particles to grow steadily while protecting them from aggregation and oxidation. In effect, trialkylphosphine reversibly coordinates the metal surface, slowing but not stopping the growth. Oleic acid used alone is an excellent stabilizing agent, but it binds so tightly to the surface during synthesis that it impedes particle growth.

Reduction by dihydrogen of the organometallic compound, Co(η^3-C$_8$H$_{13}$)(η^4-C$_8$H$_{12}$) in THF in the presence of polyvinyl pyrolidone (PVP) forms very small Co nanoparticles with fcc crystalline structure, around 1 nm and 1.5 nm depending on the temperature, 0 and 20°C, respectively [135]. The decomposition at 150°C under H$_2$ of the same organometallic precursor, but in anisole and in the presence of oleic acid and oleylamine (1:1 Co/oleic acid and Co/amine molar ratios) produces initially (3 hours of reaction) spherical 3 nm Co nanoparticles. After 48 hours, nanorods 9 × 40 nm are obtained [136]. The role of surfactants is very important in controlling the shape of particles. Increasing the concentration by a factor of 2 of oleic acid produces very long (micron range) nanowires of 4 nm in diameter, whereas the nature of the amine changes drastically the dimensions of nanorods having an aspect ratio varying between 1.7 and 22. The combination of lauric acid and hexadecylamine produces monodisperse 5 × 85 nm nanorods forming spontaneously 2D crystalline superlattices [137].

Cobalt nanoparticles have also been obtained from micellar media. Reverse micelles are water droplets dispersed in oil and stabilized by a surfactant monolayer, typically sodium bis-(2-ethylhexyl)sulfosuccinate [Na(AOT)]. The size of the micelles depends on the amount of solubilized water and varies from 0.5 to 1.8 nm. In the liquid, collisions between droplets induce exchange between water pools [138]. By mixing two micellar solutions, one containing Co(OAT)$_2$ and the other one NaBH$_4$, well-crystallized fcc Co nanoparticles are formed with a mean size around 6.4 nm. Extraction of nanoparticles with TOP forms a protective layer against aggregation and oxidation.

Pyrolysis around 180°C of cobalt carbonyl Co$_2$(CO)$_8$ dissolved in anhydrous o-dichlorobenzene containing surfactant (mixtures of oleic acid, lauric acid and TOP) produces monodisperse ε-Co nanoparticles ranging

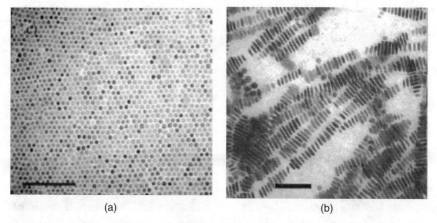

(a)   (b)

Figure 3.31   (a) Spherical Co nanocrystals synthesized in the presence of both oleic acid and TOPO and (b) Co nanodisks synthesized in the presence of amine. The disks are stacked in columns because of magnetic interactions. Bars are 100 nm [139].

in size from 3–17 nm. The particle size is tuned by the reaction temperature and the composition of surfactant [138]. In the presence of linear amines, hcp-Co nanodisks are formed, coexisting with ε-Co nanospheres (Figure 3.31) [139]. The length and diameter of the disks are controlled by variation of the reaction time following nucleation as well as by variation of the precursor to amine surfactant ratio. It has been observed that the length of the linear amine carbon chain controls the dimensions of Co disks (lower disks are obtained with shorter chains) whereas tri-substituted amines R_3N hinder the formation of disks. This suggests the $R\text{-}NH_2$ function is responsible for disk formation by selective adsorption, and steric interactions among neighboring adsorbed molecules may have an impact on the growth rate of the (001) faces.

It is difficult to synthesize iron nanoparticles by the polyol process because in the conditions of the process, Fe(0) results from disproportionation of Fe(II) whereas Co(II) and Ni(II) are quantitatively reduced [140]. Other methods to synthesize iron nanoparticles have been used. For instance, reduction of $FeCl_2$ in THF by $V(C_5H_5)_2$ in the presence of PVP forms PVP stabilized α-Fe nanoparticles 18 nm in mean size. Sonolysis of $Fe(CO)_5$ in solution in anisol in the presence of poly (dimethylphenylene oxide) (PPO) yields nonagglomerated spherical Fe nanoparticles with a mean size of around 3 nm. Interestingly, the smaller particles (< 2.5 nm) have the bcc structure of α-Fe and are superparamagnetic, whereas the larger ones (≥ 2.5 nm) adopt the fcc structure of γ-Fe and are antiferromagnetic or paramagnetic. These γ-Fe nanoparticles could result from thermal gradients and very high local temperatures during sonication with rapid quenching avoiding recrystallization,

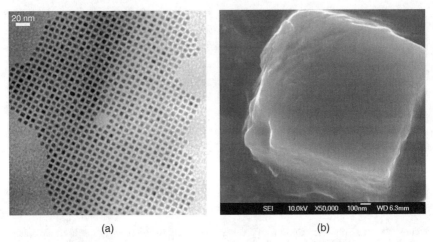

(a)   (b)

Figure 3.32   (a) TEM micrograph of 2D assembly of iron nanocubes and (b) SEM of 3D superlattice of nanocubes [141].

or from the presence of interstitial carbon inside the particles that would stabilize the γ-Fe phase.

Very beautiful iron nanocubes were obtained from the decomposition of the organometallic compound $Fe[N(SiMe_3)_2]_2$ (Figure 3.32) [141]. Heating at 150°C under dihydrogen pressure for 48 hours the solution of complex in mesitylene in the presence of hexadecylamine and oleic acid forms a black precipitate containing monodisperse 7 nm bcc-Fe nanocubes. These nanocubes are included in bigger cubes forming extended 3D superlattices.

Thermal decomposition of $Fe(CO)_5$ in TOPO at 340°C under an Ar atmosphere produces spherical 2 nm Fe nanoparticles easily dispersible in pyridine [142]. These nanoparticles can be further transformed to nanorods. $Fe(CO)_5$ solubilized in POP is added to a hot suspension (320°C) in TOPO of spherical 2 nm Fe nanoparticles. This operation yields a black solid, which is washed with acetone to remove the surfactant and then dispersed in pyridine containing didodecyl-methylammonium (DDAB). After refluxing for 12 hours, the super-natant contains 2×11 nm bcc-Fe nanorods (α-Fe). An increase in the concentration of DDAB increases the aspect ratio of rods. While diameter remains close to 2 nm, the length may be increased up to 22 nm. Such a transformation of nanospheres to nanorods seems to be caused by aggregation and by the strong binding of DDAB on the growing aggregates. After the aggregation of two particles, the third one will be bound on the top instead the central part of the aggregate where DDAB is strongly bounded. Then, further aggregation generates a unidimensional nanostructure.

Bimetallics and alloys

Bimetallic nanoparticles, composed of two different metals, are of interest in the improvement of catalytic properties [143] and in development of magnetic properties [144]. For instance, the ordered alloys Co-Pt and Fe-Pt are particularly interesting for magnetic recording because of their very high magnetocrystalline anisotropy, making these materials especially useful for practical applications such as magnetic memory devices as well as in biomedicine. The synthesis of bimetallics is generally made by coreduction of metal salts. Coreduction is the most simple preparative method and it is very similar to that used for monometallic nanoparticles. Successive reduction, carried out to prepare core-shell structured nanoparticles, is of little importance and will not be discussed here.

Au-Pt bimetallic nanoparticles have been obtained by citrate reduction of the mixture of tetrachloroauric acid, $HAuCl_4$, and hexachloroplatinic acid, H_2PtCl_6. The UV-Vis absorption spectra of the citrate stabilized sol of nanoparticles is not the simple sum of those of the two monometallic nanoparticles, indicating that the bimetallic particles have an alloy structure, as confirmed by X-ray diffraction and X-ray absorption spectroscopy [120]. A similar method produces citrate-stabilized Pd-Pt nanoparticles. Polymer-stabilized Pd-Pt nanoparticles have been prepared by simultaneous reduction of palladium chloride $PdCl_2$ and hexachloroplatinic acid by refluxing the alcohol/water (1:1 v/v) mixed solution in the presence of PVP at about 90°C for 1 hour [143]. Similarly, PVP-stabilized Pd-Rh nanoparticles were obtained by reduction in alcohol.

The polyol process [133] appears to be an efficient way to synthesize bi- or polymetallic nanoparticles of 3D metals, when heterogeneous nucleation controls the size of ferromagnetic nanoparticles. The addition of small amounts of a platinum or silver salt, which reduces at a lower temperature than 3D elements, forms nuclei for the growth of cobalt, nickel, or iron. Polymetallic spherical particles of the alloys $Co_xNi_{(100-x)}$ can be synthesized by precipitation from cobalt and nickel acetate dissolved in 1,2-propanediol with an optimized amount of sodium hydroxide. The number of nuclei depends on the relative amount of platinum, allowing the control of the particle size of $Co_xNi_{(100-x)}$ alloys over a very large range, from micrometric to nanometric [140]. The sodium hydroxide allows the precipitation of metal as hydroxides or alkoxides before the reduction. Their slow dissolution takes place at lower temperatures than the reduction, and this step likely controls the growth of metallic particles in solution. In controlling the basicity of the medium of the Pt or Ru seed-mediated polyol process, $Co_{80}Ni_{20}$ nanoparticles with surprising anisotropic shapes can be obtained [145, 146]. These particles are hcp crystallized when x > 30 (for x < 30, the particles very rich in

Figure 3.33 Particles $Co_{80}Ni_{20}$ obtained by polyol process with molar ratio HO/(Co + Ni) = 1.25 (a, b), 1.8 (c) and 2.5 (d) [146].

nickel are fcc structured and always isotropic in shape). At low hydroxide concentrations (molar ratio $[HO^-]/[Co + Ni] < 2$), growth is favored along the c-axis, forming nanowires 8 nm in diameter and 100–500 nm in length with cone-shaped ends (Figure 3.33). Theses wires appear linked to a core, forming a sea-urchin–like shape (Figure 3.33b). With molar ratios $[HO^-]/[Co + Ni] > 2$, the growth occurs preferentially perpendicular to the c-axis resulting in rods that are 20–25 nm in diameter and 75–100 nm in length. The particular shape of the ends of nanowires seen in Figure 3.33b probably results from a lower growth rate at the end of the reaction, when the Co(II) and Ni(II) concentration fall (as the result of their reduction) and when the molar ratio $[HO^-]/[Co + Ni]$ is increased. If it is assumed that the shape of wires is controlled by the growth rate, the head, formed at the end of the reaction, grows under conditions of low supersaturation inducing growth perpendicular to the c-axis. The polyol process has also be used to form Fe-Pt nanoparticles from the acetylacetonates $Fe(acac)_3$ and $Pt(acac)_2$ in ethylene glycol [146, 147].

A strong reducing agent such as hydrazine N_2H_4 has been used to reduce metal salts and to form Fe-Pt nanoparticles in water at low temperature.

H_2PtCl_6 and $FeCl_2$ together with hydrazine and a surfactant such as sodium dodecyl sulfate (SDS) or CTAB, are mixed in water. Heating at 70°C allows reduction and forms fcc-structured Fe-Pt nanoparticles [148]. Reduction of $FeCl_2$ and $Pt(acac)_2$ mixtures in diphenylether by superhydride, $LiBEt_3H$, in the presence of oleic acid, oleylamine, and 1,2-hexadecanediol at 200°C has led to 4 nm Fe-Pt nanoparticles [59]. The initial molar ratio of the metal precursor allows control of the composition of the final particles more easily than in the polyol process. However, a major drawback of using borohydride is the contamination of the final product by boron.

Another process used to produce Fe-Pt nanoparticles is the thermal decomposition of $Fe(CO)_5$ and reduction of $Pt(acac)_2$ by 1,2-alkanediol [59]. The mixture is heated to reflux (297°C). Oleic acid and oleylamine are used for surface passivation and stabilization of particles. The composition of particles is controlled by the Fe/Pt ratio and fine-tuning of particle size between 2 and 5 nm is achieved by controlling the surfactant to metal ratio. Alternatively, to make larger Fe-Pt nanoparticles, a seed-mediated growth method has been used [59].

Carbon Based Nanomaterials

Although nanomaterials had been known for many years prior to the discovery of C_{60}, the field of nanoscale science was really founded upon this seminal discovery and that of subsequent carbon nanomaterials. Part of the reason for this explosion in nanochemistry is that the carbon materials range from well-defined nano-sized molecules (i.e., C_{60}) to tubes with lengths in hundreds of microns range. Despite this range of scale, carbon nanomaterials have common reaction chemistry: that of the field of organic chemistry. This provides them with almost infinite functionality. A further advantage is that since every C_{60} molecule is like every other one, ignoring ^{12}C and ^{13}C isotope effects, C_{60} provides a unique monodispersed prototype nanostructure assembly with particle size of 0.7 nm. In no other nanomaterial system is there the ability to prepare true monodispersed material.

The previously unknown allotrope of carbon, C_{60}, was discovered in 1985 [1], and in 1996, Curl, Kroto, and Smalley were awarded the Nobel Prize in Chemistry for this discovery. The other allotropes of carbon are graphite (sp^2) and diamond (sp^3). C_{60}, commonly known as the "buckyball" or buckminsterfullerene, has a spherical shape comprising of highly pyramidalized sp^2 carbon atoms. The C_{60} variant is often compared to the typical white and black soccer ball, hence buckyball, however, this is also used for higher derivatives. Fullerenes are similar in structure to graphite, which is composed of a sheet of linked hexagonal rings, but they contain pentagonal (or sometimes heptagonal) rings that prevent the sheet from being planar. The unusual structure of C_{60} led to the introduction of a new class of molecules known as fullerenes, which now constitute the third allotrope of carbon. Fullerenes are commonly defined as "any of a class of closed hollow aromatic carbon compounds that are made up of twelve pentagonal and differing numbers of hexagonal faces." The number of carbon atoms in a fullerene range from C_{60} to C_{70}, C_{76}, and higher. Higher order fullerenes include carbon nanotubes that can be described as fullerenes that have been stretched along a rotational axis to form a tube. Given the differences in the chemistry and size of fullerenes such as C_{60} and C_{70} as compared to nanotubes, these will be dealt with separately. However, it should be appreciated that they are all part of the fullerene allotrope of carbon. In addition there have also been reports of nanohorns [149] and nanofibers [150]. Nanohorns are single-walled carbon cones that can be filled with biological material or metal oxides.

Fullerenes and carbon nanotubes are not necessarily products of high-tech laboratories; they are commonly formed in such mundane places as ordinary flames, produced by burning hydrocarbons [151, 152], and they have been found in soot from both indoor and outdoor air [153]. However, these naturally occurring varieties can be highly irregular in size and quality because the environment in which they are produced is often highly uncontrolled. Thus, although they can be used in some applications, they can lack in the high degree of uniformity necessary to meet many needs of both research and industry.

Fullerenes

The vast majority of studies have involved the chemistry of C_{60} (IUPAC name = (C_{60}-Ih)[5,6]fullerene) and given its now commercial synthesis this is the most common fullerene studied. The spherical shape of C_{60} is constructed from twelve pentagons and twenty hexagons, has I_h symmetry, and resembles a soccer ball (Figure 3.34a).

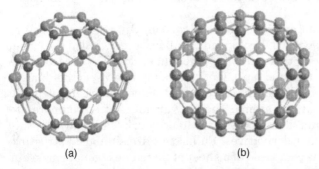

(a)   (b)

Figure 3.34 Molecular structures of (a) C_{60} and (b) C_{70}.

The next stable homologue is C_{70} (Figure 3.34b), which has D_{5h} symmetry with a shape similar to a rugby ball or American football. This is followed by C_{74}, C_{76}, C_{78}, and so on, in which an additional six membered rings are added. Mathematically and chemically, to make a stable fullerene, one has to follow two principles: Euler's theorem and the isolated pentagon rule (IPR). Euler's theorem states that for the closure of each spherical network, n (n $\geq$ 2) hexagons and 12 pentagons are required, while the IPR says no two pentagons may be connected directly with each other, as destabilization is caused by two adjacent pentagons.

Fullerenes are composed of sp^2 carbons in a similar manner to graphite, but unlike graphite's extended solid structure, fullerenes are spherical and soluble in various common organic solvents. Due to their hydrophobic nature, fullerenes are most soluble in CS_2 (C_{60} = 7.9 mg $\cdot$ mL^{-1}) and toluene (C_{60} = 2.8 mg $\cdot$ mL^{-1}), an important requirement for chemical transformation. Although fullerenes have a conjugated system, their spherical aromacity is distinctive from benzene due to the closed shell structure. In contrast to benzene, which has all C-C bonds in equal length, fullerenes such as C_{60} have two distinct classes of bonds. As determined experimentally, the shorter bonds at the junctions of two hexagons ([6, 6] bonds) and the longer bonds at the junctions of a hexagon and a pentagon ([5,6] bonds); see Figure 3.35. The alternation in bond length shows that the double bonds are located at the junction of hexagons, while there is almost no double bond character in the pentagon rings in the lowest energy state. As expected from the I_h symmetry of C_{60}, there is by symmetry only one carbon type, which can be confirmed by ^{13}C NMR spectroscopy. In agreement with its D_{5h} symmetry, C_{70} has five ^{13}C signals in the ratio of 1:2:1:2:1.

Fullerenes represent a unique category of cage molecules with a wide range of sizes, shapes, and molecular weights. Most of the effort thus far has gone into the study of C_{60} fullerenes, which can now be prepared to a purity of parts per thousand. Because of the unique icosahedral

1.45 Å [5, 6]   1.38 Å [6,6]

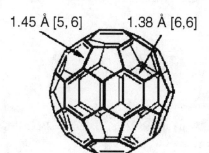

Figure 3.35 Bond lengths of [5,6] and [6,6] bonds of C_{60}.

symmetry of C_{60}, these molecules provide prototype systems for spectroscopy, optics, and other basic science investigations.

Synthesis of fullerenes.   The first observation of fullerenes was in molecular beam experiments, where discrete peaks were observed corresponding to molecules with the exact mass of 60 or 70 or more carbon atoms. In 1985, Harold Kroto (of the University of Sussex), James R. Heath, Sean O'Brien, Robert Curl, and Richard Smalley (from Rice University) published their observations along with the proposed structure for C_{60} [1]. Subsequent studies demonstrated that C_{60} was in fact ubiquitous in carbon combustion, and by 1991 it was relatively easy to produce grams of fullerene powder. Although the synthesis is relatively straightforward, fullerene purification remains a challenge to chemists and determines fullerene prices to a large extent.

The first method of production of fullerenes used laser vaporization of carbon in an inert atmosphere, but this produced microscopic amounts of fullerenes. In 1990, a new type of apparatus was developed by Krätschmer and Huffman in which carbon rods were vaporized in a helium atmosphere (approximately 100 Torr) (Figure 3.36) [154]. The subsequent black soot is collected and the fullerenes in the soot are then extracted by solvation in toluene. Pure C_{60} is obtained by liquid chromatography. The mixture is dissolved in toluene and pumped through a column of activated charcoal mixed with silica gel. The magenta C_{60} comes off first, followed by the red C_{70}.

Although many mechanisms have been described, only the "pentagon road" appears to explain high yields of C_{60}. In this proposed mechanism,

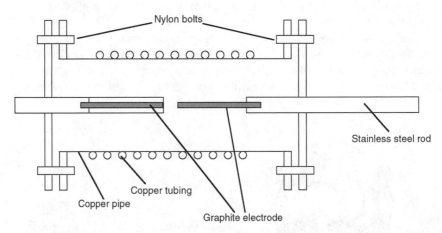

Figure 3.36   Schematic representation of the fullerene synthesis apparatus developed by Krätschmer and Huffman.

the yield is high because clustering continues in a hot enough region to permit the growing clusters to anneal to the minimum energy path: one where the graphene sheet (a) is made up solely of pentagons and hexagons, (b) has as many pentagons as possible, while (c) avoiding structures where two pentagons are adjacent. If the pentagon rule structures really are the lowest energy forms for any open carbon network, then one can readily imagine that high-yield synthesis of C_{60} may be possible. In principal, all one needs to do is adjust the conditions of the carbon cluster growth such that each open cluster has ample time to anneal into its favored pentagon rule structure before it grows further.

In the Krätschmer-Huffman (KH) experiment [154], carbon radicals are produced by the slow evaporation of the surface of a resistively heated carbon rod. After the KH method was introduced, it was found at Rice University that a simple AC or DC arc would produce C_{60} and the other fullerenes in good yield as well, and this is now the method used commercially [155]. Even though the mechanism of a carbon arc differs from that of a resistively heated carbon rod (because it involves a plasma) the He pressure for optimum C_{60} formation is very similar. Thus, it is not so much the vaporization method that matters, but rather the conditions prevailing while the carbon vapor condenses. Adjusting the helium gas pressure, the rate of migration of the carbon vapor away from the hot graphite rod is controlled and thereby the carbon radical density in the region where clusters in the size range near C_{60} are formed.

A ratio between the mass of fullerenes and the total mass of carbon soot defines fullerene yield. The yields determined by UV-Vis absorption are approximately 40 percent, 10–15 percent, and 15 percent in laser, electric arc, and solar processes. Productivity of a production process can be defined as a product between fullerene yield and flow rate of carbon soot. Interestingly, laser ablation technique has both the highest yield and low productivity and, therefore, a scale-up to a higher power is costly. Thus, fullerene commercial production is a challenging task. The world's first computer controlled fullerene production plant is now operational at the MER Corporation, which pioneered the first commercial production of fullerene and fullerene products.

Despite the commercialization of C_{60}, interest in the "rational" synthesis has continued. Scott and coworkers have reported a 12-step synthesis of C_{60} [156]. Despite the low overall yield the important step is that C_{60} is the only fullerene produced. A molecular polycyclic aromatic precursor bearing chlorine substituents at key positions forms C_{60} when subjected to flash vacuum pyrolysis at 1100°C (Figure 3.37).

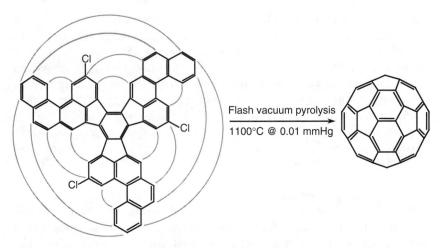

Figure 3.37   Rational synthesis of C_{60} showing the proposed C $\cdots$ C connectivity.

Endohedral fullerenes.   Endohedral fullerenes are fullerenes that have incorporated in their inner sphere atoms, ions, or clusters. Endohedral fullerenes are generally divided into two groups: endohedral metallo-fullerenes and nonmetal doped fullerenes. The first endohedral met-allofullerenes were synthesized in 1985, called $La@C_{60}$. The @ sign in the name reflects the notion of a small molecule trapped inside a shell.

Doping fullerenes with electropositive metals takes place *in situ* during the fullerene synthesis in an arc reactor or via laser evaporation. A wide range of metals have been successfully encased inside a fullerene, including Sc, Y, La, Ce, Ba, Sr, K, U, Zr, and Hf. Unfortunately, the syn-thesis of endohedral metallofullerenes is unspecific because in addition to unfilled fullerenes, compounds with different cage sizes such as $La@C_{60}$ or $La@C_{82}$ are prepared. In addition, the synthesis of $Sc_3N@C_{80}$ in 1989 demonstrated that a molecular fragment could be encapsulated within a fullerene cage.

Endohedral metallofullerenes are characterized by the fact that elec-trons will transfer from the metal atom to the fullerene cage and that the metal atom takes a position off-center in the cage. The size of the charge transfer is not always simple to determine. In most cases it is between 2 and 3 charge units (e.g., $La_2@C_{80}$) or as large as 6 electrons (e.g., $Sc_3N@C_{80}$). These anionic fullerene cages are very stable molecules and do not have the reactivity associated with ordinary empty fullerenes. For example, the Prato reaction yields only the monoadduct and not multiadducts as with empty fullerenes (see the following section). This lack of reactivity is utilized in a method to purify endohedral metallo-fullerences from empty fullerenes [157].

Saunders reported the existence of the endohedral He@C_{60} and Ne@C_{60} that form when C_{60} is exposed to a pressure of around 3 bars of the appropriate noble gases [158]. Under these conditions it was possible to dope one out of every 650,000 C_{60} cages with a helium atom. Endohedral complexes with He, Ne, Ar, Kr, and Xe as well as numerous adducts of the He@C_{60} compound have also been proven [159] with operating pressures of 3000 bars and incorporation of up to 0.1 percent of the noble gases. While the isolation of single atoms of the noble gases is not unexpected, the isolation of N@C_{60}, N@C_{70}, and P@C_{60} is very unusual. Unlike the metal derivatives, no charge transfer of the pnictide atom in the center to the carbon atoms of the cage takes place.

Chemically functionalized fullerenes. Although fullerene has a conjugated aromatic system, its reactivity is very different from planar aromatics, as all fullerene carbons are quaternary, containing no hydrogen, which renders characteristic substitution reactions of planar aromatics impossible. Therefore, only two types of primary chemical transformations exist: redox reactions and addition reactions. Among those two, addition reactions have the largest synthetic value in fullerene chemistry as they can also function as a screening probe for the chemical properties of fullerene surfaces. Another remarkable feature of fullerene addition chemistry is its thermodynamics. The sp^2 carbon atoms in a fullerene are pyramidalized. This dramatic variation from planarity draws great strain energy, especially in C_{60} fullerene. The strain energy is *ca* 8 kcal · mol^{-1}, which is about 80 percent of its heat of formation. So the relief of strain energy to fullerene cage resulting from a moderate number of addends bound to the fullerene surface is the major driving force for addition reactions as reactions leading to sp^3 hybridized C atoms strongly relieve the local strain of pyramidalization, as shown in Figure 3.38. As a consequence, most additions to C_{60} and C_{70} are exothermic reactions; however, this energy decreases as number of addends increase.

Another important feature for fullerene carbons is its rehybridization of the sp^2 σ and the p π orbitals corresponding to the derivation from planarity. Calculations have shown that average hybridization at carbon

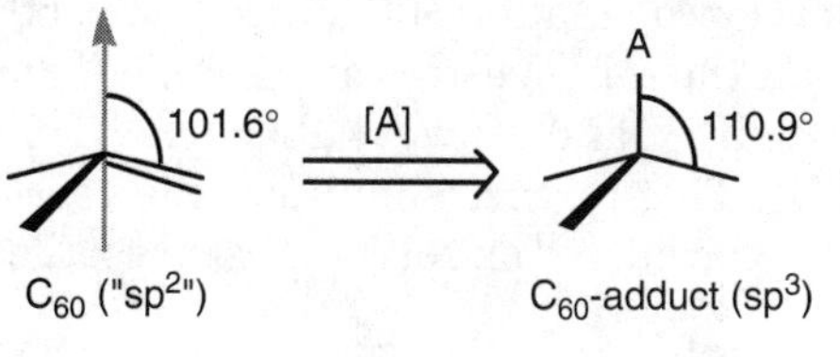

C_{60} ("sp^2") C_{60}-adduct (sp^3)

Figure 3.38 Strain release after addition of addend A to a pyramidalized carbon of C_{60}.

in C_{60} is $sp^{2.278}$ and a fractional s character of 0.085, which results in the π orbitals extending further beyond the outer surface than into the interior of fullerene. This analysis implies that fullerenes, especially C_{60}, are fairly electronegative molecules as well as having the low lying π^* orbital with considerable s character. Indeed, the reactivity of fullerene can be considered as a fairly localized, but electron deficient, polyolefin void of substitution reaction. This trend in terms of electrophilicity allows for the ready chemical reduction and nucleophilic addition to fullerenes.

Theoretical calculations show that the LUMO (t_{1u} symmetry) and LUMO+1 (t_{1g} symmetry) of C_{60} molecular orbitals exhibit a relatively low lying energy and are triply degenerated. So C_{60} was predicted to be a fairly electronegative molecule that can be reduced up to hexanion. Cyclic voltammetry (CV) studies show that C_{60} can be reduced and oxidized reversibly up to 6 electrons with one-electron transfer processes. Indeed, reduction reactions were the first chemical transformation carried out to C_{60}. Fulleride anions can be generated by electrochemical method and then be used to synthesize covalent organofullerene derivatives by quenching the anions with electrophiles. Alkali metals can chemically reduce fullerene in solution and solid state. It is with the alkali metal doped K_3C_6 that superconductivity was first found in fullerene materials. Alkaline earth metals can also be intercalated with C_{60} by direct reaction of C_{60} with alkaline earth metal vapor to form M_xC_{60} (x = 3 − 6), which possess superconductivity as well. Besides, C60 can also be reduced by less ectropositive metals such as mercury to form C_{60}^- and C_{60}^{2-}. In addition, fulleride salts can also be synthesized with organic molecules to form fullerene based charge transfer (CT) complexes. The well-known example of this type is $[TDAE^+][C_{60}^-]$, which possesses remarkable electronic and magnetic behavior.

As stated above, geometric and electronic analysis predicted that fullerene behaves like an electro-poor conjugated polyolefin. Indeed, C_{60} and C_{70} undergo various nucleophilic reactions with carbon, nitrogen, phosphorous, and oxygen nucleophiles. C_{60} reacts readily with organolithium and Grignard compounds to form alkyl, phenyl, or alkanyl fullerenes. However, one of the most widely used nucleophilic additions to fullerene is the Bingel reaction (Figure 3.39), where a carbon nucleophile was generated by deprotonation of α-halo malonate esters or ketones and added to form a clean cyclopropanation of C_{60} with a 30 ~ 60 percent yield. Later, it was found that the α-halo esters and ketones can be generated *in situ* with I_2 or CBr_4 and a weak base as 1,8-diazabicyclo[5.4.0]undec-7ene (DBU). This further simplified the reaction procedures. The Bingel reaction is considered one of the most versatile

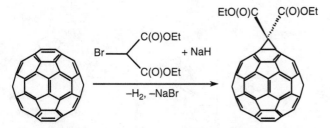

Figure 3.39 Bingel reaction of C_{60} with 2-bromoethylmalonate.

and efficient methods to functionalize C_{60}. Hundreds of fullerene derivatives have been made this way.

Cycloaddition is another powerful tool to functionalize fullerenes. The
great advantage of cycloaddition reaction is that the reaction generally
occurs at the 6,6 bonds, which limits the possible isomers (Figure 3.40).

The dienophilic feature of the [6,6] double bonds of C_{60} enables the
molecule to undergo various Diels-Alder reactions—that is a [4+2]
cycloaddition reaction. Another important feature of cycloaddition is
that monoadducts can be generated in high yields and purified by flash-
chromatography on SiO_2. However, the best studied cycloaddition reactions of fullerene are [3+2] additions with diazoderivatives (Figure 3.41)
and azomethine ylides (Prato reactions). In this reaction, azomethine
ylides can be generated *in situ* from condensation of α-amino acids with
aldehydes or ketones, which produce 1,3 dipoles to further react with
C_{60} in good yields (40–60 percent) (Figure 3.39). Hundreds of useful
building blocks have been generated by those two methods. Interestingly,
the Prato reactions have also been successfully applied to carbon nanotube and carbon onions to yield highly soluble carbon nanotube
derivatives.

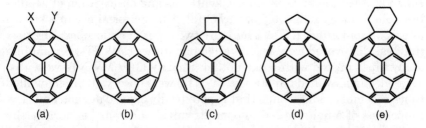

Figure 3.40 Geometrical shapes built onto a [6,6] ring junction: (a) open, (b) three-
membered ring, (c) four-membered ring, (d) five-membered ring, and (e) six-membered ring.

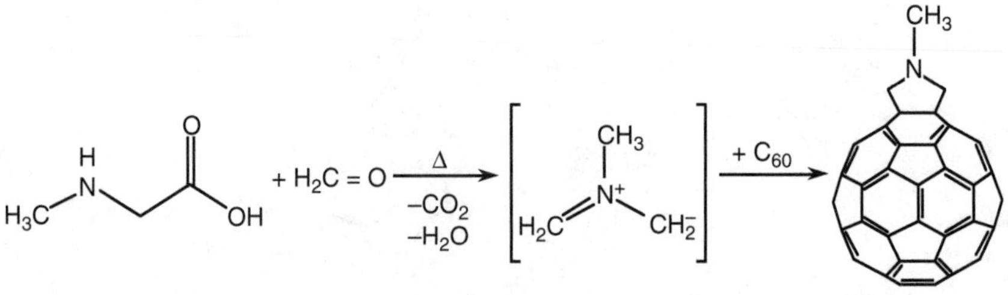

Figure 3.41   Prato reaction of C_{60} with N-methyglycine and paraformaldehyde.

Successful reactions for producing fullerenes have also involved the use of Grignard reagents [160, 161], organolithiums [162], and other nucleophiles. Current research includes the derivatization of the fullerene cage for the addition of amino acids [163] or pharmaceuticals for biological applications [164].

The oxidation of fullerenes, such as C_{60}, has been of increasing interest with regard to applications in photoelectric devices, biological systems, and possible remediation of fullerenes [165]. It has also been shown that $C_{60}O$ will undergo a thermal polymerization [166, 167], in an analogous manner to that of organic epoxides. The oxidation of C_{60} to $C_{60}O_n$ (n = 1, 2) may be accomplished by a range of methods, including photooxidation, ozonolysis, and epoxidation. With each of these methods, there is a limit to the isolable oxygenated product, $C_{60}O_n$ with n < 3. The only exception involves passing C_{60} through a corona discharge ionizer in the presence of oxygen, which allows for the detection of species formulated as $[C_{60}O_n]^-$ (n ≤ 30); however, the products were only observed in the MS [168]. Highly oxygenated fullerenes, $C_{60}O_n$ with 3 ≤ n ≤ 9, have been prepared by the Lewis base enhanced catalytic oxidation of C_{60} with $ReMeO_3/H_2O_2$ (Figure 3.42) [169].

Carbon nanotubes

Another key breakthrough in carbon nanochemistry came in 1993, when Iijima and Ichihashi reported the synthesis and observation of needle-like tubes made exclusively of carbon[3]. This material became known as carbon nanotubes (NTs). There are two types of nanotubes. The first that was discovered were multiwalled nanotubes (MWNTs) resembling many pipes nested within each other. Shortly after MWNTs were discovered, single-walled nanotubes (SWNTs) were observed. Single-walled tubes resemble a single pipe that is potentially capped at each end. The properties of single-walled and multiwalled tubes are generally the same. Though single-walled tubes are believed to have superior mechanical strength and thermal and electrical conductivity; it is also more difficult to manufacture them.

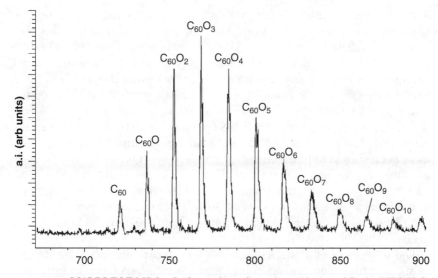

Figure 3.42   MALDI-TOF-MS for $C_{60}O_n$ products from the oxidation of C_{60} by MTO/H_2O_2 in the presence of 4-bromopyrazole.

Single-walled carbon nanotubes are by definition fullerene-based materials. Their structure consists of a graphene sheet rolled into a tube and capped by half a fullerene (Figure 3.43). The carbon atoms in a SWNT, like those in a fullerene, are sp^2 hybridized. They form a hexagonal ring

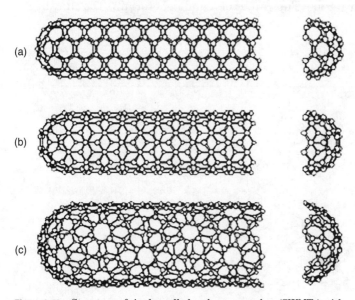

Figure 3.43   Structure of single-walled carbon nanotubes (SWNTs) with (a) armchair, (b) zigzag, and (c) chirality.

network that resembles a graphene sheet. The structure of a nanotube is analogous to taking this graphene sheet and rolling it into a seamless cylinder. The critical differentiation parameters of individual carbon nanotubes are their diameter and chirality. Most of the presently used single-wall carbon nanotubes have been synthesized by a pulsed laser vaporization method, pioneered by the Smalley group at Rice University. Their result represented a major breakthrough in the field.

The physical properties of SWNTs have made them an extremely attractive material for the manufacturing of nano-devices. SWNTs have been shown to be stronger than steel as estimates for the Young's modulus approaches 1 Tpa [170]. Their electrical conductance is comparable to copper with anticipated current densities of up to $10^{13}\,\text{A}\cdot\text{cm}^{-2}$ and a resistivity as low as $0.34\times 10^{-4}\,\Omega\cdot\text{cm}$ at room temperatures. Finally, they have a high thermal conductivity (3000–6000 $\text{W}\cdot\text{m}\cdot\text{K}^{-1}$) [171–173].

The properties of a particular SWNT structure are based on its chirality. If a tube were unrolled into a graphene sheet, vectors ($m a_2$ and $n a_1$) could be drawn starting from a carbon atom that intersects the tube axis (Figure 3.44). Then the armchair line is drawn. This line separates the hexagons into equal halves. Point B is a carbon atom that intersects the tube axis closest to the armchair line. The resultant vector of a_1 and a_2 is R and is termed the chiral vector. The wrapping angle (ϕ) is formed between R and the armchair line. If $\phi = 0°$, the tube is an armchair nanotube; if $\phi = 30°$, it is a zigzag tube. If ϕ is between 0° and 30°, the tube is called a chiral tube. The values of n and m determine the chirality or

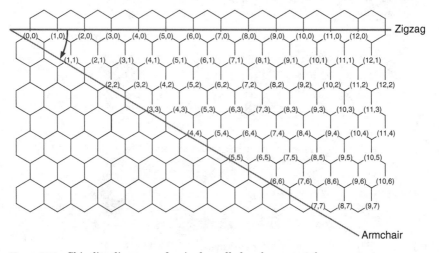

Figure 3.44   Chirality diagrams of a single-walled carbon nanotube.

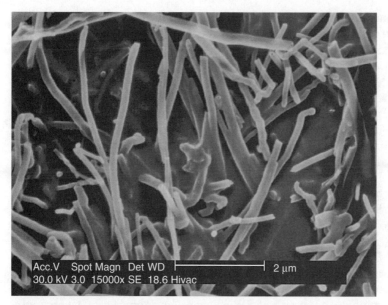

Figure 3.45   SEM image of vapor-grown carbon nanofibers.

"twist" of the nanotube. The chirality in turn affects the conductance of the nanotube, its density, its lattice structure, and other properties. A SWNT is considered metallic if the value $n - m$ is divisible by three. For example, an armchair tube is metallic in character. Otherwise, the nanotube is semiconducting [174]. Environment also has an effect on the conductance of a tube. Due to its highly delocalized π electrons, it is possible for a nanotube to accept electrons from or donate electrons to its environment [175, 176]. Molecules such as O_2 and NH_3 can change the overall conductance of a tube.

Multiwalled carbon nanotubes (MWNTs) range from double-walled NTs to carbon nanofibers. Carbon nanofibers are the extreme of multiwalled tubes (Figure 3.45). They are thicker and longer than either SWNTs or MWNTs, having a cross-sectional area of approximately 500 Å^2 and are between 10 to 100 μm in length. They have been used extensively in the construction of high-strength composites [177].

Synthesis of single-walled carbon nanotubes.   A range of methodologies have been developed to produce nanotubes in sizeable quantities, including arc discharge, laser ablation, high-pressure carbon monoxide (HiPco), and vapor-liquid-solid (VLS) growth. It is worth noting that the latter is often referred to as chemical vapor deposition (CVD), however, this is not strictly correct. All of these processes take place in a vacuum or at low pressure with process gases, although VLS growth can take place at atmospheric pressure. Large quantities of nanotubes can be synthesized by these methods; advances in catalysis and continuous growth processes are making SWNTs more commercially viable.

Nanotubes were first observed in 1991 in the carbon soot of graphite electrodes during an arc discharge that was intended to produce fullerenes. Because of the high temperatures caused by the discharge in this process, the carbon contained in the negative electrode sublimed. The fullerenes appear in the soot that is formed, while the CNTs are deposited on the opposing electrode. Tubes produced by this method were initially multiwalled tubes (MWNTs). However, in 1993 Bethune et al. reported that with the addition of cobalt to the vaporized carbon, it was possible to grow single-walled nanotubes [178]. This plasma-based process is analogous to the more familiar electroplating process in a liquid medium. This method produces a mixture of components and requires further purification to separate the CNTs from the soot and the residual catalytic metals. Producing CNTs in high yield depends on the uniformity of the plasma arc and the temperature of the deposit forming on the carbon electrode.

Higher yield and purity of SWNTs may be prepared by the use of a dual-pulsed laser. In 1995, Guo et al. reported that SWNTs could be grown through direct vaporization of a Co/Ni doped graphite rod with a high-powered laser in a tube furnace operating at 1200°C [179, 180]. By this method, it was possible to grow SWNTs in a 50 percent yield without the formation of an amorphous carbon overcoating. Samples are prepared by laser vaporization of graphite rods with a catalyst of cobalt and nickel (50:50) at 1200°C in flowing argon, followed by heat treatment in a vacuum at 1000°C to remove the C_{60} and other fullerenes. The initial laser vaporization pulse is followed by a second pulse, to vaporize the target more uniformly and minimize the amount of soot deposits. The second laser pulse breaks up the larger particle ablated by the first pulse (that would result in soot formation) and feeds the products to the growing SWNT structure. The material produced by this method appears as a mat of "ropes," 10–20 nm in diameter and up to 100 μm or more in length. Each rope consists of a bundle of SWNTs, aligned along a common axis. By varying the growth temperature, the catalyst composition, and other process parameters, the average nanotube diameter and size distribution can be varied.

Although arc-discharge and laser vaporization are currently the principal methods for obtaining small quantities of high-quality SWNTs, both methods suffer from drawbacks. The first is that they involve evaporating the carbon source, making scale-up on an industrial level difficult and energetically expensive. The second issue relates to the fact that vaporization methods grow SWNTs in highly tangled forms, mixed with unwanted forms of carbon and/or metal species. The SWNTs thus produced are difficult to purify, manipulate, and assemble for building nanotube-device architectures for practical applications.

To overcome some of the difficulties of these high-energy processes, Smalley and coworkers developed a chemical catalysis method. In 1998 Smalley and coworkers reported the use of hydrocarbons as a carbon feedstock for single-walled tube growth [181]. Here molybdenum and iron/molybdenum catalysts were heated in a tube furnace to 850°C under 1.2 atm of ethylene. Previous reports utilizing a gas-phase growth reaction had produced multiwalled tubes or single-walled tubes in very low yield [182, 183]. The use of CO as a feedstock led to the development by Smalley and coworkers of the high-pressure carbon monoxide (HiPco) procedure [184]. By this method, it was possible to produce gram quantities of SWNTs. The process involves injecting $Fe(CO)_5$ into a gas-phase reactor operating between 800–1200°C and 1–10 atm carbon monoxide. The HiPco method was better than previously reported gas-phase growth methods because it did not use a hydrocarbon as a feedstock. The key to the HiPco process is the formation of metal catalyst particles in the vapor phase, and it is thought that these are responsible for the SWNT growth, based upon an analogy with the growth by a catalyst on substrate.

The growth of SWNTs from chemical processes was first reported in 1992, and can be likened to the vapor-liquid-solid (VLS) growth of SiC wiskers [185, 186]. During VLS growth a preformed catalyst particle (most commonly nickel, cobalt, iron, or a combination thereof) is placed on a substrate. The diameters of the nanotubes that are to be grown has been proposed to be related to the size of the metal particles; however, recent work has shown that this is not necessarily true [187].

VLS growth apparatuses are usually constructed of a tube furnace that is set up so gases can flow through the tube while it is being heated to high temperatures. As with other growth systems, multiwalled tubes were the first type of tubes grown [188]. In 1999, Dai and coworkers reported the large-scale VLS growth of SWNTs using iron-impregnated silicon nanoparticles and methane [189]. Methane was chosen as the feedstock because of its high thermal stability and its ability to retard the formation of amorphous carbon in the reactor. Since these reports, a wide range of precursors have been used for the catalyzed VLS growth of SWNTs. Recent approaches have involved the use of well-defined nanoparticle or molecular precursors [187, 190]. Many different transition metals have been employed, but iron, nickel, and cobalt remain the focus of most research. SWNTs grow at the sites of the metal catalyst; the carbon-containing gas is broken apart at the surface of the catalyst particle, and the carbon is transported to the edges of the particle, where it forms the SWNTs. The catalyst particles generally stay at the tip of the growing SWNT during the growth process, although in

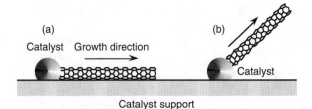

Catalyst support

Figure 3.46 Schematic representation of supported catalyst SWNT growth in which the SWNT grows parallel to the surface (a) or out from the surface (b).

some cases they remain at the SWNT base, depending on the adhesion between the catalyst particle and the substrate.

The length of the SWNTs grown in surface-supported catalyst VLS systems appears to be dependent on the orientation of the growing tube with the surface. Within particular catalyst samples there are often two classes of tubes grown: short, straight SWNTs and long, curved ones. It has been proposed that the straight SWNTs are a result of growth along the surface (Figure 3.46a) while the longer SWNTs are formed by growth out of the plane of the surface (Figure 3.46b). The growth rate of the former will be limited due to SWNT/surface interactions, while the later has unrestricted growth away from the surface [187]. Once the reaction run is complete (and the gas flow is removed) the SWNTs grown out of the surface will fall over. In the absence of additional factors, the rate of SWNT growth parallel to the surface is controlled by the frictional forces between the SWNT and the surface. By properly adjusting the surface concentration and aggregation of the catalyst particles, it is possible to synthesize vertically aligned carbon nanotubes—that is, as a carpet perpendicular to the substrate (Figure 3.47).

Of the various means for nanotube synthesis, the chemical processes show the most promise for industrial scale deposition in terms of its price/unit ratio. There are additional advantages to the VLS growth of SWNTs. Unlike the above methods, VLS is capable of growing SWNTs directly on a desired substrate, whereas the SWNTs must be collected in the other growth techniques. The growth sites are controllable by careful deposition of the catalyst. Additionally, no other growth methods have been developed to produce vertically aligned SWNTs.

Chemical functionalization of carbon nanotubes. The limitation on using carbon nanotubes in any practical applications has been their solubility; SWNTs have little to no solubility in most solvent due to the aggregation of the tubes. Aggregation is a result of the highly polarizable,

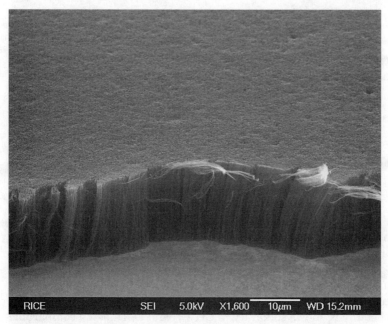

Figure 3.47 SEM image of a "carpet" of SWNTs grown perpendicular to the surface of the support.

smooth sides of the SWNTs forming bundles or ropes with a van der Waals binding energy of approximately 500 eV per μm of tube contact [191]. The van der Waals force between the tubes is so great that it takes tremendous energy to pry them apart. The insolubility of nanotubes makes it very difficult to make combinations of nanotubes with other materials, such as in composite applications. The functionalization of nanotubes—that is, the attachment of "chemical functional groups"— provides a strategy for overcoming those barriers. Functionalization can improve solubility and processibility, and will be able to link the unique properties of nanotubes to those of other materials. Through the chemical functional groups, nanotubes might take the interaction with other entities, such as solvents, polymer, nanoparticles, and other nanotubes. In functionalization of SWNTs, a distinction should be made between covalent and noncovalent functionalization. Covalent functionalization shows covalent linkage of functional groups onto the surface of nanotubes, either the sidewall or the cap of nanotubes. It is important to note that covalent functionalization methods have one problem in common: extensive covalent functionalization modifies SWNT properties by disrupting the continuous π–system of SWNTs.

Current methods for solubilizing nanotubes without covalent functionalization include highly aromatic solvents, super acids [192], DNA [193],

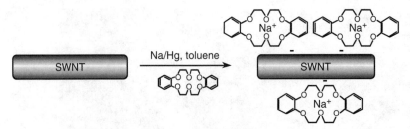

Figure 3.48 Representation of the reaction between Na/Hg amalgam, dibenzo-18-crown-6, and purified SWNTs in toluene and the formation of the [Na(dibenzo-18-crown-6)]$_n$[SWNT] complex.

polymers [194], or surfactants [195, 196]. These methods allow the sidewall of the nanotube to remain untouched, and conserve the tube's electronic structure. However, upon drying of the solution, bundles re-form. SWNTs may be made soluble in a range of organic solvents without sidewall functionalization via their reduction by Na/Hg amalgam in the presence of dibenzo-18-crown-6 (Figure 3.48) [197]. The [Na(dibenzo-18-crown-6)]$_n$ [SWNT] complex shows solubility in CH_2Cl_2 and DMF being comparable to surfactant-dispersed SWNTs; however, measurable solubilities are also observed in hexane, toluene, and alcohols.

A noncovalent functionalization is mainly based on supramolecular interaction using various adsorption forces, such as van der Waals and π-stacking interactions. Covalent functionalization relies on the chemical reaction at either the sidewall or end of the SWNT. The high aspect ratio of nanotubes, sidewall functionalization is much more important than the functionalization of the cap. Direct covalent sidewall functionalization is associated with a change of hybridization from sp^2 to sp^3 and a simultaneous loss of conjugation [198]. Defect functionalization takes advantage of chemical transformations of defect sites already present. Defect sites can be the open ends and holes in the sidewalls, and pentagon and heptagon irregularities in the hexagon graphene framework. All these functionalizations are exohedral derivatizations. Taking the hollow structure of nanotubes into consideration, endohedral functionalization of SWNTs is possible—that is, filling the tubes with atoms or small molecules [198–200].

Different application of nanotubes requires varied, specified modification to achieve processibility and accessibility of nanotubes. Thus, the covalent functionalization can provide a higher degree of fine-tuning the chemistry and physics of SWNTs than noncovalent functionalization. Until now, a variety of methods have been used to achieve the functionalization of nanotubes (Figure 3.49).

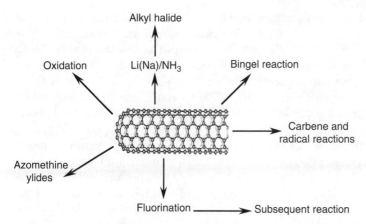

Alkyl halide

Oxidation   Li(Na)/NH$_3$ Bingel reaction

Carbene and
radical reactions

Azomethine
ylides

Fluorination   Subsequent reaction

Figure 3.49   Schematic description of various covalent functionalization
strategies for SWNTs.

Functionalization of SWNTs using 1,3 dipolar addition of azomethine ylides, a method originally developed for modification of C$_{60}$ [201].
Substituted pyrrolidine moieties were successfully introduced onto the
surface of SWNTs. The functionalized SWNTs are soluble in most
common organic solvents. The azomethine ylide functionalization
method was also used for the purification of SWNTs. In 2001, under an
electrochemical condition, a series of aryl diazonium salts were used to
react with SWNTs to achieve functionalized SWNTs. Subsequently,
SWNTs were functionalized by the diazonium ions *in situ* generated
from the corresponding aniline [202, 203]. A solvent-free reaction
appears to be the best chance for large-scale application of this method
[204]. Here, single-walled nanotubes are reacted with a para-
substituted aniline and isoamyl nitrate. This forms a diazonium salt
in situ that reacts with the tube's sidewall. It is possible to control the
amount of functionalization on the tube by varying reaction times and
the amount of aniline used. It has been reported that this method leads
to high functionalization (1 group per every 10–25 carbon atoms or
8–12 percent).

Billups and coworkers have reported organic functionalization
through the use of alkyl halides on tubes treated with lithium in liquid
ammonia [205]. The reaction occurs through a radical pathway. In this
reaction, functionalization occurs on every 17 carbons. Most success
has been found when the tubes are dodecylated. These tubes are soluble in chloroform, DMF, and THF. Besides functionalization, there is the
possibility of creating highly lithiated carbon materials. The lithium
intercollates between the SWNTs to give a C:Li ratio of approximately
1 lithium atoms per 2.2 carbons.

The addition of oxygen moieties to SWNT sidewalls can be achieved by treatment with acid or wet air oxidation and ozonolysis [206]. The direct epoxidation of SWNTs may be accomplished by the reaction with either trifluorodimethyldioxirane, formed *in situ* from trifluoroacetone and Oxone (potassium peroxymonosulfate, $KHSO_5$) in $MeCN/H_2O^4$ or 3-chloroperoxybenzoic acid (*m*-CPBA)/CH_2Cl_2 [207], or using $ReMeO_3/H_2O_2$ catalysis (Figure 3.50) [169]. Catalytic de-epoxidation (Figure 3.50) allows for the quantitative analysis of sidewall epoxide and led to the surprising result that previously assumed "pure" SWNTs actually contain approximately 1 oxygen per 250 carbon atoms. Sidewall osmylation of SWNTs has been obtained by exposing the SWNTs to OsO_4 vapor under UV photoirradiation [208]. The covalent attachment of osmium oxide increased the electrical resistance of tubes by up to several orders of magnitude. Cleavage of OsO_4 resulted in the recovery of the original resistance.

In 1999, Margrave and coworkers reported the direct fluorination of a nanotube sidewall [209]. For this method, elemental fluorine was passed over the tubes at 150–325°C. The fluorination allows for the tubes to be soluble in alcohols after brief ultrasonication (1 mg · mL^{-1} in 2-propanol). It has been ascertained that in fluorination at the optimal temperature, C:F ratios of up to 2:1 can be achieved without disruption of the tubular structure. The fluorinated SWNTs (F-SWNTs) proved to be much more soluble than pristine SWNTs in alcohols, DMF, and other selected organic solvents. Investigation of the structure of F-SWNTs has been explored by density functional theory (DFT) calculations and scanning tunneling microscopy (STM) imaging [210, 211]. STM revealed that the fluorine formed bands of approximately 20 nm [211]. Calculations on

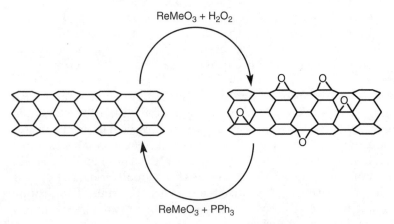

Figure 3.50 Catalytic oxidation and de-epoxidation of SWNTs.

the fluorinated (10,10) SWNTs with C_2F stoichiometry performed by using DFT revealed 1,2 addition is more energetically preferable than 1,4 addition. Recently, solid-state ^{13}C NMR has demonstrated the predominance of 1,2 addition and provided accurate quantification of the C:F ratio [212]. F-SWNTs make highly flexible synthons and subsequent elaboration has been performed with organo lithium, Grignard reagents, and amines [213–216].

Functionalized nanotubes can be characterized by a variety of techniques, such as atomic force microscopy (AFM), transmission electron microscopy (TEM), UV-vis spectroscopy, and Raman spectroscopy. Changes in the Raman spectrum of a nanotube sample can indicate if functionalization has occurred. Pristine tubes exhibit two distinct bands. They are the radial breathing mode (230 cm^{-1}) and the tangential mode (1590 cm^{-1}) [217]. When functionalized, a new band, called the disorder band, appears at approximately 1350 cm^{-1}. This band is attributed to sp^3-hybridized carbons in the tube. Unfortunately, while the presence of a significant D mode is consistent with sidewall functionalization, the relative intensity of D (disorder) mode versus the tangential G mode (1550–1600 cm^{-1}) is often used as a measure of the level of substitution. However, it has been shown that Raman is an unreliable method for determination of the extent of functionalization since the relative intensity of the D band is also a function of the substituents' distribution as well as concentration [215].

References

1. Kroto, H. W.; Heath, J. R.; O'Brien, S. C.; Curl, R. F.; Smalley, R. E. *Nature* 1985, *318*, 162.
2. Iijima, S. *Nature* 1991, *364*, 56.
3. Iijima, S.; Ichihashi, T. *Nature* 1993, *363*, 603.
4. Berry, C. R. *Phys. Rev.* 1967, *161*, 848.
5. Rossetti, R.; Nakahara, S.; Brus, L. E. *J. Chem. Phys.* 1983, *79*, 1086.
6. Efros, A. A.; Efros, A. L. *Sov. Phys. Semicond.* 1982, *16*, 1209.
7. Murray, C. B.; Norris, D. J.; Bawendi, M. G. *J. Am. Chem. Soc.* 1993, *115*, 8706.
8. Jun, Y. W.; Choi, J. S.; Cheon, J. *Angew. Chem. Int. Ed.* 2006, *45*, 2.
9. Baes, C. F.; Mesmer, R. E. *The Hydrolysis of Cations*; J. Wiley and Sons: New York, 1976.
10. Jolivet, J. P. *Metal Oxide Chemistry and Synthesis. From Solution to Solid State*; Wiley: Chichester, 2000.
11. Sillen, L. G. *Stability Constants of Metal-Ion Complexes*; The Chemical Society: London, 1964.
12. Jorgensen, C. K. *Inorganic Complexes*; Academic Press: New York, 1963.
13. Müller, A.; Beckmann, E.; Bögge, H.; Schmidtmann, M.; Dress, A. *Angew. Chem. Int. Ed.* 2002, *41*, 1162.
14. Wells, A. F. *Structural Inorganic Chemistry*; 5th ed.; Clarendon Press: Oxford, 1991.
15. Nielsen, A. E. *Kinetics of Precipitation*; Pergamon Press: Oxford, 1964.
16. Sugimoto, T. *Adv. Colloid Interface Sci.* 1987, *28*, 65.
17. Haruta, M.; Delmon, B. *J. Chim. Phys.* 1986, *83*, 859.
18. LaMer, V. K.; Dinegar, R. H. *J. Amer. Chem. Soc.* 1950, *72*, 4847.

19. Jolivet, J. P.; Chanéac, C.; Tronc, E. *Chem. Comm.* 2004, 481.
20. Euzen, P.; Raybaud, P.; Krokidis, X.; Toulhoat, H.; LeLoaer, J. L.; Jolivet, J. P.; Froidefond, C. *Handbook of Porous Materials*; Schüth, F., Sing, K. S. W., Weitkamp, J., Eds.; Wiley-VCH: Chichester, 2002, pp. 1591–1677.
21. Pottier, A.; Cassaignon, S.; Chanéac, C.; Villain, F.; Tronc, E.; Jolivet, J.-P. *J. Mater. Chem.* 2003, *13*, 877–882.
22. Lemaire, B. J.; Davidson, P.; Ferré, J.; Jamet, J. P.; Panine, P.; Dozo, I.; Jolivet, J. P. *Phys. Rev. Let.* 2002, *88*, 125507.
23. Combes, J. M.; Manceau, A.; Calas, G. *Geochimica Cosmochim. Acta* 1990, *54*, 1083.
24. Jolivet, J. P.; Tronc, E.; Chanéac, C. *Chimie* 2002, *5*, 659.
25. Jolivet, J. P.; Belleville, P.; Tronc, E.; Livage, J. *Clays Clay Miner.* 1992, *40*, 531.
26. Tronc, E.; Belleville, P.; Jolivet, J. P.; Livage, J. *Langmuir* 1992, *8*, 313.
27. Vayssières, L.; Chanéac, C.; Tronc, E.; Jolivet, J. P. *J. Colloid Interface Sci.* 1998, *205*, 205.
28. Jolivet, J. P.; Froidefond, C.; Pottier, A.; Chanéac, C.; Cassaignon, S.; Tronc, E.; Euzin, P. *J. Mater. Chem.* 2004, *14*, 3281.
29. Jolivet, J. P.; Tronc, E. *J.Colloid Interface Sci.* 1988, *125*.
30. Matijevic, E. *Pure and Appl. Chem.* 1980, *52*, 1179.
31. Cornell, R. M.; Schwertmann, U. *The iron oxides, structure, properties, reactions, occurrence and uses*; VCH Weinheim Germany Publishers, 2003.
32. Schwertmann, U.; Friedl, J.; Stanjek, H. *J. Colloid Interface Sci.* 1999, *209*, 215.
33. Bottero, J. Y.; Manceau, A.; Villieras, F.; Tchoubar, D. *Langmuir* 1994, *10*, 316.
34. Bailey, J. K.; Brinker, C. J.; Mecartney, M. L. *J. Colloid Interface Sci.* 1993, *157*, 1.
35. Sugimoto, T.; Muramatsu, A.; Sakata, K.; Shindo, D. *J. Colloid Interface Sci.* 1993, *158*, 420.
36. Matijevic, E.; Scheiner, P. *J. Colloid Interface Sci.* 1978, *63*, 509.
37. Pottier, A.; Chanéac, C.; Tronc, E.; Mazerolles, L.; Jolivet, J. P. *J. Mater. Chem.* 2001, *11*, 1116.
38. Keesmann, I. *Z. Anorg. Allg. Chem.* 1966, *346*, 31.
39. Arnal, P.; Corriu, J. P. R.; Leclercq, D.; Mutin, P. H.; Vioux, A. *Chem. Mater.* 1997, *9*, 694.
40. Bradley, D. C.; Mehrotra, R. C.; Gaur, D. P. *Metal Alkoxides*; Academic Press: London, 1978.
41. Turova, N. Y.; Turevskaya, E. P.; Kessler, V. G.; Yanoskaya, M. I. *The Chemistry of Metal Alkoxides*; Kluwer Academic Publishers: Boston, 2002.
42. Brinker, C. J.; Scherer, G. W. *Sol Gel Science*; Academic Press: San Diego, 1990.
43. Sanchez, C.; Soler-Illia, G. J. A. A.; Ribot, F.; Lalot, T.; Mayer, C. R.; Cabuil, V. *Chem. Mater.* 2001, *13*, 3061.
44. Sanchez, C.; Ribot, F. *New J. Chem.* 1994, *18*, 1007–1047.
45. Sanchez, C.; Gomez-Romero, P. *Hybrid Functional Materials*; Wiley: Chicester, 2003.
46. Stöber, W.; Fink, A.; Bohn, E. *J. Colloid Interface Sci.* 1968, *26*, 62.
47. Barringer, E. A.; Bowen, H. K. *Langmuir* 1985, *1*, 414.
48. Scolan, E.; Sanchez, C. *Chem. Mater.* 1998, *10*, 3217.
49. Rozes, L.; Steunou, N.; Fornasieri, G.; Sanchez, C. *Monatshefte Chemie* 2006, *137*, 501.
50. Vioux, A. *Chem. Mater.* 1997, *9*, 2292.
51. McMahon, C. N.; Bott, S. G.; Barron, A. R. *J. Chem. Soc., Dalton Trans.* 1997, 3129–3138.
52. Power, M. B.; Bott, S. G.; Clark, D. L.; Atwood, J. L.; Barron, A. R. *Organometallics* 1990, *9*, 3086.
53. Power, M. B.; Bott, S. G.; Atwood, J. L.; Barron, A. R. *J. Am. Chem. Soc.* 1990, *112*, 3446.
54. Power, M. B.; Nash, J. R.; Healy, M. D.; Barron, A. R. *Organometallics* 1992, *11*, 1830.
55. Niederberger, M.; Bartl, M. H.; Stucky, G. D. *J. Am. Chem. Soc.* 2002, *124*, 13642.
56. Niederberger, M.; Bartl, M. H.; Stucky, G. D. *Chem. Mater.* 2002, *14*, 4364.
57. Trentler, T. J.; Denler, T. E.; Bertone, J.; Agrawal, A.; Colvin, V. L. *J. Am. Chem. Soc.* 1999, *121*, 1613.
58. Jun, Y. W.; Casula, M. F.; Sim, J. H.; Kim, S. Y.; Cheon, J.; Alivisatos, P. *J. Am. Chem. Soc.* 2003, *125*, 15981.

59. Sun, S.; Anders, S.; Thomson, T.; Baglin, J.; Toney, M.; Hamann, H.; Murray, C.; Terris, B. *J. Phys. Chem. B* 2003, *107*, 5419–5425.
60. Zeng, H.; Rice, P.; Wang, S.; Sun, S. *J. Am. Chem. Soc.* 2004, *126*, 11458–11459.
61. Sun, S.; Zeng, H.; Robinson, D.; Raoux, S.; Rice, P.; Wang, S.; Li, G. *J. Am. Chem. Soc.* 2004 *126*, 273–279.
62. Chen, M.; Liu, J.; Sun, S. *J. Am. Chem. Soc.* 2004, *126*, 8394-8395.
63. Han, S.; Yu, T.; Park, J.; Koo, B.; Joo, J.; Hyeon, T.; Hong, S.; Im, J. *J. Phys. Chem. B* 2004, *108*, 8091–8095.
64. Song, Q.; Zhang, Z. J. *J. Am. Chem. Soc.* 2004, *126*, 6164.
65. Yin, M.; Gu, Y.; Kuskovsky, I. L.; Andelman, T.; Zhu, Y.; Neumark, G. F.; O'Brien, S. *J. Am. Chem. Soc.* 2004, *126*, 6206.
66. Jana, N. R.; Chen, Y.; Peng, X. *Chem. Mater.* 2004, *16*, 3931.
67. Sun, S.; Zeng, H.; Robinson, D. B.; Raoux, S.; Rice, P. M.; Wang, S. X.; Li, G. *J. Am. Chem. Soc.* 2004, *126*, 273.
68. Hyeon, T.; Lee, S. S.; Park, J.; Chung, Y.; Na, H. B. *J. Am. Chem. Soc.* 2001, *123*, 12798.
69. Cheon, J.; Kang, N. J.; Lee, S. M.; Lee, J. H.; Yoon, J. H.; Oh, S. J. *J. Am. Chem. Soc.* 2004, *126*, 1950.
70. Park, J.; Kang, E.; Bae, C. J.; Park, J. G.; Noh, H. J.; Kim, J. Y.; Park, J. H.; Park, H. M.; Hyeon, T. *J. Phys. Chem. B* 2004, 13594.
71. Lee, K.; Seo, W. S.; Park, J. T. *J. Am. Chem. Soc.* 2003, *125*, 3408.
72. Andrianov, K. A.; Zhadanov, A. A. *J. Polym. Sci,* 1958, *30* 513.
73. Pasynkiewicz, S. *Polyhedron* 1990, *9*, 429.
74. Barron, A. R. *Comments Inorg. Chem.* 1993, *14*, 123.
75. Harlan, C. J.; Mason, M. R.; Barron, A. R. *Organometallics* 1994, *13*, 2957.
76. Landry, C. C.; Harlan, C. J.; Bott, S. G.; Barron, A. R. *Angew. Chem., Int. Ed. Engl.* 1995, *34*, 1201–1202.
77. Apblett, A. W.; Barron, A. R. *Ceramic Transactions* 1991, *19*, 35.
78. Apblett, A. W.; Warren, A. C.; Barron, A. R. *Chem. Mater.* 1992, *4*, 167.
79. Landry, C. C.; Davis, J. A.; Apblett, A. W.; Barron, A. R. *J. Mater. Chem.* 1993, *3*, 597–602.
80. Landry, C. C.; Pappè, N.; Mason, M. R.; Apblett, A. W.; Tyler, A. N.; MacInnes, A. N.; Barron, A. R. *J. Mater. Chem.* 1995, *5*, 331–341.
81. Landry, C. C.; Pappè, N.; Mason, M. R.; Apblett, A. W.; Barron, A. R. In *Inorganic and Organometallic Polymers*; ACS Symposium Series: 1998; Vol. 572, p 149.
82. Callender, R. L.; Harlan, C. J.; Shapiro, N. M.; Jones, C. D.; Callahan, D. L.; Wiesner, M. R.; Cook, R.; Barron, A. R. *Chem. Mater.* 1997, *9*, 2418–2433.
83. Shahid, N.; Barron, A. R. *J. Mater. Chem.* 2004, *14*, 1235–1237.
84. Vogelson, C. T.; Barron, A. R. *J. Non-Cryst. Solids* 2001, *290*, 216–223.
85. Callender, R. L.; Barron, A. R. *J. Am. Ceram. Soc.* 2000, *83*, 1777.
86. Harlan, C. J.; Kareiva, A.; MacQueen, D. B.; Cook, R.; Barron, A. R. *Adv. Mater.* 1997, *9* 68.
87. Kareiva, A.; Harlan, C. J.; MacQueen, D. B.; Cook, R.; Barron, A. R. *Chem. Mater.* 1996, *8*, 2331–2340.
88. Koide, Y.; Barron, A. R. *Organometallic* 1995, *14*, 4026–4029.
89. Bethley, C. E.; Aitken, C. L.; Koide, Y.; Harlan, C. J.; Bott, S. G.; Barron, A. R. *Organometallics* 1997, *16*, 329–341.
90. Rose, J.; Cortalezzi-Fidalgo, M. M.; Moustier, S.; Magnetto, C.; Jones, C. D.; Barron, A. R.; Wiesner, M. R.; Bottero, J. Y. *Chem. Mater.* 2002, *14*, 621–628.
91. Cortalezzi-Fidalgo, M. M.; Rose, J.; Wells, G. F.; Bottero, J. Y.; Barron, A. R.; Wiesner, M. R. *Mat. Res. Soc., Symp. Proc.* 2003, *800*.
92. Rossetti, R.; Nakahara, S.; Brus, L. E. *J. Chem. Phys.* 1983, *79*.
93. Eychmüller, A. *J. Phys. Chem. B* 2000, *104*, 6514.
94. Healy, M. D.; Laibinis, P. E.; Stupik, P. D.; Barron, A. R. *J. Chem. Soc., Chem. Commun.* 1989, 359.
95. Wells, R. L.; Pitt, C. G.; McPhail, A. T.; Purdy, A. P.; Shafieezad, S.; Hallock, R. B. *Chem. Mater.* 1989, *1*, 4.
96. Manna, L.; Scher, E. C.; Alivisatos, A. P. *J. Am. Chem. Soc.* 2000, *122*, 12700.

97. Manna, L.; Milliron, D. J.; Meisel, A.; Scher, E. C.; Alivisatos, A. P. *Nature Mater.* 2003, *2*, 382.
98. Peng, Z. A.; Peng, X. *J. Am. Chem. Soc.* 2001, *123*, 183.
99. Peng, Z. A.; Peng, X. *J. Am. Chem. Soc.* 2002, *12*, 3343.
100. Peng, Z. A.; Peng, X. *J. Am. Chem. Soc.* 2001, *123*, 1389.
101. Zong, X.; Feng, Y.; Knoll, W.; Man, H. *J. Am. Chem. Soc.* 2003, *125*, 13559.
102. Jun, Y. W.; Lee, S. M.; Kang, N. J.; Cheon, J. *J. Am. Chem. Soc.* 2001, *123*, 5150.
103. Jun, Y. W.; Jung, Y. Y.; Cheon, J. *J. Am. Chem. Soc.* 2002, *12*, 615.
104. Lee, S. M.; Jun, Y. W.; Cho, S. N.; Cheon, J. *J. Am. Chem. Soc.* 2002, *124*, 11244.
105. M.B. Sigman, J.; Ghezelbash, A.; Hanrath, T.; Saunders, A. E.; Lee, F.; Korgel, B. A. *J. Am. Chem. Soc.* 2003, *125*, 16050.
106. Cho, K. S.; Talapin, D. V.; Gaschler, W.; Murray, C. B. *J. Am. Chem. Soc.* 2005, *12*, 7140.
107. MacInnes, A. N.; Power, M. B.; Barron, A. R. *Chem. Mater.* 1992, *4*, 11.
108. MacInnes, A. N.; Power, M. B.; Barron, A. R. *Chem. Mater.* 1993, *5*, 1344.
109. MacInnes, A. N.; Cleaver, W. M.; Barron, A. R.; Power, M. B.; Hepp, A. F. *Adv. Mater. Optics. Electron.* 1992, *1*, 229
110. Cleaver, W. M.; Späth, M.; Hnyk, D.; McMurdo, G.; Power, M. B.; Stuke, M.; Rankin, D. W. H.; Barron, A. R. *Organometallics* 1995, *14*, 690.
111. Okuyama, K.; Huang, D. D.; Seinfeld, J. H.; Tani, N.; Matsui, I. *Jpn. J. Appl. Phys.* 1992, *31*, 1.
112. Stoll, S. L.; Gillan, E. G.; Barron, A. R. *Chem. Vapor Deposition* 1996, *2*, 182.
113. Stoll, S. L.; Bott, S. G.; Barron, A. R. *J. Chem. Soc., Dalton Trans.* 1997, 1315
114. Power, M. B.; Barron, A. R.; Hnyk, D.; Robertson, H. E.; Rankin, D. W. H. *Adv. Mater. Optics Electron.* 1995, *5*, 177.
115. Harle, O. I.; Thomas, J. R. US, 1966.
116. Allendoy, M. D.; Hurt, R. H.; Young, N.; Reagon, P.; Robbins, M. *J. Mater. Res.* 1993, *8*, 1651.
117. Goia, D. V.; Matijevic, E. *New J. Chem.* 1998, 1203.
118. Faraday, M. *Phil. Trans.* 1857, *147*, 145.
119. Turkevitch, J.; Stevenson, P. C.; Hillier, J. *Faraday Discuss. Chem. Soc.* 1951, *11*, 55.
120. Toshima, N.; Yonezawa, T. *New J. Chem.* 1998, 1179.
121. Jolivet, J. P.; Gzara, M.; Mazières, J.; Lefebvre, J. *J. Colloid Interface Sci.* 1985, *107*, 429.
122. He, S.; Yao, J.; Jiang, P.; Shi, D.; Zhang, H.; Xie, S.; Pang, S.; Gao, H. *Langmuir* 2001, *17*, 1571.
123. Belloni, J.; Mostafavi, M.; Remita, H.; Marignier, J. L.; Delcourt, M. O. *New J. Chem.* 1998, 1239.
124. Henglein, A. *Chem. Mater.* 1998, *10*, 444.
125. Kim, F.; Connor, S.; Song, H.; Kuykendall, T.; Yang, P. *Angew. Chem. Int. Ed.* 2004, *43*, 3673.
126. Jana, N. R.; Gearheart, L.; Murphy, C. J. *J. Phys. Chem. B* 2001, *105*, 4065.
127. Pileni, M. P. *Nature Mater.* 2003, *2*, 145.
128. Jana, N. R.; Gearheart, L.; Murphy, C. J. *Adv. Mater.* 2001, *13*, 1389.
129. Nikoobakht, B.; El-Sayed, M. A. *Chem. Mater.* 2003, *15*, 1957.
130. Jana, N. R.; Gearheart, L.; Murphy, C. J. *Chem. Commun.* 2001, 617.
131. Malandrino, G.; Finocchiaro, S. T.; Fragalà, I. L. *J. Mater. Chem.* 2004, *14*, 2726.
132. Murray, C. B.; Sun, S.; Doyle, H.; Betley, T. *Mater. Res. Bull.* 2001, 985.
133. Fievet, F.; Lagier, J. P.; Figlarz, M. *Mater. Res. Bull.* 1989, *14*, 29.
134. Sun, S.; Murray, C. B. *J. Appl. Phys.* 1999, *85*, 4325.
135. Osuna, J.; deCaro, D.; Amiens, C.; Chaudret, B.; Snoeck, E.; Respaud, M.; Broto, J. M.; Fert, A. *J. Phys. Chem.* 1996, *100*, 14571.
136. Dumestre, F.; Chaudret, B.; Amiens, C.; Fromen, M. C.; Casanove, M. J.; Renaud, P.; Zurcher, P. *Angew. Chem. Int. Ed.* 2002, *41*, 4286.
137. Dumestre, F.; Chaudret, B.; Amiens, C.; Respaud, M.; Fejes, P.; Renaud, P.; Zurcher, P. *Angew. Chem. Int. Ed.* 2003, *42*, 5213.
138. Puntes, V. F.; Krishnan, K. M.; Alivisatos, P. *App. Phys. Lett.* 2001, *78*, 2187.
139. Puntes, V. F.; Zanchet, D.; Erdonmez, C. K.; Alivisatos, P. *J. Am. Chem. Soc.* 2002, *124*, 12874.

140. Toneguzzo, P.; Viau, G.; Acher, O.; Fievet-Vincent, F.; Fievet, F. *Adv. Mater.* 1998, *10*, 1032.
141. Dumestre, F.; Chaudret, B.; Amiens, C.; Renaud, M.; Fejes, P. *Science* 2004, *303*, 821.
142. Park, S. J.; Kim, S.; Lee, S.; Khim, Z. G.; Char, K.; Hyeon, T. *J. Am. Chem. Soc.* 2000, *122*, 8581.
143. Toshima, N.; Yonezawa, T.; Kushihashi, K. *J. Chem. Soc. Faraday Trans.* 1993, *89*, 2537.
144. Sun, S. *Adv. Mater.* 2006, *18*, 393.
145. Chakroune, N.; Viau, G.; Ricolleau, C.; Fievet-Vincent, F.; Fievet, F. *J. Mater. Chem.* 2003, *13*, 312.
146. Ung, D.; Viau, G.; Ricolleau, C.; Warmont, F.; Gredin, P.; Fievet, F. *Adv. Mater.* 2005, *17*, 338.
147. Iwaki, T.; Kakihara, Y.; Toda, T.; Abdullah, M.; Okuyama, K. *J. Appl. Phys.* 2003, *94*, 6807.
148. Gibot, P.; Tronc, E.; Chanéac, C.; Jolivet, J. P.; Fiorani, D.; Testa, A. M. *J. Magn. Magn. Mater.* 2005, *290-291*, 555.
149. Harris, P. J. F.; Tsang, S. C.; Claridge, J. B.; Green, M. L. H. *J. Chem. Soc., Faraday Trans.* 1994, *90*, 2799.
150. Chambers, A.; Park, C.; Baker, R. T. K.; Rodrigues, N. M. *J. Phys.Chem. B* 1998, *102*, 4253.
151. Liming, Y.; Saito, K.; Hu, W.; Chen, Z. *Chem. Phys. Lett.* 2001, *346*, 23–28.
152. Duan, H. M.; McKinnon, J. T. *J. Phys. Chem.* 1994, *98*, 12815–12818.
153. Murr, L. E.; Bang, J. J.; Esquivel, E. V.; Guerrero, P. A.; Lopez, D. A. *J. Nano. Res.* 2004, *6*, 241–251.
154. Krätschmer, W.; Lamb, L. D.; Fostiropoulos, K.; Huffman, D. R. *Nature* 1990, *347*, 354–358.
155. Smalley, R. E. *Acc. Chem. Res.* 1992, *25*, 98–105.
156. Scott, L. T.; Boorum, M. M.; McMahon, B. J.; Hagen, S.; Mack, J.; Blank, J.; Wegner, H.; deMeijere, A. *Science* 2002, *295*, 1500.
157. Ge, Z.; Duchamp, J. C.; Cai, T.; Gibson, H. W.; Dorn, H. C. *J. Am. Chem. Soc.* 2005, *127*, 16292–16298.
158. Saunders, M.; Jiménez-Vázquez, H. A.; Cross, R. J.; Poreda, R. J. *Science* 1993, *259*, 1428–1430.
159. Saunders, M.; Jimenez-Vazquez, H. A.; Cross, R. J.; Mroczkowski, S.; Gross, M. L.; Giblin, D. E.; Poreda, R. J. *J. Am. Chem. Soc.* 1994, *116*, 2193–2194.
160. Hirsch, A.; Soi, A.; Karfunkel, H. R. *Angew. Chem., Intl. Ed.* 1992, *31*, 766.
161. Hirsch, A.; Grösser, T.; Skiebe, A.; Soi, A. *Chem. Ber.* 1993, *126*, 1061.
162. Fagan, P. J.; Krusic, P. J.; Evans, D. H.; Lerke, S.; Johnston, E. *J. Am. Chem. Soc.* 1992, *114*, 9697.
163. Yang, J.; Barron, A. R. *Chem. Commun.* 2004, 2884–2885.
164. Zakharian, T.; Ashcroft, J.; Mirakyan, A.; Tsyboulski, D.; Benedict, N.; Weisman, B.; Wilson, L. J.; Mark, J. W.; Rosenblum, M. R. *Electrochem. Soc. Proceedings* 2004, *14*, 338.
165. Chikkannanavar, S. B.; Luzzi, D. E.; Paulson, S.; Jr., A. T. *J. Nano Lett.* 2005, *5*, 151.
166. III, A. B. S.; Tokuyama, H.; Strongin, R. M.; Furst, G. T.; Romanow, W. J.; Chait, B. T.; Mizra, U. A.; Haller, I. *J. Am. Chem. Soc.* 1995, *117*, 9359.
167. Britz, D. A.; Khlobystov, A. N.; Porfyrakis, K.; Ardavan, A.; Briggs, G. A. D. *J. Chem. Soc., Chem. Commun.* 2005, 37.
168. Tanaka, H.; Takeuchi, K.; Negishi, Y.; Tsukuda, T. *Chem. Phys. Lett.* 2004, *384*, 283.
169. Ogrin, D.; Barron, A. R. *J. Mol. Cat. A: Chem.* 2006, *244*, 267–270.
170. Treacy, M. M. J.; Ebbesen, T. W.; Gibson, J. M. *Nature* 1996, *381*, 678.
171. Berber, S.; Kwon, Y. K.; Tomanek, D. *Phys. Rev. Lett.* 2000, *84*, 4613.
172. Thess, A.; Lee, R.; Nikolaev, P.; Dai, H. J.; Petit, P.; Robert, J.; Xu, C.; Lee, Y. H.; Kim, S. G.; Rinzler, A. G.; Colbert, D. T.; Scuseria, G. E.; Tomanek, D.; Fischer, J. E.; Smalley, R. E. *Science* 1996, *273*, 483.
173. Girifalco, L. A.; Hodak, M.; Lee, R. S. *Phys. Rev. B* 2000, *62*, 13104.
174. Bachilo, S. M.; Strano, M. S.; Kittrell, C.; Hauge, R. H.; Smalley, R. E.; Weisman, R. B. *Science* 2002, *298*, 2361.
175. Kong, J.; Franklin, N. R.; Zhou, C. W.; Chapline, M. G.; Peng, S.; Cho, K. J.; Dai, H. J. *Science* 2000, *28*, 622.

176. Collins, P. G.; Bradley, K.; Ishigami, M.; Zettl, A. *Science* 2000, *287*, 1801.
177. Schadler, L. S.; Giannaris, S. C.; Ajayan, P. M. *Appl. Phys. Lett.* 1998, *73*, 26.
178. Bethune, D. S.; Klang, C. H.; deVries, M. S.; Gorman, G.; Savoy, R.; Vazquez, J.; Beyers, R. *Nature* 1993, *363*, 605.
179. Guo, T.; Nikolaev, P.; Rinzler, A. G.; Tománek, D.; Colbert, D. T.; Smalley, R. E. *J. Phys. Chem.* 1995, *99*, 10694.
180. Guo, T.; Nikolaev, P.; Thess, A.; Colbert, D. T.; Smalley, R. E. *Chem. Phys. Lett.* 1995, *243*, 49.
181. Hafner, J. H.; Bronikowski, M. J.; Azamian, B. R.; Nikolaev, P.; Rinzler, A. G.; Colbert, D. T.; Smith, K. A.; Smalley, R. E. *Chem. Phys. Lett.* 1998, *296*, 195.
182. Peigney, A.; Laurent, C.; Dobigeon, F.; Rousset, A. *J. Mater. Res.* 1997, *12*, 613.
183. Dai, H.; Rinzler, A. G.; Nikolaev, P.; Thess, A.; Colbert, D. T.; Smalley, R. E. *Chem. Phys. Lett.* 1996, *260*, 471.
184. Nikolaev, P.; Bronikowski, M. J.; Bradley, R. K.; Rohmund, F.; Colbert, D. T.; Smith, K. A.; Smalley, R. E. *Chem. Phys. Lett.* 1999, *313*, 91.
185. Bootsma, G. A.; Gasson, H. J. *J. Cryst. Growth* 1971, *10*, 223.
186. Westwater, J.; Gosain, D. P.; Tomiya, S.; Usui, S.; Ruda, H. *J. Vac. Sci. Technol. B* 1997, *15*, 554.
187. Ogrin, D.; Jr., R. C.; Maruyama, B.; Pender, M. J.; Smalley, R. E.; Barron, A. R. *Dalton Trans.* 2006, 229–233.
188. Fonseca, A.; Hernadi, K.; Nagy, J. B.; Bernaerts, D.; Lucas, A. A. *J. Mol. Catal. A: Chem.* 1996, *107*, 159.
189. Cassell, A. M.; Raymakers, J. A.; Kong, J.; Dai, H. *J. Phys. Chem. B* 1999, *103*, 6484.
190. Anderson, R. E.; Jr., R. C.; Crouse, C.; Ogrin, D.; Maruyama, B.; Pender, M. J.; Edwards, C. L.; Whitsitt, E.; Moore, V. C.; Koveal, D.; Lupu, C.; Stewart, M.; Tour, J. M.; Smalley, R. E.; Barron, A. R. *Dalton Trans.* 2006, 3097–3107.
191. Thess, A.; Lee, R.; Nikolaev, P.; Dai, H. J.; Petit, P.; Robert, J.; Xu, C.; Lee, Y. H.; Kim, S. G.; Rinzler, A. G.; Colbert, D. T.; Scuseria, G. E.; Tomanek, D.; Fischer, J. E.; Smalley, R. E. *Science* 1996, *273*, 483.
192. Davis, V. A.; Erickson, L. M.; Parra–Vasquez, A. N. G.; Ramesh, S.; Saini, R. K.; Kittrell, C.; Billups, W. E.; Adams, W. W.; Hauge, R. H.; Smalley, R. E.; Pasquali, M. *Macromolecules* 2004, *37*, 154.
193. Zheng, M.; Jagota, A.; Strano, M. S.; Santos, A. P.; Barone, P.; Chou, S. G.; Diner, B. A.; Dresselhaus, M. S.; McLean, R. S.; Onoa, G. B.; Samsonidze, G. G.; Semke, E. D.; Usrey, M.; Walls, D. J. *Science* 2003, *302*, 1545.
194. Tang, B. Z.; Xu, H. *Macromolecules* 1999, *32*, 2569.
195. Moore, V. C.; Strano, M. S.; Haroz, E. H.; Hauge, R. H.; Smalley, R. E. *Nano Lett.* 2003, *3*, 1379.
196. Moore, V. C.; Strano, M. S.; Haroz, E. H.; Hauge, R. H.; Smalley, R. E. *Nano Lett.* 2003, *3*, 1379.
197. Anderson, R. E.; Barron, A. R. *J. Nanosci. Nanotechnol.* 2006 in press.
198. Hirsch, A. *Angew. Chem. Int. Ed.* 2002, *40*, 4002.
199. Bahr, J. L.; Tour, J. M. *J. Mater. Chem.* 2002, *12*, 1952.
200. Banerjee, S.; Hermraj-Benny, T.; Wong, S. S. *Adv. Mater.* 2005, *1*, 17.
201. Georgakila, V.; Kordatos, K.; Prato, M.; Guldi, D. M.; Holzinger, M.; Hirsch, A. *J. Am. Chem. Soc.* 2002, *124*, 760.
202. Bahr, J. L.; Yang, J.; Kosynkin, D. V.; Bronikowski, M. J.; Smalley, R. E.; Tour, J. M. *J. Am. Chem. Soc.* 2001, *123*, 6536.
203. Bahr, J. L.; Tour, J. M. *J. Mater. Chem.* 2001, *12*, 3823.
204. Dyke, C. A.; Tour, J. M. *J. Am. Chem. Soc.* 2003, *125*, 1156.
205. Liang, F.; Sadana, A. K.; Peera, A.; Chattopadhyay, J.; Gu, Z.; Hauge, R. H.; Billups, W. E. *Nano Lett.* 2004, *4*, 1257.
206. Mawhinney, D. B.; Naumenko, V.; Kuznetsova, A.; Yates Jr., J. T.; Liu, J.; Smalley, R. E. *J. Am. Chem. Soc.* 2000, *122* 2383.
207. Ogrin, D.; Chattopadhyay, J.; Sadana, A. K.; Billups, E.; Barron, A. R. *J. Am. Chem. Soc.* 2006, *128*, 11322–11323.
208. Cui, J.; Burghard, M.; Kern, K. *Nano Lett.* 1993, *3*, 613.
209. Mickelson, E. T.; Huffman, C. B.; Rinzler, A. G.; Smalley, R. E.; Hauge, R. H.; Margrave, J. L. *Chem. Phys. Lett.* 1998, *296*, 188.

210. Kudin, K. N.; Bettinger, H. F.; Scuseria, G. E. *Phys. Rev. B* 2001, *63*, 45413.
211. Kelly, K. F.; Chiang, I. W.; Mickelson, E. T.; Hauge, R. H.; Margrave, J. L.; Wang, X.; Scuseria, G. E.; Radloff, C.; Halas, N. *Chem. Phys. Lett.* 1999, *313*, 445.
212. Alemany, L. B.; Zeng, L.; Zhang, L.; Edwards, C. L.; Barron, A. R. Solid state NMR analysis of fluorinated single-walled carbon nanotubes; assessing the extent of fluorination, *Chem. Mater.*, 2007, *19*, 735–744.
213. Boul, P. J.; Liu, J.; Mickelson, E. T.; Huffman, C. B.; Ericson, L. M.; Chiang, I. W.; Smith, K. A.; Colbert, D. T.; Hauge, R. H.; Margrave, J. L.; Smalley, R. E. *Chem. Phys. Lett.* 1999, *310*, 367.
214. Saini, R. K.; Chiang, I. W.; Peng, H.; Smalley, R. E.; Billups, W. E.; Hauge, R. H.; Margrave, J. L. *J. Am. Chem. Soc.* 2003, *125*, 3617.
215. Zhang, L.; Zhang, J.; Schmandt, N.; Cratty, J.; Khabashesku, V. N.; Kelly, K. F.; Barron, A. R. *Chem. Commun.* 2005, 5429–5430.
216. Zeng, L.; Zhang, L.; Barron, A. R. *Nano Lett.* 2005, *5*, 2001.
217. Rao, A. M.; Richter, E.; Bandow, S.; Chase, B.; Eklund, P. C.; Williams, K. A.; Fang, S.; Subbaswamy, K. R.; Menon, M.; Thess, A.; Smalley, R. E.; Dresselhaus, G.; Dresselhaus, M. S. *Science* 1997, *275*, 187.

Methods for Structural and Chemical Characterization of Nanomaterials

Jérôme Rose   *CNRS, University of Aix-Marseille, Aix-en-Provence, France*
Antoine Thill   *CEA, Saclay, France*
Jonathan Brant   *Duke University, Durham, North Carolina, USA*

Introduction

In this chapter, we survey several methods for characterizing physical-chemical properties of nanoparticles. Among these properties are particle size, charge, structure, shape, and chemical composition. Particle size can influence an array of material properties and is therefore of concern when studying nanoparticles. For example, the crystal properties of the material such as the lattice symmetry and cell parameters may change with size due to changes in surface free energy [see, for example, Zhang and Banfield, 1998]. This is particularly true for particles where the number of atoms at the surface represents a significant fraction of the total number of atoms (e.g., below 10 to 20 nm). Size may also affect the electronic properties of the materials due to the confinement of electrons, commonly discussed in terms of the quantum size effect, and the existence of discrete electronic states that give rise to properties such as the size-dependent fluorescence of CdS and CdSe nanoparticles or the electrical properties of carbon nanotubes.

The surface properties of nanomaterials play an important role in determining nanoparticle toxicity. The toxicity of CdS and CdSe nanoparticles, for instance, is completely controlled by their surface coating. The intracellular oxidation of bare CdSe nanoparticles results in the release of Cd^+ ions, the toxicity of which is well known. However, when CdSe nanoparticles are covered with organic molecules, the cytotoxicity is reduced (see Chapters 11 and 12). Adsorptive interactions involving nanomaterials (Chapter 10), the effects of nanoparticle surface chemistry on particle stability and mobility (Chapter 7), and photocatalytic properties (Chapter 5) are also covered in greater detail later in this book.

Principles of Light-Material Interactions, Atomic Force Microscopy, and Scanning Tunnel Microscopy

Phenomena resulting from the interaction between electromagnetic radiation and matter can be interpreted to yield a great deal of information on nanomaterials. Light (e.g., X-ray, UV-vis, infrared) is an oscillating electromagnetic field. It is a wave that is characterized by a specific frequency ($\omega_0 = 2\pi\nu_0$) and a radiation wave vector ($\vec{k}_0$). However, light is also a particle called a photon, where each photon carries a packet of energy that is proportional to its frequency. Light can also be associated to particulate beams like electron beams. An electron beam also interacts with matter in a similar fashion to that of photons. The interaction between light and matter can occur in a variety of different scenarios, which can be summarized as follows (see also Figure 4.1):

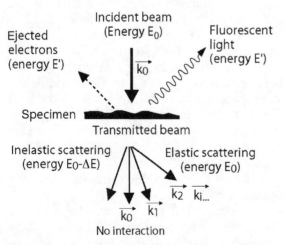

Figure 4.1   Examples of the various manners of interactions between light and electron beams with matter.

- No interaction occurs and the light is transmitted through the material with its initial characteristics ($\vec{k}_0$, ω_0).

- The wave vector is affected, but there is no change in frequency. The incident beam is dispersed over a range of "angles" via elastic scattering/diffraction processes. Modification of the wave vector is related to the spatial aspects of the matter (crystal structure, size, shape). These are the classical diffractions (X-ray diffraction [XRD], electron and neutron diffraction) and scattering techniques (small angle x-ray scattering [SAXS], light scattering, neutron scattering).

- The frequency of light is affected by internal excitation processes (electronic, nuclear, etc., transitions) that lead to absorption of the incident beam at the frequency ω_0. Examples of techniques that measure changes in the frequency of light are the absorption spectroscopy techniques such as X-ray absorption spectroscopy (XAS), Raman, Fourier transform infrared spectroscopy (FTIR), and nuclear magnetic resonance (NMR) imaging. The absorption of the incident beam, depending on the nature and structure of the sample matter, can also be used in microscopy by analyzing the absorption contrast as is done in electronic, light, and X-ray microscopy.

- The excited matter can relax through several processes causing the emission of fluorescent photons (X-ray and light) or electrons (auger and secondary electrons) having different frequencies. Examples of characterization techniques utilizing these measurement principles are X-ray fluorescence spectroscopy (XRF), luminescence spectroscopy, energy loss spectroscopy (ELS), and X-ray photoelectron spectroscopy (XPS).

In contrast with methods that rely on the interaction of materials with electromagnetic radiation, two techniques that were developed in the 1980s, atomic force microscopy (AFM) and scanning tunneling microscopy (STM), take an entirely different approach in which a probe "feels" its way along a surface at a resolution that may approach the atomic scale. AFM and STM are advanced microscopy techniques developed to measure surface topography with angstrom level resolution. In both cases the operating principle consists of scanning a probe in close proximity to the surface. The location between the probe and the surface is determined by the change in cantilever deflection and tunnel current in AFM and STM, respectively.

Structural Characterization

In the following parts we will detail different techniques enabling the characterization of the structure from the atomic scale to the size and shape of the nanoparticles.

Characterization of atomic structure

The characterization of the atomic arrangement of nanoparticles can be performed to determine both the long-range (i.e., the crystal parameters) and short-range order of a material's atomic structure. Of particular interest is the arrangement of atoms at the material's surface. The various characterization techniques that are available for these purposes may or may not be element specific. Therefore, care must be taken to select an appropriate characterization method. We will therefore discuss first the nonspecific techniques followed by the element-specific ones.

Nonspecific techniques

X-ray diffraction

Operating principles.   X-ray scattering was first used to study the long-range order of the atomic arrangement in crystals in the early 20th century (Wyckoff, 1964). More recently, it has been applied to characterize nanomaterials. The pattern, position, intensity, and shape of the peaks in X-ray diffraction (XRD) are all influenced by the atomic structure—that is, atomic interatomic distances and crystal cell parameters. The comparison of the diffraction pattern of unknown samples with the diffraction pattern of reference compounds from the database is necessary to identify the nature of minerals present in mixtures. XRD is a common technique used in mineralogy, and details to determine or refine atomic position can be found elsewhere.

Furthermore, information regarding the size of nanoparticles may also be derived from the XRD pattern, as particle size strongly affects the peak width. Particle size may be calculated based on the Sherrer formula, which states:

$$S = \lambda/w\cos\theta \qquad (1)$$

where S is the particle size, λ the wavelength of the beam, θ the diffraction angle and w is the width of the peak at half-maximum.

Sample preparation.   X-ray diffraction instruments may have a number of different configurations resulting in different types of sample holders. Powder diffraction requires a fine and homogeneous powder. Diffraction setup for "single" crystal analysis is not really adapted to study small objects since the size of the beam is much larger than individual nanoparticles. For nanoparticles, the sample needs to be dry and deposited either on a sample holder or in capillary tubes.

Application in the particular case of nanoparticles.   For nanoparticles, as for other types of materials, it is essential to determine both the crystallinity and polymorph type (i.e., two or more minerals having the

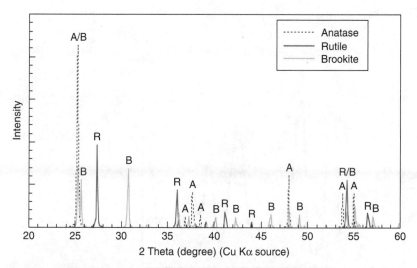

Figure 4.2   Theoretical X-ray diffraction patterns of the three polymorphs of TiO_2 particles. A: anatase, B: Brookite, R: Rutile.

same chemical composition, but different crystal structures). One of the best examples for illustrating this point is titanium dioxide (TiO_2). Due to the unique physical and chemical properties of TiO_2, it is widely used in industrial applications, particularly as a photocatalyst. Three different TiO_2 polymorphs exist—rutile, anatase, and brookite—however, only anatase is generally accepted to have significant photocatalytic activity. The respective XRD diffraction patterns of the three TiO_2 polymorphs (Figure 4.2) represent a simple and easy way to distinguish between them based on peak location and height.

Limitations.   One of the main limitations of XRD when characterizing nanoparticles is that if the amorphization process occurs without changing particle size, it will affect both the intensity and diffraction peak width in the same fashion. Moreover, in certain cases the position of the peaks for different minerals can overlap, leading to ambiguous identification of mineral phases. Because peak width increases with decreasing particle size, the overlapping of peaks is particularly problematic for very small particles.

Total scattering.   XRD is used to probe the periodic structure of minerals (periodicity over distances above 100 Å), but a method has been recently "rediscovered" as a result of synchrotron light sources. This technique is based on the total scattering of particles and is called the pair distribution function (PDF). This method essentially fills the gap between XAS (see below) and XRD. The PDF technique has long been

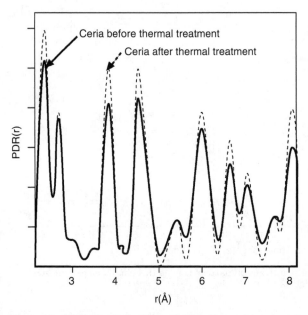

Figure 4.3 PDF of ceria before and after thermal treatment (adapted from Mamontov and Egami, 2000).

used for studying the nonperiodic structure of matter in noncrystalline materials (glasses) [Warren, 1990]. The local structure determined by the pair distribution function is the probability of finding an atom at a distance r from a reference z atom. The PDF transforms the signal obtained in the reciprocal space (wave vector space $Q = 4\pi \sin\theta/\lambda$ with θ the scattering angle and λ wavelength of the incident beam) to the real space (interatomic distance space). The spatial resolution is directly linked to the Q range scanned (greater than 20 Å^{-1}). Therefore, to obtain a high interatomic size resolution it is necessary to measure data using high X-ray energy (> 100 KeV), that is to say, low wavelength (<0.12 Å). The high X-ray energy is a strong limitation of this technique, since it requires a synchrotron radiation source. The PDF analysis of ceria nanoparticles [Mamontov and Egami, 2000] illustrates the interest of such a tool.

Using neutron diffraction, the authors have shown that the nanoscale ceria had Fenkel-type oxygen defects. The defects disappeared after a thermal treatment as shown in the post-thermal treatment PDF curve, which exhibits a higher intensity compared to the ceria before thermal treatment, for example, a higher number of interatomic distances (Figure 4.3).

Raman spectroscopy

Operating principles. Raman spectra result from the scattering of electromagnetic radiation by the molecules in solid bulk materials. The energy of the incident light beam (usually in the visible region of the spectrum and sometimes in the ultraviolet zone) is slightly lowered or raised by inelastic interactions with the vibrational modes. It is a powerful tool for investigating the structural and morphological properties of solids at a local level [see, for example, Ferraro and Nakamoto, 1994]. In a simple approach, Raman spectra can fingerprint the nature or type of crystal phases. Raman can in some cases assess the crystal, or amorphous nature, of minerals. These properties of Raman spectroscopy, however, are not particular to nanoparticles.

Sample preparation and limitations. One very interesting point concerning Raman spectroscopy is that the sample can be solid or in solution. A limitation of Raman spectroscopy, however, is that the sensitivity of the technique is dependent upon the material being characterized.

Application in the particular case of nanoparticles. Raman spectroscopy is a powerful tool to identify the nature of nanoparticles. Like XRD, the position of Raman peaks can be considered as a fingerprint for different minerals. More than XRD, overlapping of peaks can lead to difficulties in identifying the constituent minerals in a complex matrix. In some cases like TiO_2, for which the Raman peaks are particularly intense, it is possible to distinguish between the different polymorphs. Taking for example anatase and rutile, both of which are polymorphs of TiO_2, the Raman spectroscopy may be used to differentiate between them. For anatase, six different peaks are recorded at 144 cm^{-1} (*Eg*), 197 cm^{-1} (*Eg*), 397 cm^{-1} (*B*1g), 518 cm^{-1} (*A*1g and *B*1g, unresolved), and 640 cm^{-1} (*Eg*). On the other hand, for rutile, three different peaks are detected at 144 cm^{-1} (*B*1g), 448 cm^{-1} (*Eg*), and 613 cm^{-1} (*A*1g) (a fourth very weak band corresponding to the *B*2g mode also exists at 827 cm^{-1}) [Robert et al., 2003]. In the particular case of nanoparticles, the signal is strongly affected by particle size as well as shape. This point will be detailed further later in this chapter.

Element-specific techniques

X-ray absorption spectroscopy (XAS)

Operating principles. X-ray absorption spectroscopy (XAS) is one of the most powerful techniques for probing the local atomic structure in a vast array of materials. XAS is a short-range order method that can be used regardless of the sample's physical state (crystalline, amorphous, in solution, or in a gas phase). Another important property of XAS is that it is an element-specific technique, which is in contrast to other spectroscopic methods such as Raman spectroscopy. The operating principle of this method requires that the incident X-ray beam energy be scanned from below to above, the binding energy of the core shell electrons of the target atom. By doing so, one observes an abrupt increase in the absorption coefficient corresponding to the characteristic absorption edge of the selected

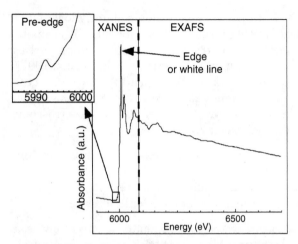

Figure 4.4 The Cr K edge absorption spectrum of chromite ($CrFe_2O_4$) showing the XANES (X-ray absorption near edge structure) and EXAFS (extended X-ray absorption fine structure) parts.

element (Figure 4.4). Depending on the electron that is excited, the absorption edges are named K for 1s , L_I for 2s, L_{II}, L_{III} for 2p, and so on.

The theory of XAS is described in detail elsewhere [Koningsberger and Prins, 1988, Fontaine, 1993, Rehr and Albers, 2000]. Two characterization methods involving X-ray absorption are XANES and EXAFS, which focus on different portions of the absorption spectra. XANES yields information on chemical bonds and symmetry, while EXAFS provides information on coordination number, chemical species, and distances. In the next section we will briefly introduce the theoretical basis of XAS and detail its use in characterizing the atomic arrangement of nanoparticles.

The energy position of the absorption edge and its shape reflect the excitation energy of the inner-shell electrons. The transition is always from core level to unoccupied states. The resulting excited photoelectron has generally enough kinetic energy to move through the material and this phenomenon can even occur in insulators. The presence of neighboring atoms around the central and excited atom leads to a modulation of the absorption coefficient due to interferences between outgoing and backscattered photoelectron waves. These modulations are present in the EXAFS zone. Oscillations can be extracted as a function of the photoelectron wave vector $k = \sqrt{2m_e(E - E_0)/\hbar^2}$, where m_e is the mass of electron, E is the energy, and E_0 is the binding energy of the photoelectron. The conventional EXAFS analysis based on single scattering was developed by Sayers et al. [1971]:

$$\chi(k) = \frac{\mu(E) - \mu_0(E)}{\Delta\mu_0(E)}$$

$$= -S_0^2 \sum_i \frac{N_i}{kR_i^2} |f_i(k)| e^{-2\sigma^2 k^2} e^{-\frac{2R_i}{\lambda(k)}} \sin[2kR_i + \phi_{ij}(k)] \qquad (2)$$

where $\mu(E)$ is the measured absorption coefficient, $\mu_0(E)$ is a background function representing the absorption of an isolated atom, $\Delta\mu_0(E)$ is the jump in the absorption coefficient at the energy of the edge, S_0^2 is the amplitude reduction factor due to multielectronic effects. N_i is the coordination number, R_i is the interatomic distance between the central atom and the neighboring atom of type i, σ_i is a Debye-Waller factor describing the static and dynamic disorder in a Gaussian approximation, $|f_i(k)|$ is the amplitude of the backscattering wave from the neighbor of type i, $\lambda(k)$ is the free mean path of the photoelectron, that accounts for inelastic losses, and $\phi_{ij}(k)$ is the phase shift between the central ion j and its neighbors i. From Eq. 2 it is possible to extract from EXAFS oscillations information such as the interatomic distances and the number and nature of surrounding atoms.

The pioneering work of Sayers et al. [1971] revolutionized the way EXAFS data is analyzed. Because of the sinusoidal nature of EXAFS spectra, Sayer et al. used a Fourier transform to visualize the various electronic shells surrounding the central absorber. A pseudoradial distribution function (RDF) is obtained that provides the position of the different scatterers (Figure 4.5).

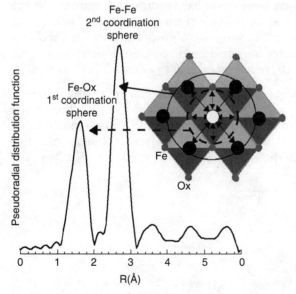

Figure 4.5 Radial distribution function of iron in lepidocrocite (γ-FeOOH).

XANES analysis is more sensitive than EXAFS to the site geometry and the oxidation state of the target atom. Therefore, the combination of XANES and EXAFS is very powerful for probing the speciation of elements.

Sample preparation. XAS has the ability to measure samples under various physical forms (solid, liquid, and gas) with little preparation, in contrast to other techniques such as TEM. In transmission mode, it is crucial to prepare pellets (for solid samples) with no pinholes and constant and appropriate thickness. It is also important that the size of the particles in the sample not be much larger than one absorption length (which is always the case for nanoparticles). Solutions produce the best transmission samples. For nanoparticles in suspension it is essential to prevent any settling during the measurement.

Limitations. The interpretation of EXAFS is limited in that it does not provide any information about bond angles between atoms. Moreover, it is not practically possible to distinguish between atoms that are in the same line in the periodic table ($Z \pm 2$). For example, it is not possible to distinguish between O and N, whereas the differences between O and S allow for their identification. This quite strong limitation is due to the energy dependence of $|f_i(k)|$ and the phase shift function $\phi_{ij}(k)$. These two functions contain the information characteristic of the nature of the scattering atoms. Unfortunately, the differences of $|f_i(k)|$ and $\phi_{ij}(k)$ for two atoms like O and N are not great enough to easily distinguish between them (Figure 4.6).

In XANES analysis the theory is not as yet fully quantitative as is the case for EXAFS and requires different physical considerations. Here, X-rays from a synchrotron source may induce significant chemical changes within the sample (e.g., oxydo-reduction processes), which has obvious consequences for accurately characterizing the material of interest.

Application in the particular case of nanoparticles. XAS can be helpful in determining the structural evolution of nanoparticles as a function of the nature of the ligands capping them. For example, Chemseddine et al. [1997] demonstrated that the presence of acetate or thiolate modifies the symmetry of CdS in the surface layer. Whereas in the bulk on the

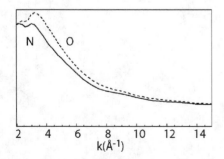

Figure 4.6 Backscattering amplitude factor $|f_i(k)|$ for oxygen and nitrogen.

particle the Cd is tetrahedraly coordinated and CdS octahedra appeared at the surface. XAS can also probe the surface layer structure of particles, but only if objects are within the nanometer size range. Indeed, XAS and EXAFS give average information concerning all crystallographic sites that exist in a mineral. Unfortunately, XAS is only sensitive to the largest fraction of sites when there are several sites present. In the case of large particles the surface sites represent a limited number of atoms that cannot be detected by XAS. As soon as the fraction of surface atoms becomes higher than 15–20 percent, XAS is sensitive enough to determine the modification of the surface site. For example, Auffan et al. [2006] were able to determine that iron atoms at the surface of nanomaghemites (γ-Fe_2O_3) recovered by DMSA (di-mercaptosuccinic acid) were highly asymmetric due to ligand exchange between OH and SH groups from the DMSA (Figure 4.7). The resulting EXAFS spectra were then the combination of two Fe sites.

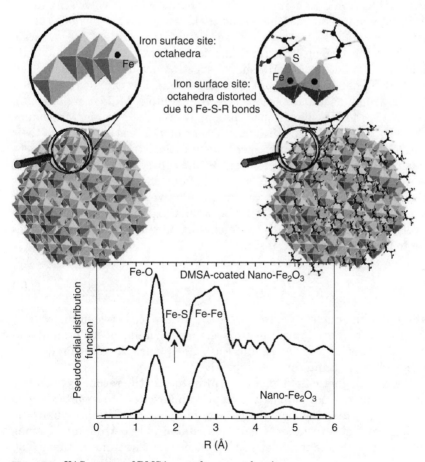

Figure 4.7 XAS spectra of DMSA-coated nanomaghemites.

EXAFS can be a powerful tool for characterizing nanomaterials when used along with XRD. Indeed, the decrease in the size of crystalline particles leads to an increase in the FWHM of the XRD peaks. But a decrease in particle crystallinity without any change in size leads to the same effect. For very small particles (<3 to 4 nm) the XRD spectra are quite noisy, and accurate information concerning the minerals is difficult to extract. Using EXAFS in combination with XRD can help in solving this limitation. The Debye-Waller factor determined by EXAFS modeling is related to the disorder of the particles. For example Choi et al. [2005] observed an increase of the static disorder as TiO_2 particle size decreased. Moreover, a volume contraction as particle size decreases has been highlighted by a decrease of the Ti-Ti interatomic distances.

Mössbauer spectroscopy

Operating principles. While XAS spectroscopy is based on the measurement of electronic transitions, the Mössbauer effect involves the interaction of γ radiation (i.e., the resonant absorption) with the nuclei of the atoms of a solid. Here, γ-rays are used to probe the nuclear energy levels related to the local electron configuration and the electric and magnetic fields of the solid. To date, Mössbauer spectroscopy has been mainly used to study Fe nanoparticles, but Au and Pt materials can also be studied by following nuclear transitions [see, for example, Mulder et al., 1996]. Like XAS, the Mössbauer spectroscopy is element specific. Mössbauer spectra consist of plotting the transmission of γ rays as a function of their source velocity. A Mössbauer spectrometer consists of a vibrating mechanism that imparts a Doppler shift to the source energy and then to a source. In the absence of any magnetic field, the Mössbauer spectrum consists of one or two absorption maxima between I1/2 and I3/2 nuclear levels (Figure 4.8). The difference between the ground and excited state levels is called the chemical or isomer shift, δ, which is described according to the following relationship:

$$\delta = \frac{4\pi}{5} Ze^2 R^2 \left(\frac{\delta R}{R}\right) \left[|\psi(0)|^2_{ABS} - |\psi(0)|^2_{SOURCE}\right] \qquad (3)$$

where Z is the nuclear charge, δR is the difference between the radii of the ground and excited states, R is the mean radius of the ground and excited states, and $|\psi(0)|^2_{ABS}$ and $|\psi(0)|^2_{SOURCE}$ are the electron density of the absorbant and source, respectively.

When a magnetic field exists it will influence the resonant nuclei by splitting the nuclear spin of the ground and excited states into various new states. This phenomenon leads to multiple transitions, where the position and absorption intensity are related to hyperfine interactions between the resonant nuclei and the electrons surrounding them. The

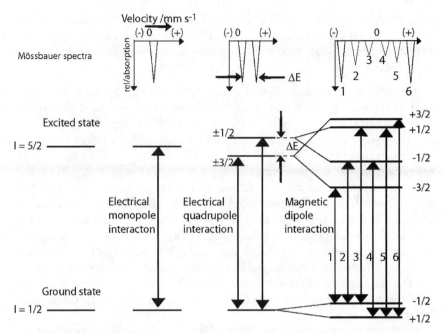

Figure 4.8 Resonant Mössbauer absorption in the presence and absence of a magnetic field.

different interactions that may occur between the nuclei and accompanying electrons and the information that may be derived from each can be summarized as follows:

- The shift due to monopole interactions provides information about the coordination number, valency and spin state of the studied atom.
- The quadrupole shift provides information about the site distortion.
- The magnetic hyperfine field provides information about the valence and magnetic properties of the compound.

Accordingly, inspection of Mössbauer spectra can provide information regarding the crystallographic nature of minerals and the local scale evolution of a target atom. For iron nanoparticles Mössbauer spectroscopy is a very efficient technique for probing the Fe^{2+}/Fe^{3+} ratio and the nature and size of the different oxide phases.

Sample preparation. In performing Mössbauer spectroscopy, samples must be prepared either as powders or as homogeneous thin-films (e.g., 5 mg/cm^2 of Fe). As with the case of powder-XRD it is important to avoid any textural effect of the powder. The samples do not require any vacuum. However, measurements are generally performed at low temperature (i.e., in the presence of liquid nitrogen). The samples must therefore not change phase and/or be resilient in such conditions. With regards to sensitivity, the matrix or support of the studied element is of high importance. But it seems difficult to detect elements at concentrations lower than 1–2 percent (w/w).

Application in the particular case of nanoparticles. For nanoparticles some specificity exists when applying Mössbauer spectroscopy. When the grain size of fine particles is smaller than a critical grain size, Dc, they are composed of many single magnetic domains even though there is no external magnetic field. Therefore, the magnetic fields will not be stationed in a fixed direction like for a bulk material, but will rather jump from one easy-magnetization direction to another; this is the superparamagnetic phenomenon. The Dc varies from one mineral to another. For instance, the Dc of α-Fe_2O_3 is 20 nm. If the grain size of iron oxide in composites is smaller than Dc, there appear superparamagnetic doublet lines in the Mössbauer spectra. Using this property, Liu et al. [2005] combined XRD and Mössbauer spectroscopy on an Fe_2O_3-Al_2O_3 nanocomposite, and found that below 1373 K the average grain size is below 20 nm (i.e., it exhibited a superparamagnetic behavior).

As Mössbauer spectroscopy probes the local structure of a target atom, the crystallographic sites of doped nanoparticles can also be studied. For example, Zhu et al. [2005] found that Fe can form superparamagnetic α-Fe_2O_3 phases in iron-doped TiO_2 if the Fe concentration was higher than 4 percent (w/w). This phenomena is also a consequence of the fact that Fe does not substitute Ti in the TiO_2 structure.

In the case of iron, Mössbauer spectroscopy is capable of differentiating redox states and also atomic sites. The chemical shifts of ^{57}Fe in various solid systems are summarized in Figure 4.9.

Like XAS, Mössbauer spectroscopy is sensitive to the atomic arrangement of the surface layer in a bulk solid. For metallic clusters of Au and Pt, Mulder et al. [1996] showed that for nanoparticles composed of 55,

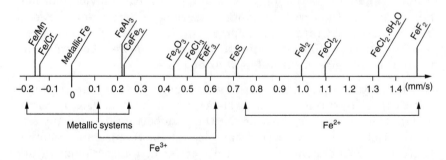

Figure 4.9 Chemical shifts of various iron species.

147, 309 atoms (2, 3, or 4 atom shells) with cuboctahedral (fcc.) arrangement and for a given n-shell compound, the "surface" metal atoms in the outer n^{th}-shell of the metal core do not show purely "metallic" behavior. Furthermore, experiments with two shell Au-55 nanoparticles recovered by different types of ligands (triphenylphosphine, tri-para-tolylphosphine, metasulfonatophenyldiphenylphosphine, and trianisylphosphine) demonstrated that they do not exhibit a perfect metallic character due to the influence of the surface layer. The charge transfer between the surface atoms and the ligand influences all the particles even at the core of the Au-55 particles. For Pt-308 nanoparticles charge transfer between the ligand and the surface layer did not influence the particle core. In summary, Mössbauer spectroscopy provides a sensitive local probe to measure these charge densities and therefore assess the influence of ligands on the surface layer structure.

Nuclear magnetic resonance (NMR)

Operating principles. Many nuclei have spin, and all nuclei are charged. The number of energy levels of a nuclide is 2I+1, where I is the spin quantum number. When a magnetic field is applied, nuclei can be excited to a higher energy level that corresponds to a change in spin. When the spin on the nuclide returns to its base level, it emits energy at a wavelength that corresponds to the energy transfer that excited the nuclide. By measuring the resonance frequency of the nuclides, information on chemical structure can be obtained. The interaction between the field and the energy levels leads to energy differences between the different levels. The number of energy levels is 2 for I = 1/2 and 6 for I = 5/2 (Figure 4.10).

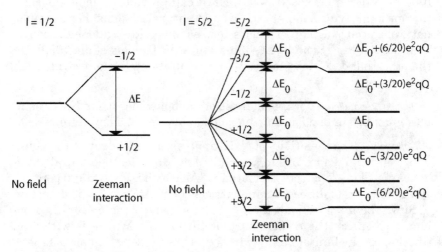

Figure 4.10 Nuclear spin energy level diagram for a spin = 1/2 and 5/2 nuclide.

The differences between the various energy states is given by: $\Delta E = |\gamma \hbar H|$ where H is the magnetic field at the nucleus. The Boltzmann distribution $\frac{N_a}{N_b} = \exp(-\Delta E/kT)$ gives the probability of higher and lower energy levels (N_a and N_b). It is important, however, to recall that the difference between high and low energy levels tends to be very low. The increase or decrease of nucleus energy is due to absorption or emission of photons with a frequency v, given by:

$$v = \frac{\gamma}{2\pi}H \cdot \qquad (4)$$

NMR measures the frequency of this radiation, which is in the radio-frequency range. Nuclei with various atomic arrangements absorb and emit photons of slightly different frequencies due to shielding:

$$H = H_0(1 - \sigma) \qquad (5)$$

and

$$v = \gamma \frac{H_0}{2\pi}(1 - \sigma) \qquad (6)$$

The resonance frequencies are generally reported as chemical shifts (δ, which are described according to the following expression):

$$\delta = \left(\frac{v_{\text{sample}} - v_{\text{standard}}}{v_{\text{standard}}}\right) \times 10^6 \qquad (7)$$

As with Mössbauer spectroscopy, NMR can provide valuable information for identifying solid phases even in natural systems. For example, naturally occurring nanoparticles such as imogolite or allophane can be identified using ^{29}Si and ^{27}Al solid-state NMR [Denaix et al., 1999]. On the other hand, XRD is severely limited in its ability to provide such information.

Sample preparation. For the analysis of nanoparticles using NMR, samples must be supplied as a homogeneous powder in a dry state.

Limitations. The sensitivity of NMR analysis is dependent on the abundance of the given isotope in a sample. For example, the natural abundance of ^{29}Si is 4.7 percent compared to ^{27}Al, which is close to 100 percent. With mixtures of different nanoparticles (e.g., Al and Si) the detection of Si will be more difficult than Al. One possible solution to this problem is to enrich the Si nanoparticle during the synthesis with ^{29}Si. Another strong limitation to NMR is that the presence of paramagnetic elements strongly affects the signal. For example, the presence of iron in a system leads to a strong decrease of the peak intensity and therefore limits the application of NMR in these cases.

Application in the particular case of nanoparticles. For nanoparticles, NMR is of high interest since this element-specific technique does not require long-range order and may be used to characterize nanoparticle surface properties like XAS. For example, ^{7}Li NMR of polycrystalline nano-LiMn$_2$O$_4$ indicated that the lithium ion occupies the 8a position, however, it also revealed that it has two different distances to neighboring manganese and oxide ions [Hon et al., 2002]. Such information is difficult to extract from XRD due to the small size of the product. More specifically, in the case of quantum tunneling in nanomagnets such as Mn, ^{55}Mn NMR [Kubo et al., 2001] can provide unique information on the magnetic behavior of these nanoparticles. This is evidence of phonon-activated resonant quantum tunneling.

Microscopy

SEM and TEM. Spectroscopic techniques can provide detailed information about the structure and size of nanoparticles. The most common examples of electron microscopy techniques used for characterizing nanoparticles include scanning electron microscopy (SEM) and transmission electron microscopy (TEM). A potential drawback to these techniques is that in some cases particle shape can induce indirect modification of the spectroscopic signal and is thus a source of error in these types of measurements. Nevertheless, these microscopy techniques allow for direct visualization of nanoparticles, and thereby provide information about particle size, shape, and structure. With this in mind, both SEM and TEM imaging are highly versatile and powerful techniques for characterizing nanoparticles.

Operating principle. The general operating principles and the components that make up the respective instruments of TEM and SEM imaging are summarized in Figure 4.11.

The interaction between an electron beam and a solid surface results in a number of elastic or inelastic scattering processes (backscattering or reflection, emission of secondary electrons, X-rays or optical photons, and transmission of the undeviated beam along with beams deviated as a consequence of elastic—single atom scattering, diffraction—or inelastic phenomena). The operational principle for a scanning electron microscope (SEM) is based on the scanning of finely focused beams of electrons onto a surface. When using an SEM, there are a number of different visualization techniques that can be used. During scanning, the incident electrons are completely backscattered, reemerging from the incident surface

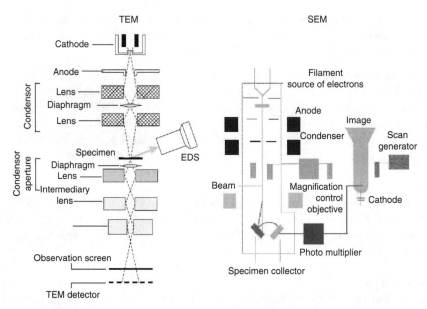

Figure 4.11 Schematics illustrating the operating principles of SEM and TEM microscopes.

of the sample. Since the scattering angle is strongly dependent on the atomic number of the nucleus involved, the primary electrons arriving at a given detector position can be used to yield images containing both topological and compositional information. The backscattering mode is generally used on a polished section to minimize the effects of local topology and therefore obtain information on the composition of the sample.

The high-energy incident electrons can also interact with loosely bound conduction band electrons in a sample. The amount of energy given to these secondary electrons as a result of these interactions is small, and so they have a very limited range in the sample (a few nanometers). Because of this, only secondary electrons that are emitted within a very short distance of the surface are able to escape from the sample. This means that the detection mode boasts high-resolution topographical images, making this the most widely used of the SEM modes. SEM can provide both morphological information at the submicron scale and elemental information at the micron scale. Recent developments in terms of electron source (field emission) have led to the development of high-resolution SEM. Using a secondary or backscattering electron image one can look at particles as small as 10–20 nm (Figure 4.12A). Chemical information using EDX, however, is obtained at the micron scale and not for individual particles.

In contrast with SEM, transmission electron microscopy (TEM) analyzes the transmitted or forward-scattered electron beam. Here the

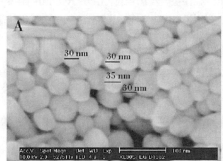
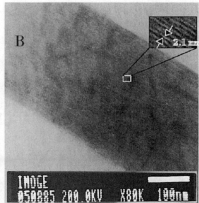

Figure 4.12   A. An FESEM secondary electron image of Ag nanoparticles where the particle size is determined using the appropriate scale (courtesy of Vladimir Tarabara). B. A TEM image of imogolite (single-walled aluminosilicate nanotube) (courtesy of Clément Levard).

electron beam is passed through a series of lenses to determine the image resolution and obtain the magnified image (Figure 4.12B). The highest structural resolution possible (point resolution) is achieved upon use of high-voltage instruments (acceleration voltages higher than 0.5 MeV). Enhanced radiation damage, which may have stronger effects for nanostructured materials, must however be considered in these cases. With corrections it is possible to achieve sub-angstrom resolution with microscopes operating at lower voltages (typically, 200 keV), allowing the oxygen atoms to be resolved in oxides materials. On the other hand, as high resolution is achieved in TEM as the result of electron wave interference among diffracted peaks and not only to the transmitted beam in the absence of deflection, a limitation to structural resolution can arise from nanoparticles with a very low number of atoms. Nevertheless conventional TEM is the most common tool used to investigate the crystal structure of materials at the sub-nanometer scale. There exist a number of different TEM techniques that may be used to obtain structural images with atomic level resolution; two of these techniques are detailed below: high-resolution TEM (HRTEM) and high angle annular dark field (HAADF) scanning transmission electron microscopy (STEM).

HRTEM images are formed by the interference of coherent electron waves. The object transmits the (nearly) planar incident electron wave, interacts with it, and the resulting electron wave ψ_e at the exit plane of the object carries information about the atom arrangement in the object. The ψ_e corresponds to a set of "diffracted" coherent plane waves. The electron optics transfers these waves to the image plane, and the intensity distribution of their interference pattern constitutes the HRTEM image.

With the HAADF-STEM techniques, images are formed by collecting electrons that have forward scatter at high angles, typically a few degrees or more, using high angle annular dark field (HAADF) scanning transmission electron microscopy (STEM). Unlike normal dark-field imaging, where the signal comes from elastic (Bragg) scattering of electrons typically to smaller angles, the HAADF-STEM signal is the result of inelastic electron scattering typically to larger angles. At high angles, elastic and inelastic interactions between the incident electrons and the columns of atoms within the specimen produce the image contrast. Since inelastic scattering depends on the number of electrons in an atom, the strength of scattering varies with atomic number. Spatial resolution is determined by the size of the focused incident electron probe. With electron beam sizes of less than 3 angstroms, imaging at atomic level resolution is possible. In a HAADF image, brighter spots represent the heavier atomic columns while the less intense spots indicate the lighter atomic columns.

Coupled to EDS (energy dispersive spectrometry) or to EELS (energy electron loss spectroscopy), TEM can provide information about the elemental distribution at a very low spatial scale (several nanometers). The energy resolution of EELS is 0.2 eV, while for EDX it is > 140 eV at 6 keV. In addition to elemental information, EELS can be used to determine the electronic structure, bonding, and nearest neighbor distribution of the specimen atoms. The high-loss energy is related to electrons that have interacted with inner-shell or core electrons of the specimen atoms. Thus, the information obtained is similar to that given by XAS (see the above discussion)—that is, K-, L-, M-, and such ionization edges of the elements present in the sample appear in the EELS spectrum (near edge and extended energy-loss fine structure, ELNES and EXELFS, regions being defined within them). Therefore, EELS can provide information on the speciation of elements. However, the very low intensity of the EELS signal at energies higher than 1500–2000 eV is a strong limitation to study the K edge of elements with an atomic number higher than the silicon. Therefore EELS is well adapted to study elements in a 100–1500 eV energy range corresponding to the K edge of low atomic number elements like carbon and oxygen or the l edge of transition metals like Fe or Cu.

Samples preparation.   For conventional SEM measurements the sample must be dry and conductive. If the sample material is not conductive it must be coated with some material, usually carbon or gold, using a sputter coating device. In some instances it may also be advantageous to coat already conductive materials to improve image contrast. The conductive coatings are usually applied at a thickness of about 20 nm, which is too thin to interfere with dimensions of surface features. Prior to coating the sample is mounted on a ring stand using nonconducting carbon tape. In many instances, as with most common forms of SEMs, it is necessary that the measurement be done under high vacuum. This requires that the sample be dry in order to prevent off-gassing during the measurement. However, recent advances in the design of SEM measurement chambers has led to the development of environmental (ESEM) and cryogenic SEMs, for imaging wet and frozen or fixed samples, respectively. As opposed to conventional SEM, in ESEM the sample may be both wet and does not need to be conductive. It is therefore desirable for delicate biological samples. In cryo-SEM and cryo-TEM the sample is frozen or fixed using liquid nitrogen and transferred to a cryo-preparation chamber that is held in vacuum. In the case of SEM measurements, a thin conductive coating is usually applied to allow high-resolution imaging or microanalysis in the SEM. Transfer to the SEM/TEM chamber is via an interlocked airlock and onto a cold stage module fitted to the SEM/TEM stage.

Materials that are appropriate for TEM analysis are constrained to very thin samples (1000–2000 Å), but do not require the presence of a conducting layer as in conventional SEM. However, a high vacuum is required and is accompanied by the aforementioned constraints. Nanoparticles are particularly viable for study using TEM as they are appropriately thin and may be imaged using TEM support grids. TEM support grids are fine mesh supports that are commonly made of copper, which may be covered with a range of materials (e.g., carbon, Formvar, holey, SiO_2, etc.). Particles are either deposited through evaporation or through electrostatic attraction using positively charged grids. The grid/sample must be allowed to dry prior to imaging to prevent off-gassing once the sample is placed in the vacuum. When examining particle samples it is important to avoid aggregation during the drying step, which will inhibit analysis of individual particles.

Application in the particular case of nanoparticles. High-resolution TEM (HRTEM) may be employed to provide extremely valuable information about the atomic structure of nanoparticles. For instance, Marın-Almazo et al. [2005] used HRTEM to determine the atomic arrangement of rhodium nanoparticles (d = 1.8 nm). The authors found that for these nanoparticles the [111] and [200] inter-planar distances corresponded to large minerals indicating that no distortion of the network existed. However, for the smaller clusters (below 1.5 nm) some range stacking faults, dislocations, and twins were identified (as illustrated in Figure 4.13).

In another study by Yan et al. [2005] it was found that by coupling XRD analysis (Sherer equation) and TEM imaging, it was possible to determine the structure and size of ultra-small gold nanoparticles (d = 1 nm). TEM can also be a powerful characterization technique for studying the dispersion and chemistry of nanoparticles in the environment. For

55 atoms Rh fcc cluster   Planar defect

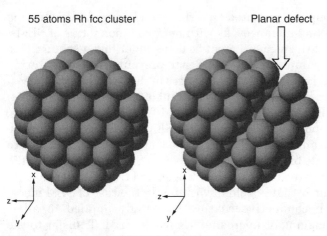

Figure 4.13 Images of a stable structure composed of 55 atoms and one with a planar defect that can be identified with HRTEM images. These two pictures show a hard-ball model of the structure (adapted from Marın-Almazo et al., 2005).

instance, using TEM, Hiutsunomiya and Ewing [2003] found that airborne particulates (d < 2 μm) from coal-fired power plants contained 1 to 10 ppm of uranium using HAADF-STEM images and that the uranium was located in nanoparticles (d < 10 nm) as uraninite (UO_2). These nanoparticles were encapsulated in graphite, which may retard oxidation of the tetravalent uranium to the more mobile hexavalent form.

Spatially resolved EELS has been used to study the morphology of carbon nanotubes [Stephan et al., 2001]. EELS results demonstrated that even for tiny nanotubes the covalent nature of the chemical bonds is preserved, whereas near-field EELS pointed out the specific character of the surface valence electron excitation modes in nanotubes in relation with their curved anisotropy.

AFM/STM

Operating principles. The invention of the atomic force microscope (AFM) in 1982 is considered one of the most important instrumental breakthroughs in the development of nanoscience. The AFM provides a means both to characterize the physical properties of materials at the atomic scale and to measure forces between surfaces with piconewton resolution. The operating principles of both the AFM and the scanning tunneling microscope (STM) may be described in terms of an optical lever acting as a sensitive spring. The optical lever operates by reflecting a laser beam off the end of a cantilever, typically made of silicon or silicon nitride, at the end of which is attached a tip or probe. Angular deflection of the tip causes a twofold larger angular deflection of the laser beam. The reflected laser beam strikes a position-sensitive photo-detector consisting of two side-by-side photodiodes. The difference between the two photodiode signals indicates the position of the laser spot on the detector and thus the angular deflection of the tip. Because the tip-to-detector distance generally measures thousands of times the length of the cantilever, the optical lever greatly magnifies the motions of the tip. Because of the approximately 2000-fold magnification in the measured deflection, the optical lever detection can theoretically obtain a noise level as low as 10^{-14} m/Hz$^{1/2}$.

The advantage of the AFM/STM over electron microscopes is that it is possible to measure in the z-axis in addition to both the x-axis and y-axis. In this way it is possible to get a three-dimensionally resolved image of a surface. Furthermore, using the AFM it is possible to measure interfacial forces between surfaces in both gaseous and liquid environments. For example, Brant et al. [2002 and 2004] used an AFM to characterize the surface morphology of water-treatment membranes and to subsequently measure the interfacial forces between the membrane surface and various nanoparticles. This information may then be used to either optimize or prevent particle attachment to a given surface as in groundwater transport processes or engineered systems.

Although originally conceived as an imaging device, because the operating principle of the AFM is based on the measurement of force between a small tip and a surface with piconewton sensitivity, this method can also be used to characterize interactions between surfaces and nanomaterials. Force is measured by recording deflection of the free end of a cantilever as its fixed end approaches and is subsequently retracted from a sample surface. The interaction force occurring between the AFM probe and the sample surface is then calculated according to Hooke's law ($F = k\Delta z$) where F is the force; k is the cantilever spring constant; and Δz is the vertical deflection of the cantilever. A positive vertical deflection indicates repulsion while a negative one indicates attraction. Figure 4.14 represents a typical force curve generated by an AFM, where force is plotted as a function of separation distance (h). Initially the colloid probe is far from the sample surface and no force is detected (a). As the probe approaches the surface it encounters some type of force (repulsive or attractive) before contacting the surface (b). The probe is then pressed against the sample surface until a preset loading force is reached, and then the probe is retracted (c). The pull-off force is measured as the force required to separate the two surfaces (maximum negative deflection) (d) and is used to approximate the strength of adhesion between the two surfaces.

A significant advantage of the AFM over other force measuring techniques, is its ability to operate in either air or water [33]. AFM force measurements may be carried out using either a standard silicon nitride tipped cantilever or a probe with attached material such as a colloid or

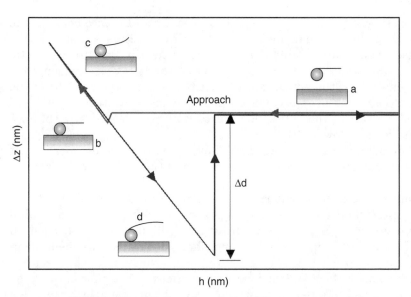

Figure 4.14   AFM force as a function of separation distance on approach (a-c) and retraction (c-d) from the surface.

nanoparticle [31, 32, 34]. By attaching nanomaterials such as a single-wall nanotube, information on interacting forces between the nanomaterial and an approaching surface may be measured directly. Alternatively, a colloid probe can be attached to the AFM tip and used to measure interactions with a lawn of deposited nanomaterials. Measured AFM force curves are then related to hypothetical interfacial forces arising from hydration, electrostatic repulsion, Lifshitz van der Waals energies (Hamaker constant), and acid-base interactions. This is accomplished by modifying the solution chemistry to isolate the specific interactions or properties of interest. For example, the Lifshitz van der Waals surface energy component of the nanoparticles can be determined by performing AFM force measurements in purely apolar solvents (e.g., cyclohexane) [36]. By measuring the interfacial interaction in a nonpolar solvent, both acid-base and electrostatic interactions are eliminated, thus isolating the van der Waals component. The Hamaker constant for the system (see Chapter 7) is proportional to the AFM measured force and the square of separation distance. Therefore, the Hamaker constant may be calculated as the gradient of a plot of the square root of Force versus *separation distance* for the sphere-plate geometry represented, for example, by a colloid probe approaching a nanomaterial lawn.

Sample Preparation.   For AFM-type measurement, no special sample preparation is required other than fixation to a sample stage, typically a small circular metal disk. Nonconducting carbon tape is commonly used for these purposes. One of the distinct advantages of AFM is that measurements may be carried out in both air and water. For analysis of powder materials they must be deposited onto a surface and fixated to prevent movement during imaging.

Limitations. Samples for AFM analysis are generally constrained to rather smooth samples with average roughness values of less than around several microns. This constraint results from the need to maintain contact between the scanning probe and the surface and limitations in the change in height that the probe can accommodate. Particles in size from around 1 μm to several 10s of microns may be attached to AFM cantilevers for force measurements. The minimum particle size continues to decrease with improving methods of particle attachment. For example, nanotubes are now being used as tethers for attaching particles to cantilevers serving to reduce the minimum particle size. The principle constraint here results from the use of epoxy to fix the particle to a tipless cantilever. Below a critical size the epoxy covers the sample particle and thus alters the interaction chemistry.

Form and size characterization

Form and size of nanoparticles can be determined using various techniques, but scattering techniques are particularly well adapted for this purpose.

Scattering experiments. Scattering experiments may be used to characterize particle suspension *in situ*. Light or X-ray scattering experiments may be used to measure the size, shape, agglomeration state, and dynamic properties (diffusion coefficient) of nanoparticles in a sample. These properties are averaged over the whole scattering volume, and thus over a large number of nanoparticles or nanoparticle aggregates. The statistical relevance and the nondestructive *in situ* nature of scattering experiments represent significant advantages of these techniques over the electron scattering techniques previously discussed. The operating principles of a typical scattering apparatus are illustrated in Figure 4.15. One monochromatic beam (photons or neutrons) is incident on a sample. Typically, both the incident and scattered beam are shaped by optics adapted to the radiation characteristics (e.g., apertures, slits, and lens). Part of the incident beam crosses the sample unaffected and some is scattered. A detector measures at an angle θ the scattered intensity $I(\theta, t)$. The volume illuminated by the incident beam and analyzed by the detector is the scattering volume V.

From the general experimental setup shown in Figure 4.15, three different types of measurements can be performed: dynamic, static, and

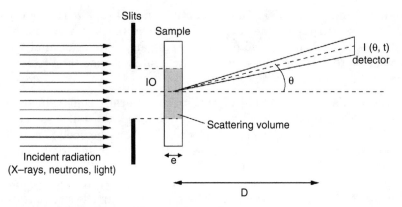

Figure 4.15 Illustration of the scattering principle upon which most scattering techniques and apparatuses are based.

calibrated light scattering. Static scattering measures the value of $I(\theta, t)$ averaged over a period of time t, typically as a function of θ. Static scattering measurements can be used to obtain structural information such as size, shape, and agglomeration state. Dynamic scattering measures the instantaneous values of $I(t)$ over time at a fixed angle, θ. The variability in scattering over time gives information on the diffusion coefficient due to Brownian motion of the scattering nanoparticles in suspension. Dynamic scattering is thus an indirect measurement of particle size, shape, and interactions. The third type of information given by scattering experiments is obtained by the calibration of the measured intensity to obtain the quantitative average number of scattered photons per unit solid angle [Thill et al., 2002]. This method yields, with almost no approximations or model assumptions, physical quantities such as the volume or specific surface averaged over the whole sample. Different radiation sources used for scattering experiments can yield complementary information on the sample. Light scattering is associated with variations in dielectric properties (or refractive index) [Berne and Pecora, 1976; Berne, 1996], X-rays are scattered by electrons [Berne, 1996; Glatter and Kratky, 1982], and neutrons are scattered by nuclei [Higgins and Benoit, 1994].

Static scattering experiment

Absolute scattered intensity. When a flux N_0 (counts/s) of photons or neutrons is incident on a sample (volume V and thickness e_s) then a part of the flux $\Delta N(u, \Delta\Omega)$ is scattered in a direction u with a solid angle $\Delta\Omega$:

$$\Delta N(u,\Delta\Omega) = N_0 T(e_s/V)d\sigma/d\Omega(u)\Delta\Omega \qquad (8)$$

where $d\sigma/d\,\Omega(u)$ is the differential scattering cross-sectional area (m^2). This cross section is characteristic of the interaction between the material and the incident beam. The scattering intensity by a sample is the differential scattering cross section per unit volume and is expressed as the inverse of a length (m^{-1}).

$$I = 1/V\,d\sigma/d\,\Omega(u) = \Delta N/(N_0 T e_S \Delta \Omega) \tag{9}$$

This quantity is experimentally accessible, provided that the thickness e_s is known and the transmission T is measured with the scattering properties $\Delta N/(N_0 \Delta \Omega)$. If the exact composition of the sample is known, e_s can be deduced from the transmission measurement according to the following relationship: $T = \Delta N(0)/N_0 = e^{-\mu e}$, where $\mu*$ is the absorption coefficient, which depends only on the scattering volume composition.

The experimental setup directly measures a differential scattering cross section per unit volume if the detector geometry is well defined ($\Delta \Omega$) and if the same detector is used to measure the direct beam (N_0) and the scattered beam (ΔN). Some experimental configurations have become standard, as for example the Bonse-Hart Ultra Small Angle X-ray Scattering apparatus, where the same punctual detector is scanned from the direct beam to the scattered beam and because $\Delta \Omega$ is precisely defined by the optics before the detector. Generally, however, a standard must be used to calibrate the instrument. For X-ray scattering, a practical standard is pure water. Indeed, the scattering of pure water is related to its isothermal compressibility χ_T and is well defined ($I = 0.016$ cm^{-1}). The transmission of pure water also gives the thickness as the mass attenuation coefficient for this medium, which is known ($\mu*/\rho$(water) $= 9,91$ cm^2/g at 8 keV). Using water as a standard, however, requires highly sensitive instruments. Other standards with known $d\sigma/d\Omega$ values may also be used. For light scattering, pure liquids can be employed (benzene is often used for these purposes) or for less sensitive low angle light scattering instruments, calibrated pinholes can be used as a standard [Thill et al., 2002].

General expression of the scattering intensity. For an incident radiation source of wavelength λ the incident beam has a wave vector k_i of amplitude $|k_i| = k_i = 22\pi/\lambda$. The scattered wave vector k_d making an angle θ with k_i has the same amplitude for an elastic scattering process and defines a scattering wave vector $q = k_d - k_i$. Thus, the amplitude of the scattering wave vector is $q = 4\pi/\lambda \sin(\theta/2)$. A static scattering experiment yields structural information on the dispersed phase at a typical spatial scale of $1/q$. For nanoparticles, the interesting range of scattering wave vectors is 10^{-2} to 1 nm^{-1}. This range is observable for reasonable angles for very small wavelengths that are typical of X-rays and neutrons ($\lambda \sim 0.1$ nm). Thus, the following notations, also very general, are more adapted to X-ray scattering.

The scattering amplitude, $A(q)$, from a scattering volume, V is:

$$A(q) = \int_v \rho(r) e^{-iqr} dr \tag{10}$$

where $-qr$ gives the phase shift between two scatterers separated by the vector r and $\rho(r)$ is the density of scattering length. The density of scattering length is $\rho(r) = \sum_i \rho_i(r) b_i \rho_i(r)$ $\rho_i(r)$ being the local density of scatterers of type i and b_i is the scattering length. In the case of x-rays, the photons interact with all the electrons in the sample. Thus, the scattering length is the Thomson scattering length of a single electron $b_e = e^2/(4\pi\varepsilon_0 mc^2) = 0.282\ 10^{-14}$ m and $\rho_i(r)$ is simply the local density of electrons in the scatterers $\rho_e(r)$. The scattering intensity per unit volume is given by the following expression:

$$I(q) = \frac{A(q)A'(q)}{V} \tag{11}$$

Sample preparation.   No particular limitations exist for the type of media that can be analyzed by SAXS. The sample can be in a solid, liquid, or gaseous phase. For liquids or gases, the sample must be put in a container that has windows to allow for a beam path. It is further necessary that the container walls that compose the window do not interact with the X-rays. Thin plastic sheets, thin mica sheets, or beryllium are some typical materials used as windows. The thickness of the sample is the most critical parameter. It has to be large enough so that the X-rays can interact with the matter and thin enough so that both the incident beam and scattered beam can cross the sample. A good criterion for choosing the sample thickness is when it is possible to detect a transmitted beam. Indeed, from Eq. 9, it can be shown that the quantity of scattered photons is proportional to eT, where e is the sample thickness and T is the transmission. As $e \sim \ln(1/T)$, the optimal transmission is obtained for the maximum value of $T^*\ln(1/T)$ which is obtained for $T = 0.36$. For example, with an 8 keV incident X-ray beam, in the case of water ($\mu^* r = 10$ cm^2/g, $r = 1$ g/cm^3), the maximum scattered signal will be measured for a thickness of $e = \ln(1/0.36) / (10 * 1.) \sim 0.1$ cm; for cerium oxides ($\mu^* r = 290$ cm^2/g, $r = 6.5$ g/cm^3), optimum thickness falls down to $e = \ln(1/0.36)/(290^*6.5) = 5$ μm.

Scattering by a nanoparticle dispersion.   We now consider calculation of scattering for a solvent of homogeneous scattering length density *rsol* containing N nanoparticles with a homogeneous scattering length density *rnp*. For a single nanoparticle, we can rewrite Eq. 11 as:

$$A(q) = \int_{Vnp} \rho_{np} e^{(-iqr)} dr + \int_{Vsol} \rho_{sol} e^{(-iqr)} dr \qquad (12)$$

The scattering amplitude of this system can be rewritten in terms of the scattering length density contrast between the nanoparticles and solvent $\Delta\rho = \rho_{np} - \rho_{sol}$:

$$A(q) = \int_{Vnp} \rho_{np} e^{(-iqr)} dr + \int_{Vsol} \rho_{sol} e^{(-iqr)} dr \qquad (13)$$

The last term of Eq. 12 is only present at $q = 0$ and can be removed for practical purposes. The scattering amplitude of a single nanoparticle is then only dependent on the electronic density contrast between the nanoparticles and solvent. The intensity can be written as:

$$A(q) = A(q)A'(q) = \int\int_{Vnp} \Delta\rho^2 e^{(-iq(r-v))} dr dv \qquad (14)$$

For dilute suspensions of nanoparticles, the intensity per unit volume is obtained by summation of the scattering from each nanoparticle and is often written as:

$$I(q) = \frac{N}{V} V_{np}^2 P(q) = \phi VnpP(q) \qquad (15)$$

where $P(q)$ is the normalized form factor which is defined as:

$$P(q) = \frac{1}{V_{np}^2} \int\int_{Vnp} \Delta\rho^2 e^{(-iq(r-v))} dr dv \qquad (16)$$

The form factor $P(q)$ is characteristic of the shape and scattering length density contrast of the nanoparticle. For example, the form factor of a spherical particle can be obtained as a function of the scattering wave vector amplitude by:

$$P(q) = \Delta\rho^2 \left(\frac{3[\sin(qr) - qr\cos(qr)]}{(qr)^3} \right)^2 \qquad (17)$$

When particles are interacting through long-range forces in dilute suspensions or simply by collision in a concentrated suspension, the scattering of the different nanoparticles is no longer independent and interferences between nanoparticles must be accounted for. It can be shown that the scattering intensity of the unstable suspension can be rewritten as:

$$I(q) = \phi V_{np} P(q) S(q) \qquad (18)$$

where $S(q)$ is the structure factor accounting for the correlation between the nanoparticles.

$$S(q) = 1 + \frac{1}{N}\left(\sum_i \sum_{j \neq i} e^{(iq(r_j - r_i))}\right) \tag{19}$$

A detailed quantitative analysis of the structure factor goes beyond the scope of this book. However, a qualitative description of the structure factor can give useful information on the state of the nanoparticle dispersion. The value of $S(0)$ is proportional to the average mass of the nanoparticle aggregates and the colloidal liquid compressibility. For a dilute suspension, $S(q) = 1$. For stable nanoparticle suspensions, the limit of $S(q)$ when q approaches zero is less than 1. For unstable nanoparticle suspensions, $S(q)$ is larger than 1 when q approaches zero.

Limiting behaviors of the form factor.   In the limit of forward scattering and for dilute nanoparticles suspensions, the form factor can be approximated as an exponential:

$$I(q) \approx \phi V_{part} e^{\left(\frac{-(qR_0)^2}{3}\right)} \tag{20}$$

where R_G is the radius of gyration of the nanoparticle. This regime of forward scattering is called the Guinier regime and is valid for $qR_G < 1$. In the limit of very large q, and provided that a perfect interface exists between the nanoparticles and the solution (i.e., no surface rugosity), then the following limit is observed:

$$\lim_{q \Rightarrow \infty} I(q) = \frac{2\pi(\Delta\rho)^2}{q^4} \frac{S}{V} \tag{21}$$

where $\Delta\rho$ is the scattering length density contrast between the nanoparticles and the solvent, S, is the total surface area of the nanoparticles, and V is the scattering volume. This approximation allows the specific surface area of a nanoparticle to be measured. When the suspension is concentrated or if the solution is bicontinuous, the same formula holds. This large angle regime is called the Porod regime. The specific surface of the particles is conveniently extracted using a $I(q)q^4$ vs. q plot as shown in Figure 4.3.

Example SAXS experiment.   As an illustration of these principles, let us consider an example of the information that may be obtained through small angle X-ray scattering (SAXS) by a suspension of CeO_2 nanoparticles. Our goal is to characterize concentration, size, and shape of the nanoparticles in the suspension. The initial nanoparticle suspension is made by peptisation of 38 g of a powder with raw formula CeO_2 $(NO_3)_{0.5}$ $(H_2O)_x$ mixed with 30 g of poly acrylic acid $(C_3H_4O_2)_n$ and a molar mass of

2000 g/mol (i.e., n~ 28) in 1 L. The density of the resulting solution is 1.06 g/cm^3. First, the solution is mounted in a variable light path X-ray cell for measuring its transmission as a function of its thickness. The transmission T of a monochromatic X-ray beam obeys the following relationship:

$$\frac{I}{I_o} = e^{-\mu^*/\rho e} \tag{22}$$

where μ/ρ is the mass absorption coefficient (cm^2/g), $e = d\tilde{\rho}\, d$ is the thickness, and ρ the density. The mass absorption coefficient is tabulated for each atom at all X-ray incident energies. (The mass absorption coefficient can be found on the NIST website http://www.nist.gov). For a mix of compounds, the average mass attenuation coefficient is obtained with the following formulation:

$$\frac{\mu^*}{\rho} = \sum x_i(\mu/\rho)_i \tag{23}$$

where x_i is the mass fraction of compound i. For the cerium solution, we have $x_{powder} = 38/1060 = 3.58\ 10^{-2}$; $x_{PAA} = 30/1060 = 2.83\ 10^{-2}$; and $x_{water} = 1 - x_{powder} = 0.9242$. The mass attenuation coefficient of the powder and water are respectively 179.88 and 9.83 cm^2/g. Thus, according to Eq.17, the solution has a mass attenuation coefficient of 15.9 cm^2/g. Figure 4.16 shows the comparison between the experimental measurement of the

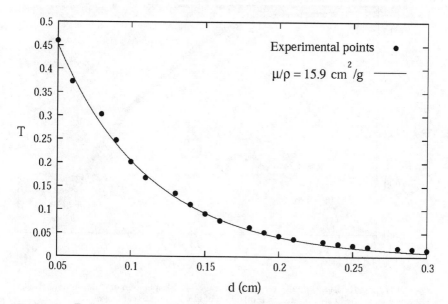

Figure 4.16 Experimental and theoretical transmission through a 38 g/l CeO$_2$ nanoparticle suspension.

transmission obtained by dividing the measured transmitted intensity I by the incident beam intensity I_0 with the result of Eq. 22 using $\mu*/\rho = 15.9 \, cm^2/g$ and $\rho = 1.06 \, g \cdot cm^{-3}$.

The scattered intensity as a function of the scattering vector q obtained for this NPs solution is shown in Figure 4.17. The plain line in this curve is a fit using Eq. 24 of the scattering intensity using the form factor of an ellipsoid of revolution.

$$I(q) = \phi V \Delta \rho^2 P(q) \qquad (24)$$

where $P(q)$ is the form factor of an ellipsoid, V is the volume of the ellipsoid, ϕ is the volumetric fraction of the ellipsoid in solution, and $\Delta \rho$ is the scattering length density contrast between CeO_2 and water (= 4.26 $10^{11} cm^{-2}$). The form factor of the ellipsoid is calculated according to the following expression:

$$\int_0^{\pi/2} \frac{3[\sin(qR^e) - qR^e \cos(qR^e)]}{(qR^e)^3} \sin(\alpha) d\alpha \qquad (25)$$

where $R^e = R\sqrt{\sin^2\alpha + eps \cos^2\alpha}$ and R, R and εR are the semi-axes of the ellipsoid.

Figure 4.17 clearly shows a very good agreement with the experimental data. The model is very sensitive to variation in size, aspect

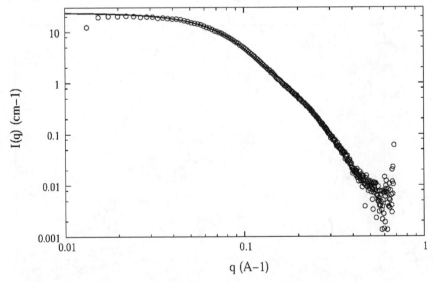

Figure 4.17   Scattering intensity measured for a 38 g/l suspension of CeO_2 nanoparticles. The red line is the best fit using an ellipsoid model with the parameters: $R = 3.5$ nm, $\varepsilon = 0.26$ and $\rho_{NP} = 7.13$ g/cm^3.

ratio of the ellipsoid and agglomeration stage. Even slightly agglomerated nanoparticles would have displayed a strong increase of the scattering intensity in the low q region. In summary, using small angle X-ray scattering, information on the size, shape, and interaction between nanoparticles in solution can be obtained, even at high concentration suspensions. However, this method is not well-adapted for characterization of complex mixtures or very small nanoparticle concentration.

Light scattering.   The same experiment as for small angle X-ray scattering can be performed by light scattering. However for the visible light wavelength, the particles that can be successfully analyzed by static light scattering are of course much larger (more than 100 nm). Figure 4.18 shows the effect of adsorbed nanoparticles on *Synecocystis* bacteria. We see that once a specific coverage of the bacteria by CeO_2 nanoparticles is reached, the bacteria rapidly aggregate.

To observe nanoparticles in suspension, a different approach has to be performed. The Brownian motion of nanoparticles can be used to measure their radius by dynamic light scattering (DLS). When a monochromatic light source interacts with one nanoparticle whose size is small compared to the light wavelength, the beam is scattered in all directions. When two nanoparticles are scattering the same incident beam, the total scattering detected in one point of space also depends on the interference between the scattered light of the two nanoparticles and therefore of their respective distance. For an ensemble of nanoparticles,

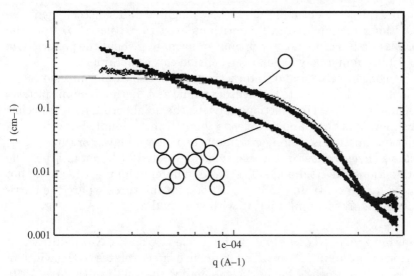

Figure 4.18 Small angle light scattering measurements of *Synecocystis* bacteria before and after contact with 50 ppm CeO_2 nanoparticles. The bacteria in contact with the nanoparticles form aggregates. (Courtesy of O. Zeyons)

the scattered intensity observed on a screen shows an ensemble of bright and dark points, which are commonly referred to as speckles. If the nanoparticles are immobile in the solution, the speckles are fixed on the screen. However, as a result of Brownian motion, the intensity on the screen fluctuates due to variations of the interference part of the scattered intensity. These fluctuations are directly related to the diffusion coefficient of the nanoparticles.

The signal is analyzed by a correlator to obtain the intensity autocorrelation function $C(\tau) + \langle I(t)I(t + \tau)\rangle$. When $\tau = 0$, $C(\tau)$ is simply $<I(t)^2>$, but when τ is sufficiently large, $I(t)$ and $I(t + \tau)$ are completely uncorrelated and $C(\tau) = <I(t)>^2$. The characteristic time required for the autocorrelation function to go from $<I(t)^2>$ to $<I(t)>^2$ is related to the diffusion coefficient of the nanoparticles. Therefore, for relatively dilute suspensions of noninteracting and monodisperse nanoparticles the following relationship is true:

$$C(\tau) \cong e^{-\Gamma\tau} \qquad (26)$$

where $\Gamma = Dq^2$ and D is the Stokes-Einstein diffusion coefficient $D = \frac{kT}{6\pi\eta R}$. For a polydisperse nanoparticle suspension, the autocorrelation function is then:

$$C(\tau) = \int P(D)e^{-Dq^2\tau}dD \qquad (27)$$

where $P(D)$ is the normalized intensity weighted diffusion coefficient distribution function. Several algorithms exist to extract $P(D)$ from the correlation function. So, in principle, information about the particle size distribution in a polydisperse suspension can be obtained.

For concentrated and/or interacting nanoparticles, the diffusion coefficient is modified due to the hydrodynamic and thermodynamic (attraction and repulsion) interactions between the neighboring nanoparticles. The motion of the nanoparticles is then strongly coupled and the correlation function is no longer a simple exponential law. For typical particle size measurements, the suspension is diluted such that interparticle interactions are negligible. Therefore, dynamic light scattering is not suited for concentrated samples or complex mixtures where the particles are likely to be interacting with one another.

X-ray diffraction.   The influence of particle size on the X-ray diffraction pattern is quite important. The main size effect relates to the width of the diffraction peaks for particles smaller than 50 nm (Figure 4.19). The size of the crystallites can then be determined from the diffraction pattern using the Sherrer formula (Eq. 1).

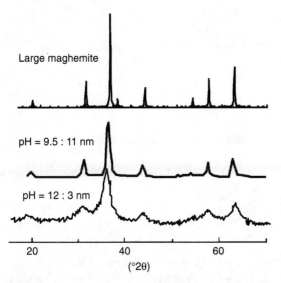

Figure 4.19   X-ray diffraction patterns ($\lambda_{Cu\,K\alpha} = 1.5406$ Å) of maghemite particles synthesized at different pHs with $I = 1$ mol l^{-1} (T = 25°C; 8-day aged suspensions) compared with large crystals. (Adapted from Jolivet et al., 2002)

Raman.   The Raman spectra for nanoparticles is modified, compared to that measured for larger particles, as a result of phonon confinement [see, for example, Richter et al., 1981]. In the model developed to analyze the modification of the Raman peaks for nanocrystalline materials, the nanoparticles are considered as an intermediate case between a perfect infinite crystal and an amorphous material. The development of this model indicates that the Raman line of a perfect crystal is modified for nanoparticles by producing asymmetric broadening and peak shifts. For example, for TiO_2, the Raman peak at 142 cm^{-1} measured for large crystals shifts to 146 cm^{-1} for 8 nm particles and to 148 cm^{-1} for 5 nm particles. Simultaneously, the FWHM increases from 10 to 18 cm^{-1} [Kelly et al., 1997]. In separate experiments, Choi et al. [2005] observed a similar effect when examining TiO_2 anatase nanoparticles on the peak at 142 cm^{-1}.

XAS.   For a detailed description of the application of XAS for characterizing materials as a function of size, the reader is referred to the literature [see, for example, Greegor and Lytle, 1980; Jentys, 1999]. XAS can be used to characterize the size and shape of metallic nanoparticles. The parameter primarily reflecting the size and shape of metal particles is the average coordination number, since when small clusters are examined by EXAFS the apparent average coordination number is

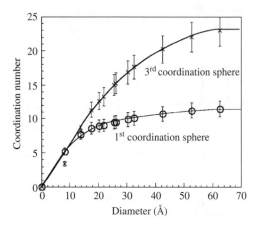

Figure 4.20 Average coordination of the first and third coordination spheres as a function of the size of the fcc particles. (Adapted from Jentys, 1999)

smaller than that observed in the bulk metal because of the high proportion of surface atoms. This effect is dependent on the size and shape of the metal cluster. However, the estimation of particle geometry relies on an accurate description of the relation between particle size/shape and the average coordination number [Greegor and Lytle, 1980; Jentys, 1999]. In the case of *fcc* metal it has been shown that the coordination number for the first and second coordination spheres are sensitive to the size of the cluster up to 20 nm (Figure 4.20). For larger particles the uncertainty on the coordination sphere determined by EXAFS is too high to accurately determine the size. The average coordination number of the third coordination sphere (N3) is more sensitive to the size of larger objects, but determining accurately the third coordination sphere requires a scan of EXAFS spectra at a high k value (up to 20 $\mathring{A}^{-1}$) [Frenkel et al., 2001], which is not always possible.

Because metal oxides display a multitude of different crystal structures, no general correlation between EXAFS coordination numbers and average particle size/morphology has been published [Fernandez-Garcia et al., 2004]. In fact, information can only be obtained for specific minerals (e.g., lepidocrocite (FeOOH) [Rose et al., 2002].

XANES can in some cases be used to determine the nanoparticle size. Indeed, the LIII white line for 5d metals (Pt . . .) is at the center of electronic charge transfer between metals that are present inside the cluster. Bazin [Bazin et al., 1997] showed that a strong correlation exists between the intensity of the white line and the size of the Pt cluster. But the quantitative correlation between size and XANES shape remains difficult.

EXAFS has been successfully applied to determine the size and structure of polycations during metal hydrolysis and the formation of gels or nanoparticles. Additional examples include iron [Combes

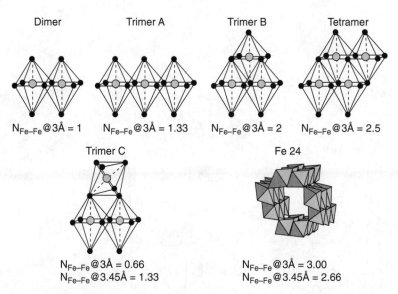

Dimer   Trimer A   Trimer B   Tetramer

$N_{Fe-Fe}@3Å = 1$ $N_{Fe-Fe}@3Å = 1.33$ $N_{Fe-Fe}@3Å = 2$ $N_{Fe-Fe}@3Å = 2.5$

Trimer C   Fe 24

$N_{Fe-Fe}@3Å = 0.66$ $N_{Fe-Fe}@3Å = 3.00$
$N_{Fe-Fe}@3.45Å = 1.33$ $N_{Fe-Fe}@3.45Å = 2.66$

Figure 4.21 Examples of 3D Fe clusters and their corresponding coordination numbers (N_{Fe-Fe} @3 Å = number of edge sharing, N_{Fe-Fe} @ 3.45 Å = number of double corner sharing).

et al., 1989; Bottero et al., 1994; Rose et al., 1996; Rose et al., 1997], chromium, [Jones et al., 1995; Roussel et al., 2000], gallium [Michot et al., 2000], titanium [Chemseddine, 1999], and zirconium [Turillas et al., 1993; Helmerich et al.; 1994, Peter et al., 1995]. In the case of iron chloride, for instance, the very first steps of iron octahedra polymerization due to a pH increase (i.e., evolution of the hydrolysis ratio = [OH]/[Fe]) were determined. In fact, the Fe-Fe interatomic distances can be associated to iron octahedra linkages (face, edge, double corner, single corner sharing). Moreover, the number of neighbors at a given distance is associated to the length of iron octahedra polymers. Theoretical values for N for a number of different iron clusters are reported in Figure 4.21.

The use of XAS at the Fe K-edge can be used to characterize the different early polymerization steps of a metal salt (Figure 4.22). Using this tool it is possible to visualize the arrangement of iron clusters as they undergo hydrolysis.

Microscopy. Both SEM and TEM imaging may be used to determine particle size distributions and shapes. There exists a large body of literature on the use of SEM and TEM in particle characterization. However, because the particles in any given SEM or TEM image only

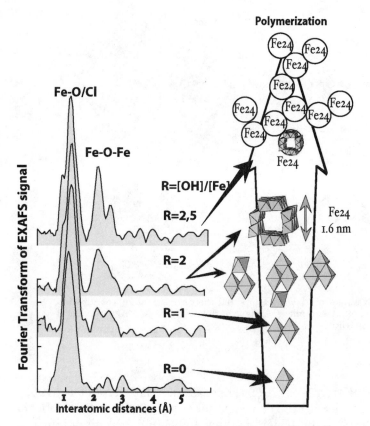

Figure 4.22   Nucleation steps during the hydrolysis of $FeCl_3.6H_2O$ solutions as a function of the hydrolysis ratio R.

represent a small fraction of the total particle population, these measurements are not statistically accurate. Nevertheless, various software packages dedicated to image analysis can be used to automatically measure and analyze the size and shape of a high number of particles in a single image as illustrated in Figure 4.23.

For example, Marın-Almazo et al. [2005] used TEM to determine the size distribution of rhodium nanoparticles. The image analysis indicated that the nanoparticle size distribution followed a Gaussian form with a mean particle size in a range of 1 nm to 3 nm, however with a high preference for the very small nanoparticles, around 1.5 nm. The advantage of scattering techniques is the level of detail that they can provide regarding particle size and structure.

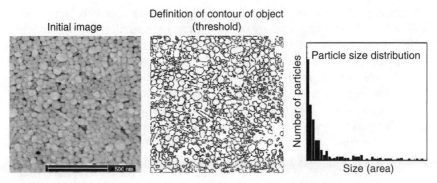

Initial image

Definition of contour of object
(threshold)

Number of particles

Particle size distribution

Size (area)

Figure 4.23 SEM image and the associated contrast picture and size distribution of a particle suspension.

Surface Physico-Chemical Properties

Physical and chemical properties such as particle charge, size, chemical composition, and surface functionality are often linked. These characteristics may be modified, intentionally or unintentionally, through adsorption of species on the particle surface (Chapter 10) and particle aggregation (Chapter 7).

Surface charge characterization

Operating principle. Surface charge is a key parameter controlling the stability of nanoparticle dispersions. In polar solvents, surfaces may have a charge of a specified density σ that can be approximated by a potential $\tilde{\psi}$. Depending on the type of material, the surface charge can vary as a function of pH as is the case for oxide minerals. For some minerals (like clays), fixed charges exist that are added with the pH dependent charges to give the total surface charge. Sources of pH dependant charge are the protonation and deprotonation of functional groups, while fixed charges result from crystal lattice defects and atomic substitution. In an aqueous medium, a diffuse electrical double layer will form at the solid-liquid interface as ionic species are attracted to the charged surface. Ions with a same charge compared to the nanoparticle surface are repelled from the surface, while those ions having the same charge are attracted to it. This effect decreases from the charged interface with a characteristic length κ^{-1} (the Debye screening length) depending on the solution ionic strength and composition. Various techniques can be used to determine particle surface charge. First the charge can be assessed using a proton as an atomic probe, by titrating a suspension of particles.

The charge can also be determined indirectly by determining the zeta potential (ζ) of the particles. The zeta potential is the electrical potential that exists at the "shear plane" at some distance from the particle surface. It is derived from measuring the electrophoretic mobility distribution of a dispersion of charged particles as they are subjected to an electric field. The electrophoretic mobility is defined as the velocity of a particle per electric field unit and is measured by applying an electric field to the dispersion of particles and measuring their average velocity. Depending on the concentration of ions in the solution, either the Smoluchowski (for high ionic strengths) or Huckel (for low ionic strengths) equations are used to calculate the zeta potential from the measured mobilities. For very small particles the mobility is generally determined using laser Doppler velicometry.

It is essential to determine the modifications of the surface charge (or zeta potential) as a function of pH and ionic strength. By doing so, it is possible to determine the point of zero charge (PZC) or isoelectric point (pH_{iep}) for a sample. This is significant because nanoparticle suspensions are generally stable above and below the pH point of zero charge <pH_{pzc}> while the relationship with ionic strength is somewhat similar.

Application in the particular case of nanoparticles.   For biological tests, knowledge of the particle PZC is crucial as it relates to the stability or propensity to aggregate of the sample. The importance of this concept may be illustrated by considering cytotoxicology experiments conducted with fibroblasts cells and nano-maghemite. Here, the PZC of the nano-maghemites is around pH 7, which is close to the solution pH used during experiments looking at nanoparticle-cell interactions (DMEM, for example). Adding 8 nm nano-maghemite to a DMEM nutritive medium results in aggregation of the nano-maghemite, with the particle size approaching several tens of microns (Figure 4.24).

In order to increase the stability of nanoparticles, their surface chemistry is often modified through functionalization or polymer encapsulation to modify their surface charge. In the case of nano-maghemites, the adsorption of DMSA (dimercaptosuccinic acid) at the maghemite surface strongly modified the surface charge leading to a stable dispersion in the nutritive medium. The DMSA is fixed through the SH group at the maghemite surface. Therefore, the external part of the DMSA-covered maghemite corresponds to the carboxylic group of the DMSA. At pH 7, the carboxylic groups are negatively charged resulting in repulsive electrostatic forces between the various particles (Figure 4.25).

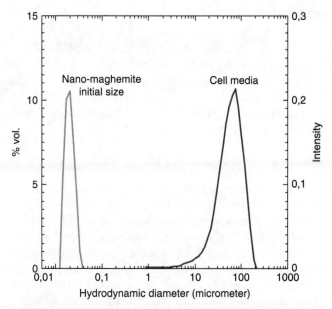

Figure 4.24 Evolution of the size of nano-maghemite nanoparticles in DMEM nutritive medium (pH = 7.4) measured by Photon Correlation Spectroscopy (PCS). (Courtesy of Mélanie Auffan)

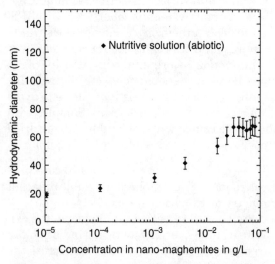

Figure 4.25 Stability of a DMSA-coated nano-Fe_2O_3 suspension as a function of their concentration in an abiotic supplemented DMEM (contact time = 48 hr). (Courtesy of Mélanie Auffan)

Evolution of the oxidation state
at the surface

For metallic and some oxide nanoparticles, the oxidation state of elements composing the objects might evolve at the surface. Many of the techniques previously described in this chapter may be used to probe the evolution of the elemental oxidation state. Some techniques, however, like X-ray photoelectron spectroscopy (XPS), can be used to specifically probe the surface or near-surface region.

X-ray photoelectron spectroscopy (XPS)

Operating principle. XPS is based on the measurement of the energies of photoelectrons emitted as a result of the interaction between an incident X-ray beam with matter. As it measures the energies emitted from photoelectrons, XPS is classified as an electron spectroscopy technique. The main difference between electron and X-ray spectroscopies is the depth to which the surface is characterized. For example, electrons travel through a extremely short distance in a solid before losing their energy, while X-rays penetrate deeper into the solid matrix. Therefore, XPS is a surface or near-surface sensitive technique, as it is sensitive to a depth of between 1 to 5 nms. On the other hand, X-rays are used to characterize the bulk structure. However, for nanoparticles smaller than 20 nm, XPS can also be considered as a bulk sensitive technique!

Sample preparation. XPS is classified as a surface sensitive technique, but only for relatively large surfaces. For analysis of nanoparticles with XPS, the best preparation method is to deposit them on a clean and flat surface. More than for TEM, XPS requires an ultra-high vacuum chamber. It is therefore important to dry the system before analyzing it using XPS. When examining interactions between nanoparticles and living cells, sample preparation can be quite complicated as it is imperative to avoid any chemical modifications during drying (e.g., oxidation). Samples with large surface areas, or with volatile components, should ideally be dried and placed in a vacuum chamber prior to insertion into the XPS equipment.

Application to nanoparticles. The ejected electrons correspond to core photoemission. The core level peaks for a given atom can exhibit different binding energies due to symmetry or oxidation state effects. In particular, core-level photoemission can be very sensitive to changes in the oxidation state of an element. As an example, for TiO_2 optical irradiation can lead to the formation of charge carriers by optical absorption across the band gap. These charge carriers can directly participate in redox processes on the TiO_2 surface. XPS can distinguish the different Ti oxidation states—that is, the $Ti2p_{3/2}$ photoemission varies from 455.3 eV for TiO (Ti^{2+}) to 456.7 eV for Ti_2O_3 (Ti^{3+}), 457.6 eV for Ti_3O_5 ($2xTi^{3+}$, Ti^{4+}), and

458.8 eV for TiO$_2$ (Ti^{4+}) [(Song et al., 2005]. In the case of Mo metallic nanoparticles, XPS can probe for surface oxidation. Indeed, the two peaks of the 3D photoemission vary from 231.6, 228.3 eV to 235.8 and 232.7 eV for metallic and 6+ Mo oxidation state. The presence of the 235.8 and 232.7 eV peaks fingerprint the oxidation of Mo nanoparticles [Song, 2003].

Bulk redox sensitive spectroscopies. X-ray absorption spectroscopy and Mössbauer spectroscopy are redox sensitive techniques. The information is not specific from the surface or near surface region. Therefore if a modification of the oxidation state of the nanoparticles occurs it is almost impossible to attribute it to a bulk or surface oxidation. Moreover, if oxidation occurs at the surface, since both techniques provide information from all atoms of the particles, if the fraction of surface atoms is low (large particles) the signal will not be affected. XAS and Mössbauer can only determine the evolution of surface oxidation state for small particles (<20–50 nm) as soon as the fraction of surface atoms is higher than 15–20 percent.

For example, Raj Kanel et al. [in press] have determined the kinetics of iron oxidation during arsenic(V) removal from groundwater using nanoscale zerovalent iron (ZVI). The average particle size measured using TEM images was 30 nm. Mössbauer spectroscopy results confirmed that even before any contact between ZVI and As(V) solution, that 19 percent in mass were in zerovalent state with a coat of 81 percent iron oxides [maghemite + Fe(OH)$_2$]! (Figure 4.26) The interaction with As(V) solutions leads to an increase of the iron zerovalent state as it was identified by Mössbauer.

Using XAS it is also possible to identify the evolution of oxidation states for nanoparticles. For example, Thill et al. [2006] were able to determine the surface reduction of CeO$_2$ nanoparticles when reacted with a suspension of *Escherichia coli*. In this study the nanoparticles had

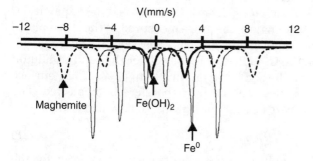

Figure 4.26 The Mössbauer individual component spectra of ZVI illustrating the maghemite, Fe(OH)$_2$, and Fe^0 constituents. (Adapted from Raj Kanel et al., in press)

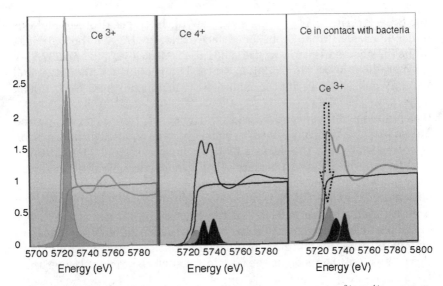

Figure 4.27 Results from XANES analysis at the Ce LIII edge for Ce^{3+}, Ce^{4+}, and CeO_2 in contact with bacteria. (Adapted from Thill et al., 2006)

a diameter of around 8 nm with 30 percent of Ce atoms at the surface. Using XAS spectroscopy at the Ce LIII edge, it has been demonstrated that 30 to 40 percent of the cerium was reduced in the biological system. Indeed, the $Ce^{(4+)}$ and $Ce^{(3+)}$ XANES spectra are very different (Figure 4.27), allowing for their relatively easy differentiation. The XANES spectra can be decomposed using Gaussian individual components that can lead to a quantification of the $Ce^{3+}/(Ce^{4+} + Ce^{3+})$ ratio.

Interactions with water

Hydrophobicity describes the strength of the interaction between the nanoparticle and water. Hydrophobicity may be directly quantified using contact angle measurements or qualitatively described using partitioning experiments. Contact angle is the angle measured at the three-phase interface made between solid, liquid, and gas phases. Essentially the contact angle depends on the free energy of cohesion for the solid and liquid and the free energy of adhesion between these two phases. The surface energy of air is essentially zero and may therefore be neglected. The energy balance is illustrated by the Young equation, which states:

$$\gamma_L \cos\theta = \gamma_s - \gamma_L \qquad (28)$$

where θ is the contact angle, γ_L is the total surface free energy for the liquid phase, γ_s is the total surface free energy for the solid, and γ_{SL} is the free energy of interaction between the solid and liquid.

For measuring nanoparticle contact angle two different techniques are available. Nanoparticles may be deposited onto a surface to form a lawn of particles. Here the deposited nanoparticles form a surface onto which the contact angle with different liquids may be measured. This technique requires that the surface formed be smooth and even in order to get reproducible results and to minimize the influence of roughness on the measured contact angle. Furthermore, this technique gives an average contact angle as the surface energies of many particles are contributing to the measurement. However, there is no limitation with regards to minimum particle size in using this technique, so the contact angle for very small nanoparticles may be measured. More precise information may be obtained using an AFM where the contact angle for a single nanoparticle with a liquid may be measured.

Nanoparticle contact angle may be measured based on the equilibrium position (i.e., zero net pressure) of a nanoparticle probe from AFM force curves [Yakubov et al., 2000]. The nano-probe is formed by attaching or gluing a single nanoparticle to the end of a tipless AFM cantilever. The receding contact angle is measured by placing an air bubble (diameter ~ 1–1.5 mm) on the bottom of a liquid cell. The nano-probe is then brought toward the air bubble. Once the nano-probe makes contact with the bubble, a three-phase interface is formed (air-liquid-solid) and the capillary force pulls the colloid into the air bubble. As the probe is extended further into the bubble, a point is reached where the net force acting on the nano-probe is equal to zero (Figure 4.28). This zero-force equilibrium position of the nanoparticle is characterized by a penetration depth D_b, which can be directly determined from the AFM force curve. Penetration depth is defined as the difference between the jump in point and the zero-force position. The receding contact angle may then be calculated according to the following relationship:

$$\cos \theta_r = 1 - \frac{D_b}{R} \tag{29}$$

where R is the nanoparticle radius. The procedure for determining the advancing contact angle is similar to that for the receding contact angle. For the advancing contact angle a drop of water (diameter ~ 1–1.5 mm) is placed on the AFM stage and the nano-probe is extended into it. From the penetration depth of the colloid into the zero-force position of the drop surface D_d the advancing contact angle may be calculated according to:

$$\cos \theta_a = \frac{D_d}{R - 1} \tag{30}$$

The limitation of this method for determining nanoparticle contact angle falls on the ability to successfully attach the nanoparticle to the end of an AFM cantilever. Attaching increasingly smaller particles to AFM cantilevers becomes problematic as the epoxy used to glue the two surfaces together may cover or alter the nanoparticle surface chemistry. However, as expertise in this area continues to develop and the use of other probes such as carbon nanotubes progresses, smaller nanoparticles (d < 50 nm) may be investigated using this technique.

The relative hydrophobicity of nanoparticles may be determined by measuring their propensity to partition into hydrophobic solvents. In these measurements a particle suspension of known concentration is prepared by dispersing the nanoparticles in water. This suspension is then mixed with a volume of a hydrophobic solvent such as octanol or dodecane. The two phases are mixed and subsequently allowed to phase separate. The particle concentration in the water phase is measured. The relative hydrophobicity may then be expressed in terms of the percentage of the initial nanoparticle concentration that has partitioned into the hydrophobic phase. For octanol-water partitioning tests, the relative hydrophobicity is expressed in terms of the logarithm of the octanol-water partition coefficient (log P), which is defined as:

$$\log P = \log\left(\frac{C_0}{C_w}\right) \tag{31}$$

where C_0 and C_w are the nanoparticle concentrations in the octanol and water phases after phase separation, respectively.

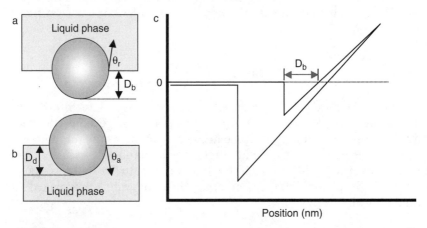

Figure 4.28 Illustration of the methods used to determine the (a) receding and (b) advancing contact angle for a nanoparticle. (c) Determination of the zero-force position in an AFM force plot.

TABLE 4.1 Applicability of Analytical Techniques to Providing Specific Information on Nanoparticles. (Adapted from Oberdorster et al., 2005)

	XRD	Raman	XAS	Mössbauer	NMR	SAXS, SANS	DLS	TEM	SEM	AFM	Zeta potential	XPS
Size	▼	▼			▼	▼▼	▼	▼▼	▼▼	▼		
Shape	▼					▼▼	○	▼▼	▼			
Surface area						▼						
Chemical composition		▼		▼	▼			▼	▼			
Speciation of elements		▼	▼▼	▼▼	▼▼							
Redox-state of elements			▼▼	▼▼								▼▼
Surface chemistry			▼		▼							▼▼
Surface charge						•					▼▼	
Crystal structure	▼▼	▼▼	▼▼	▼▼								
Agglomeration state						▼▼	○	▼	▼			
Heterogeneity								▼▼	▼			

▼▼ Highly applicable, gives quantitative information

▼ Applicable in some cases, gives quantitative information

• Applicable in some cases, gives quantitative information with validation with another technique

○ Applicable in some cases, gives qualitative information

151

Adsorption

As it has been previously discussed, the chemical reactivity of nanoparticle surface sites is an important issue for many industrial applications as well as for environmental concerns. Chemical reactivity is strongly influenced by the large ratio of surface atoms to nonsurface atoms. Adsorption properties of nanoparticles from an environmental perspective are discussed in Chapter 10. The adsorption mechanisms of molecules or atoms on the nanoparticle surface can be described using many of the techniques previously covered and using other techniques such as infrared spectroscopy and NMR.

References

Auffan M., Decome L., Rose J., Orsiere T., Demeo M., Briois V., Chaneac C., Olivi L., Berge-Lefranc J.-L., Botta A., Wiesner M. R., and Bottero J.-Y. (2006) In Vitro Interactions between DMSA-Coated Maghemite Nanoparticles and Human Fibroblasts: A Physicochemical and Cyto-Genotoxical Study. *Environmental Science and Technology.*

Bazin D., Sayers D., Rehr J. J., and Mottet C. (1997) Numerical Simulation of the Platinum LIII Edge White Line Relative to Nanometer Scale Clusters. *Journal of Physical Chemistry B.*(101), 5332–5336.

Berne B. J. and Pecora R. (1976) *Dynamic Light Scattering.* Wiley.

Berne W. (1996) Light Scattering: Principles and Development.

Bottero J., Manceau A., Villieras F., and Tchoubar D. (1994) Structure and Mechanisms of Formation of FeOOH(Cl) Polymers. *Langmuir* 10(1), 316–319.

Chemseddine A., Fieber-Erdmann M., Holub-Krappe E., and Boulmaaz S. (1997) XAFS Study of Functionalized Nanoclusters and Nanocluster Assemblies. *Zeitschrift für Physik D Atoms, Molecules and Clusters* 40(1–4), 566–569.

Chemseddine A. M., Moritz T. (1999) *European Journal of Inorganic Chemistry* 2, 235.

Choi H. C., Jung Y. M., and Kim S. B. (2005) Size Effects in the Raman Spectra of TiO2 Nanoparticles. *Vibrational Spectroscopy* 37, 33–38.

Combes J.-M., Manceau A., Calas G., and Bottero J.-Y. (1989) Formation of Ferric Oxides from Aqueous Solutions: A Polyhedral Approach by X-ray Absorption Spectroscopy. 1. Hydrolysis and Formation of Ferric Gels. *Geochimica and Cosmochimica Acta* 53(3), 583–594.

Denaix L., Lamy I., and Bottero J.-Y. (1999) Structure and Affinity Towards Cd2+, Cu2+, Pb2+ of Synthetic Colloidal Amorphous Aluminosilicates and Their Precursors. *Colloids and Surfaces A: Physicochemical and Engineering Aspects* (158), 315–325.

Fernandez-Garcia M., Martinez-Arias A., Hanson J. C., and Rodriguez J. A. (2004) Nanostructured Oxides in Chemistry: Characterization and Properties. *Chemical Review* 104, 4063–4104.

Ferraro J. R. and Nakamoto K. (1994) *Introductory Raman Spectroscopy.* Academic Press.

Fontaine A. (1993) Interactions of X-rays with Matter: X-ray Absorption Spectroscopy. In *Neutron and Synchrotron Radiation for Condensed Matter Studies,* (eds. Baruchel J., Itodeau J.-L., Lehmann M. S., Regnard J. R., and Schlenker C.). Les Editions de Physique—Springer Verlag.

Frenkel A. I., Hills C. W., and Nuzzo R. G. (2001) A View from the Inside: Complexity in the Atomic Scale Ordering of Supported Metal.

Nanoparticles, Dextran and Albumin Derivatised Iron Oxide. *The Journal of Physical Chemistry B* 105(51), 12689–12703.

Glatter O. and Kratky O. (1982) *Small-Angle X-ray Scattering.* London: Academic Press.

Greegor R. and Lytle F. (1980) Morphology of Supported Metal-Clusters—Determination by EXAFS and Chemisorption. *Journal of Catalysis* 63(2), 476–486.

Helmerich A., Raether F., Peter D., and Bertagnoly H. (1994) Structural Studies on an ORMOCER System Containing Zirconium. *Journal of Material Science* 29, 1388–1389.

Higgins J. S. and Benoit H. C. (1994) *Polymers and Neutrons Scattering*. Oxford: Clarendon Press.

Hiutsunomiya S. and Ewing R. (2003) Application of High-Angle Annular Dark Field Scanning Transmission Electron Microscopy, Scanning Transmission Electron Microscopy-Energy Dispersive X-ray Spectrometry, and Energy-Filtered Transmission Electron Microscopy to the Characterization of Nanoparticles in the Environment. *Environmental Science and Technology* (37), 786–791.

Hon Y. M., Lin S. P., Fung K. Z., and Hon M. H. (2002) Synthesis and Characterization of Nano-LiMn2O4 Powder by Tartaric Acid Gel Process. *Journal of the European Ceramic Society* 22, 653–660.

Jentys A. (1999) Estimation of Mean Size and Shape of Small Metal Particles by EXAFS. *Physical Chemistry Chemical Physics* 1(17), 4059–4063.

Jolivet J.-P., Tronc E., and Chanéac C. (2002) Synthesis of Iron Oxide-Based Magnetic Nanomaterials and Composites. *C. R. Chimie* 5, 659–664.

Jones D., Roziere J., Maireles-Torres P., Jimenez-Lopez A., Olivera-Pastor P., Rodriguez-Castellon E., and Tomlinson A. A. G. (1995) *Inorganic Chemistry* 34, 4611.

Kelly S., Pollak F. H., and Tomkiewicz M. (1997) Raman Spectroscopy as a Morphological Probe for TiO2 Aerogels. *Journal of Physical Chemistry B* 101, 2730–2734.

Koningsberger D. C. and Prins R. (1988) *X-Ray Absorption: Principles, Applications, Techniques of EXAFS, SEXAFS and XANES*. John Wiley.

Kubo T., Koshiba T., Goto T., Oyamada A., Fujii Y., Takeda K., and Awaga K. (2001) The Observation of Magnetization Behavior of the Nano-Scale Cluster Magnet Mn ac by 55Mn NMR. *Physica B* 294–295, 310–313.

Liu M., Li H., Xiao L., Yu W., Lu Y., and Zhao Z. (2005) XRD and Mössbauer Spectroscopy Investigation of Fe2O3–Al2O3 Nano-Composite. *Journal of Magnetism and Magnetic Materials* 294, 294–297.

Mamontov E. and Egami T. (2000) Structural Defects in a Nano-Scale Powder of CeO2 Studied by Pulsed Neutron Diffraction. *Journal of Physics and Chemistry of Solids* 61(8), 1345–1356.

Marın-Almazo M., Ascencio J. A., Perez-Alvareza M., Gutierrez-Wing C., and Jose-Yacaman M. (2005) Synthesis and Characterization of Rhodium Nanoparticles Using HREM Techniques. *Microchemical Journal* 81, 133–138.

Michot L. J., Montargès-Pelletier E., Lartiges B. S., d'Espinose de la Caillerie J.-B., Briois V. (2000) *Journal of American Chemical Society* 122, 6048.

Mulder E., Thiel R. C., de Jongh L., and Gubbens P. C. M. (1996) Size-Evolution Towards Metallic Behavior in Nano-Sized Gold And Platinum Clusters as Revealed by 197Au Mössbauer Spectroscopy. *NanoStructured Materials* 7(3), 269–292.

Peter D., Ertel T. S., and Bertagnolli H. (1995) EXAFS Study of Zirconium as Precursor in the Sol-Gel Process: II. The Influence of the Chemical Modification. *Journal of Sol-gel Science and Technology* 5, 5–14.

Raj Kanel S., Greneche J.-M., and Choi H. (in press) Arsenic(V) Removal from Groundwater Using Nano Scale Zero-Valent Iron as a Colloidal Reactive Barrier Material. *Environmental Science and Technology.*

Rehr J. J. and Albers R. C. (2000) Theoretical Approaches to X-ray Absorption Fine Structure. *Reviews of Modern Physics* 72, 621–654.

Richter H., Wang Z. P., and Ley L. (1981) *Solid State Communication* 39, 625.

Robert T., Laude L., Geskin V., Lazzaroni R., and Gouttebaron R. (2003) Micro-Raman Spectroscopy Study of Surface Transformations Induced by Excimer Laser Irradiation of TiO2. *Thin Solid Films* 440(1–2), 268–277.

Rose J., Fidalgo M. M., Moustier S., Magnetto C., Jones C. D., Barron A. R., Wiesner M. R., and Bottero J.-Y. (2002) Synthesis and Characterisation of Carboxylate-FeOOH Nanoparticles (Ferroxane) and Ferroxane-Derived Ceramics. *Chemistry of Materials* 14, 621–628.

Rose J., Flank A. M, Masion A, Bottero J. Y, and Elmerich P. (1997) Nucleation and Growth Mechanisms of Fe(III) Oxy-hydroxide in the Presence of PO4 Ions. 2. P-K Edge EXAFS Study. *Langmuir* 13(6), 1827–1834.

Rose J., Manceau A., Bottero J. Y., Masion A., and Garcia F. (1996) Nucleation and Growth Mechanisms of Fe(III) Oxy-Hydroxide in the Presence of PO4 Ions. 1. Fe-K edge EXAFS study. *Langmuir* 12(26), 6701–6707.

Roussel H., Briois V., Elkaim E., de Roy A., and Besse J. P. (2000) *Journal of Physical Chemistry B* 104(25), 5915.

Sayers D. A., Stern E. A., and Lytle F. W. (1971) New Technique for Investigating Noncrystalline Structures: Fourier Analysis of the Extended X-Ray—Absorption Fine Structure. *Physical Review Letter* 27, 1204–1207.

Song Z., Hrbek J., and Osgood R. (2005) Formation of TiO2 Nanoparticles by Reactive-Layer-Assisted Deposition and Characterization by XPS and STM. *Nano Letters,*5(7), 1327–1332.

Song Z., Cai T., Chang Z., Liu G., Rodriguez J. A., and Hrbek J. (2003) J. Am. Chem. Soc. (125), 8060.

Stephan O., Kociak M., Henrard L., Suenaga K., Gloter A., Tence M., Sandre E., and Colliex C. (2001) Electron Energy-Loss Spectroscopy on Individual Nanotubes. *Journal of Electron Spectroscopy and Related Phenomena* 114–116, 209–217.

Thill A., Desert S., and Delsanti M. (2002) Small Angle Static Light Scattering: Absolute Intensity Measurement. *The European Physical Journal, Applied Physics* 17, 201–208.

Thill A., Zeyons O., Spalla, Rose J., Auffan M., and Flank A.-M., (2006) Adsorption of CeO2 Nanoparticles on Escherishia Coli. Assessment of Nanoparticle Localization and Cytotoxicity, *Environmental Science and Technology*, 40 (19), 6151–6156.

Turillas X., Barnes P., Dent A. J., Jones S. L., and Norman C. J. (1993) 'Hydroxyde' Precursor to Zirconia: Extended X-ray Absorption Fine Structure Study. *Journal of Materials Chemistry* 3(6), 583–586.

Warren B. E. (1990) *X-Ray Diffraction*. Dover Publications.

Yan W., Petkov V., Mahurin S. M., Overbury S. H., and Dai S. (2005) Powder XRD Analysis and Catalysis Characterization of Ultra-Small Gold Nanoparticles Deposited on Titania-Modified SBA-15. *Catalysis Communications*(6), 404–408.

Zhang H. and Banfield J. (1998) Thermodynamic Analysis of Phase Stability of Nanocrystalline Titania. *Journal of Materials Chemistry* 8(9), 2073–2076.

Zhu S., Li Y., Fan C., Zhang D., Liu W., Sun Z., and Wei S. (2005) Structural Studies of Iron-Doped TiO2 Nano-composites by Mössbauer Spectroscopy, X-ray Diffraction and Transmission Microscopy. *Physica B* 374, 199–205.

Reactive Oxygen Species Generation on Nanoparticulate Material

Michael Hoffmann   *California Institute of Technology, California*

Ernest M. Hotze   *Duke University, Durham, North Carolina*

Mark R. Wiesner   *Duke University, Durham, North Carolina*

Background

Photoactive nanomaterials can be grouped roughly into two different classes: metal oxide or metal chalocogenide photocatalysts such as titanium dioxide or cadmium sulfide, respectively; and materials that can be photosensitized such as chromophores and certain types of fullerenes.

Semiconductor photochemistry behavior is governed by the band gap that can be described as the energy difference between the valence band (fermi level highest energy electrons) and the conduction band (lowest energy unoccupied molecular orbitals). Light absorbed at wavelength less than or equal to the band-gap energy ($\lambda \leq \lambda_{bg}$) of a semiconductor will result in the promotion of an electron to the conduction band, and as a consequence, a hole or vacancy in the HOMO state of the molecular orbitals is created. Both the promoted electrons and the holes can migrate to the surface of the semiconductor and subsequently react in aqueous solution to form reactive oxygen species such as superoxide ($O2^-$) and hydroxyl radical ($\cdot OH$).

Photosensitizers are sensitized by light, and their electrons are excited within the molecular orbitals. Sensitized electrons can behave according to two mechanisms: type I electron transfer involving a donor molecule and type II energy transfer involving no reaction. Both of these pathways will potentially produce reactive oxygen species in solution. The net effect of both these processes is to convert light energy into oxidizing chemical energy.

In engineered systems, oxidation serves many purposes, from elimination of potential hazards to improving aesthetic qualities. Chemical oxidation may be used in a wide range of applications, such as the breakdown of organic compounds such as TCE and atrazine, the oxidation of the reduced states of metals such as iron(II) and arsenic(III), or the inactivation of pathogenic organisms in water treatment.

Oxidation is an electron transfer process in which electrons from a reductant (i.e., an electron donor) are transferred to an oxidant and electron acceptor. The thermodynamic constraints or boundary conditions for electron transfer in aqueous solution are given by the electrochemical potentials or half-cell potentials as illustrated in Eq. 1:

$$O_2 + 4H^+ + 4e^- \rightleftharpoons 2H_2O \quad E_H^0 = +1.23 \text{ V} \tag{1}$$

where oxygen, O_2, is the electron acceptor in this half-reaction leading to the formation of water. The redox potential E_H, under standard conditions, is given as 1.29 volts relative to the standard hydrogen electrode (NHE). In the reduction-oxidation (i.e., redox) scale, reduction potential values (E_H^0) are positive for oxidizing species (i.e., oxidants) negative for reducing species (reductants). The redox potential for any set of conditions other than the standard conditions (i.e., concentrations are fixed at 1.0 M at a temperature of 298.15 K) can be determined with the Nernst equation as follows:

$$E_H = +1.23 - \frac{2.3RT}{nF} \log \frac{[H_2O]^2}{[O_2][H^+]^4} \tag{2}$$

for any set of concentration conditions where R is the universal gas constant, T is temperature, n is the number of electrons transferred, and F is the Faraday constant. For example, at pH 7 in water, the reduction potential of O_2 given as follows:

$$E_H^0(W) = +1.23 - \frac{2.3RT}{nF} \log \frac{[a_{H_2O}]^2}{P_{O_2}[10^{-7}]^4} = +0.81 \text{ V} \tag{3}$$

where the activity of water, $a_{H_2O} = 1$ by definition, and the partial pressure of oxygen, P_{O_2}, gas are by definition equal to 1 for standard state conditions.

Key reduction potentials for the important oxygen-containing species are given in Table 5.1.

TABLE 5.1   Standard State Reduction Potentials of Important Oxygen Species Where T = 298.15 K, P = 1.0 atm and All Concentrations and Activities (by Definition) Are Constant at 1.0 M

Oxygen species half-reaction	EH (pH 0)	$\Delta G/n$ (kJ mol^{-1})
$O_2 + 4\,H^+ + 4\,e^- \rightleftharpoons 2\,H_2O$	+1.23	−118.56
$O_2 + 2\,H^+ + 2\,e^- \rightleftharpoons H_2O_2$	+0.70	−67.47
$^3O_2 + e^- \rightleftharpoons O_2^-\cdot$	−0.16	+15.42
$^3O_2 + H^+ + e^- \rightleftharpoons HO_2^{\cdot}$	+0.12	−11.57
$^1O_2 + e^- \rightleftharpoons O_2^-\cdot$	+0.83	−80.01
$O_2^+ + e^- \rightleftharpoons O_2$	+3.20	−308.45
$\cdot OH + e^- \rightleftharpoons {}^-OH$	+1.90	−183.14
$\cdot OH + H^+ + e^- \rightleftharpoons H_2O$	+2.72	−262.19
$O^-\cdot + H^+ + e^- \rightleftharpoons HO^-$	+1.77	−170.61
$HO_2 + e^- \leftrightarrow \rightleftharpoons HO_2^-$	+0.75	−72.29
$HO_2 + H^+ + e^- \rightleftharpoons H_2O_2$	+1.50	−144.59
$H_2O_2 + 2\,H^+ + 2\,e^- \rightleftharpoons 2\,H_2O$	+1.77	−170.61
$H_2O_2 + e^- \rightleftharpoons \cdot OH + H_2O$	+0.72	−69.40
$O_3 + 2\,H^+ + 2\,e^- \rightleftharpoons O_2 + H_2O$	+2.08	−200.50
$O_3 + e^- \rightleftharpoons O_3^-\cdot$	+1.00	−96.39
$O_3 + H^+ + e^- \rightleftharpoons O_2 + \cdot OH$	+1.34	−129.17

For an overall redox reaction, coupling the half-reactions for an oxidant with a reductant, we can write the following simple equation:

$$Ox_1 + Red_2 = Red_1 + Ox_2 \qquad (4)$$

The corresponding Nernst equation can be expressed in terms of the overall redox potential, E_{rxn}, for a given set of nonstandard conditions as follows:

$$E_{rxn} = E_H^0 - \frac{2.3RT}{nF} \log \frac{[Red_1][Ox_2]}{[Ox_1][Red_2]} \qquad (5)$$

The highest occupied molecular orbitals (HOMO) of ground state oxygen contain unpaired electrons with parallel spins (Figure 5.1). The parallel spins are characterized by triplet signal response in an applied magnetic field, while anti-parallel spins have a characteristic singlet response in a magnetic field.

As a consequence of the unpaired, parallel spins in the ground-state ($^3O_2$ ($^3\Sigma_g^-$)) oxygen is paramagnetic. Thus, molecular oxygen (i.e., dioxygen) has a triplet spin state. However, most ground molecules in the ground electronic state are spin-paired and singlet state. Given "Woodward-Hoffmann"

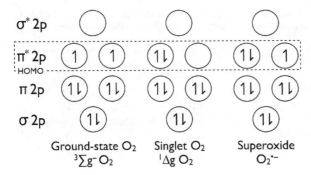

Figure 5.1 The 2p molecular orbitals of the dioxygen molecule. Electron spins in the HOMO states for ground-state molecular oxygen are compared to spin-paired singlet oxygen and the one-electron reduction product, superoxide.

symmetry constraints, a kinetically favorable reaction occurs when the reacting molecules have the same spin state. As a consequence, triplet-state oxygen, $^3O_2$, is unlikely to be highly reactive with a singlet-state molecule such as propane, $CH_3CH_2CH_3$, due to the symmetry considerations (i.e., the reaction is spin-forbidden), even though the overall thermodynamic driving force for a four-electron transfer is highly favorable. In order to oxidize propane rapidly (i.e., rapid kinetics), ground state oxygen must undergo activation or spin-pairing, which can be achieved in a number of ways. For example, ground-state dioxygen can be activated simply by near IR light as shown in Eq. 6. In contrast, reactive oxygen species (ROS, Table 5.1) are substantially more reactive since they have overcome symmetry restrictions.

$$O_2(^3\Sigma_g^-) + h\nu \longrightarrow O_2(^1\Delta_g) \{\lambda < 1000\,nm\} \tag{6}$$

A primary example of the difference in reactivity between ($O_2(^1\Delta_g)$) and $O_2(^3\Sigma_g^-)$ is illustrated by the difference in rate constants for the same reductant.

$$O_2(^3\Sigma_g^-) + NO_2^- \xrightarrow{k_7} NO_2 + O_2^{-\cdot} \tag{7}$$

$$O_2(^1\Delta_g) + NO_2^- \xrightarrow{k_8} NO_2 + O_2^{-\cdot} \tag{8}$$

where in water $k_8 \gg k_7$ ($k_8 = 3.1 \times 10^6\,M^{-1}s^{-1}$; $k_7 = 0.036\,M^{-1}s^{-1}$).

Reactive oxygen species (ROS) include but are not limited to singlet oxygen ($O_2(^1\Delta_g)$), hydroperoxyl radical (HO_2), superoxide ($O_2^{-\cdot}$), dihydrogen trioxide (H_2O_3, HO_3^-), and hydroxyl radical ($\cdot OH$). As shown in Table 5.1, reduction potentials for most ROS species are substantially more favorable than $O_2(^3\Sigma_g^-)$.

The lifetime of singlet oxygen in water is on average 3 μs due to deactivation by collision with H_2O according to Eq. 9.

$$O_2(^1\Delta_g) + H_2O \xrightarrow{k_d} O_2(^3\Sigma_g^-) + H_2O^* \qquad (9)$$

The first-order decay constant for the deactivation of singlet oxygen in water is $k_d = 3.2 \times 10^5 \, M^{-1}s^{-1}$ where the characteristic lifetime is given by $1/k_d$.

Singlet oxygen can be generated in nature by sensitization (type II reaction) of dye compounds such as rose bengal or C_{60} molecules, and also by irradiation of naturally occurring humic acids in lakes and rivers [1, 2]. Singlet oxygen reacts with molecules of biological significance such as nucleic acids, lipids, and amino acids [3] with toxic consequences. For example, rhodopsin reacts with $O_2(^1\Delta_g)$ at pH 8.0 with a second-order rate constant of $1.1 \times 10^9 \, M^{-1}s^{-1}$.

Superoxide, the one-electron reduction product of dioxygen (Table 5.1), and its protonated conjugate acid, HO_2, have the following equilibrium relationship:

$$HO_2 \xrightleftharpoons{K_a} O_2^{-\cdot} + H^+ \qquad (10)$$

where $pK_a = 4.8$. Superoxide is readily formed by electron transfer from sensitized dyes (type I reactions) or by sensitized oxidation of secondary alcohols. In addition, $O_2^{-\cdot}$ is formed in aqueous suspensions of semiconductor photocatalysts (e.g., ZnO or TiO_2) where oxygen is reduced by the photo-excited conduction-band electrons on the surface of the metal oxide or metal sulfide semiconductors. However, the characteristic lifetime of superoxide in aqueous solution is short due to competition from its self-reaction (i.e., dismutation) into oxygen and hydrogen peroxide.

$$HO_2^\cdot + O_2^{-\cdot} \xrightarrow{H^+} H_2O_2 + O_2 \qquad (11)$$

$$HO_2^\cdot + HO_2^\cdot \longrightarrow H_2O_2 + O_2 \qquad (12)$$

$$O_2^{-\cdot} + O_2^{-\cdot} \xrightarrow{2H^+} H_2O_2 + O_2 \qquad (13)$$

However, there is a pronounced difference in the rates of Eqs. 11 and 12. For example, $k_{11} = 9.7 \times 10^7 \, M^{-1}s^{-1}$; this can be compared to $k_{12} = 8.3 \times 10^5 \, M^{-1}s^{-1}$ and $k_{13} < 2 \, M^{-1}s^{-1}$. As a consequence of the relative slowness of Eq. 13, most living cells employ a protein "superoxide dismutase" (SOD) to catalyze the reaction under biological pH conditions (e.g., pH 7.8).

Hydroxyl radical, the three-electron reduction product of dioxygen, is the most highly reactive oxygen species in terms of redox potential and typical electron transfer rates, which are most often in the diffusion controlled regime (k = 10^9 to 10^{10} $M^{-1}s^{-1}$). Hydroxyl radical reacts by three pathways: hydrogen atom extraction, HO radical addition, or by direct electron transfer as illustrated by the following three reactions:

$$CH_3CH_2OH + \cdot OH \longrightarrow CH_3\overset{\cdot}{C}HOH + H_2O \tag{14}$$

$$\tag{15}$$

$$SO_3^{2-} + \cdot OH \longrightarrow SO_3^- \cdot + H_2O \tag{16}$$

The array of oxygen-containing radicals known as ROS in solution is known to damage cell membranes, cellular organelles, and nucleic acids contained within RNA and DNA. Moreover, the oxidizing properties of ROS can also be generated on the surface of nanomaterials.

Reactive oxygen species are highly reactive with low selectivity. In addition, ROS species present many challenges for direct time-resolved detection due to their short lifetimes and relatively low concentrations under steady-state conditions. This limitation can be overcome by trapping the free radicals with appropriate chemical trapping agents (e.g., chemical compounds that readily react with ROS and stabilize their unpaired electrons).

Electron paramagnetic resonance (EPR/ESR) detects the small changes that an unpaired electron exerts on an applied magnetic field. The unpaired electron, in a spin state of +1/2 or −1/2, responds to a magnetic field by aligning either parallel or anti-parallel to the applied field. Both spin states have distinct energy levels, which are determined by the magnetic field strength. In order to detect these energy states, the sample is exposed to electromagnetic radiation with sufficient energy to excite the electrons from the lower state to the upper state. This energy gap is given by

$$\Delta E = g\beta H \tag{17}$$

where ΔE is the energy gap between the +1/2 and −1/2 state, H is the applied magnetic field, β is the Böhr magneton constant, and g is the splitting factor for the free electron. This splitting factor depends on the atoms within the radical compound being detected. An adsorption spectrum can be obtained by using the applied magnetic field to detect the changes. EPR detectors take the first derivative of initial spectrum.

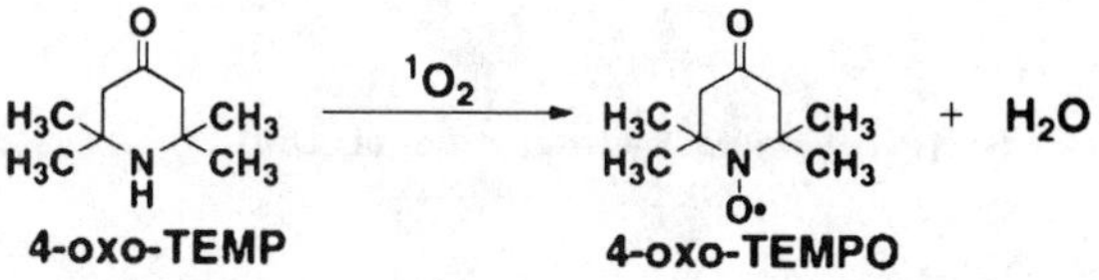

Figure 5.2 Reaction of TEMP with singlet oxygen.

The spin-trapping molecules affect the signal of the unpaired electron on the free radical species that is detected; this effect is known as splitting and is most commonly caused by adjacent hydrogen atoms. Often the splitting leads to a unique pattern for the EPR signals. The combination of the hyperfine structure (number of lines), line shape, and hyperfine splitting (the distance between peaks) give a radical its unique imprint. An ideal spin-trapping molecule will react quickly and specifically with the radical species of interest and produce a characteristic signal. Two spin traps, which are most frequently used for oxygen radical detection, are 5,5-dimethyl-1-pyrolline-N-oxide (DMPO) and 2,2,6,6-tetramethylpiperidine-N-oxyl (TEMP/TEMPO). Figures 5.2, 5.3, and 5.4 give the reaction of DMPO with superoxide and hydroxyl radical and TEMP with singlet oxygen.

In each case, the product of the reaction is a nitroxide compound, which is stabilized by charge delocalization between the nitrogen and oxygen atom. Figures 5.5, 5.6, and 5.7 illustrate typical EPR spectra for TEMPO, DMPO-OOH, and DMPO-OH.

The TEMPO spectrum is a 1:1:1 hyperfine structure that results from the interaction of the unpaired electron with the nitrogen nucleus. Both DMPO-OH and DMPO-OOH have the same interaction, but splitting occurs due to the presence of adjacent hydrogen and oxygen atoms. Hydroxyl radical reacts with DMPO about nine orders of magnitude faster than superoxide, so the DMPO-OH signal will predominate unless hydroxyl radical is quenched (*vide infra*). As a result, superoxide detection with DMPO requires much higher concentrations than hydroxyl detection would. DMPO-OOH can also decompose to DMPO-OH giving a false positive for hydroxy radical, but there are ways to avoid this [4, 5]. Many other spin traps are available, and new ones are developed on a regular basis.

Another common option for detection of reactive oxygen is chemical reduction. Two examples are Cytochrome c [6, 7] or nitroblue tetrazolium

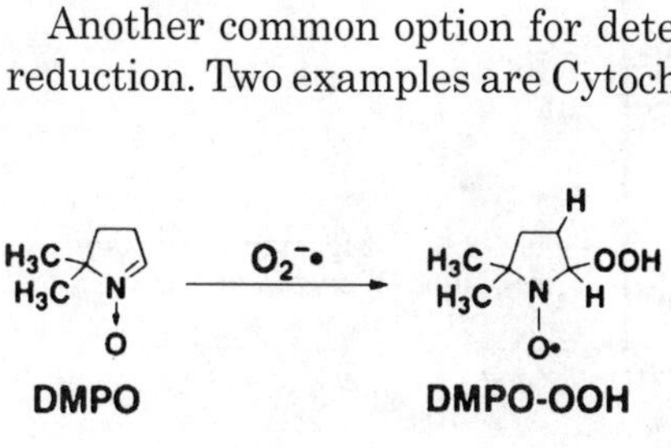

Figure 5.3 Reaction of DMPO with superoxide.

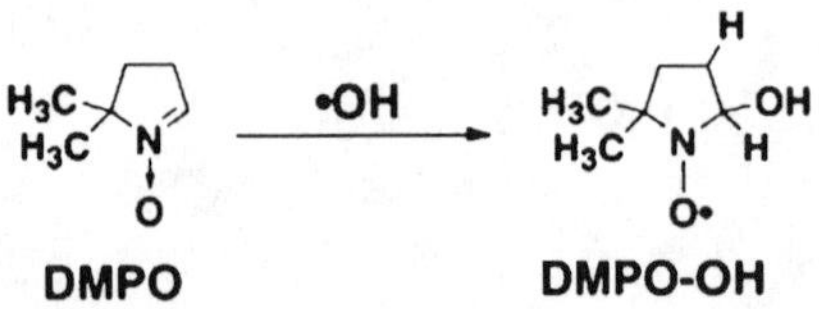

H₃C — •OH → H₃C—OH

DMPO   **DMPO-OH**

Figure 5.4 Reaction of DMPO with hydroxyl radical.

(NBT) [8–10] reduction. Both of these compounds are reduced by super-oxide. Cycochrome c reacts with a rate constant of $2.6 \times 10^5 \, M^{-1}s^{-1}$ and NBT with a constant of $6 \times 10^4 \, M^{-1}s^{-1}$. While Cytochrome c has a relatively simple reaction pathway, the pathway for NBT is more complex [8]. A simplified version is depicted in Figure 5.8.

This reduction is not selective, so the presence of the reduction products does not assure superoxide activity. Thus, superoxide dismutase [6] (*vide infra*) must be added to quench superoxide and create a baseline reduction level. These reductions can be followed by simple spectrophotometry, which provides a significant experimental advantage to using these compounds to quantify and study ROS production. Another advantage is the ability to analyze reduction data and determine rates of ROS production. However, concentration detection limits are higher than typically found using EPR methods. As in the case of EPR, there are other chemical reductants available (e.g. XTT).

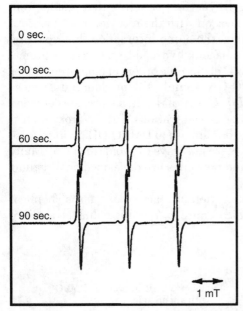

Figure 5.5 EPR signal of the TEMP-singlet oxygen adduct.

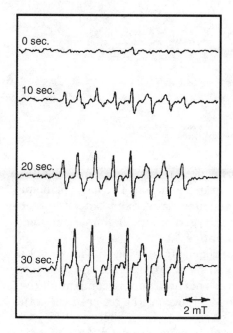

Figure 5.6 EPR signal of the DMPO-OOH adduct.

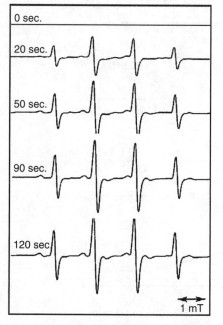

Figure 5.7 EPR signal of DMPO-OH adduct.

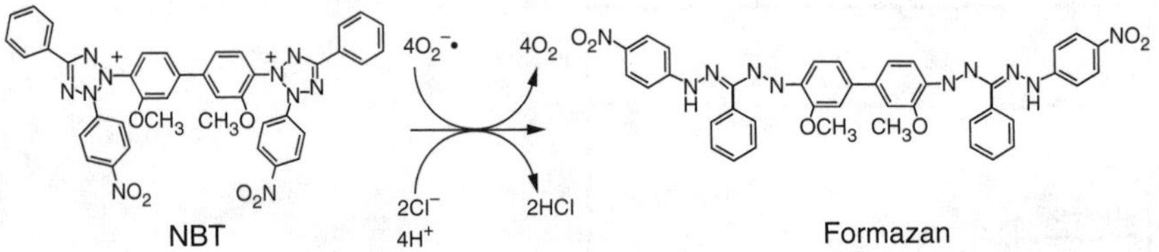

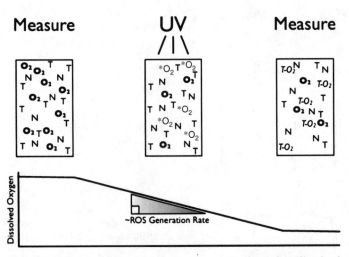

Figure 5.8   NBT reduction by superoxide.

In a less common approach, ROS is detected by the simple monitoring of dissolved oxygen concentration in a solution. The basic principle is shown below in Figure 5.9. A compound irreversibly traps the reactive oxygen and a drop of dissolved oxygen occurs. A simple handheld meter can be used to monitor this change.

One major advantage of the approach illustrated in Figure 5.9 is that the rate of ROS generation can be obtained from the dissolved oxygen (DO) loss by adding an excess of the trapping compound; theoretically every ROS will be trapped. However, the free radical traps must react with ROS at much higher rates than observed for ROS decay. In previous research, Zepp and coauthors used 2,5-dimethylfuran (DMF) to trap ROS [1, 2], while Haag et al. used furfuryl alcohol (FFA) as the trapping agent [11]. However, FFA reacts with singlet oxygen at relatively high rates (e.g., $k = 1.2 \times 10^8 \ M^{-1}s^{-1}$). It is often

Figure 5.9   Dissolved oxygen ROS measurement method. O_2 = dissolved oxygen; T = ROS trap; N = nanoparticle; $*O_2$ = ROS; $T\text{-}O_2$ = trapped ROS.

TABLE 5.2 A Brief Summary of ROS Detection Methods

Common ROS Detection Methods			
Type of ROS (decreasing oxidative power)	Electron Paramagnetic Resonance (EPR)	Spectrophotometric	Dissolved Oxygen Measurement
Hydroxyl	**DMPO** [Yamakoshi 2003] [Dugan 1997] **DMSO**	N/A	N/A
Superoxide	**DMPO** [Yamakoshi 2003] [Dugan 1997] **DEPMPO** [Frejaville 1995]	**NBT** [Saito 1983] **Cytochrome C** [You 2003] **XTT** [Okado-Matsumoto 2001]	**FFA** [Haag 1984] [Pickering 2005] **DMF** [Zepp 1985]
Singlet Oxygen	**TEMP** [Wilkinson 1993]	**Dipropionic acid** [Lindig 1981]	**FFA** [Haag 1984] [Pickering 2005] **DMF** [Zepp 1985]

difficult to detect low concentrations of ROS when compared with EPR methods (Table 5.2).

Chemical or biochemical quenchers can also be added to target specific ROS. For example, superoxide dismutase is routinely used to quench superoxide in a reaction medium [7]. Beta carotene and azide ion (N_3^-) are known quenchers for singlet oxygen [12]. These compounds serve to eliminate the response seen from any of the above detection methods. If a response is eliminated by the addition of a quencher, it is easier to assume that the specific ROS generated the signal. However, most quenchers lack complete specificity for a single ROS component. For example, superoxide dismutase seems to have an effect on singlet oxygen production despite its supposed specificity [13] for superoxide.

Alternative methods for detection of ROS include fingerprinting of reaction products and direct chemiluminescence detection (e.g., singlet oxygen luminescence can be quantified at 1270 nm [14]).

Nanoparticulate Semiconductor Particles and ROS Generation

Semiconducting metal oxides and metal chalcogenides have been used as catalysts for a wide variety of chemical reactions in the gas-solid phase and liquid-solid reactions over a broad range of temperatures from < 0 to > 500°C [15]. Semiconducting oxides and sulfides can be activated by an applied electrical potential, by the absorption of photons, or by elevated temperatures. The energies required for activation of some common metal oxide and metal chalocogenide semiconductors are given in Tables 5.3 and 5.4, respectively.

TABLE 5.3   Bandgap Energies for Metal Oxide and Mixed-Metal Oxide Semiconductors [15]

Semiconductor	E_g (eV)	E_g (nm)
TiO_2-anatase	3.2	385
TiO_2-rutile	3.0	413
ZnO	3.3	376
$SrTiO_3$	3.2	387
WO_3	2.8	443
Nb_2O_5	3.4	365
Bi_2O_3	3.2	387
CeO_2	3.4	365
In_2O_3	3.1	403
SnO_2	3.7	330
$MnTiO_3$	2.8	443
$FeTiO_3$	3.2	390
$BaTiO_3$	3.2	385
$CaTiO_3$	2.8	448
$PbTiO_3$	3.4	365
$Bi_4Ti_3O_{12}$	3.1	403
$Bi_2Ti_2O_7$	3.0	420
$Bi_{12}TiO_{20}$	3.1	386
t-ZrO_2	5.0	248
m-ZrO_2	5.2	238
Fe_2O_3	2.2	564
$YFeO_3$	2.58	481
CeO_2	3.4	365
In_2O_3	3.6; 3.1	344, 400
IrO_2	3.12	397
MoO_3	3.24	383
Bi_2WO_6	2.69	461
$K_3Ta_3Si_2O_{13}$	4.1	302
$LiTaO_3$	4.7	264
$NaTaO_3$	4.0	310
$KTaO_3$	3.3	376
$KNbO_3$	3.8	326
γ-Bi_2O_3	2.8	443
$K_4Nb_6O_{17}$	3.3	376
$CaTa_2O_6$	4.0	310
$SrTa_2O_6$	4.4	282
$BaTa_2O_6$	4.1	302
$Sr_2Ta_2O_7$	4.6	270
K_2PrTaO_{15}	3.8	326
$Sr_2Nb_2O_7$	3.9	318
$AgNbO_3$	2.9	428
$BiVO_4$	2.4	517
Ag_3VO_4	2.0	620
Bi_2WO_4	2.8	443
$La_2Ti_2O_7$	2.8	443
Bi_2YNbO_7	2.0	620
Bi_2CeNbO_7	2.10	590
Bi_2GdNbO_7	2.13	582
Bi_2SmNbO_7	2.21	561
Bi_2NdNbO_7	2.25	551
Bi_2PrNbO_7	2.26	549

TABLE 5.3   (*Continued*)

Semiconductor	E_g (eV)	E_g (nm)
Bi_2LaNbO_7	2.38	521
Ga_2O_3	4.98	249
SnO_2	3.5	354
In_2O_3	2.7	459
Cu_2O	2.6, 2.0	477, 620
$CdWO_4$	2.94	422
$CdMoO_4$	2.43	510

*1 eV = 4.42 × 10^{14} Hz (s^{-1}); $\lambda = \dfrac{hc}{E_g(eV)\nu(Hz\,eV^{-1})}$; C = 299792458 m s^{-1}; h = 4.14 × 10^{-15}

Activation can occur by the input of energy (i.e., eV or nm), which is sufficient to promote an electron from the valence band to the conduction band or from energy states that lie within the bandgap (trapped or doped states), as illustrated in Figures 5.10 and 5.11. For example, absorption of photons (with $E_{hv} > E_{bg}$) by a semiconductor leads to promotion of an electron to the conduction band, e_{cb}^-, and at

TABLE 5.4   Bandgap Energies for Metal Sulfide and Chalcogenide Semiconductors [15]

Semiconductor	E_g (eV)	E_g (nm)
ZnS	3.66	339
ZnSe	2.90	428
ZnTe	2.25	551
CdS	2.45	506
CdSe	1.74	713
CdTe	1.45	855
HgS	2.2	564
$CuAlS_2$	3.4	365
$CuAlSe_2$	2.7	459
$AgGaS_2$	2.5	496
$CuGaS_2$	2.4	517
$AgInS_2$	1.8	689
GaS_2	3.4	365
Ga_2S_3	2.8	443
GaS	2.5	496
As_2S_3	2.5	496
Gd_2S_3	2.55	486
La_2S_3	2.91	426
MnS	3.0	413
Nd_2S_3	2.7	459
Sm_2S_3	2.6	477
ZnS_2	2.7	459
ZrS_2	1.82	681
$Zn_3In_2S_6$	2.81	441
Pr_2S_3	2.4	517

Energy

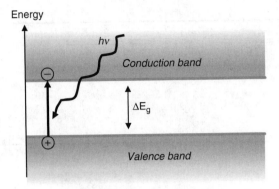

Figure 5.10   Schematic representation of the energy gap between the HOMO states or valence band of a semi-conductor and the LUMO states or conduction band where ΔE_g corresponds to the bandgap energy in either energy (eV) or wavelength (nm).

the same time creation of a vacancy, h_{vb}^+, in the valence band as illustrated in Figure 5.11.

The relative energy levels of the conduction band and valence band states for some of the representative semiconductors listed in Tables 5.3 and 5.4 are shown in Figure 5.12. In the specific case of TiO_2, a semiconductor used in numerous commercial products, the conduction band edge has a relative energy level at pH 0 of -0.4 V and the valence band hole has an electrochemical potential of $+2.8$ V. Thus, in the case of TiO_2 the conduction band electron is mildly reducing although with sufficient potential to reduce O_2 to superoxide, O_2, while the valence band hole at the surface of TiO_2 is a powerful oxidizing agent.

For example, with use of the appropriate semiconductor catalysts, it is possible to drive the photoelectrolysis of water into hydrogen and oxygen as follows:

$$2H_2O \xrightarrow{\text{UV \& visible light}} 2H_2 + O_2 \qquad (18)$$

The overall multi-electron redox potentials for the multi-electron reactions at pH 7 suggest that H_2 production with the input of UV-visible light energy is feasible.

$$H_2O \leftrightarrow H^+ + OH^- \qquad (19)$$

$$2H_2O + 2\,\bar{e} \leftrightarrow H_2 + 2^-OH \qquad (20)$$

$$2H_2O \leftrightarrow O_2 + 4H^+ + 4\bar{e} \qquad (21)$$

The redox potential for equation 21 at pH 7 is $E_H = -1.23$ V (NHE) with the corresponding half-reactions of -0.41 V (Eq. 20) and 0.82 V (Eq. 21),

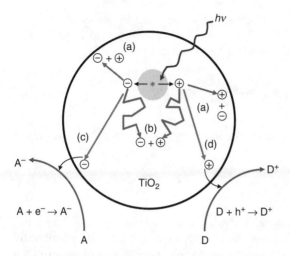

Figure 5.11 Presented above is a graphic depiction of a spherical semiconductor particle, which has absorbed a photon of sufficient energy to promote and electron from the conduction band to the valence band. After charge separation, the electron, e_{cb}^-, and the hole, h_{vb}^+, may simply recombine either direct bandgap recombination (a) or indirect recombination from trapped states (b) e_{tr}^- or h_{tr}^+. Some of mobile electrons and holes can migrate to the surface of the nanoparticulate semiconductor where they can undergo electron transfer reactions (c and d). The trapped electron can be transferred to an electron acceptor, A, and the trapped hole can accept an electron from an electron donor, D.

which gives a $\Delta G^\circ = +237$ kJ/mole). However, with the input of light at wavelengths ≤ 1000 nm (i.e., 1.23 eV ≈ 1000 nm), the overall energy requirement for the photosynthetic splitting of water can be met with solar radiation in principle. On the other hand, the rate of reaction in the normal Marcus regime should depend on the overall driving force (i.e., lower wavelength irradiation is preferable kinetically) and the thermodynamics of the initial or sequential one-electron transfer processes at the semiconductor surfaces. Moreover, the one-electron transfers are much less favorable thermodynamically than the overall two-electron transfer reactions as shown below:

$$H^+ + e_{aq}^- \xrightleftharpoons{E_H = -2.5 \text{ V (pH 7)}} H_{aq}^\cdot \tag{22}$$

$$H_2O \xrightleftharpoons{E_H = -2.3 \text{ V (pH 7)}} {}^\cdot OH + H^+ + e_{aq}^- \tag{23}$$

$$^-OH \xrightleftharpoons{E_H = -1.8 \text{ V (pH 0)}} {}^\cdot OH + e_{aq}^- \tag{24}$$

In the presence of oxygen, the reductive process in water leading to superoxide ion, O_2^- has the following half-reaction and a reduction potential of $E_H = -0.33$ V: at pH 7.

$$O_2 + \bar{e} \underset{pH=7}{\overset{E_H = -0.33V}{\rightleftharpoons}} O_2^- \qquad (25)$$

On the oxidative side, the water or hydroxyl radical has corresponding potential at pH 0:

$$^-OH \underset{pH=7}{\overset{E_H = -1.8V}{\rightleftharpoons}} {}^\cdot OH + e_{aq}^- \qquad (26)$$

However, when working with hydrated metal oxide surfaces one must take into account the nature of the reactive surface sites that normally involve metal hydroxyl functionalities. In the specific case of dehydrated TiO_2 (see Figure 5.13), we must consider that when exposed to water either in a humid atmosphere or in aqueous suspension the surface titanium-oxygen bonds in the crystallites undergo hydrolysis to produce surface hydroxyl groups as follows [16−27]:

$$> Ti\text{-}O\text{-}Ti < + H_2O \underset{k_d}{\overset{k_h}{\rightleftharpoons}} 2 > TiOH \qquad (27)$$

The metal hydroxyl surface sites (e.g., >TiOH or >FeOH) exist under ambient conditions in an open atmosphere or in an aqueous suspension. Once hydrolyzed, the surface hydroxyl groups either gain or lose a proton

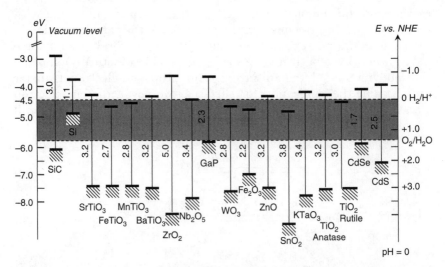

Figure 5.12   Comparison of the bandgap energies for an array of semiconductor relative to the reduction and oxidation potentials of water at pH 0.

depending on the intrinsic acidity of the metal oxide [24, 28]. A simple formalism for characterizing the surface acidity versus pH, for quantifying the surface buffering capacity, the surface ion-exchange properties, and surface complexation capacity for cations, anions, and ligands is presented in Eqs. 28 and 29. The pH dependent changes in terms of the acid-base chemistry of surface hydroxyl functionalities (e.g., $>MOH$, $>TiOH$, $>FeOH$) can be treated as a conventional diprotic acid, although there may be more than one type of surface site undergoing protonation and deprotonation (i.e., a distribution of surface acidity constants, K_a^s.

In the case of nanoparticulate TiO_2, the titration of a colloidal suspension with NaOH gives a classical titration curve for a diprotic acid as shown in Figure 5.14. Using the titration data of Figure 5.14, the surface acidity of TiO_2 can be characterized is terms of two surface acidity constants as follows:

$$< TiOH_2^+ \underset{k_{-1}}{\overset{k_1}{\rightleftharpoons}} < TiOH + H^+$$

$$\text{(28)}$$

$$K_{al}^s = \frac{k_1}{k_{-1}} \qquad pK_{al}^s = 2.4$$

$$< TiOH \underset{k_{-2}}{\overset{k_2}{\rightleftharpoons}} < TiO^- + H^+$$

$$\text{(29)}$$

$$K_{a2}^s = \frac{k_2}{k_{-2}} \qquad pK_{a2}^s = 8.0$$

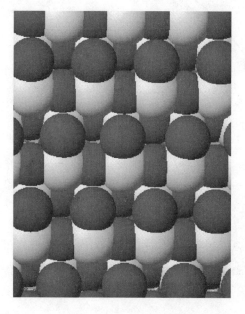

Figure 5.13 The anhydrous TiO_2 surface looking at the predominant 101" crystalline face of TiO_2 (anatase) showing oxygen in red (O^{2-}) and titanium in white (Ti^{4+}).

$$pH_{zpc} = \frac{(pK_{a1}^s + pK_{a2}^s)}{2} = 5.25 \qquad (30)$$

In the simplest case, at low ionic strength in the absence of added cations or anions, the isoelectric point or "point of zero charge" is described as follows in terms of the concentrations of the relevant surface species as a function of pH:

$$[> TiOH_2^+] = [> TiO^-] \qquad (31)$$

At a fixed ionic strength, the surface charge on TiO_2 is a function of the solution pH as follows [24, 28]:

$$\left.\sigma_0\right|_{[\mu]=const} = \left.\frac{\partial\sigma_0}{\partial pH}\right|_{pH\neq pH_{zpc}} (pH - pH_{zpc}) \qquad (32)$$

The above acidity constants and the pH of zero point of charge, pH_{zpc}, are given for quantum-sized TiO_2 in the particle size range of 1.0 to 3.0 nm. The titration data can be plotted with respect to the surface site density as shown in Figure 5.15 to get more reliable estimates to the pH_{zpc} for most semiconductors listed in Tables 5.3 and 5.4.

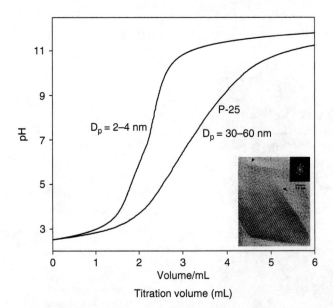

Figure 5.14 A typical titration curve for two different size fractions of colloidal TiO_2. This data can be used to determine the surface acidity constants and provide an estimate of the pH of zero point of charge.

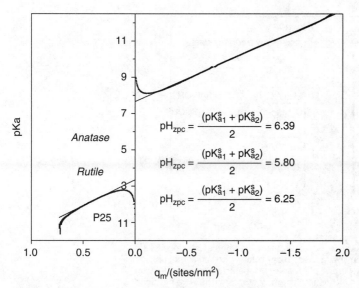

Figure 5.15 Using titration data similar to that presented in Figure 5.14, the individual surface acidity constants and the surface site density can be determined. The pH_{zpc} for anatase, rutile, and a mixture of anatase and rutile (Degussa P25) are shown here.

The number of moles of surface hydroxyl groups {>MOH} in an aqueous suspension of volume V can be estimated as follows:

$$\{>MOH\} \approx \frac{(m_{MO}/\rho_{MO}V_{MO})A_{MO}d_{OH}(10^{18}nm^2/1m^2)}{N} \tag{33}$$

where A_{MO} is the surface area of the metal oxide particle (m^2), d_{OH} is the site density of hydroxyl groups for MO (number per nm^2), N is Avogadro's number, m_{MO} is the mass of the metal oxide, MO, in a suspension (g) of a given volume, V is the volume of an MO particle (cm^3), and ρ is the density of the MO (g/cm^3). The density of surficial >MOH groups, d_{OH}, ranges from 4 to 10 per nm^2.

Irradiation of TiO$_2$ (as a primary example) generates electrons and holes on a 100 femtosecond to picosecond time scale. The approximate energy level positions for TiO$_2$ in the anatase crystalline form are shown in Figure 5.12 with an equivalent scale of redox potentials versus NHE (normal hydrogen electrode). After photoexcitation (Figure 5.16) some of the electrons and holes migrate to the surface where they can be trapped within the nanosecond time frame on the surface of TiO$_2$ in the form of >Ti(III)OH and >Ti(IV)OH$^-$.

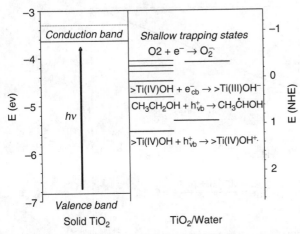

Figure 5.16   Energy level positions for the photoexcitation of TiO$_2$ ($\Delta E_g = 3.2$ eV) in the anatase form relative to the solid-solution interface redox potentials for key steps and possible electron transfer reactions. Surface trapping states within the bandgap energy domain are indicated.

The surface hydration and dehydration process and photoexcitation can be followed with DRIFT (diffuse reflectance infrared Fourier transform) spectroscopy [29–31]. In Figure 5.18, evidence for the reversible hydration (Figure 5.17) and dehydration of TiO$_2$ is shown where the

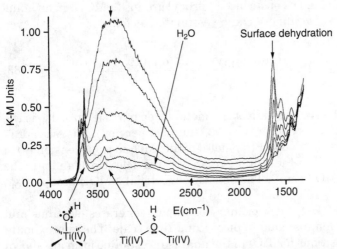

Figure 5.17   DRIFT spectra of TiO$_2$. The broadband spanning 2500–3900 cm^{-1} is due to >TiOH stretching vibrations in different atomic environments. With progressive dehydration, this characteristic feature disappears, and discrete stretches within 3400–3800 cm^{-1} arise. Complete dehydration required thermal treatment for 12 hours at 623 K under a ~1 μTorr vacuum. Dehydrated TiO$_2$ is reversibly rehydrated with water vapor. Surface trapping states clearly indicated in the dehydrated spectra appear at 3716 cm^{-1}.

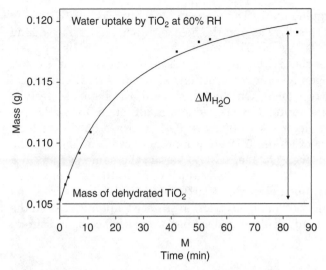

Figure 5.18   The rehydration of anhydrous TiO_2 is a relatively slow process which takes place over several hours.

presence of a humid atmosphere with water peak in the IR spectrum is clearly visible near 3500 cm^{-1}.

In a typical photolysis experiment (Figure 5.19), there is convincing FTIR evidence for the formation of >Ti(III)OH (e_{tr}^-) and >Ti(IV)OH$^{\cdot}$ (h_{tr}^+)

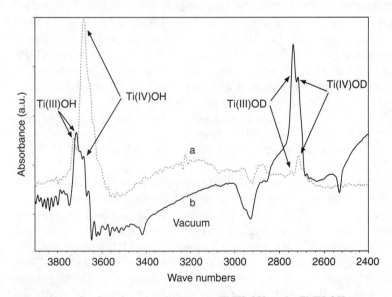

Figure 5.19   Surface functional groups, >Ti(III)OH and >Ti(IV)OH$^{\cdot}$ groups reflecting the trapping of electrons and holes after illumination under a vacuum.

as the surface trapping state. The electron is trapped as $>$Ti(III)OH (a reduced Ti site on the surface) and the hole is trapped as surface-bound hydroxyl radical, $>$Ti(IV)OH$\cdot$. The trapped hole, $>$Ti(IV)OH$\cdot$, is a powerful oxidizing species that is able to undergo either direct electron transfer reactions or hydrogen abstraction reactions, depending on the chemical nature of the electron donor. On the other hand, the trapped electron [$>$Ti(III)OH] is a moderate to weak reductant, although it is still capable of transferring electrons to dioxygen (O_2) adsorbed on the surface. The DRIFT spectra of Figure 5.19 show a pronounced peak shift in the presence of oxygen (i.e., $O_2 + e_{tr}^- \rightarrow O_2^-$) versus the same system in a vacuum. For example, the band that appears at 3716 cm^{-1} has been identified as the trapped electron, $>$Ti(III)OH, which results from the localization of conduction band electrons in surface trapping sites. In the presence of O_2, the trapped electron disappears and surface-bound superoxide, $O_2^{-\cdot}$, is formed.

$$O_2 + e_{tr}^- \rightleftharpoons O_2^{-\cdot} \qquad (34)$$

As an alternative, Eq. 34 can be written as

$$O_2 + {}>Ti(III)OH^- \rightleftharpoons O_2^{-\cdot} + {}>Ti(IV)OH \qquad (35)$$

The trapped electron, $>$Ti(III)OH$^-$, is completely removed by exposure to Br_2 in the gas phase. The trapped electron has an ESR (electron spin resonance) signal with $g_\perp = 1.957$, and $g_\parallel = 1.990$. However, after irradiation the 3716 cm^{-1} band persists under a 1.0 atmosphere of dry (0 percent RH) O_2 at 300 K, whether in the dark or under illumination, whereas the TiO_2 surface must be exposed to water vapor before the trapped electron band at 3716 cm^{-1} disappears. *ab initio* calculations of oxygen-deficient TiO_2 in the rutile form indicate that excess charge in the bulk remains spin-paired and localized at vacant oxygen sites. Inter-bandgap states on reduced TiO_2 surfaces are associated with spin-polarized Ti(III) 3d^1 and Ti(II) 3d^2 configurations.

The conduction band electrons in the trapped state electrons have been observed using a variety of laser-based pump-probe photolysis experiments. The so-called blue electron in either a deep or swallow state trap—that is, an internal Ti(III) site or a surface $>$Ti(III)OH site—has a characteristic spectrum in the visible with a band peak at 600 nm. The appearance and disappearance of the "blue electron" can be followed kinetically by rapid-scan spectroscopy in order to get an estimate of the actual lifetimes of mobile electrons that are available for electron transfer.

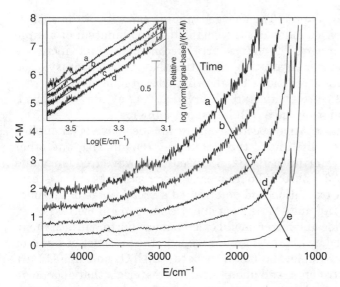

Figure 5.20 DRIFT spectra of TiO_2 (a) 13 minutes after UV irradiation, (b) 52 minutes after, (c) 94 minutes after, (d) 135 minutes after, and (e) following full relaxation of the excited-state signal. Inset: $\log(S_{norm})$ vs. $\log(\bar{\nu}(cm^{-1}))$ where S_{norm} is the background subtracted signal normalized to $S = 1$ at $2500\ cm^{-1}$.

The trapped electron and hole can recombine over a broad range of timeframes from nanoseconds to hours depending on the energy of the trapped state. In the absence of external electron donors or acceptors, the shallow trapped electrons can persist for up to 40 hours before eventual recombination as shown in Figure 5.20.

$$> Ti(IV)OH^{+\cdot} + > Ti(III)OH^- \xrightarrow{\text{recombination}} 2 > Ti(IV)OH \quad (36)$$

In the presence of O_2, recombination can occur through reversible electron transfer from $> Ti{:}O_2^-\cdot$ back to a bound surface hydroxyl radical as follows:

$$> Ti(IV)OH^{+\cdot} + > Ti(IV)OH{:}O_2^-\cdot \xrightarrow{\text{recombination}} 2 > Ti(IV)OH + O_2 \quad (37)$$

However, the majority of electrons in bandgap-excited TiO_2 exist in a free state, giving rise to a broad, featureless absorption with intensity proportional to $(\lambda/\mu m)^{1.73}$ as shown in Figure 5.20. These electrons decay according to a saturation kinetic mechanism that is limited by the density of trapped states. Kinetic observations suggest that free charge carriers are relatively stable in the bulk phase, that surface charge trapping is a reversible process, and that recombination of trapped states does not necessarily occur rapidly, even in the presence of an opposing charge acceptor. The charge trapping capacity of the surface eventually decreases following the complete recombination of a large number of free carriers. The charge carrier recombination rate increases with increasing surface hydration, indicating that surface hydroxylation assists annihilation reactions by allowing irreversible electron trapping. A surface electron trap state observed at 0.42 eV may be responsible for the mediating the annihilation mechanism.

A number of alternative spectroscopic techniques have been applied to characterize the lifetime of the mobile electrons in TiO_2 and other semiconductors. Martin et al. [32, 33] used microwave frequency and Herrmann et al. [34] used radio frequency spectroscopy as a probe technique after laser excitation and determined a broad range of lifetimes for the various trapping states from microseconds to milliseconds.

Most metal oxide and mixed-metal oxide semiconductor surface chemistry is dominated by hydroxyl groups when in the presence of water or humid air. For example, the metal niobates, $LiNbO_3$ and $KNbO_3$, are widely used electro-optic and photorefractive materials that depend on the activation of surface protons (i.e., protons bound in hydroxyl ions, ^-OH). The hydroxyl bound protons have activation energies in the range of 1 eV for mobility in $LiNbO_3$ and $KNbO_3$ crystals. The corresponding surface hydration in $KNbO_3$ leads to the following reactions:

$$> Nb(V)ONb(V) < + H_2O \rightleftharpoons 2 > Nb(V)OH \tag{38}$$

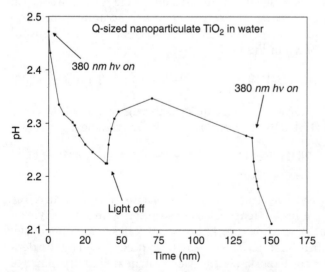

Figure 5.21 Irradiation of TiO_2 quantum-sized colloids ($D_p = 2$ nm) in water also produces trapped electrons and holes, which lead to shifts in the measured pH in the presence and absence of light with no added electron donor or acceptor except H_2O.

The photochemistry and photophysics of quantum-sized nanoparticles [16, 22, 24, 35] (1 nm $< D_p <$ 10 nm) in contrast with the larger, bulk-phase particles ($D_p \approx$ 100 nm), can be quite different. Many nanoparticulate semiconductors, which are often described as "quantum-sized particles" or "quantum dots" depending on their applications, exhibit a characteristic blue shift in the UV or visible absorption spectrum. Along with the blue shift in the absorption spectrum, there is a corresponding increase in the bandgap energy, ΔE_g which can be described in terms of a simple solution to the Schrödinger equation with an appropriate Hamiltonian.

$$\Delta E_g \simeq \left(\frac{\pi^2 \bar{h}^2}{2R^2} \frac{1}{\mu} \right) - \frac{1.8 e^2}{\varepsilon R} \qquad (39)$$

where R is the particle radius and μ is the reduced mass of the exciton or the electron-hole pair.

$$\frac{1}{\mu} = \left(\frac{1}{m_{e^-}^*} + \frac{1}{m_{h^+}^*} \right) \qquad (40)$$

where $m_{e^-}^*$ is the effective mass of the electron and $m_{h^+}^*$ is the effective mass of the hole, and ε is the dielectric constant of TiO_2, and $\bar{h}$ is Planck's constant.

According to Eq. 39, as R decreases the bandgap energy, ΔE_g, increases (i.e., R↓ = ΔE_g↑). As an example, Kormann et al. [24] prepared Q-sized TiO_2 with a characteristic blue shift from the bulk state bandgap of 385 nm for anatase TiO_2 down to 350 nm (i.e., ΔE_g = 3.2 to 3.35 eV). The steady-state particle size ranged from 2.0 nm to 2.5 nm depending on the preparation conditions and the Ti(IV) reagent used in the synthesis (e.g., $TiCl_4$ or Ti(IV)-isopropoxide). The corresponding cluster size (oligomer) for the nanoparticles ranged from 120 to 220 monomers. In an earlier study, Bahnemann et al. [16] reported that Q-ZnO exhibited bandgap increases as large as 1 eV or $\Delta E_{g,\,Q-ZnO}$ = 4.2 eV).

An increase in ΔE_g often enhances the reactivity of the photocatalyst by increasing its reduction/oxidation potential and thus the driving force for electron transfer in the normal Marcus regime; thereby ROS (reactive oxygen species) should also be a function of particle size.

In a subsequent study, Hoffman et al. [36] investigated the photochemical production of H_2O_2 on irradiated Q-ZnO over the wavelength range of 320 $\leq \lambda \leq$ 370 nm in the presence of carboxylic acids and oxygen. Steady-state concentrations up to 2 mM H_2O_2 were formed. Maximum H_2O_2 concentrations were obtained only with added electron donors (i.e., hole scavengers). The order of photochemical efficiency for H_2O_2 production with carboxylic acids as electron donors was HCO_2^- > $C_2O_4^{2-}$ > $CH_3CO_2^-$ > citrate. Isotopic labeling of the electron acceptor,

O_2, with ^{18}O verified that H_2O_2 was produced directly by the reduction of adsorbed oxygen by conduction band electrons. Quantum yields were as high as 30 percent for H_2O_2 production at low photon fluxes. At the same time, the quantum yield was shown to vary with the inverse square root of absorbed light intensity [i.e., $\phi \propto (\sqrt{I_{abs}})^{-1}$], with the wavelength of excitation $\phi \propto (\lambda)^{-1}$, and with the diameter of the Q-sized colloids (i.e., $\phi \propto D_p^{-1}$). For example, d[H_2O_2]/dt is 100 to 1000 times faster on Q-sized ZnO particles ($D_p = 2 - 4$ nm) than with bulk-phase ZnO particles ($D_p = 100$ nm).

Hydrogen peroxide production proceeds, after initial photoactivation, by electron transfer from the conduction band to dioxygen adsorbed on the surface of the excited-state metal oxide as follows:

$$2[e_{cb}^- + O_2 \longrightarrow O_2^{-\cdot}] \tag{41}$$

$$O_2^{-\cdot} + H^+ \underset{}{\overset{pK_a = 4.8}{\rightleftharpoons}} HO_2^{\cdot} \tag{42}$$

$$2HO_2^{\cdot} \longrightarrow H_2O_2 + O_2 \tag{43}$$

Hoffmann and coworkers [25, 36, 37] observed a tenfold increase in the measure quantum yield for H_2O_2 production upon reduction of the mean particle diameter from 40 to 23 nm for ZnO, where O_2 was the electron acceptor and small molecular organic compounds (e.g., carboxylic acids and alcohol) the electron donor. Similar effects were reported by Hoffmann and coworkers [35, 38, 39] for photo-polymerization reactions catalyzed by Q-sized CdS, Q-ZnO, and Q-TiO$_2$ and for SO$_2$ oxidation in the aqueous phase.

In addition to ROS generated from surface hydroxyl species and from adsorbed O_2, there are other oxygen-containing free radical species that are generated on the surface of photoactivated semiconductors. For example, S(IV) ([S(IV)] $\equiv$ [SO$_2 \cdot$ H$_2$O] + [HSO$_3^-$] + [SO$_3^{2-}$]) is readily photooxidized [28] in the presence of colloidal suspensions of nanoparticulate α-Fe$_2$O$_3$.

$$O_2 + 2 \, HSO_3^- \xrightarrow[\alpha\text{-Fe}_2O_3]{h\nu \leq 520 \text{ nm}} 2 \, SO_4^{2-} + 2H^+ \tag{44}$$

Quantum yields ranged from 0.08 to 0.3 with a maximum yield found at pH 5.7. The primary initiation pathway involved irradiation at wavelengths equal to or less than the nominal bandgap of hematite, which is 2.2 eV or 560 nm. Upon bandgap illumination, conduction-band electrons and valence-band holes are separated; the trapped electrons are transferred either to surface-bound dioxygen or to Fe(III) sites on or near the surface, while the trapped holes accept electrons from adsorbed SO$_3^{2-}$ to produce surface-bound SO3⁻·. The relatively high quantum yields are attributed in part to the desorption of SO3⁻· from the α-Fe$_2$O$_3$ surface and subsequent initiation of a homogeneous aqueous-phase free radical chain oxidation of S(IV) to S(VI). The following photochemical rate expression describes the observed kinetics over a broad range of conditions:

$$-\frac{d[S(IV)]}{dt} = \phi I_0 (1 - 10^{-\varepsilon[\alpha\text{-Fe}_2O_3]\ell})\left(\frac{K_s[HSO_3^-]}{1 + K_s[HSO_3^-]}\right) \tag{45}$$

where the quantum yield ϕ is defined as follows [40, 41]:

$$\phi_i(\lambda) \equiv \frac{\text{\# of molecules reacting via pathway } i}{\text{total number of photons absorbed by reacting molecule}} \tag{46}$$

or

$$\phi_r(\lambda)(\text{mol einstein}^{-1}) = \frac{\text{moles of compound transformed}}{\text{moles of photons absorbed}} \tag{47}$$

where

$$\sum_i \phi_i = 1 \tag{48}$$

A similar kinetic expression [38] was observed for the photocatalytic oxidation of S(IV) on TiO$_2$. In this case, for $\lambda \le 385$ nm, quantum yields in excess of unity (e.g., $0.5 \le \phi \le 300$) were observed and attributed also to desorption of the SO$_3^-$· radical anion from the TiO$_2$ surface leading to the initiation of homogeneous free radical chain reactions. These chain reactions have an amplified effect on the measured quantum efficiency. Depending on the free radical chain length, the measured ϕ values can be greater than one. In addition, the observed quantum yields depend on the concentration and nature of free radical inhibitors present in the heterogeneous suspension.

For SO$_2$ in water, the free radical chain reactions involve the formation of sulfur radical species such as SO$_3^-$·, SO$_4^-$·, and SO$_5^-$· that are alternative forms of ROS with similar reactivity to superoxide and hydroxyl radicals.

Iron oxides and iron oxide polymorphs initiate the chain reaction as follows:

$$O_2 + 2HSO_3^- \xrightarrow[\alpha\text{-Fe}_2O_3]{h\nu} 2HSO_4^- \tag{49}$$

$$>FeOH + HSO_3^- \rightleftharpoons >FeSO_3^- + H_2O \tag{50}$$

$$> Fe(III)SO_3^- + h_{vb}^+ \rightleftharpoons > Fe(III) + SO_3^- \cdot \qquad (51)$$

$$SO_3^- \cdot + O_2 \longrightarrow SO_5^- \cdot \qquad (52)$$

$$SO_5^- \cdot + SO_3^{2-} \longrightarrow SO_4^{2-} + SO_4^- \cdot \qquad (53)$$

$$SO_5^- \cdot + SO_3^{2-} \longrightarrow SO_5^{2-} + SO_3^- \cdot \qquad (54)$$

$$SO_4^- \cdot + SO_3^{2-} \longrightarrow SO_4^{2-} + SO_3^- \cdot \qquad (55)$$

$$SO_5^{2-} + SO_3^{2-} \longrightarrow 2SO_4^{2-} \qquad (56)$$

The other iron(III) oxide polymorphs (e.g., γ-FeOOH, $\Delta E_g = 2.06$ eV; β-FeOOH, $\Delta E_g = 2.12$ eV; α-FeOOH, $\Delta E_g = 2.10$ eV; δ-FeOOH, $\Delta E_g = 1.94$ eV; γ-Fe$_2$O$_3$, $\Delta E_g = 2.03$ eV) are photocatalytic for certain reactions [42]. In general, the observed order of relative photochemical reactivity toward SO_2 oxidation [43] is γ-FeOOH > α-Fe$_2$O$_3$ > γ-Fe$_2$O$_3$ > δ-FeOOH > β-FeOOH > α-FeOOH, while in the case of $C_2O_4^{2-}$ oxidation [44–46] the relative catalytic order is ferrihydrite (am-Fe(OH)$_3$), γ-Fe$_2$O$_3$ > γ-FeOOH > α-Fe$_2$O$_3$ > α-FeOOH > β-FeOOH.

$$2 > FeOH + hv \xrightarrow{e^-/h^+} \underset{e_{tr}^-}{> Fe(II)OH^-} + \underset{h_{tr}^+}{> Fe(IV)OH^+} \qquad (57)$$

$$> Fe(IV)OH^+ + C_2O_4^{2-} \longrightarrow > Fe(III)OH + CO_2 + CO_2^- \cdot \qquad (58)$$

$$> Fe(II)OH^- + O_2 \longrightarrow > Fe(III)OH + O_2^- \cdot \qquad (59)$$

Similar reactions occur with other carboxylic and dicarboxylic acids on the surface of the iron oxides and oxyhydroxides leading to superoxide and hydrogen peroxide formation. Pehkonen et al. [46] found that the rates of H_2O_2 formation on a series of iron oxide polymorphs depended on the chemical nature of the carboxylate electron donor where HCO_2^- > $CH_3CO_2^-$ > $CH_3CH_2CH_2CO_2^-$.

Metal Sulfide Surface Chemistry and Free Radical Generation

Metal sulfide and related chalocogenide semiconductors are used in a variety of electronic applications. For example, CdS, which is an n-type semiconductor with $E_g = 2.4$ eV, has been shown to have photocatalytic activity for H_2 production under visible light irradiation under anoxic conditions in water in the presence of electron donors such as C_2H_5OH, HS^- and SO_3^{2-}.

The electronic levels of nanoparticulate or quantum-sized CdS (Q-CdS) can be tuned by changing or controlling particle size without changing the chemical composition. For example, Hoffman et al. [35] found an increase in quantum efficiency for photo-polymerization of methylmethacrylate with a corresponding decrease in particle size using Q-CdS.

The surface chemistry [47−53] of nanoparticulate CdS in the Q-size domain has some similarities, which are initiated by the hydrolysis of the surface of CdS to form surface functionalities [54] that are dominated by the cadmium mercapto group, >CdSH, and cadmium hydroxyl, >CdOH, functionalities as follows:

$$\left[S_{Cd^{2+}}^{2-} > (CdS) \right]_2 + H_2O \xrightleftharpoons{K_H} S_{Cd^{2+}}^{2-} > Cd(II)SH$$

$$\tag{60}$$

$$+ S_{Cd^{2+}}^{2-} > S(-II)Cd(II)OH$$

The variable surface charges arise from protonation and deprotonation of surface sulfhydryl and hydroxyl groups as depicted in following equations:

$$S_{Cd^{2+}}^{2-} > CdSH + H^+ \xrightleftharpoons{} S_{Cd^{2+}}^{2-} > CdSH_2^+ \tag{61}$$

$$S_{Cd^{2+}}^{2-} > CdSH_2^+ \xrightleftharpoons{K_{a1}^s} S_{Cd^{2+}}^{2-} > CdSH + H^+ \tag{62}$$

$$S_{Cd^{2+}}^{2-} > CdSH \xrightleftharpoons{K_{a2}^s} S_{Cd^{2+}}^{2-} > CdS^- + H^+ \tag{63}$$

$$S_{Cd^{2+}}^{2-} > CdOH + H^+ \xrightleftharpoons{} S_{Cd^{2+}}^{2-} > CdOH_2^+ \tag{64}$$

$$S_{Cd^{2+}}^{2-} > CdOH_2^+ \xrightleftharpoons{K_{a1.1}^s} S_{Cd^{2+}}^{2-} > CdOH + H^+ \tag{65}$$

$$S_{Cd^{2+}}^{2-} > CdOH \xrightleftharpoons{K_{a2.1}^s} S_{Cd^{2+}}^{2-} > CdO^- + H^+ \tag{66}$$

In the simplest case, at low ionic strength in the absence of added cations or anions, the isoelectric point or "point of zero charge" is described as follows in terms of the concentrations of the relevant surface species as a function of pH:

$$\left[{}^{S^{2-}}_{Cd^{2+}} > CdSH_2^+ \right] + \left[{}^{S^{2-}}_{Cd^{2+}} > CdOH_2^+ \right] = \left[{}^{S^{2-}}_{Cd^{2+}} > CdS^- \right] + \left[{}^{S^{2-}}_{Cd^{2+}} > CdO^- \right]$$

$$(67)$$

A likely set of surface chemical reactions that take place upon illumination of metal sulfide particles under anoxic conditions are given below for the case of an aqueous suspension of CdS in the presence of dissolved ethanol [55–60]:

$$ {}^{S^{2-}}_{Cd^{2+}} > Cd(+II)S(-II)H_2^+ \xrightarrow{hv} \xrightarrow[h_{vb}^+]{\bar{e}_{cb}} {}^{S^{2-}}_{Cd^{2+}} > Cd(+I)S(-I)H_2^+$$

$$(68)$$

$$ {}^{S^{2-}}_{Cd^{2+}} > Cd(+I)S(-I)H_2^+ \xrightarrow{hv} \xrightarrow[h_{vb}^+]{\bar{e}_{cb}} {}^{S^{2-}}_{Cd^{2+}} > Cd(0)S(0)H_2^+ \quad (69)$$

$$ {}^{S^{2-}}_{Cd^{2+}} > Cd(0)S(0)H_2^+ \rightleftharpoons {}^{S^{2-}}_{Cd^{2+}} > Cd(+II)S(0)^+ + H_2 \quad (70)$$

$$ {}^{S^{2-}}_{Cd^{2+}} > Cd(+II)S(0)^+ + CH_3CH_2OH \rightleftharpoons$$

$$(71)$$

$$ {}^{S^{2-}}_{Cd^{2+}} > Cd(+II)S(-I)H^+ + CH_3\dot{C}HOH$$

$$ {}^{S^{2-}}_{Cd^{2+}} > Cd(+II)S(-I)H^+ + CH_3CH_2OH \rightleftharpoons$$

$$(72)$$

$$ {}^{S^{2-}}_{Cd^{2+}} > Cd(+II)S(-II)H_2^+ + CH_3\dot{C}HOH$$

$$2H\cdot \longrightarrow H_2 \quad (73)$$

$$2CH_3\dot{C}HOH \longrightarrow 2CH_3CHO + H_2 \quad (74)$$

Similar photoreactions can occur at neutral >CdSH and >CdOH sites and protonated surface sites involving >CdSH$_2^+$ and >CdOH$_2^+$.

$$\underset{Cd^{2+}}{S^{2-}} > Cd(+II)S(-II)H \xrightarrow{\ hv\ } \xrightarrow[h^+_{vb}]{\bar{e}_{cb}} \underset{Cd^{2+}}{S^{2-}} > Cd(+I)S(-I)H \quad (75)$$

$$\underset{Cd^{2+}}{S^{2-}} > Cd(+I)S(-I)H \xrightarrow{\ hv\ } \xrightarrow[h^+_{vb}]{\bar{e}_{cb}} \underset{Cd^{2+}}{S^{2-}} > Cd(0)S(0)H \quad (76)$$

$$\underset{Cd^{2+}}{S^{2-}} > CdOH_2^+ \xrightarrow{2\,\bar{e}_{cb}} \underset{Cd^{2+}}{S^{2-}} > CdO^- + H_2 \quad (77)$$

$$\underset{Cd^{2+}}{S^{2-}} > CdO^- + 2H^+ \xrightarrow{2\,\bar{e}_{cb}} \underset{Cd^{2+}}{S^{2-}} > CdOH_2^+ \quad (78)$$

$$\underset{Cd^{2+}}{S^{2-}} > CdOH \xrightarrow{h^+_{vb}} \underset{Cd^{2+}}{S^{2-}} > CdOH^{+\cdot} \xrightarrow{CH_3CH_2OH}$$
$$(79)$$
$$\underset{Cd^{2+}}{S^{2-}} > CdOH_2^+ + CH_3\dot{C}HOH$$

In the presence of oxygen (i.e., under oxic conditions), O_2 would immediately react with the carbon-centered ethanolic radical, $CH_3\dot{C}HOH$, to form the corresponding peroxy radical ($RO_2\cdot$), which is an alternative form of ROS.

Fullerene Photochemistry and ROS Generation Potential

Similar to the cases of the metal oxide and sulfide semiconductors, the photochemical properties of fullerenes can be viewed in the context of excitation of electrons across a bandgap. For example, the bandgap of pure C_{60} has been reported to be 2.3 eV, which is comparable to that of iron oxide polymorphs (Table 5.3). The bandgap for carbon nanotubes (CNTs) depends on its chirality and is inversely proportional to the diameter of the nanotube. In the case of ROS generation by fullerenes, they can act either as a photosensitizers or an electron shuttle.

Two distinct pathways are recognized for the photosensitization of fullerenes. Both pathways involve the initial excitation of the photosensitizing molecule (i.e., a fullerene). Type I sensitization involves electron transfer and depends upon the presence of a donor molecule that can reduce the triplet state of the sensitizer. The triplet state is more susceptible to electron donation than is the ground-state singlet molecule. In the presence of oxygen, superoxide radical anion can be formed by direct electron transfer from this excited radical to molecular oxygen.

The type II photosensitization pathway involves the transfer of excited spin-state energy from the sensitizer to another molecule. Type II sensitization does not depend upon the presence of a donor molecule, but only requires a long-lived triplet excited state. In the presence of oxygen, this triplet state is quenched by ground-state oxygen, which is transformed into singlet oxygen. As noted previously, singlet oxygen is a reactive oxygen species that can participate in reactions in solution that are spin-forbidden in the case of ground-state molecular oxygen. Singlet oxygen formation via type II photosensitization and quenching has been reviewed extensively by Wilkinson et al.

A typical photosensitizing molecule in the ground state is represented by S_0 with S_1 and T_1 representing the lowest energy singlet and triplet states, respectively. Figure 5.22 is a graphic representation of their main photosensitized pathways.

Light within the absorbance range of the photosensitizer is absorbed and promotes electrons into the excited singlet state (Eq. 80).

$$S_0 + h\nu \xrightarrow{k_w} S_1 \qquad (80)$$

The intensity and wavelength of the light will govern the rate of S_1 formation (k_W). The singlet state then decays via at least three different pathways: fluorescence, internal conversion, or intersystem crossing

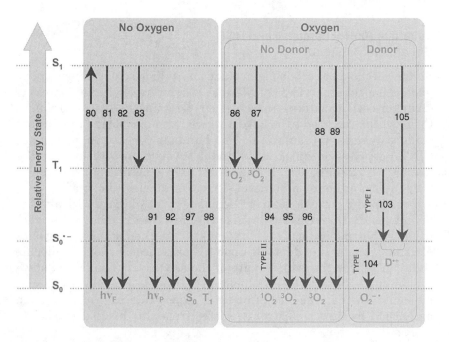

Figure 5.22   General photosensitization kinetic scheme.

(Eqs. 81–83). Intersystem crossing is a spontaneous nonradiative transition from the singlet to the triplet electronic state.

$$S_1 \xrightarrow{k_F} S_0 + hv_F \tag{81}$$

$$S_1 \xrightarrow{k_{ic}} S_0 \tag{82}$$

$$S_1 \xrightarrow{k_{isc}} T_1 \tag{83}$$

The rate of singlet decay is taken as the sum of these rates of singlet decay, k_{SD}, (Eq. 84) and the quantum yield, ϕ_T, is defined by the fraction of the singlet state that decays to the triplet state (Eq. 85).

$$k_{SD} = k_F + k_{ic} + k_{isc} = \frac{1}{\tau(S_1)} \tag{84}$$

$$\phi_T = \frac{k_{isc}}{k_{SD}} \tag{85}$$

The singlet form of a photosensitizer can also be quenched by oxygen to form the triplet state of the sensitizer, generating either singlet oxygen (Eq. 86) or ground-state oxygen (Eq. 87). It can also be quenched to the ground state (Eq. 88) or the sensitizer may be altered through a reaction to form a new product (Eq. 89).

$$S_1 + O_2 \xrightarrow{k_{S\Delta}^{O_2}} T_1 + {}^1O_2^* \tag{86}$$

$$S_1 + O_2 \xrightarrow{k_{isc}^{O_2}} T_1 + {}^3O_2 \tag{87}$$

$$S_1 + O_2 \xrightarrow{k_{Sd}^{O_2}} S_0 + {}^3O_2 \tag{88}$$

$$S_1 + O_2 \xrightarrow{k_{Sr}^{O_2}} \text{reaction product} \tag{89}$$

$$k_{SQ}^{O_2} = k_{S\Delta}^{O_2} + k_{isc}^{O_2} + k_{Sd}^{O_2} + k_{Sr}^{O_2} \tag{90}$$

The triplet state in turn decays by phosphorescence (Eq. 91) or intersystem crossing (Eq. 92). Phosphorescence releases light energy as electrons fall to a lower energy level, and internal conversion involves the same drop in energy without the associated energy release. The rate of triplet decay is the sum of these processes (Eq. 93).

$$T_1 \xrightarrow{k_{Tp}} S_0 + hv_p \tag{91}$$

$$T_1 \xrightarrow{k_{Td}} S_0 \tag{92}$$

$$k_{TD} = k_{Tp} + k_{Td} = \frac{1}{\tau(T_1)} \tag{93}$$

Photosensitizers in the triplet state can also be quenched by oxygen (Eqs. 94–97) or by sensitizers in the ground state. Quenching by sensitizer in the ground state is known quite simply as self-quenching (Eq. 97). Finally, a sensitizer in the triplet state may also be quenched when a triplet collides with another triplet in a process known as triplet-triplet annihilation (Eq. 98). The summation of these processes is represented in Eq. 99.

$$T_1 + O_2 \xrightarrow{k_{T\Delta}^{O_2}} S_0 + {}^1O_2^* \text{ (Type II reaction)} \tag{94}$$

$$T_1 + O_2 \xrightarrow{k_{Td}^{O_2}} S_0 + {}^3O_2 \tag{95}$$

$$T_1 + O_2 \xrightarrow{k_{Tr}^{O_2}} \text{product} \tag{96}$$

$$T_1 + S_0 \xrightarrow{k_{SQ}^{S_0}} 2S_0 \tag{97}$$

$$T_1 + T_1 \xrightarrow{k_{AN}^{T_1}} S_0 + T_1 \tag{98}$$

$$k_{TQ}^{O_2} = k_{T\Delta}^{O_2} + k_{Td}^{O_2} + k_{Tr}^{O_2} \tag{99}$$

Often the triplet state is more energetically stable than the singlet state, so the products of the above reactions are favored over those resulting from reactions with the singlet state in photosensitizing systems. The triplet state is quenched by oxygen by three different pathways; the fraction of triplet that participates in the Type II reaction to form singlet oxygen is represented by Eq. 100. Oxygen also competes with triplet-triplet annihilation and self-quenching as potential pathways for triplet quenching. Consequently, the proportion of the triplet state quenched with oxygen can be calculated using Eq. 101. The quantum yield for singlet oxygen expresses how efficiently the photons are used to produce singlet oxygen (Eq. 102).

$$f_\Delta^{II} = \frac{k_{T\Delta}^{O_2}}{k_{TQ}^{O_2}} \tag{100}$$

$$P_T^{O_2} = \left. k_{TQ}^{O_2}\left[O_2\right] \middle/ \left(k_{TD} + k_{TQ}^{O_2}\left[O_2\right] + k_{SQ}^{S_0}\left[S_0\right] + k_{AN}^{T_1}\left[T_1\right] \right) \right. \tag{101}$$

$$\phi_\Delta = \phi_T f_\Delta^{II} P_T^{O_2} \tag{102}$$

The nonproductive pathways regarding ROS production such as triplet quenching by oxygen along with triplet-triplet annihilation and self-quenching represent inefficiencies in converting light energy to chemical energy. Increasing the lifetime of the triplet state and minimizing the effect of the nonproductive pathways will result in higher quantum yields for ROS production; these pathways are controlled by the concentration and proximity of the photosensitizer in the solution.

Type I reactions can occur in parallel with type II reactions, when the photosensitizers are in the presence of electron donors. Type I reactions are initiated by the reduction of the triplet state by an electron donor (Eq. 103). The donor in this case has a reduction potential lower than either the ground state (S_0) or the excited state (T_1 or S_1) of the sensitizer (Eqs. 2 to 5). Once the reduction occurs, the sensitizer takes the form of a radical anion (S_0^-) that has the possibility of reducing oxygen to superoxide (Eq. 104) and subsequently returning to the ground state. Reduction of the singlet state is more thermodynamically favorable but is kinetically limited because of the short lifetime of the singlet excited state (Eq. 105). As a result, singlet-state reactions with the electron donor are not likely to occur.

$$T_1 + D \xrightarrow{k_{T_1}^D} S_0^- + D^{\cdot+} \tag{103}$$

$$S_0^- + O_2 \xrightarrow{k_{S_0^-}^{O_2}} O_2^- + S_0 \tag{104}$$

$$S_1 + D \xrightarrow{k_{S_1}^D} S_0^- + D^{\cdot+} \tag{105}$$

Thus, the proportion of oxygen reacting with the triplet-state sensitizer must be modified to express the reactions between the triplet state and the electron donor (Eq. 106).

$$P_T^{O_2} = \left. k_{TQ}^{O_2}[O_2] \middle/ \left(k_{TD} + k_{TQ}^{O_2}[O_2] + k_{SQ}^{S_0}[S_0] + k_{AN}^{T_1}[T_1] + k_{T_1}^D[D] \right) \right. \tag{106}$$

In order to determine the quantum yield for superoxide formation, the following assumptions are made: anion radical sensitizers react with oxygen to form superoxide (i.e., $f_{sup}^{II} = 1$); anion radicals are unlikely to be formed (Eq. 105) due to the short lifetime of S_1; and the donor reacts only with the triplet-state molecule, T_1, of Eq. 103.

Given these assumptions, the proportion of the triplet-state molecules that react with a donor is given by Eq. 107 with the corresponding quantum yield given by Eq. 108.

$$P_T^D = \left. k_{T_1}^D[D] \middle/ \left(k_{TD} + k_{TQ}^{O_2}[O_2] + k_{SQ}^{S_0}[S_0] + k_{AN}^{T_1}[T_1] + k_{T_1}^D[D] \right) \right. \tag{107}$$

$$\phi_{sup} = \phi_T P_T^D \qquad (108)$$

The overall quantum yield for the production of ROS ($^1O_2$ and O_2^{-}) in such a system can then be given as the sum of the quantum yields:

$$\phi_{ROS} = \phi_\Delta + \phi_{sup} \qquad (109)$$

Kinetically, the triplet state is a key intermediary in photosensitizing processes. The quantum yield for ROS (Eq. 109) cannot be maximized unless triplet decay pathways (k_{TD}), oxygen quenching that does not lead to singlet oxygen ($k_{TQ} - k_{T\Delta}^{O_2}$), self quenching ($k_{SQ}^{S_0}$), and triplet-triplet annihilation ($k_{AN}^{T_1}$) are minimized as potential pathways for removal of the triplet-state sensitizer. Based on these kinetic limitations, the ideal photosensitizer has three properties: the absorbance of low energy light to create the singlet excited state efficiently; preferred conversion of the singlet state to the triplet state due to intersystem crossing (Eq. 83); and a low occurrence of non-ROS forming triplet removal pathways. The nanomaterial class known as fullerenes holds great promise due to properties that correspond to each of these desirable traits.

ROS production by fullerenes

Carbon-based nanomaterials such as fullerenes have been known to be photoactive as photosensitizers from the first studies of their physical properties [61]. Fullerenes, and in particular C_{60}, have been studied intensively for applications in fields such as photodynamic therapy [62], photovoltaics [63], and materials [64].

An advantage of using fullerenes, and in particular C_{60}, as photosensitizers in an engineered system is that they are highly stable. For example, the carbon cage making up C_{60} appears to be nearly impervious to degradation by oxidation or susceptible to enzymatic attack. However, fullerenes may be modified in aqueous environments such as in the formation of epoxide derivatives of C_{60} in the presence of UV light [65] or on the surface of a metal oxides such as TiO_2 [66].

When fullerenes are illuminated under the appropriate wavelength, the electrons are excited from the ground state ($^0C_{60}$) to the singlet state (Eq. 80). The singlet state ($^1C_{60}$) can decay in three main manners: fluorescence (Eq. 81), internal conversion (Eq. 82), and intersystem crossing (ISC) (Eq. 83). The first two result in the ground state while the latter leads to the relaxation of singlet C_{60} to the triplet state ($^3C_{60}$). Interaction of the singlet state with oxygen can also result in the triplet state (Eqs. 86 and 87). Eq. 86 results in the production of singlet oxygen via type II photosensitization. The triplet state, $^3C_{60}$, has a significantly longer lifetime than $^1C_{60}$ in solution, allowing it to participate in type II formation of singlet oxygen to a greater extent than does the

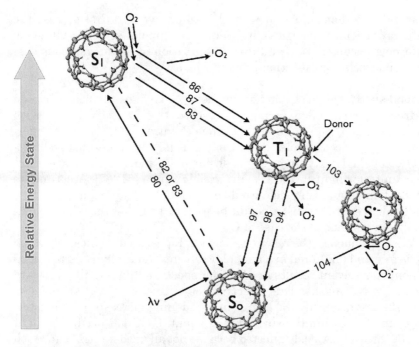

Figure 5.23   Major photosensitization pathways for C_{60}. Numbers *vide supra* and *vide infra* refer to the primary reaction pathways shown in this figure and Figure 5.22.

$^1C_{60}$ (Eq. 94). The triplet state is also susceptible to self-quenching (Eq. 97) via interaction with the ground state $(^0C_{60})$ and triplet-triplet annihilation (Eq. 98) via interaction with another triplet $(^3C_{60})$. Type I sensitization (Eq. 103) occurs when the triplet state comes in contact with a donor molecule that has a more negative reduction potential than $(^3C_{60})$. The resulting radical $(C_{60}^{\cdot-})$ can then pass the electron to ground-state oxygen to form superoxide (Eq. 104). An illustration of these main fullerene photosentization pathways is shown in Figure 5.23.

In general, conditions that lead to ROS generation can be grouped into four categories in which reaction pathways are most directly related and in many cases have been investigated together:

1. Ground-state Fullerene: Excitation and Decay (Eqs. 80, 81, 82)

2. Intersystem Crossing: Fullerene Triplet-state Formation (Eqs. 83, 86, 87)

3. Fullerene Triplet Quenching: Type II Photosensitization (Eqs. 97, 98, 94)

4. Fullerene Triplet Reduction: Type I Photosensitization (Eqs. 103, 104).

Each of these groups of reaction pathways are affected by changes in fullerene functionality and aggregation state, such as those produced to suspend fullerene in water. In each section, we will first discuss the pathway specific properties of C_{60} found in nonpolar solvents, following with how encapsulation and functionalization of C_{60} for aqueous suspension affects that reaction group pathway.

Ground-state fullerene: Excitation and decay. The absorbance spectrum of free C_{60} suspended in nonpolar solvents has sharp peaks with absorbance in the UV and visible range (Figure 5.24). This has important consequences because quantum yields for the photosensitized production of singlet oxygen by a suspension of C_{60} in nonpolar benzene are near unity for light in the UV and visible range [61, 67]. According to Eq. 85, triplet quantum yield would equal singlet oxygen quantum yield under ideal energy transfer conditions, indicating that the triplet quantum yield is unity in the case of C_{60}.

As a consequence, the reaction shown in Eq. 80 occurs efficiently when C_{60} is free and unaltered in a nonpolar solvent. Generally, C_{60} suspended in nonpolar solvent results in a broader spectrum that is shifted toward the red wavelengths. This effect varies with the type of suspension [68–72], but generally the degree of broadening increases with clustering of the C_{60} within the surrounding agent. C_{60} cages excited by light (Eq. 80) are deleteriously affected because not all incident light can reach the surface of the C_{60}. Fluorescence (Eq. 81) and internal conversion (Eq. 82) have not been identified as important contributors to the decay of the $^{1}C_{60}$ back to $^{0}C_{60}$. However, addends decrease triplet quantum yield by promoting non-triplet forming singlet decay pathways such as florescence (Eq. 81) and most likely internal conversion (Eq. 82) [73–75].

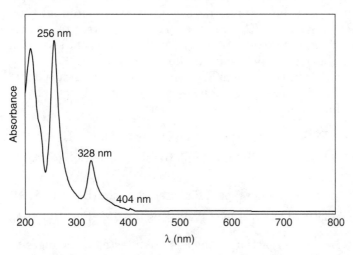

Figure 5.24 Typical UV/Vis absorbance of C_{60} suspended in nonpolar solvent.

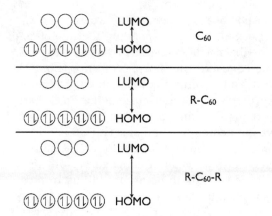

Figure 5.25 Increasing the amount of addends enlarges the gap between the HOMO and LUMO.

The cage structure can be altered by various chemical reactions that perturb the extended π-bonding network and subsequently raise the energy of the LUMO electrons due to loss of conjugation [76] (Figure 5.25). This increases the energy required to excite electrons across the band gap and into their excited states; thus requiring higher energy light (Eq. 80) and contributing to a reduction in quantum yield (Eq. 85) [73].

The photophysical properties of C_{70}, a higher order fullerene cage, are influenced by its structure, which may have an oblong shape in order to maintain the cagelike carbon structure. The singlet oxygen quantum yield (Eq. 102) (a measure of the lower limit of $^3C_{70}$ quantum yield [Eq. 85]) was found to be around 0.81, indicating light conversion was not as efficient in this molecule [67]. This decrease is partially attributed to deactivation of $^1C_{70}$ by alternative pathways such as internal conversion (Eq. 82) that do not produce the triplet state. Thus, it appears that either increasing the size of the fullerene cage or decreasing fullerene symmetry—or both—may lead to a decrease in quantum yield (Eq. 85).

Intersystem crossing: Fullerene triplet-state formation. The characteristic rates of ISC (k_{isc}) for C_{60} and C_{70} had been determined to be $3.0 \times 10^{10}\,s^{-1}$ and $8.7 \times 10^9\,s^{-1}$, respectively [77]. Decay of the singlet-excited C_{60} is predominantly ISC to the triplet state (Eq. 83) [61, 67]. This phenomenon can be explained in terms of small energy splitting between $^1C_{60}$ and $^3C_{60}$, low fluorescence (Eq. 81), and large spin-orbital interaction. The large diameter and spherical nature of C_{60} promote these properties by lowering electron repulsion and the extended π-bonding network, respectively. In C_{70}, the extended π-bonding network seems to be perturbed enough to promote internal conversion (Eq. 82) rather than ISC (Eq. 83); this results in reduced $^3C_{60}$ quantum yields (Eq. 85). Addends reduce the ISC rate (Eq. 83) in the same way by reducing the amount of π bonds on the surface of the C_{60} cage. Since the singlet state cannot be as easily relaxed, its lifetime is noticeably longer but still on the order of nanoseconds [78–80].

Fullerene triplet quenching: Type II photosensitization. After excitation of C_{60} to the triplet state via ISC (Eq. 83), the corresponding lifetime of the triplet state is microseconds in absence of quenching by oxygen [61, 77, 81]. However, the lifetime also depends on phosphorescence (Eq. 91), internal conversion (Eq. 92), self-quenching (Eq. 97), and triplet-triplet annihilation (Eq. 98). Therefore, the triplet lifetime depends on concentration of C_{60}. By measuring triplet lifetime at various concentrations, the effect of these alternative quenching mechanisms (Eqs. 97 and 98) can be eliminated from lifetime calculations, allowing for an estimation of the intrinsic triplet lifetime around 133 µs in nonpolar solvent [82]. More generally, the triplet lifetime tends to be around 40 µs in most studies involving nonpolar solvents because a single concentration is used to measure the triplet lifetime. These triplet lifetimes are exceptional, but along with type II photosensitization rates lifetimes change dramatically when C_{60} is suspended in the aqueous environment.

Encapsulating agents such as γ-cyclodextrin (γ-CD), Triton-X, and poly(vinylpyrrolidine) (PVP) increase the lifetime of the triplet state (Eq. 93) up to 130 µs regardless of concentration [68–70, 83]. This is likely due to the encapsulating agent's ability to reduce contact between fullerenes making self-quenching (Eq. 97) and triplet-triplet annihilation (Eq. 98) less frequent (i.e., γ-CD encapsulation reduced triplet-triplet annihilation by four times as compared with free C_{60} in toluene [69]). Photosensitization rates benefit from the increased lifetime of the triplet state, but the same encapsulation effects reduce the rate of type II singlet oxygen formation (Eq. 94). In the case of γ-CD, the triplet quenching by oxygen was determined to be half that of free C_{60} in toluene after correction for oxygen diffusion rates [69]. In addition, illuminated PVP/C_{60} was monitored for the characteristic singlet oxygen emissions band at 1270 nm. However, the IR emission was not observed [68] and thus, it was concluded that type II sensitization (Eq. 94) was not occurring or that it was taking place at a very low rate.

A more specific example, "mechanically entrapped C_{60}," recently has been developed for C_{60} suspension [84]. The carbon cage is entrapped in an all-silica zeolite Y supercage (Figure 5.26).

In the absence of O_2, the triplet lifetime (Eq. 93) is extended into the order of minutes. Presumably, this is due to the complete lack of quenching mechanism activity and indicates the C_{60} molecules are more than likely entrapped individually and not as clusters. Otherwise, triplet-triplet annihilation (Eq. 98) and self-quenching (Eq. 97) would lower the triplet lifetime significantly. Despite this, in the presence of oxygen, the type II (Eq. 94) mechanism occurs effectively but at a slight rate decrease from the diffusion-controlled quenching of free C_{60}.

Increasing the number of addends on the C_{60} cage decreases the quantum yield of the triplet state (Eq. 85) and, as a result, the quantum yield

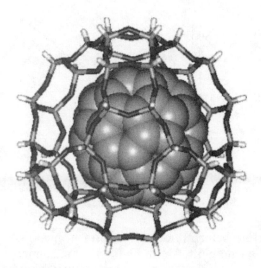

Figure 5.26 A molecular model representation of C_{60} trapped in a zeolite cage [84].

of singlet oxygen (Eq. 102) [85]. The fused core area can be correlated to the triplet C_{60} and singlet oxygen quantum yields (Figures 5.27 and 5.28) [86].

However, since the ratio of quantum yields ($\phi\Delta/\phi_T$) is approximately unity for all different functional groups, the nature of the addend does not hinder the ability of oxygen to quench the $^3C_{60}$ via a type II mechanism (Eq. 94) [85, 86].

All the photosensitization properties of functionalized fullerene were first measured in nonpolar solvent in order to accurately gauge the induced changes with increasing addends. Nonetheless, these modifications were originally intended for increased water solubility

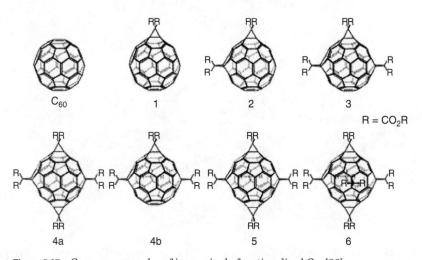

Figure 5.27 Common examples of increasingly functionalized C_{60} [86].

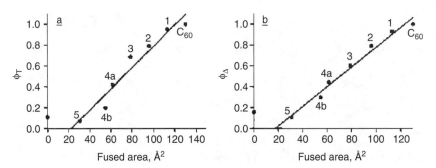

Figure 5.28 $^3C_{60}$ quantum yield and singlet oxygen quantum yield correlated with functionalized cage area [86].

so it was only a matter of time before photosensitizing properties were tested in the aqueous environment. An immediate consequence of placing functionalized C_{60} in the aqueous system is clustering. Monofunctionalized malonic acid C_{60} ($C_{60}C(COOH)_2$) is a good example. The nonpolar ends of these molecules are thought to group together facing the negatively charged carboxyl groups out into the polar environment [71] (Figure 5.29).

Clustering reduces the lifetime of the triplet-state two orders of magnitude [71] by promoting triplet-triplet annihilation (Eq. 98) and self-quenching (Eq. 97). Therefore, the triplet-excited state does not survive long enough to participate in photosensitization reactions. In order to alleviate this problem γ-CD can be added to cap the exposed nonpolar ends and reduce clustering. This increases triplet lifetime comparable to that observed for γ-CD encapsulated C_{60} [69].

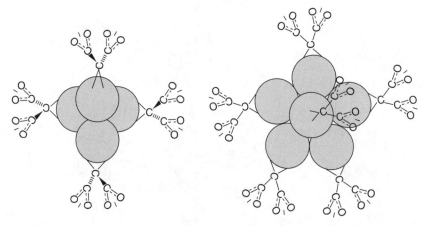

Figure 5.29 Drawing of possible cluster formations of monofunctionalized malonic acid C_{60} derivatives in the aqueous environment [71].

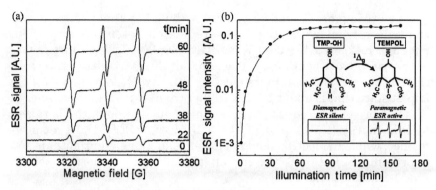

Figure 5.30 EPR signal growth of the TEMP-singlet oxygen adduct in the presence of fullerol suspended in an aqueous system.

Type II quenching by oxygen (Eq. 94) for the encapsulated mono-functionalized compounds has a rate constant on the same order of magnitude as that of γ-CD encapsulated C_{60} [71]. However, bi-functionalized, tri-functionalized, and poly-functionalized compounds can be suspended in the aqueous environment without the use of an encapsulating agent (Figure 5.27 provides examples of multi-functionalized compounds). Bi-functional and tri-functional C_{60} singlet oxygen formation rates can be compared with free C_{60} and they show only a slight slowdown in singlet oxygen formation, whereas the poly-functionalized fullerol ($C_{60}(OH)_x$) is about an order of magnitude slower. The slowdown is attributed to the extremely perturbed π bonding system caused by the addition of hydroxyl groups [71].

Suspensions of fullerol have been observed to exhibit two distinct triplet lifetimes. A shorter time constant for triplet decay appears to be concentration dependent, while triplet lifetime simultaneously decays with a longer time constant that is concentration independent [87]. Annihilation (Eq. 98) and self-quenching (Eq. 97) play an important role in the shorter decaying component, while the longer decaying component is probably associated with the presence of individually suspended fullerol. EPR spin-trapping methods can be used to monitor the type II formation of singlet oxygen (Eq. 94), and the singlet oxygen signal is both time and concentration dependent in the presence of UV illumination (Figure 5.30) [3, 88].

Fullerol exhibits lower singlet oxygen quantum yields due to its perturbed π-bond system, but the hydroxyl groups increase the triplet lifetime by reducing cage contact so that the molecule can participate in type II reactions in water.

Fullerene triplet reduction: Type I photosensitization.   In the presence of appropriate donor compounds, the high electron affinity of C_{60} results in type I reactions. C_{60} can be reduced with up to five electrons in benzonitrile solvent with progressive reduction potentials of (-0.36, -0.83, -1.42, -2.01, -2.60 V vs. SCE) [89, 90]. This affinity is due to the extended π-bonding that can spread the extra electrons across the surface. In addition to its electron affinity, C_{60} also has a stable triplet state that is about 1.56 eV higher than $^0C_{60}$. The higher energy $^3C_{60}$ is more easily reduced because the reduction potential is raised by this energy (1.56 V $-$ 0.42 V = 1.14 V vs. SCE) [61]. As a consequence, when an electron donor of lower reduction potential than $^3C_{60}$ is present, excitation to the triplet state plays an important role in type I reactions (Eq. 103).

The electron transfer capabilities of γ-CD encapsulated C_{60} as opposed to free C_{60} in propan-2-ol can be compared in terms of their bi-molecular rate constants (Eq. 103). Interestingly enough, the rate constant is about a factor of 2 slower in the encapsulating agent [72]. This is consistent with the rate of oxygen quenching by γ-CD/C_{60} [69]. Similar C_{60} micellular suspensions formed with the non-ionic surfactant Triton-X 100 form monomeric or colloidal suspensions of C_{60} depending on the preparation method. The bi-molecular rate constant for reduction by a donor (Eq. 103) is three orders of magnitude less than the free C_{60} in toluene [83]. Independent measurement confirms that triton X encapsulation slows reduction by one order of magnitude compared with γ-CD [70]. The inability of the donor molecule to approach the surface of C_{60} is likely due to steric and charge repulsion effects [72]. PVP is another encapsulating agent that has been used extensively to suspend C_{60} in aqueous solution at concentrations of up to 400 mg/L [68]. In the presence of adenosine 5'-(trihydrogen diphosphate) (NADH), a C_{60} suspension has been shown to damage DNA; concurrently EPR and NBT detection confirms superoxide formation via type I reaction (Eq. 104) but at reduced rates from free C_{60} [9, 14, 91]. As noted previously for type II reactions, encapsulation represents a tradeoff between triplet lifetime (Eq. 93) and quantum yield of type I (Eq. 108) reactions.

As discussed earlier, the LUMO increases with the addition of addends, and because C_{60} is fully occupied in the HOMO, a reducing electron must jump a larger and larger gap in order to complete the reduction. This translates into increasingly more negative reduction potentials that drop about 0.1 to 0.15 V for each additional addend (Figure 5.31) [76, 78–90, 92]. Concurrently, the triplet energy increases with the addition of addends (Figure 5.32) [86]. However, this energy increase is not as dramatic as the decrease in reduction potential, and upon the summation of these two effects a net decrease in reduction potential for the triplet state of the increasingly functionalized C_{60} cage occurs. As $^3C_{60}$ is increasingly functionalized it takes on electrons less readily than nonfunctionalized $^3C_{60}$ (Figure 5.33).

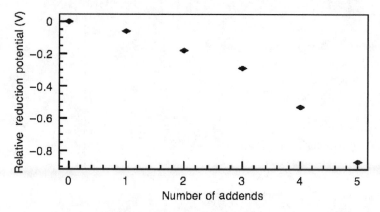

Figure 5.31 Reduction potential of increasing number of functional groups relative to the ground state C_{60}.

However, the energy of $^3C_{60}$ is approximately 1.5 V higher than C_{60} and as a result all functionalized triplet states remain easier to reduce than ground state C_{60} despite the negative trend in their reduction potentials.

As a result, the change in reduction potential for monofunctionalized/encapsulated forms of C_{60} have lower rates of reduction than the unfunctionalized/encapsulated C_{60} [80]. The type of addend does seem to make a significant difference. Cages with positively charged amino functional groups are more susceptible to reduction than the negatively charged carboxyl groups.

In the case of fullerol, the reduction potential is estimated between -0.358 and -0.465 V vs. NHE (-0.600 and -0.707 vs. SCE), indicating the reduction potential decrease with addition of addends (Figure 5.33) has a limit. Fullerol has been shown to produce type I (Eq. 103) superoxide in

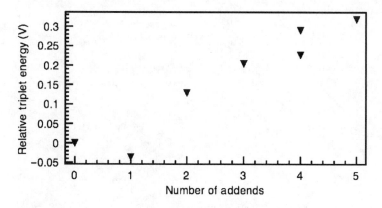

Figure 5.32 Relative triplet energy with increasingly functionalized C_{60} cage.

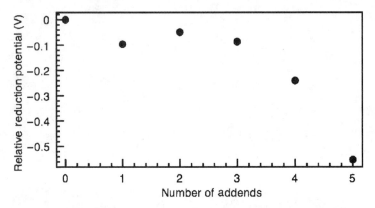

Figure 5.33 Relative reduction potential of increasingly functionalized $^3C_{60}$.

the presence of UV light by selectively quenching with azide. Reduction of triplet-state fullerol is possible in neutral aqueous solution because electron approach is not hampered by the presence of hydroxyl groups, but acidity may influence the photochemical reaction via increased donor access to the surface at low pH due to protonation of surface hydroxyl groups [13].

Dark "type I" pathway.   In some cases, a donor may be present in solution with ground-state C_{60} or a derivative thereof. If the donor has a lower reduction potential than the C_{60} (-0.42 V vs. SCE), there is no need for light to excite the fullerene into the higher energy triplet state. This can be categorized as a dark type I reaction (Figure 5.34). In this case, the donor can reduce ground-state C_{60}. Furthermore, the product $C_{60}^{\cdot-}$ radical anion can transfer its electron to molecular oxygen with the formation of superoxide.

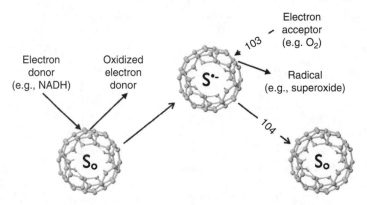

Figure 5.34 Dark "type I" reaction reduction pathway.

References

1. Zepp, R. G.; Braun, A. M.; Hoigne, J.; Leenheer, J. A. *Environmental Science & Technology* 1987, *21*, 485.
2. Zepp, R. G.; Schlotzhauer, P. F.; Sink, R. M. *Environmental Science & Technology* 1985, *19*, 74.
3. Vileno, B.; Lekka, M.; Sienkiewicz, A.; Marcoux, P.; Kulik, A. J.; Kasas, S.; Catsicas, S.; Graczyk, A.; Forro, L. *Journal of Physics-Condensed Matter* 2005, *17*, S1471.
4. Halliwell, B.; Gutteridge, J. M. C. *Free Radicals in Biology and Medicine*; Oxford University Press: Oxford, England, 1999; Vol. 3rd Ed.
5. Frejaville, C.; Karoui, H.; Tuccio, B.; Lemoigne, F.; Culcasi, M.; Pietri, S.; Lauricella, R.; Tordo, P. *Journal of Medicinal Chemistry* 1995, *38*, 258.
6. McCord, J. M.; Fridovic, I. *Journal of Biological Chemistry* 1968, *243*, 5753.
7. McCord, J. M.; Fridovic, I. *Journal of Biological Chemistry* 1970, *245*, 1374.
8. Bielski, B. H. J.; Shiue, G. G.; Bajuk, S. *Journal of Physical Chemistry* 1980, *84*, 830.
9. Yamakoshi, Y.; Umezawa, N.; Ryu, A.; Arakane, K.; Miyata, N.; Goda, Y.; Masumizu, T.; Nagano, T. *Journal of the American Chemical Society* 2003, *125*, 12803.
10. Saito, I.; Matsuura, T.; Inoue, K. *Journal of the American Chemical Society* 1983, *105*, 3200.
11. Haag, W. R.; Hoigne, J. *Environmental Science & Technology* 1986, *20*, 341.
12. Schmidt, R. *Journal of Physical Chemistry A* 2004, *108*, 5509.
13. Pickering, K. D.; Wiesner, M. R. *Environmental Science & Technology* 2005, *39*, 1359.
14. Nakanishi, I.; Ohkubo, K.; Fujita, S.; Fukuzumi, S.; Konishi, T.; Fujitsuka, M.; Ito, O.; Miyata, N. *Journal of the Chemical Society-Perkin Transactions* 2002, 2, 1829.
15. Hoffmann, M. R.; Martin, S. T.; Choi, W. Y.; Bahnemann, D. W. *Chemical Reviews* 1995, *95*, 69.
16. Bahnemann, D. W.; Kormann, C.; Hoffmann, M. R. *Journal of Physical Chemistry* 1987, *91*, 3789.
17. Carraway, E. R.; Hoffman, A. J.; Hoffmann, M. R. *Environmental Science & Technology* 1994, *28*, 786.
18. Choi, W.; Termin, A.; Hoffmann, M. R. *Angewandte Chemie-International Edition in English* 1994, *33*, 1091.
19. Choi, W. Y.; Hoffmann, M. R. *Environmental Science & Technology* 1995, *29*, 1646.
20. Choi, W. Y.; Hoffmann, M. R. *Journal of Physical Chemistry* 1996, *100*, 2161.
21. Choi, W. Y.; Hoffmann, M. R. *Environmental Science & Technology* 1997, *31*, 89.
22. Choi, W. Y.; Termin, A.; Hoffmann, M. R. *Journal of Physical Chemistry* 1994, *98*, 13669.
23. Choi, W. Y.; Termin, A.; Hoffmann, M. R. *Angewandte Chemie-International Edition in English* 1994, *33*, 1091.
24. Kormann, C.; Bahnemann, D. W.; Hoffmann, M. R. *Journal of Physical Chemistry* 1988, *92*, 5196.
25. Kormann, C.; Bahnemann, D. W.; Hoffmann, M. R. *Environmental Science & Technology* 1988, *22*, 798.
26. Kormann, C.; Bahnemann, D. W.; Hoffmann, M. R. *Journal of Photochemistry and Photobiology a-Chemistry* 1989, *48*, 161.
27. Kormann, C.; Bahnemann, D. W.; Hoffmann, M. R. *Environmental Science & Technology* 1991, *25*, 494.
28. Faust, B. C.; Hoffmann, M. R.; Bahnemann, D. W. *Journal of Physical Chemistry* 1989, *93*, 6371.
29. Szczepankiewicz, S. H.; Moss, J. A.; Hoffmann, M. R. *Journal of Physical Chemistry B* 2002, *106*, 7654.
30. Szczepankiewicz, S. H.; Moss, J. A.; Hoffmann, M. R. *Journal of Physical Chemistry B* 2002, *106*, 2922.
31. Szczepankiewicz, S. H.; Colussi, A. J.; Hoffmann, M. R. *Journal of Physical Chemistry B* 2000, *104*, 9842.
32. Martin, S. T.; Herrmann, H.; Choi, W. Y.; Hoffmann, M. R. *Journal of the Chemical Society-Faraday Transactions* 1994, *90*, 3315.
33. Martin, S. T.; Herrmann, H.; Hoffmann, M. R. *Journal of the Chemical Society-Faraday Transactions* 1994, *90*, 3323.

34. Herrmann, H.; Martin, S. T.; Hoffmann, M. R. *Journal of Physical Chemistry* 1995, *99*, 16641.
35. Hoffman, A. J.; Mills, G.; Yee, H.; Hoffmann, M. R. *Journal of Physical Chemistry* 1992, *96*, 5546.
36. Hoffman, A. J.; Carraway, E. R.; Hoffmann, M. R. *Environmental Science & Technology* 1994, *28*, 776.
37. Hong, A. P.; Bahnemann, D. W.; Hoffmann, M. R. *Journal of Physical Chemistry* 1987, *91*, 2109.
38. Hong, A. P.; Bahnemann, D. W.; Hoffmann, M. R. *Journal of Physical Chemistry* 1987, *91*, 6245.
39. Hoffman, A. J.; Yee, H.; Mills, G.; Hoffmann, M. R. *Journal of Physical Chemistry* 1992, *96*, 5540.
40. Cornu, C. J. G.; Colussi, A. J.; Hoffmann, M. R. *Journal of Physical Chemistry B* 2001, *105*, 1351.
41. Cornu, C. J. G.; Colussi, A. J.; Hoffmann, M. R. *Journal of Physical Chemistry B* 2003, *107*, 3156.
42. Frank, S. N.; Bard, A. J. *Journal of Physical Chemistry* 1977, *81*, 1484.
43. Leland, J. K.; Bard, A. J. *Journal of Physical Chemistry* 1987, *91*, 5076.
44. Siefert, R. L.; Pehkonen, S. O.; Erel, Y.; Hoffmann, M. R. *Geochimica Et Cosmochimica Acta* 1994, *58*, 3271.
45. Erel, Y.; Pehkonen, S. O.; Hoffmann, M. R. *Journal of Geophysical Research-Atmospheres* 1993, *98*, 18423.
46. Pehkonen, S. O.; Siefert, R.; Erel, Y.; Webb, S.; Hoffmann, M. R. *Environmental Science & Technology* 1993, *27*, 2056.
47. Davis, A. P.; Huang, C. P. *Langmuir* 1990, *6*, 857.
48. Davis, A. P.; Huang, C. P. *Water Research* 1991, *25*, 1273.
49. Davis, A. P.; Huang, C. P. *Langmuir* 1991, *7*, 709.
50. Davis, A. P.; Hsieh, Y. H.; Huang, C. P. *Chemosphere* 1994, *28*, 663.
51. Hsieh, Y. H.; Huang, C. P. *Colloids and Surfaces* 1991, *53*, 275.
52. Hsieh, Y. H.; Huang, C. P.; Davis, A. P. *Chemosphere* 1992, *24*, 281.
53. Hsieh, Y. H.; Huang, C. P.; Davis, A. P. *Chemosphere* 1993, *27*, 721.
54. Park, S. W.; Huang, C. P. *Journal of Colloid and Interface Science* 1987, *117*, 431.
55. Borgarello, E.; Kalyanasundaram, K.; Graetzel, M.; Pelizzetti, E. *Helvetica Chimica Acta* 1982, *65*, 243.
56. Buehler, N.; Meier, K.; Reber, J. F. *Journal of Physical Chemistry* 1984, *88*, 3261.
57. Darwent, J. R.; Porter, G. *Journal of the Chemical Society, Chemical Communications* 1981, 145.
58. De, G. C.; Roy, A. M.; Bhattacharya, S. S. *International Journal of Hydrogen Energy* 1995, *20*, 127.
59. De, G. C.; Roy, A. M.; Bhattacharya, S. S. *International Journal of Hydrogen Energy* 1995, *20*, 127.
60. De, G. C.; Roy, A. M. *Journal of Surface Science and Technology* 1999, *15*, 147.
61. Arbogast, J. W.; Darmanyan, A. P.; Foote, C. S.; Rubin, Y.; Diederich, F. N.; Alvarez, M. M.; Anz, S. J.; Whetten, R. L. *Journal of Physical Chemistry* 1991, *95*, 11.
62. Wang, S. Z.; Gao, R. M.; Zhou, F. M.; Selke, M. *Journal of Materials Chemistry* 2004, *14*, 487.
63. Kamat, P. V.; Haria, M.; Hotchandani, S. *Journal of Physical Chemistry B* 2004, *108*, 5166.
64. Echegoyen, L.; Echegoyen, L. E. *Accounts of Chemical Research* 1998, *31*, 593.
65. Creegan, K. M.; Robbins, J. L.; Robbins, W. K.; Millar, J. M.; Sherwood, R. D.; Tindall, P. J.; Cox, D. M.; Smith, A. B.; McCauley, J. P.; Jones, D. R.; Gallagher, R. T. *Journal of the American Chemical Society* 1992, *114*, 1103.
66. Ziolkowski, L.; Vinodgopal, K.; Kamat, P. V. *Langmuir* 1997, *13*, 3124.
67. Arbogast, J. W.; Foote, C. S. *Journal of the American Chemical Society* 1991, *113*, 8886.
68. Yamakoshi, Y. N.; Yagami, T.; Fukuhara, K.; Sueyoshi, S.; Miyata, N. *Journal of the Chemical Society, Chemical Communications* 1994, 517.
69. Andersson, T.; Nilsson, K.; Sundahl, M.; Westman, G.; Wennerstrom, O. *Journal of the Chemical Society, Chemical Communications* 1992, 604.

70. Guldi, D. M.; Huie, R. E.; Neta, P.; Hungerbuhler, H.; Asmus, K. D. *Chemical Physics Letters* 1994, *223*, 511.
71. Guldi, D. M.; Hungerbuhler, H.; Asmus, K. D. *Journal of Physical Chemistry* 1995, *99*, 13487.
72. Hungerbuhler, H.; Guldi, D. M.; Asmus, K. D. *Journal of the American Chemical Society* 1993, *115*, 3386.
73. Anderson, J. L.; An, Y. Z.; Rubin, Y.; Foote, C. S. *Journal of the American Chemical Society* 1994, *116*, 9763.
74. Sayes, C. M.; Fortner, J. D.; Guo, W.; Lyon, D.; Boyd, A. M.; Ausman, K. D.; Tao, Y. J.; Sitharaman, B.; Wilson, L. J.; Hughes, J. B.; West, J. L.; Colvin, V. L. *Nano Letters* 2004, *4*, 1881.
75. Bensasson, R. V.; Bienvenue, E.; Janot, J. M.; Leach, S.; Seta, P.; Schuster, D. I.; Wilson, S. R.; Zhao, H. *Chemical Physics Letters* 1995, *245*, 566.
76. Boudon, C.; Gisselbrecht, J. P.; Gross, M.; Isaacs, L.; Anderson, H. L.; Faust, R.; Diederich, F. *Helvetica Chimica Acta* 1995, *78*, 1334.
77. Wasielewski, M. R.; Oneil, M. P.; Lykke, K. R.; Pellin, M. J.; Gruen, D. M. *Journal of the American Chemical Society* 1991, *113*, 2774.
78. Guldi, D. M. *Journal of Physical Chemistry A* 1997, *101*, 3895.
79. Guldi, D. M.; Maggini, M. *Gazzetta Chimica Italiana* 1997, *127*, 779.
80. Guldi, D. M.; Asmus, K. D. *Journal of Physical Chemistry A* 1997, *101*, 1472.
81. Haufler, R. E.; Wang, L. S.; Chibante, L. P. F.; Jin, C. M.; Conceicao, J.; Chai, Y.; Smalley, R. E. *Chemical Physics Letters* 1991, *179*, 449.
82. Fraelich, M. R.; Weisman, R. B. *Journal of Physical Chemistry* 1993, *97*, 11145.
83. Eastoe, J.; Crooks, E. R.; Beeby, A.; Heenan, R. K. *Chemical Physics Letters* 1995, *245*, 571.
84. Galletero, M. S.; Garcia, H.; Bourdelande, J. L. *Chemical Physics Letters* 2003, *370*, 829.
85. Hamano, T.; Okuda, K.; Mashino, T.; Hirobe, M.; Arakane, K.; Ryu, A.; Mashiko, S.; Nagano, T. *Chemical Communications* 1997, 21.
86. Prat, F.; Stackow, R.; Bernstein, R.; Qian, W. Y.; Rubin, Y.; Foote, C. S. *Journal of Physical Chemistry A* 1999, *103*, 7230.
87. Mohan, H.; Palit, D. K.; Mittal, J. P.; Chiang, L. Y.; Asmus, K. D.; Guldi, D. M. *Journal of the Chemical Society-Faraday Transactions* 1998, *94*, 359.
88. Vileno, B.; Marcoux, P. R.; Lekka, M.; Sienkiewicz, A.; Feher, I.; Forro, L. *Advanced Functional Materials* 2006, *16*, 120.
89. Dubois, D.; Kadish, K.; Flanagan, S.; Haufler, R. E.; Chibante, L. P. F.; Wilson, L. J. *Journal of the American Chemical Society* 1991, *113*, 4364.
90. Dubois, D.; Kadish, K. M.; Flanagan, S.; Wilson, L. J. *Journal of the American Chemical Society* 1991, *113*, 7773.
91. Yamakoshi, Y.; Sueyoshi, S.; Fukuhara, K.; Miyata, N. *Journal of the American Chemical Society* 1998, *120*, 12363.
92. Guldi, D. M.; Prato, M. *Accounts of Chemical Research* 2000, *33*, 695.

Principles and Procedures to Assess Nanomaterial Toxicity

Michael Kovochich *University of California, Los Angeles, California*

Tian Xia *University of California, Los Angeles, California*

Jimmy Xu *Brown University, Providence, Rhode Island*

Joanne I. Yeh *University of Pittsburgh School of Medicine, Pittsburgh, Pennsylvania*

Andre E. Nel *University of California, Los Angeles, California*

Introduction

By some estimates, nanotechnology promises to far exceed the impact of the Industrial Revolution and is projected to develop into a $1 trillion market by 2015. Manufactured nanomaterials (NM, see list of abbreviations at the end of this chapter) are already being used in sporting goods, tires, stain-resistant clothing, sunscreens, cosmetics, and electronics, and they will also be increasingly used in medicine for purposes of diagnosis, imaging, and drug delivery. The unique physico-chemical properties of engineered NM are attributable to their small size, large surface area, durability, chemical composition, crystallinity, electronic properties, surface reactivity, surface groups, surface coatings, solubility, shape, and aggregation. Although impressive from a physico-chemical viewpoint, the novel properties of NM raise the possibility that they could interact with and cause damage to biological components or systems. Indeed, a number of studies have suggested that not all NM are benign and some have the ability to cause adverse biological effects at cellular, subcellular, and molecular levels [1–12]. These potentially harmful effects could be enhanced by the ability of nanoparticles to be taken up and travel through the body, deposit in target organs, penetrate cell membranes, lodge in mitochondria, and trigger injurious responses.

Against this background, it should be our goal to develop appropriate methods to assess the safety of NM and, by so doing, help to safeguard the future of nanotechnology without the fear of negative public perception, government overregulation, and potentially costly litigation. It is essential that we adopt standardized test methods to assess NM safety and to generate an online databank that is accessible to all users and producers of NM. In this chapter, we propose an approach to the assessment of NM toxicity that uses a test paradigm proven useful for studying the toxicity of ambient air particles. This approach attempts to predict which NM are dangerous based on the material characteristics that predispose them to ROS production and the generation of oxidative stress. We propose that this test paradigm be developed into a high throughput screening system that can be used to predict NM toxicity *in vivo*.

Paradigms for Assessing NM Toxicity

Air pollution and mineral dust particles have been implicated in a number of adverse biological effects and disease outcomes. Major disease outcomes include the exacerbation of airway inflammation, asthma, interstitial pulmonary fibrosis, atherosclerosis, ischemic cardiovascular events, and cardiac arrhythmias. Fortunately, no clinically recognizable disease outcomes have so far been reported for manufactured NM to date. Although a household-cleaning product Magic Nano was recently implicated in respiratory symptoms, closer investigation failed to reveal a link to NM. However, we are just entering the nano-revolution and it is quite possible that clinically relevant NM toxicity could emerge. While such outcomes will no doubt launch intensive investigations into NM toxicity, a retroactive approach could be disastrous in terms of public perception and possibly harm the nanotechnology industry. It makes far more sense to instigate preventative measures to avert such a disaster.

Is it possible to formulate a preemptive approach to the potential danger(s) of NM? In our opinion the answer is yes, since the potential mechanism(s) of injury can be studied by a science-based approach. One of the key mechanisms by which ambient particulate matter (PM) causes tissue injury and cardiopulmonary disease is through the generation of the reactive oxygen species (ROS) and oxidative stress. Since the oxidative stress paradigm has evolved into a comprehensive disease model, it illustrates the type of approach that could be used to develop a predictive paradigm for the NM toxicity testing.

Biological systems are generally able to integrate multiple pathways of injury into a limited number of pathological outcomes, including

TABLE 6.1   NM Effects as the Basis for Pathophysiology and Toxicity

Experimental NM Effects	Possible Pathophysiological Outcomes
ROS generation*	Protein, DNA & membrane injury,* oxidative stress[†]
Oxidative stress*	Phase II enzyme induction, inflammation[†], mitochondrial perturbation*
Mitochondrial perturbation*	Inner membrane damage,* PTP opening,* energy failure,* apoptosis,* apo-necrosis, cytotoxicity
Inflammation*	Tissue infiltration with inflammatory cells[†], fibrosis[†], granulomas[†], atherogenesis[†], acute phase protein expression (e.g., C-reactive protein)
Uptake by reticulo-endothelial system*	Asymptomatic sequestration and storage in liver,* spleen, lymph nodes[†], possible organ enlargement and dysfunction
Protein denaturation, degradation*	Loss of enzyme activity,* auto-antigenicity
Nuclear uptake*	DNA damage, nucleoprotein clumping,* autoantigens
Uptake in neuronal tissue*	Brain and peripheral nervous system injury
Perturbation of phagocytic function,* "particle overload," mediator release*	Chronic inflammation[†], fibrosis[†], granulomas[†], interference in clearance of infectious agents[†]
Endothelial dysfunction, effects on blood clotting*	Atherogenesis,* thrombosis,* stroke, myocardial infarction
Generation of neo-antigens, breakdown in immune tolerance	Autoimmunity, adjuvant effects
Altered cell cycle regulation	Proliferation, cell cycle arrest, senescence
DNA damage	Mutagenesis, metaplasia, carcinogenesis

Adapted from [11].
*Limited experimental evidence
[†]Limited clinical evidence

inflammation, apoptosis, necrosis, fibrosis, hypertrophy, metaplasia, and carcinogenesis (Table 6.1). Although oxidative stress feeds into most of these outcomes, it is important to mention that it is by no means the only injury mechanism that will cause such pathological outcomes. Other forms of injury include the disruption of biological membranes, protein denaturation, DNA damage, immune reactivity, and the formation of foreign body granulomas. In fact, one of the first biological interactions that take place when a nanoparticle penetrates or enters a tissue is contact with the surface membrane of the target cell. This interaction can lead to membrane damage based on particle properties such as hydrophobicity, cationic charge, or detergent activity that allow the particle contact, penetration, or disruption of membrane integrity. The cell may respond by leakage of intracellular content, intracellular Ca^{2+} release, and the induction of apoptosis. To mention but one example, some cationic dendrimers are capable of disrupting cell membranes by being able to pull off lipid molecules, leading to the formation of

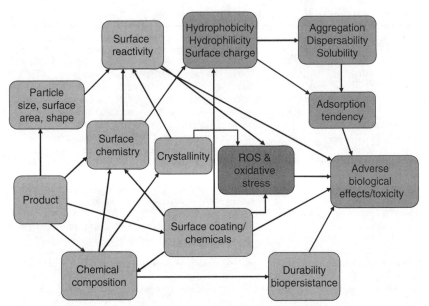

Figure 6.1 Particle-Related Characteristics Module. Nanoparticle physico-chemical characterization can be divided into modules to clarify how primary, secondary, and tertiary particle characteristics may lead to adverse biological effects. Adapted from [42].

membrane defects. This leads to the release of cellular enzymes that can be assayed for [13]. Protein denaturation or degradation at the organic/inorganic interphase can lead to functional or structural changes in proteins at the primary, secondary, or tertiary level; this could manifest as interference in enzyme function or exposure to antigenic sites [14]. This damage may result from splitting of covalent bonds that are responsible for protein structure, such as disulfide bonds (Figure 6.1). There is also some evidence that certain nanoparticle characteristics facilitate cellular uptake and access to the nucleus, where DNA damage can result [15, 16]. In addition to this list of potential NM injuries, it is conceivable that new material properties may emerge that could lead to a new mechanism of toxicity.

Oxidative stress as a predictive paradigm for ambient ultrafine particle toxicity

Although the manufacture of NM is a relatively new scientific development, particle toxicity in response to ambient PM or mineral dust particles is a mature science. Animal exposure to these particles has shown linkage of pulmonary inflammation to the ability of the particles to induce ROS production and biologically relevant levels of oxidative stress [17, 18]. More recently, PM exposure has also been linked to systemic inflammatory effects in the cardiovascular system; this could be due to either the ability of ambient PM to induce pulmonary inflammation or the ability of the ultrafine particles (aerodynamic diameter <100 nm) to gain access to the systemic circulation. Cellular studies have generally supported the role of oxidative stress, pro-inflammatory cytokines, and programmed cell death as relevant mechanisms of PM injury [19, 20]. Ambient ultrafine particles have a higher pro-oxidative potential than ambient particles of larger size [19].

In contrast to the heterogeneous characteristics of ambient PM, manufactured NM are more homogeneous in shape, size, and form. In spite of these differences, research into ambient PM has helped to establish a number of principles that can be used to study NM toxicity. These include recognition that small particle size, large surface area, chemical composition, and ability to catalyze ROS production are important properties that determine PM-induced oxidant injury and inflammation [6,7]. Additional NM properties may contribute to ROS generation and oxidant injury and will be discussed throughout this chapter.

Oxidative stress refers to a state in which cellular GSH is depleted while oxidized glutathione (GSSG) accumulates. Under normal coupling conditions, ROS are generated at low frequency, mostly in mitochondria, and are easily neutralized by antioxidant defense mechanisms such as the glutathione (GSH)/glutathione disulfide (GSSG) redox couple. Ambient nanoparticles can elicit further ROS production in mitochondria, in addition to ROS generation by catalytic conversion pathways and NADPH oxidase activation [19, 21]. PM-induced ROS production is dependent on the particles themselves, as well as the redox cycling organic chemicals and transition metals that coat the particle surface [19, 21]. In addition to intrinsic redox cycling capabilities, the metabolic transformation of these chemicals and their ability to elicit intracellular calcium flux, disrupt electron flow in the mitochondrial inner membrane, perturb the permeability transition pore, and deplete cellular GSH content could contribute to cellular ROS generation. While small amounts of ROS could be buffered by the antioxidant defense pathways in the cell, excess amounts of ROS could lead to a drop in the GSH/GSSG ratio. This elicits additional cellular responses.

Only a limited number of manufactured NM have so far been shown to exert toxicity in tissue culture and animal experiments, and usually at high doses. A recent study shows that the biological response of BV2 microglia to noncytotoxic concentrations of TiO_2 induced a rapid (<5 minutes) and sustained (120 minutes) release of reactive oxygen species [22]. The kinetics of ROS production suggests that TiO_2 stimulates immediate oxidative burst activity in microglia and can also interfere in mitochondrial energy production. Carbon nanostructures represent another material type that has undergone some toxicity testing [11]. Water-soluble, monodisperse or colloidal fullerene aggregates induce O_2-anions, lipid peroxidation in cells and tissue as well as the ability to affect GSH depletion and cytotoxicity [23]. *In vitro* incubation of keratinocytes and bronchial epithelial cells with relatively high doses of single-wall carbon nanotubes (SWNT) results in ROS generation, lipid peroxidation, oxidative stress, mitochondrial dysfunction, and changes in cell morphology [24,25]. MWNT also elicits pro-inflammatory effects in keratinocytes [26].

Oxidative stress elicits quantifiable cellular responses that can be used to study NM toxicity

Cells respond to oxidative stress by mounting a number of protective and injurious responses that, depending on the stress level, can lead to a protected or an injurious outcome [7, 11]. The protective responses are elicited by even minor changes in the cellular redox equilibrium. This sensitivity is rooted in the behavior of the cap 'n' collar transcription factor, Nrf-2, that operates on the antioxidant response element (ARE) in the promoter of > 200 phase II enzymes [27]. These phase II enzymes exert antioxidant, detoxification and anti-inflammatory effects that are responsible for preventing or slowing the effects of oxidant injury [27]. Examples of phase II enzymes include HO-1, glutathione-S-transferase, NADPH quinone oxidoreductase, catalase, superoxide dismutase, glutathione peroxidase, and UDP-glucoronosyltransferase [27, 28]. A reduced or compromised phase II response promotes susceptibility to oxidant injury. In studies conducted in epithelial cells and macrophages, phase II enzyme expression is the most sensitive oxidative response parameter and has also been referred to as Tier 1 of the hierarchical oxidative stress response [29–31]. A higher level of oxidative stress can lead to a Tier 2 response, which is characterized by pro-inflammatory effects that follow the activation of intracellular signaling cascades, including the MAP kinase and NF-κB signaling cascades [30–32]. These signaling cascades are responsible for the activation of a number of cytokines, chemokines, and adhesion molecules that play a role in local and systemic inflammatory responses [32–34]. The dynamic equilibrium between the protective (Tier 1) and pro-inflammatory responses (Tier 2) determines the outcome of the oxidative stress response and the likelihood that this will lead to injury.

Tier 3 responses involve mitochondrial perturbation, Ca^{2+} flux, and activation of apoptosis pathways [21, 30, 32, 35]. Although the *in vivo* significance of the mitochondrial pathway is still under investigation, it has been demonstrated in tissue culture cells that ambient PM interferes in mitochondrial electron transfer, thereby leading to dissipation of the mitochondrial membrane potential, increased ROS production, and the induction of programmed cell death [30, 32, 35]. The basis for the mitochondrial response may be direct oxidant injury by free radicals, as well as the effect of free ionized calcium, which acts as an intracellular pathway by which oxidative stress can indirectly impact mitochondrial function. Research on ambient UFP also reveals the interesting possibility that the mitochondrion could be directly targeted by nanoparticles [19, 21, 36]. Ambient UFP lodge in the damaged mitochondria of exposed macrophages and epithelial cells [19, 21, 37]. Other nano-sized materials have been demonstrated to target mitochondria. As early as 1970, de Lorenzo found that colloidal gold particles (50 nm) delivered intra-nasally in the squirrel monkey cross the olfactory nerve/mitral cell synapse and have the capability to lodge in mitochondria of the mitral cells [38]. A fullerene derivative, $C_{61}(CO_2H)_2$, has been shown capable of crossing the surface membrane with preferential localization in mitochondria [39]. The same observation was made with block copolymers, which are water-soluble biocompatible nanocontainers that can be used for drug delivery. Fluorescent-labeled block copolymer micelles localize in a variety of cytoplasmic organelles, including mitochondria [40]. In summary, mitochondrial targeting and damage could constitute an important mechanism of NM toxicity. Changes in mitochondrial membrane potential, calcium uptake, O_2^- production, cardiolipin integrity, and induction of cellular apoptosis represent testable responses.

Need for standardized materials

Well-characterized NM that have undergone rigorous biological testing to show reproducible biological effects are required as standard reference materials to compare the effects of newly introduced NM. These benchmark materials will help to prevent the discrepancy and paradoxical findings that arise from the study of NM types in different hands. The choice of such standards should be based on the physicochemical properties of the material, frequency of use, volume of production, and likelihood of release as a singlet substance to which humans and the environment may be exposed. Carbon black is a bulk-manufactured material that is in widespread use, including as a powder that can be inhaled. In a highly purified form, carbon black is incapable of ROS generation and devoid of cytotoxicity [41]. In their unadulterated form these could serve as a benchmark material that does not engage in ROS production and oxidant injury. TiO_2 nanoparticles, also widely used and bulk-produced, can be considered as a representative material capable of ROS production under abiotic conditions [41]. Carbon nanotubes could be used as a NM that, due to their large aspect ratios, could act as biopersistent fibers. A fullerene could also be considered for the development as a carbon-based reference standard NM. No doubt as we undertake a more deliberate exploration of NM properties that lead to adverse biological events, we may be able to develop reference materials with predictive toxicity. The establishment of benchmark standards will rapidly expand our knowledge of NM toxicity.

Overall Considerations in the Assessment of NM Toxicity

A rational approach to NM toxicity begins with an assessment of the physico-chemical properties of the materials that allow them to interact with and possibly damage biological systems. Another set of tests involves *in vitro* cellular assays that reflect the response of a number of different cell types that may be targeted at the portal of entry or systemic sites after NM uptake and spread. Finally, any test strategy should include appropriate use of animal models, potentially the most difficult assay to consider. While some experts regard animal testing as obligatory, it is important to understand that this mode of experimentation is costly and time-consuming. In our opinion, animal testing should be done judiciously and possibly reserved for those materials that show a toxic potential in the *in vitro* studies. Moreover, where such testing is used, it is likely to be more informative if used in conjunction with a known mechanism of injury rather than a descriptive approach. This aspect will be discussed later.

The decision about an appropriate test strategy for NM should consider cost and time. A sobering thought is that among the 80,000 chemicals that are currently registered for commercial use in the US, only slightly more than 500 have undergone long-term and only 70 short-term testing by the National Toxicology Program (NTP). Moreover, the resource-intensive nature of these studies puts the cost of each bioassay at $2–4 million, and tests generally take more than three years to complete. Thus, the test strategy must be pragmatic. While it may be optimal to conduct comprehensive *in vitro* and *in vivo* toxicological assessment in some instances, it is probably more practical to develop *in vitro* decision matrices that avoid unnecessary *in vivo* testing, such as when *in vitro* testing fails to show any evidence of toxicity. Accordingly, the construction of a database that is premised on an *in vitro* predictive test result could help to determine under what circumstances *in vivo* testing should be performed.

Predictive assays should ideally be premised on a mechanism of injury that is directly related to *in vivo* toxicological or pathological outcomes [11]. Elucidation of NM properties that are responsible for ROS generation, with the possibility of generating oxidative stress responses *in vitro* and *in vivo*, is an example of a predictive test paradigm. This is also compatible with the preferred NTP approach to chemical toxicity, that is, using predictive scientific models that focus on target-specific, mechanism-based biological outcomes rather than descriptive approaches (http://ntp-server.niehs.nih.gov). We propose that the same guidelines be followed for engineered NM.

Physico-chemical characterization of NM

Characterization should be done at the time of NM administration as well as at the conclusion of biological studies. This would allow one to interpret the physico-chemical changes that take place in the presence of proteins, surfactants, or biological fluids. Any test paradigm should attempt to characterize the test material with respect to size distribution, chemical composition, surface area, crystallinity, electronic properties, shape, inorganic/organic coatings, hydrophobicity, and aggregation (Table 6.2). The possible relationship of these physical and chemical properties to biological outcomes could be premised on an interactive model such as shown in Figure 6.1. Particle size, shape, surface area, chemical composition, and surface coatings are primary material characteristics that are often provided by the manufacturer. These primary characteristics, which are usually acquired under dry conditions, determine intermediary material properties such as surface reactivity, catalytic properties, crystallinity, and biopersistance (Figure 6.1). Surface chemistry and surface coating/chemicals could, in turn, determine the hydrophobicity and hydrophilicity of the particles, their surface charge in aqueous solution,

TABLE 6.2 Nanomaterial Characterization

Parameters	Methods
Size distribution (primary particles)	TEM, SEM, XRD
Shape	TEM, SEM
Surface area	BET
Composition	Mass spectrometry, spectroscopy (UV, Vis, Raman, IR, NMR)
Hydrophobicity	MATH
Surface charge—suspension/solution	Zeta potential
Crystal structure	TEM, XRD
Agglomeration state	TEM, SEM, DLS
Porosity	MIP
Heterogeneity	TEM, SEM, spectroscopy
ROS generation capacity	DTT, FFA assay, nanosensors

Adapted from [12].
TEM, transmission electron microscopy; SEM, scanning electron microscopy; XRD, X-ray diffraction; BET, Brunauer, Emmett and Teller; MATH, microbial adhesion to hydrocarbons; DLS, dynamic light scattering; MIP, mercury intrusion porosimeter; FFA, furfuryl alcohol

as well as their state of aggregation, dispersability, and solubility. These properties under aqueous conditions play important roles in particle interactions with cells, binding to the cell membrane, cellular uptake, and subcellular distribution. A sufficient number of particles need to be taken up to induce adverse biological effects. Material composition, reactive surface groups/chemicals, and crystallinity are key determinants in the ability of nanoparticles to generate ROS and oxidant injury [42].

Cell-free assays to determine ROS production

There are numerous assays that can be performed to test the inherent properties of NM to produce ROS under abiotic conditions (Table 6.3). Electron spin resonance (ESR) detects unpaired electrons in any given sample. Ascorbate and spin-trapping agents such as 5,5-dimethyl-1-pyrroline-N-oxide (DMPO) can be used to detect free oxygen radicals [43]. ESR assessment is usually performed at room temperature with a quartz flat cell and a spectrometer that records the ESR spectra. The spectra are characterized by the shape of the absorption curve, the position of the resonance field, the line width, and the area under the absorption curve. Software programs, such as EPRWare, can help to superimpose and quantify the ratio of superoxide and hydroxyl radicals on their DMPO spin adducts, thereby allowing one to calculate the free radical concentration [43].

The dithiothreitol (DTT) assay can be used to measure the presence of redox cycling organic chemicals such as quinones on the surface of ambient ultrafine particles [44]. Quinones are capable of capturing

TABLE 6.3   Hierarchical Oxidative Stress Responses

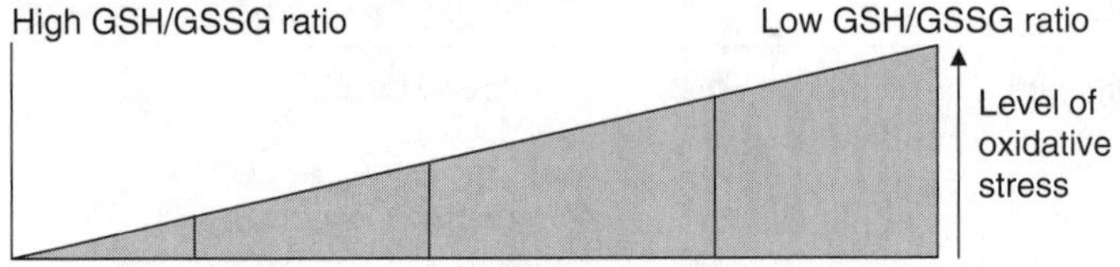

Methods	Normal	Tier 1	Tier 2	Tier 3
		Cell defense	Pro-inflammation	Mitochondria effects, apoptosis
in vitro		HO-1 WB	Cytokine ELISA	Mitochondrial $\Delta\Psi$m (DiOC$_6$)
		Phase II enzymes	Phospho-JNK WB	$[Ca^{2+}]_i$: Fluo-3; $[Ca^{2+}]_m$: Rhod-2
		(Real-time PCR)		Apoptosis: Annexin V/PI
		Nrf-2 WB		Caspase 3 activation
				DNA fragmentation(BrdU-FITC)
in vivo		HO-1 Luc Mice	BAL cytokines	Cell damage (LDH)
			Inflammatory cells	Immunohistochemistry
			Histology	(active caspase-3)

(Graph above table: "High GSH/GSSG ratio" at left, "Low GSH/GSSG ratio" at right, with arrow labeled "Level of oxidative stress")

Adapted from [31].

electrons from DTT to form metastable semiquinones that, in turn, can transfer these electrons to molecular oxygen to form the superoxide radical [45]. We have successfully implemented this assay to study the ability of ambient ultrafine particles, which are collected at different sites in the Los Angeles Basin, to generate ROS [19].

ROS production by NM can also be quantitated by using the ROS quencher, furfuryl alcohol (FFA) [41, 46]. Furfuryl alcohol chemically reacts with ROS with a reaction rate constant three orders of magnitude higher than the rate of physical quenching. Because FFA functions as an ROS quencher, the quantities of ROS that are being produced by the particles can be measured as the decrease in dissolved oxygen, corrected for an appropriate blank sample. Results are plotted as the log of the ratio of the instantaneous to initial concentrations of oxygen measured over time [41].

Unfortunately this indirect method of ROS measurement does not distinguish between individual oxygen radicals [41, 46]. This can be accomplished by ESR or nanobiosensor technology.

All things considered, cell-free systems reflect the intrinsic abilities of NM to generate oxygen radicals. It is important to remark, however, that these tests do not automatically lead to cellular ROS production or the ability of the NM to induce oxidant injury or toxicity. For that to happen, the material needs to be taken up by the target cell and must overcome the antioxidant defenses that are capable of removing ROS or restoring redox equilibrium [41]. It is also possible that coating of the NM surface with biological components such as proteins may passivate the material surface, leading to decreased ROS generation. Some materials may lack the ability to generate ROS but could do so biologically due to their ability to perturb mitochondrial electron transduction, induce the assembly of NADPH oxidase, or engage metabolic pathways that lead to ROS production [41]. Thus, although useful for predicting the intrinsic capabilities of NM to produce ROS, the acellular tests need to be interpreted in the context of the biological response outcome.

Predictive *In Vitro* Toxicological Assays in Tissue Culture Cells that Are Premised On the Hierarchical Oxidative Stress Paradigm

In vitro assays for NM toxicity should consider the portal-of-entry as well as possible systemic cellular targets (Table 6.4). Different NM may necessitate different test strategies depending on the exposure risk. With this sense of physiological relevance in mind, we can select cell type, dosage, and endpoints according to the demands of the situation. For instance, if a particular NM is found in skin-care products, it would be more logical to study its effects on keratinocytes. If, on the other hand, the NM is being produced as singlet particles that can be easily

TABLE 6.4 *In vitro* Systems Defined by Portal of Entry/Potential Target Organs

Portal of Entry/Organs	Cell/Tissue Type	Cellular Response/Pathology
Lung	Epithelium	Toxicity, inflammation, translocation, carcinogenesis
	Macrophages	Toxicity, chemotaxis, phagocytosis, inflammation
Skin	Keratinocytes	Cytotoxicity, inflammation
Mucosa	Buccal/intestinal epithelium	Cytotoxicity, inflammation, translocation
Cardiovascular system	Endothelial cells (e.g., HUVEC)	Cytotoxicity, homeostasis, translocation
Blood	Red blood cells, platelets, bone marrow (megakaryocytes)	Inflammation/immune response
Liver	Hepatocytes	Toxicity
	Kupffer cells	Inflammation, coagulation
Spleen	Lymphocytes	Immune response
Central and peripheral nervous system	Neuronal cells	Toxicity, inflammation, translocation
	Astroglial, microglial cells	Inflammation
Heart	Cardiomyocytes	Toxicity, inflammation, function
Kidney	Cell (e.g., HK-2, MDCK, LCC-PK1)	Toxicity, inflammation, translocation

Adapted from [12].

spread as a powder or a dust, it would be more appropriate to test its effects in bronchial and alveolar epithelial cells and/or pulmonary alveolar macrophages. Thus, the closer the test scenario is to the real-life exposure conditions, the better the predictive value of the test. The predictive value is further enhanced by using a test strategy that reflects a mechanism of injury, in particular, if that mechanism also leads to *in vivo* pathology. Since the hierarchical oxidative stress paradigm has proven useful to understand the generation of lung and cardiovascular injury by ambient ultrafine particles, we will briefly discuss this approach for *in vitro* NM testing.

As an initial step, we recommend rapid screening procedures such as the MTT and LDH assays. The MTT [3-(4,5-dimethylthiazol-2-yl)-2,5-diphenyltetrazolium bromide] assay, also known as the MTT cellular proliferation assay, is based on the ability of dehydrogenase enzymes in viable cells to cleave the tetrazolium rings of pale yellow MTT to form dark blue formazan crystals that are impermeable to the surface membrane of viable cells [47]. This oxidation takes place in cells with active reductases and therefore reflects cell viability. The lactate dehydrogenase (LDH) assay measures LDH release after damage to the surface membrane. This colorimetric assay is dependent on the ability of this enzyme to reduce a tetrazolium salt to water-soluble formazan, which can be measured by absorption at 492 nm [47]. The major advantage of these screening assays is the ease with which they can be performed and the ability to compare a large number of NM simultaneously.

ROS generation in cells can be detected by fluorescent dyes such as dichlorofluorescein-diacetate (DCFH-DA) and MitoSOX™ Red [41]. DCFH-DA is a nonpolar compound that readily diffuses into cells, where it is hydrolyzed to the nonfluorescent polar derivative, DCFH, which is trapped in the cell. In the presence of a H_2O_2 and hydroxyl (OH) radicals, DCFH is oxidatively modified into a highly fluorescent derivative, 2,7-dichlorofluorescein (DCF). DCF fluorescence can be detected in cells by a flow cytometry procedure. MitoSOX Red is a novel fluorogenic indicator offering direct measurement of superoxide (O_2^-) production in live cells. This cell-permeant dye is rapidly and selectively concentrated in mitochondria, where oxidation by O_2^- but not other oxygen or nitrogen species leads to the formation of a fluorescent product that interacts with nucleic acids. MitoSOX Red fluorescence is also detectable by flow cytometry. Both flow cytometry procedures can be adapted to study the kinetics of ROS generation as well as dose response relationships.

Whether a biological response follows cellular ROS production is dependent on the magnitude of the response, as well as the antioxidant buffering capacity [11]. If the antioxidant defense capability of the cell is overwhelmed, excess ROS production leads to cellular damage as well as GSH depletion. The GSH content of the cell is an important determinant of redox equilibrium and if perturbed could elicit further cellular responses. Cellular thiol levels can be assessed with a thiol-interactive fluorescent dye, monobromobimane (MBB), using a flow cytometry procedure [41]. Since GSH makes the biggest contribution towards the thiol content of the cell, this assay is helpful in predicting whether NM elicit oxidative stress.

According to the hierarchical oxidative stress hypothesis, cells respond to even minimal levels of oxidative stress with a protective antioxidant response [31]. This pathway is dependent on transcriptional activation of phase II gene promoters by the transcription factor, Nrf2 [27]. Heme oxygenase 1 (HO-1) is a prime example of a phase II enzyme that mediates antioxidant, anti-inflammatory, and cytoprotective effects, and it is useful as a marker for particle-induced oxidative stress effects [28]. HO-1 and phase II enzyme expression can be assessed by immunoblotting, real-time PCR, or reporter gene assays that reflect the activation of the antioxidant response element by Nrf2 [27, 28]. The value of these assays is the sensitivity of the phase II response to oxidative stress. It is possible to discern phase II enzyme expression at minimal levels of oxidative stress that are not necessarily related to oxidant injury. Not only is phase II enzyme expression indicative of the ability of the cell to defend against oxidative stress, but it also acts as a bio-marker that reflects subtle levels of oxidative stress that might otherwise be overlooked. Thus, the presence of phase II responses could alert one to the possibility that the material elicits more harmful oxidative stress effects, when, for instance, a higher material dose is delivered or when exposure takes place in a cell type that lacks the ability to synthesize its own GSH, such as neuronal cells.

When antioxidant defenses fail to restore redox equilibrium, escalation of the level of oxidative stress can lead to cellular injury [11, 33]. One mechanism of injury is the activation of pro-inflammatory cascades [11, 33]. The Jun kinase (JNK) and NF-κB cascades are redox-sensitive signaling cascades that are capable of inducing the expression of pro-inflammatory cytokines and chemokines. The activation of these cascades can be detected by a variety of techniques, including IκB immunoblotting, phosphopeptide immunoblotting, flow cytometry assays, and electrophoretic mobility shift assays (EMSA) [32]. The expression of pro-inflammatory cytokines (e.g., TNF-α, IL-8, IL-6, and GM-CSF) can be detected by enzyme-linked sorbent assays or real-time PCR. Adhesion molecule expression (e.g., VCAM-1 and ICAM-1) can be discerned by flow cytometry as well as real-time PCR. The utility of these assays is the ease with which they can be performed on a number of samples. It is also possible to combine the individual pro-inflammatory response markers into multiplex assays that are capable of measuring several cytokines or chemokines simultaneously. Many of the same inflammation markers play a role in the pathogenesis of PM-induced disease *in vivo* and are also detectable in body fluids and tissues, e.g., bronchoalveolar lavage fluid.

High levels of oxidative stress can induce changes in the intracellular free calcium concentration, $[Ca^{2+}]_i$, or may perturb Ca^{2+} compartmentalization in the cell, with the potential to induce toxicity [48]. The endoplasmic reticulum and mitochondria play important roles in buffering sudden increases in $[Ca^{2+}]_i$. ROS participate in this increase through inhibition of the sarco/endoplasmic reticulum Ca^{2+} ATPase (SERCA) or by inducing inositol 1,3,5-trisphosphate release [48, 49]. This, in turn, can lead to increased mitochondrial Ca^{2+} levels, $[Ca^{2+}]_m$. Saturation of the mitochondrial Ca^{2+} buffering capacity triggers further mitochondrial responses that can produce additional ROS production and more cellular damage. The mitochondrial permeability transition pore (PTP) is a Ca^{2+}-regulated channel that can be triggered by a persistent increase in $[Ca^{2+}]_m$ [50]. Large-scale PTP opening leads to cytochrome c release and apoptosis [50]. $[Ca^{2+}]_i$ and $[Ca^{2+}]_m$ levels can be followed by using fluorescent dyes such as Fluo-3 or Rhod-2 and flow cytometry [41]. These assays can be performed on several samples simultaneously. Their biological significance is the elucidation of cellular responses that result in cell death.

Toxic oxidative stress can be studied by following changes in mitochondrial function or structural integrity. We have demonstrated that the induction of oxidative stress by ambient UFP or manufactured NP induces the dissipation of the mitochondrial membrane potential ($\Delta\Psi$m), large-scale opening of the PTP, and the release of pro-apoptotic factors [41]. A change in the $\Delta\Psi$m can be followed by the use of the cationic dye, $DiOC_6$, which is highly concentrated in the negatively charged mitochondrial matrix. Dissipation of the $\Delta\Psi$m leads to a decrease in $DiOC_6$ fluorescence that is also quantifiable by flow cytometry [41]. Apoptosis can be assessed by detecting annexin V and propidium iodide (PI) fluorescence [36, 41]. Annexin V is a calcium-dependent phospholipid binding protein with high affinity for phosphatidylserine (PS), a membrane phospholipid component normally localized in the inner layer of the cell membrane. During early apoptosis, PS translocates to the outer surface of the cell membrane where this phospholipid becomes accessible to annexin V binding. At a later stage of apoptosis, the loss of membrane integrity allows PI to be taken up in the nucleus where its presence can be followed by flow cytometry. More formal demonstration of mitochondrial damage can be accomplished by ultrastructural studies and confocal microscopy, which are time-consuming and expensive methods that may be better used as a research tool.

Use of Cellular Assays to Study Other Responses that Are Relevant to NM Toxicity, Including Cellular Uptake and Subcellular Localization

A detailed investigation of NM toxicity should assess NM uptake and subcellular localization (Figure 6.1). The most common method to visualize NM uptake is transmission electron microscopy (TEM). This approach is relatively expensive and time-consuming. Fluorescent-labeled NM can help to determine cellular uptake and localization by using fluorescent or confocal microscopy. Using fluorescent dyes that localize in the cell membrane, mitochondria, endoplasmic reticulum, nucleus, Golgi apparatus, or endocytic vesicles can help to further refine the subcellular localization of the NM. It is important to demonstrate that labeling procedures do not alter the size, state of aggregation, charge, hydrophobicity, or other physico-chemical characteristics of the NM being tested.

In vivo testing

While *in vitro* techniques allow one to focus on specific biological and mechanistic pathways that can be studied under controlled conditions, the ultimate proof of NM toxicity necessitates the performance of *in vivo* studies. *In vivo* studies traditionally require a major route of exposure (inhalation, dermal uptake, ingestion, or systemic injection) that reflects how the material is produced or used in the workplace and/or marketplace. The ideal approach would be to use an animal model in which a disease process or pathological event is linked to a mechanism of injury that can be followed by cellular studies. If toxicity does occur *in vivo*, this could act as the departure point to perform more detailed evaluation of NM dosimetry, toxicokinetics, and toxicodynamics. These are expensive and labor-intensive studies that should be used under select circumstances, for example, materials that show definitive *in vitro* toxicity as well as *in vivo* pathological outcomes. Finally, a number of animal models are available to allow one to study mechanistic pathways by live imaging procedures [51]. These animal models are further discussed below.

While comprehensive coverage of *in vivo* study methods falls outside the scope of this chapter, suffice it to mention that the ideal approach would be to use animal models that address disease outcome in terms of a specific mechanism of injury. Rats and mice are currently the best animal models for which an extensive database has been developed to assess chemical toxicity. For particle exposures, the best protocols are for pulmonary exposure, using inhalation or intratracheal instillation methods. Inhalation is more physiologically relevant and the preferred approach for hazard identification and generating dose-response data. However, intratracheal instillation or pharyngeal aspiration are acceptable procedures to gather proof-of-principal evidence for an *in vivo* toxicological outcome. Relevant endpoints are the assessment of airway and interstitial inflammation, as well as the assessment of oxidative stress markers in the lung by bronchoalveolar lavage and histopathology. It is also possible to assess cardiovascular pathology by inhalation exposure, such as assessing the effect of ambient PM on atherosclerosis, blood clotting, and the generation of cardiac arrhythmias [52]. The performance of these cardiovascular and pulmonary studies can be further enhanced by susceptible animal models, for example, allergen-sensitized mice to perform asthma studies, apoE knockout animals to conduct atherosclerosis studies, or animals impaired in one or more aspects of antioxidant defense to show accelerated disease development either in the lung or cardiovascular system. The approach to the performance and assessment of dermal and gastrointestinal toxicity is described elsewhere (see reference 12).

In vivo imaging could be used to study the hierarchical oxidative stress paradigm. One example is the use of a transgenic mouse model in which the HO-1 promoter has been linked to a luciferase reporter [51]. The HO-l-luc Tg mouse was developed by Dr. Christopher Contag at Stanford University by injecting a full-length HO-1 promoter-luciferase construct into FVB mouse pronuclei [51, 53–55]. Transgenic pups were screened by bioluminescent imaging and by HO-1 luc-specific PCR.

A transgenic line was established that provides a rapid and noninvasive way to characterize *in vivo* HO-1 expression in response to systemic agents. Agents that elicit oxidative stress (e.g., $CdCl_2$, heme, and other metalloporphyrins) have been used in these mice for *in vivo* assays [53, 54]. A variety of HO-1 expression profiles have been demonstrated based on the stimulus, method of administration, and time since administration [54]. For instance, intravenous administration of $CdCl_2$ results in a luciferase signal originating primarily from the liver, while peritoneal administration of heme yields a response in the liver, spleen, and other abdominal sites [55]. Both $CdCl_2$ and heme induce this response based on their abilities to generate ROS [55]. We have successfully tested the ROS generating ability of ambient UFP using this animal model and believe it could be helpful for launching nanotoxicology studies. Similar imaging models exist to study the activation of the NF-κB signaling cascade.

Nanosensors: Sensitive Probes for the Biodetection of ROS

In the context of testing, a nanobiosensor can involve a biological molecule that serves as a detector, linker, or mediator, and nanoelectrodes. Various components can be equated with the electronic elements of a sensor, as every component has to transduce the signal generated at the source (biomolecule) to the detector (electrode) (Figure 6.2). Consequently,

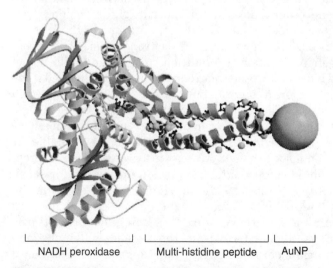

| NADH peroxidase | Multi-histidine peptide | AuNP |

Figure 6.2 Nanobiosensor assembly based on the atomic coordinates of the NADH peroxidase and MHP (multi-histidine peptide). The peptide coordinates cobalts (small sphere) through histidine residues at every i, i + 4 positions. AuNP was modeled in as a sphere, to scale with the biological molecules, with a diameter of 14 Å.

as in enzyme systems, rate improvements can occur from proximity and geometric effects, with potential enhancements of 10^2 to 10^3 at each junction. Additional advantages arise from the dimensionality at which detection is conducted. In nanoscale structures, electrons no longer behave like physical objects that flow in a continuous stream but take on wave mechanical and quantum properties and have the ability to tunnel through structures that would ordinarily be insulators. As single molecule measurements become more feasible with the advent of methods sensitive enough to study single molecule kinetics, thermodynamic, and electronics, significant deviations from ensemble measurements have been found. With the removal of ensemble averaging, distributions and fluctuations of molecular properties can be characterized, transient intermediates identified, and catalytic mechanisms elucidated. To facilitate single molecule measurements, nanoelectrode platforms have been investigated as nanosensors for enhancing detection.

We have produced nanobiosensors utilizing various redox enzymes aligned on nanoelectrode arrays [56]. One of the systems is comprised of the enzyme, NADH peroxidase, as the specific detector of hydrogen peroxide, and converts a biological binding event into an electronic signal [57, 58]. These results demonstrate the use of biosensors to investigate the ability of nanoparticles to change the redox status of the cell, as could happen due to the ability of these materials to induce ROS species such as H_2O_2 and O_2^-. Although this system used an oxidative metabolism enzyme, other redox proteins, such as glucose oxidase, can be substituted as the bioelement. The detection event in these redox enzyme systems is based on generation of electrons as one of the products of an endogenous reaction. In addition to the traditional use of redox enzymes in biosensors, other nonredox proteins can be used if the binding of a ligand triggers a conformational change that can be detected by an induced electronic event or via optical, thermal, or other detectable physical changes. Alternatively, a virion or particle can theoretically be the bioelement of a sensor, as structural information is available for many of these macromolecules. Our strategy integrates desirable properties of the individual components: the protein machinery for sensitivity and specificity of binding, peptide chemistry for aligning the various electron transducing units, and the nanoelectrodes for gain sensitivity in electronic detection (Figure 6.2). Using these NADH peroxidase biosensors has allowed us to detect the presence of ROS in ambient and commercial nanoparticle samples [59]. Comparison to standard hydrogen peroxide curves permits elucidation of amounts of peroxides generated. These results highlight the feasibility of utilizing nanobiosensors for detection and, ultimately, quantification of ROS, calcium, and other fingerprints of activation of specific pathways, thus allowing confirmation of induction of specific cellular response elements due to exposure to various NM.

Nanoelectrodes

Electrodes made at the nanoscale can display altered properties simply by the nature of their dimensionality [56, 59, 60]. Reducing something to a smaller size can result in increased available reactive surface area, as well as other effects such as focusing of the electric field due to the rod-like geometric configuration [61–63]. These are desirable effects that enhance the system through their physical aspects. Carbon nanotube (CNT) and metallic arrays have been used as electrodes to promote electron transfer in redox reactions [59–64].

Carbon nanotubes are unique and possess many interesting properties in the electronic, mechanical, chemical, and optical domains [65]. Particularly relevant to the work covered in this chapter is that they have shown great promise as a biochemically compatible, nanoelectronic interface to biomolecules. CNTs used as a bioelectronic interface with a protein immobilized at one end offer a number of advantages. It maximally maintains the protein's native conformation, hence its functional properties, by minimizing the contact surface area in much the same way as a straw holds a soap bubble at its tip. Its high electrical conductivity provides an efficient conduit for electronic signals from or to the protein, and its sharp curvature at the tip greatly enhances the electrical field and thereby increases exponentially the electron transfer between the protein and the nanotube via quantum mechanical tunneling whose rate is an exponential function of the local field. Functional activated carboxyl groups are readily produced via acid-treatment at the open end of the nanotube. This facilitates its conjugation with biomolecules localized at the tip and, therefore, enables site and molecule specific immobilization. Recent results have demonstrated the viability of linking an enzyme's site specifically with enhancement of electron transfer rates from the immobilized protein of two orders of magnitude over the known rate in the solution phase [56, 59].

Arrays of vertically aligned MWNTs are grown from a hexagonally patterned array of nanopores in an anodized aluminum oxide (AAO) template that are virtually identical in length, diameter, and spacing (Figure 6.3). Their dense packing, on the order of 10^{10} cm^{-2}, and uniformity produce a tight, even plane of nanotube tips highly suitable for interfacing with biomolecules. Within the array, each individual tube is physically separated and electrically insulated by the insulating AAO template, and a direct electrical contact of each tube can be made to a macro-scale electrode by sputtering the backside of the array with a layer of metal (e.g., gold). Details of the fabrication process have been

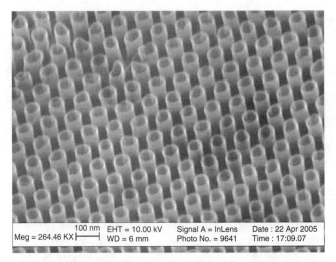

| | 100 nm | EHT = 10.00 kV | Signal A = InLens | Date : 22 Apr 2005 |
| Meg = 264.46 KX | ⊢————⊣ | WD = 6 mm | Photo No. = 9641 | Time : 17:09.07 |

Figure 6.3 Scanning electron microscopy (SEM) of platform of nanotube ensemble, in which the carbon nanotubes (CNTs) are vertically oriented, periodically spaced, and readily accessible from the top and the bottom. The multiwalled carbon nanotubes are 50 nm in diameter, have walls of 3 nm thickness, and exhibit an exposed length of 60 nm, a total length of 10 μm, and a center-to-center spacing of 100 nm between adjacent tubes.

published elsewhere [66]. This unique configuration has allowed the specific covalent linkage of glucose oxidase, as well as NADH peroxidase, at the nanotube tips and noncovalently to the sidewalls, with high coverage (Figure 6.4). These electrodes have been used to characterize the electrochemical profile of the immobilized enzymes by cyclic voltametry and chronoamperometry measurements, attesting to the active state of the enzymes after immobilization [56, 59, 67].

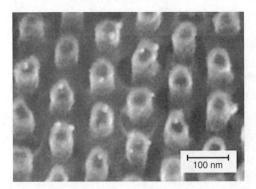

100 nm

Figure 6.4 SEM of glucose oxidase labeled with gold nanoparticles, covalently linked to the tips of highly ordered carbon nanotubes (CNTs). In the array, the specific localization of the enzyme to the tips is possible through reaction with carboxyl groups formed by acid etching, then activated by conventional EDC/NHS chemistry (description in text).

Online Data Bank

With an increasing number of NM being produced, it is essential that the scientific community have an easily accessible data bank. This online bank should hold safety information of all NM already tested and provide frequent updates on NM in the process of being tested. This format is already in use by the Hazardous Substances Data Bank (HSDB), which can be accessed at www.nlm.nih.gov/pubs/factsheets/hsdbfs.html. Another example is the protein data bank (PDB) found at www.rcsb.org/pdb. With these formats in mind, many different modes of information about NM, such as NM characterization, laboratory methods, environmental fate/exposure, safety and handling, *in vitro* studies, *in vivo* studies, and human health effects, can be provided to the interested party. This type of data bank promotes standardization and allows open access to up-to-date information. It is essential that we use these methods when communicating with a worldwide audience, especially given the rapid growth in the number of new NM.

Once a NM database has been established, the scientific community can further refine the material characteristics leading to toxicity (e.g., size/charge or hydrophobicity). It will also be possible to work with batches of NM to develop a toxicological grading system.

In summary, the demand for a predictive and pragmatic approach to nanotoxicity is compelling. While it is optimal to collect data at different tiers of toxicity, some flexibility is required to develop decision matrixes for *in vitro* and *in vivo* testing. Ultimately, the goal of the predictive approach would be to develop a series of toxicity assays that limit the demand for large-scale *in vivo* studies, both from a cost as well as an animal-use perspective. It is important to mention the potential significant difficulties that may be encountered in conducting *in vitro* and *in vivo* studies with engineered NM. This has largely to do with lack of knowledge of real-world exposures to NM, including dosage, complexity of working/living environment, aggregation status, and so on. As nanotechnology develops, it is essential that the toxicological approach also evolves and stays up to date. This will provide an important safeguard for the continued expansion of the nanotechnology industry.

Abbreviations

ARE, antioxidant response element; ESR, electron spin resonance; GPx, glutathione peroxidese; GSH, glutathione; GSSG, glutathione disulfide; GST, glutathione S-transferase; JNK, NH_2-terminal Jun kinase; HO-1, heme oxygenase 1; MAPK, mitogen-activated protein kinase; NAC, N-acetylcysteine; O_2^-, superoxide; OH, hydroxyl radical; PM, particle matter; PT, permeability transition; ROS, reactive oxygen species; SOD, superoxide dismutase; UFP, ultrafine particles; UGT, UDP-glucoronosyltransferase; $\Delta\psi$m, mitochondrial membrane potential.

Acknowledgements

The work described in this chapter was supported by the US Public Health Service Grants, U19 A1070453 (funded by National Institute of Allergy and Infectious Diseases and National Institute of Environmental and Health Science), RO1 ES10553 (National Institute of Environmental and Health Science) R01 ES 015498 and RO1 ES10253, as well as the US Environmental Protection Agency STAR award to the Southern California Particle Center RD-83241301. This work has not been subjected to the Environmental Protection Agency for peer and policy review and therefore does not necessarily reflect the views of the agency; no official endorsement should be inferred. We also thank DARPA (F49620-03-1-0365; JX, JIY) and NIH (GM066466; JIY) for funding.

References

1. Oberdorster G., Utell M.J. Ultrafine Particles in the Urban Air: To the Respiratory Tract—and Beyond? *Environmental Health Perspectives* 2002;110(8):A440–A441.
2. Donaldson K., Tran C.L. Inflammation Caused by Particles and Fibers. *Inhalation Toxicology* 2002;14(1):5–27.
3. Service R.F. Nanotoxicology. Nanotechnology Grows Up. *Science* 2004;304(5678): 1732–1734.
4. Donaldson K., Stone V., Clouter A., Renwick L., MacNee W. Ultrafine Particles. *Occupational & Environmental Medicine* 2001;58(3):211–216.
5. Oberdorster G., Oberdorster E., Oberdorster J. Nanotoxicology: An Emerging Discipline Evolving from Studies of Ultrafine Particles. *Environmental Health Perspectives* 2005;113(7):823–839.
6. Donaldson K., Stone V., Tran C.L., Kreyling W., Borm P.J. Nanotoxicology. *Occupational & Environmental Medicine* 2004;61(9):727–728.
7. Nel A. Atmosphere. Air Pollution-Related Illness: Effects of Particles. *Science* 2005;308(5723):804–806.
8. Oberdorster G., Sharp Z., Atudorei V., Elder A., Gelein R., Kreyling W., et al. Translocation of Inhaled Ultrafine Particles to the Brain. *Inhalation Toxicology* 2004;16(6–7):437–445.
9. Donaldson K., Stone V. Current Hypotheses on the Mechanisms of Toxicity of Ultrafine Particles. *Annali dell'Istituto Superiore di Sanita.* 2003;39(3):405–410.
10. Donaldson K., Stone V., Seaton A., MacNee W. Ambient Particle Inhalation and the Cardiovascular System: Potential Mechanisms. *Environmental Health Perspectives* 2001;109 Suppl 4,523–527.
11. Nel A., Xia T., Madler L., Li N. Toxic Potential of Materials at the Nanolevel. *Science* 2006;311(5761):622–627.
12. Oberdorster G., Maynard A., Donaldson K., Castranova V., Fitzpatrick J., Ausman K., et al. Principles for Characterizing the Potential Human Health Effects from Exposure to Nanomaterials: Elements of a Screening Strategy. *Particle and Fibre Toxicology* 2005;28.
13. Mecke A., Majoros I.J., Patri A.K., Baker J.R., Jr., Holl M.M., Orr B.G. Lipid Bilayer Disruption by Polycationic Polymers: The Roles of Size and Chemical Functional Group. *Langmuir* 2005;21(23):10348–10354.
14. Vertegel A.A., Siegel R.W., Dordick J.S. Silica Nanoparticle Size Influences the Structure and Enzymatic Activity of Adsorbed Lysozyme. *Langmuir* 2004;20(16): 6800–6807.

15. Dunford R., Salinaro A., Cai L., Serpone N., Horikoshi S., Hidaka H., et al. Chemical Oxidation and DNA Damage Catalysed by Inorganic Sunscreen Ingredients. *FEBS Letters* 1997;418(1–2):87–90.
16. Schins R.P., Duffin R., Hohr D., Knaapen A.M., Shi T., Weishaupt C., et al. Surface Modification of Quartz Inhibits Toxicity, Particle Uptake, and Oxidative DNA Damage in Human Lung Epithelial Cells. *Chemical Research in Toxicology* 2002;15(9):1166–1173.
17. Warheit D.B., Laurence B.R., Reed K.L., Roach D.H., Reynolds G.A., Webb TR. Comparative Pulmonary Toxicity Assessment of Single-Wall Carbon Nanotubes in Rats. *Toxicological Sciences* 2004;77(1):117–125.
18. Zhang Q., Kusaka Y., Sato K., Nakakuki K., Kohyama N., Donaldson K. Differences in the Extent of Inflammation Caused by Intratracheal Exposure to Three Ultrafine Metals: Role of Free Radicals. *Journal of Toxicology and Environmental Health*. Part A 1998;53(6):423–438.
19. Li N., Sioutas C., Cho A., Schmitz D., Misra C., Sempf J., et al. Ultrafine Particulate Pollutants Induce Oxidative Stress and Mitochondrial Damage. *Environmental Health Perspectives* 2003;111(4):455–460.
20. Brown D.M., Donaldson K., Borm P.J., Schins R.P., Dehnhardt M., Gilmour P., et al. Calcium and ROS-Mediated Activation of Transcription Factors and TNF-Alpha Cytokine Gene Expression in Macrophages Exposed to Ultrafine Particles. *American Journal of Physiology. Lung Cellular and Molecular Physiology* 2004;286(2): L344–L353.
21. Hiura T.S., Li N., Kaplan R., Horwitz M., Seagrave J.C., Nel A.E. The Role of a Mitochondrial Pathway in the Induction of Apoptosis by Chemicals Extracted from Diesel Exhaust Particles. *Journal of Immunology* 2000;165(5):2703–2711.
22. Long T.C., Saleh N., Tilton R.D., Lowry G.V., Veronesi B. Titanium Dioxide (P25) Produces Reactive Oxygen Species in Immortalized Brain Microglia (BV2): Implications for Nanoparticle Neurotoxicity. *Environmental Science & Technology* 2006;40(14):4346–4352.
23. Oberdorster E. Manufactured Nanomaterials (Fullerenes, C60) Induce Oxidative Stress in the Brain of Juvenile Largemouth Bass. *Environmental Health Perspectives* 2004;112(10):1058–1062.
24. Shvedova A.A., Castranova V., Kisin E.R., Schwegler-Berry D., Murray A.R., Gandelsman V.Z., et al. Exposure to Carbon Nanotube Material: Assessment of Nanotube Cytotoxicity Using Human Keratinocyte Cells. *Journal of Toxicology and Environmental Health*. Part A 2003;66(20):1909–1926.
25. Shvedova A.A., Kisin E.R., Mercer R., Murray A.R., Johnson V.J., Potapovich A.I., et al. Unusual Inflammatory and Fibrogenic Pulmonary Responses to Single Walled Carbon Nanotubes in Mice. *American Journal of Physiology. Lung Cellular and Molecular Physiology* 2005;289(5):L698–708.
26. Monteiro-Riviere N.A., Nemanich R.J., Inman A.O., Wang Y.Y., Riviere J.E. Multi-Walled Carbon Nanotube Interactions with Human Epidermal Keratinocytes. *Toxicological Letters* 2005;155(3):377–384.
27. Li N., Alam J., Venkatesan M.I., Eiguren-Fernandez A., Schmitz D., Di S.E., et al. Nrf2 Is a Key Transcription Factor that Regulates Antioxidant Defense in Macrophages and Epithelial Cells: Protecting Against the Proinflammatory and Oxidizing Effects of Diesel Exhaust Chemicals. *Journal of Immunology* 2004;173(5):3467–3481.
28. Li N., Venkatesan M.I., Miguel A., Kaplan R., Gujuluva C., Alam J., et al. Induction of Heme Oxygenase-1 Expression in Macrophages by Diesel Exhaust Particle Chemicals and Quinones via the Antioxidant-Responsive Element. *Journal of Immunology* 2000;165(6):3393–3401.
29. Li N., Wang M., Oberley T.D., Sempf J.M., Nel A.E. Comparison of the Pro-oxidative and Proinflammatory Effects of Organic Diesel Exhaust Particle Chemicals in Bronchial Epithelial Cells and Macrophages. *Journal of Immunology* 2002;169(8):4531–4541.
30. Li N., Kim S., Wang M., Froines J., Sioutas C., Nel A. Use of a Stratified Oxidative Stress Model to Study the Biological Effects of Ambient Concentrated and Diesel Exhaust Particulate Matter. *Inhalation Toxicology* 2002;14(5):459–486.
31. Xiao G.G., Wang M., Li N., Loo JA., Nel A.E. Use of Proteomics to Demonstrate a Hierarchical Oxidative Stress Response to Diesel Exhaust Particle Chemicals in a Macrophage Cell Line. *Journal of Biological Chemistry* 2003;278(50):50781–50790.

32. Wang M., Xiao G.G., Li N., Xie Y., Loo J.A., Nel A.E. Use of a Fluorescent Phosphoprotein Dye to Characterize Oxidative Stress-Induced Signaling Pathway Components in Macrophage and Epithelial Cultures Exposed to Diesel Exhaust Particle Chemicals. *Electrophoresis* 2005;26(11):2092–2108.
33. Li N., Hao M., Phalen R.F., Hinds W.C., Nel A.E. Particulate Air Pollutants and Asthma. A Paradigm for the Role of Oxidative Stress in PM-Induced Adverse Health Effects. *Clinical Immunology* 2003;109(3):250–265.
34. Nel A.E., Diaz-Sanchez D., Ng D., Hiura T., Saxon A. Enhancement of Allergic Inflammation by the Interaction Between Diesel Exhaust Particles and the Immune System. *The Journal of Allergy and Clinical Immunology* 1998;102(4 Pt 1):539–554.
35. Hiura T.S., Kaszubowski M.P., Li N., Nel A.E. Chemicals in Diesel Exhaust Particles Generate Reactive Oxygen Radicals and Induce Apoptosis in Macrophages. *Journal of Immunology* 1999;163(10):5582–5591.
36. Xia T., Korge P., Weiss J.N., Li N., Venkatesen M.I., Sioutas C., et al. Quinones and Aromatic Chemical Compounds in Particulate Matter Induce Mitochondrial Dysfunction: Implications for Ultrafine Particle Toxicity. *Environmental Health Perspectives* 2004;112(14):1347–1358.
37. Upadhyay D., Panduri V., Ghio A., Kamp D.W. Particulate Matter Induces Alveolar Epithelial Cell DNA Damage and Apoptosis: Role of Free Radicals and the Mitochondria. *American Journal of Respiratory Cell and Molecular Biology* 2003;29(2):180–187.
38. de Lorenzo A.J. 1970. The Olfactory Neuron and the Blood-Brain Barrier. In *Taste and Smell in Vertebrates* (Wolstenholme G., Knight J., eds). London: Churchill, 151–176.
39. Foley S., Crowley C., Smaihi M., Bonfils C., Erlanger B.F., Seta P., et al. Cellular Localisation of a Water-Soluble Fullerene Derivative. *Biochemical and Biophysical Research Communications* 2002;294(1):116–119.
40. Savic R., Luo L., Eisenberg A., Maysinger D. Micellar Nanocontainers Distribute to Defined Cytoplasmic Organelles. *Science* 2003;300(5619):615–618.
41. Xia T., Kovochich M., Brant J., Hotze M., Sempf J., Oberley T., et al. Comparison of the Abilities of Ambient and Manufactured Nanoparticles to Induce Cellular Toxicity According to an Oxidative Stress Paradigm. *Nano Letters* 2006;6(8):1794–1807.
42. Morgan K. Development of a Preliminary Framework for Informing the Risk Analysis and Risk Management of Nanoparticles. *Risk Analysis* 2005;25(6):1621–1635.
43. Kagan V.E., Tyurina Y.Y., Tyurin V.A., Konduru N.V., Potapovich A.I., Osipov A.N., et al. Direct and Indirect Effects of Single Walled Carbon Nanotubes on RAW 264.7 Macrophages: Role of Iron. *Toxicological Letters* 2006.
44. Cho A.K., Sioutas C., Miguel A.H., Kumagai Y., Schmitz D.A., Singh M., et al. Redox Activity of Airborne Particulate Matter at Different Sites in the Los Angeles Basin. *Environmental Research* 2005;99(1):40–47.
45. Kumagai Y., Koide S., Taguchi K., Endo A., Nakai Y., Yoshikawa T., et al. Oxidation of Proximal Protein Sulfhydryls by Phenanthraquinone, a Component of Diesel Exhaust Particles. *Chemical Research in Toxicology* 2002;15(4):483–489.
46. Pickering K., Wiesner M. Fullerol-Sensitized Production of Reactive Oxygen Species in Aqueous Solution. *Environmental Science & Technology* 2005;39(5):1359–1365.
47. Sayes C., Fortner J., Lyon D., Boyd A., Ausman K., Guoh W., et al. The Differential Cytotoxicity of Water-Soluble Fullerenes. *Nano Letters* 2004;1881–1887.
48. Brookes P.S., Yoon Y., Robotham J.L., Anders M.W., Sheu S.S. Calcium, ATP, and ROS: A Mitochondrial Love-Hate Triangle. *American Journal of Physiology Cell Physiology* 2004;287(4):C817–C833.
49. Adachi T., Matsui R., Xu S., Kirber M., Lazar H.L., Sharov V.S., et al. Antioxidant Improves Smooth Muscle Sarco/Endoplasmic Reticulum Ca(2+)-ATPase Function and Lowers Tyrosine Nitration in Hypercholesterolemia and Improves Nitric Oxide-Induced Relaxation. *Circulation Research* 2002;90(10):1114–1121.
50. Green D.R., Kroemer G. The Pathophysiology of Mitochondrial Cell Death. *Science* 2004;305(5684):626–629.
51. Contag C.H., Stevenson D.K. In Vivo Patterns of Heme Oxygenase-1 Transcription. *Journal of Perinatology* 2001;21 Suppl 1S119–S124.

52. Brook R.D., Franklin B., Cascio W., Hong Y., Howard G., Lipsett M., et al. Air Pollution and Cardiovascular Disease: A Statement for Healthcare Professionals from the Expert Panel on Population and Prevention Science of the American Heart Association. *Circulation* 2004;109(21):2655–2671.
53. Hajdena-Dawson M., Zhang W., Contag P.R., Wong R.J., Vreman H.J., Stevenson D.K., et al. Effects of Metalloporphyrins on Heme Oxygenase-1 Transcription: Correlative Cell Culture Assays Guide in Vivo Imaging. *Molecular Imaging* 2003;2(3): 138–149.
54. Zhang W., Feng J.Q., Harris S.E., Contag P.R., Stevenson D.K., Contag C.H. Rapid in Vivo Functional Analysis of Transgenes in Mice Using Whole Body Imaging of Luciferase Expression. *Transgenic Research* 2001;10(5):423–434.
55. Zhang W., Contag P.R., Hardy J., Zhao H., Vreman H.J., Hajdena-Dawson M., et al. Selection of Potential Therapeutics Based on in Vivo Spatiotemporal Transcription Patterns of Heme Oxygenase–1. *Journal of Molecular Medicine* 2002;80(10):655–664.
56. Withey G.D., Lazareck A.D., Tzolov M.B., Yin A., Aich P., Yeh J.I., et al. Ultra-High Redox Enzyme Signal Transduction Using Highly Ordered Carbon Nanotube Array Electrodes. *Biosensors & Bioelectronics* 2006;21(8):1560–1565.
57. Poole L.B., Claiborne A. Interactions of Pyridine Nucleotides with Redox Forms of the Flavin-Containing NADH Peroxidase from Streptococcus Faecalis. *Journal of Biological Chemistry* 1986;261(31):14525–14533.
58. Yeh J.I., Claiborne A., Hol W.G. Structure of the Native Cysteine-Sulfenic Acid Redox Center of Enterococcal NADH Peroxidase Refined at 2.8 A Resolution. *Biochemistry* 1996;35(31):9951–9957.
59. Yeh J.I., Zimmt M.B., Zimmerman A.L. Nanowiring of a Redox Enzyme by Metallized Peptides. *Biosensors & Bioelectronics* 2005;21(6):973–978.
60. Shim M., Shi Kam N.W., Chen R.J., Li Y., Dai H. Functionalization of Carbon Nanotubes for Biocompatibility and Biomolecular Recognition. *Nano Letters* 2002;2(4):285–288.
61. Bradley K., Briman M., Star A., Gruner G. Charge Transfer from Adsorbed Proteins. *Nano Letters* 2004;4(2):253–256.
62. Gruner G. Carbon Nanotube Transistors for Biosensing Applications. *Analytical and Bioanalytical Chemistry* 2006;348(2), 322–335.
63. Star A., Bradley K., Gabriel J.-C., Gruner G. Nanoelectronic Sensors: Chemical Detection Using Carbon Nanotubes. *Polymeric Materials Science and Engineering* 2003;89204.
64. Kohli P., Harrell C.C., Cao Z, Gasparac R., Tan W., Martin C.R. DNA-Functionalized Nanotube Membranes with Single-Base Mismatch Selectivity. *Science* 2004;305(5686): 984–986.
65. Iijima S. Helical Microtubules of Graphitic Carbon. *Nature* 1991;354(6348):56–58.
66. Li J., Papadopoulos C., Xu J.M., Moskovits M. Highly Ordered Carbon Nanotube Arrays for Electronics Applications. *Applied Physics Letters* 2006;75(3):367–369.
67. Xiao Y., Patolsky F., Katz E., Hainfeld J.F., Willner I. Plugging into Enzymes: Nanowiring of Redox Enzymes by a Gold Nanoparticle. *Science* 2003;299(5614): 1877–1881.

Nanoparticle Transport, Aggregation, and Deposition

Jonathan Brant *Duke University, Durham, North Carolina, USA*
Jérôme Labille *CNRS-Université Aix-Marseille, Aix-en-Provence, France*
Jean-Yves Bottero *CNRS-Université Aix-Marseille, Aix-en-Provence, France*
Mark R. Wiesner *Duke University, Durham, North Carolina, USA*

Introduction

This chapter explores physical-chemical factors that govern two key processes controlling the transport and fate of nanomaterials in aquatic environmental systems, aggregation and deposition. There is a well-developed body of work addressing the behavior of particles in water. In particular, we consider applications of the principles of colloid chemistry to specific nanomaterials and possible insights that theory provides for the case of very small colloidal particles. However, the format of nanomaterials that is environmentally relevant (e.g., as pure materials versus composites), the quantities and concentrations likely to be present, and the patterns of production, use, and disposal are, at the time of writing, largely speculative.

Particle aggregation and deposition can be thought of as occurring in two steps. First, particles are transported to the vicinity of a surface and then, if conditions allow, particles attach to the surface. In aggregation, the surface may be that of another particle or a growing aggregate. In deposition, the surface is an immobile "collector" where particles accumulate. Much of this chapter addresses the role of surface chemistry and associated interfacial relationships, as they affect the attachment step and ultimately nanoparticle mobility in aquatic environments. Basic concepts from colloid chemistry are reviewed and interpreted in the content of particles with at least one characteristic dimension measuring less than 100 nm. We will begin with a discussion of the relevant interfacial chemical properties and relationships between particles in aqueous media.

Physicochemical Interactions

In aqueous environments, the nature of particle surfaces is intimately linked with the solution conditions (pH, ionic strength, temperature, etc.). Characterization of nanoparticles is therefore relevant largely within the context of the characteristics of the solution in which the nanoparticle is suspended. Those properties of particular interest for environmental analyses include surface charge, the presence of surface functional groups, Hamaker constant, and interaction energy with water (wettability). However, the characterization of materials in the nanometer regime is particularly challenging due to fact that materials in this size range fall into a gray area where they may, in many cases, be considered as either small particles or large solutes.

Brownian motion

Particles, molecules, and ions in fluid environments, regardless of size, possess a Brownian energy (BR) equal to 1.5 kT [3], where k is the Boltzmann constant (1.38 10^{-23} J) and T is the absolute temperature in degrees Kelvin. Brownian potential energy decays exponentially from this value (1.5 kT) with distance, with a decay length equal to the respective particle's radius of gyration, R_g. (The radius of gyration is the root mean square of mass-weighted distances of all subvolumes in a particle from its center of mass. It has special interest in particle science because it can be applied to irregularly shaped particles.) Because the Brownian energy imparted to a given particle originates from collisions with (primarily) solvent molecules (e.g., water), it becomes more significant with decreasing particle size. In contrast, forces such as gravity or those originating from fluid flow (e.g., drag forces), increase with increasing particle size. Thus, as a transport mechanism, Brownian motion has particular significance in determining nanoparticle stability and mobility in aqueous systems through its influence on the collision frequency between particles and with stationary surfaces.

DLVO theory

Particle interactions in aqueous environments are generally assessed within the context of the classical Derjaguin-Landau-Verwey-Overbeek (DLVO) theory [4]. The DLVO theory expresses the total interaction energy between two surfaces as the sum of Lifshitz-van der Waals (LW) and electrostatic (EL) interactions. Other forces, defined as non-DLVO forces, have also been found to be significant for surfaces in aqueous environments [3, 5] and have thus been included in the form of an extended DLVO (XDLVO) approach. Here, the total interaction energy between two surfaces in water may be written as:

$$U_{123}^{\text{XDLVO}} = U_{123}^{\text{LW}} + U_{123}^{\text{EL}} + U_{123}^{\text{AB}} + U_{123}^{\text{BO}} \tag{1}$$

where U^{XDLVO} is the total interaction energy between two surfaces immersed in water; U^{LW} is the Lifshitz-van der Waals interaction term; U^{EL} is the electrostatic interaction term; U^{AB} is the acid-base interaction term; and U^{BO} is the interaction energy due to Born repulsion. The subscripts 1, 2, and 3 correspond to surfaces 1 and 3 separated by an aqueous medium 2. Other interactions, such as steric interactions, are also likely and should be considered, though they are not included in the energy balance presented here. Steric interactions generally result from the adsorption of polymers or other long-chained molecules and can act to either stabilize or destabilize a particle suspension. This topic is addressed later in this chapter.

When plotted as a function of separation distance, the total interaction energy shows the evolution of the magnitude and type of interaction (repulsive or attractive) that occurs as two surfaces approach each other (Figure 7.1). Three cases arise, depending on the relative

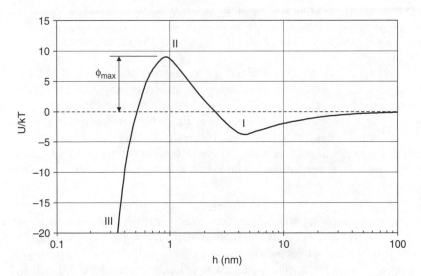

Figure 7.1 Example of DLVO interaction energy curve between a sphere and a flat surface illustrating the three characteristic regions of an energy plot: (I) attractive secondary minimum, (II) repulsive barrier, and (III) attractive primary minimum. The energy barrier height is defined as ϕ_{max}.

magnitudes of the attractive (negative energy potential) and repulsive (expressed as a positive quantity) phenomena involved. In the first case, the interaction between surfaces is attractive at all separation distances. In the second case, the interaction is repulsive as the surfaces approach one another until a repulsive energy barrier is overcome, at which point the interactions become attractive. In the third case (depicted in Figure 7.1), interactions are at first attractive, then repulsive, and finally attractive once again (as in the second case) as the approach distances became smaller. The features of the potential energy curve in this last case are designated as an attractive secondary minimum (I), a repulsive barrier (II), and an attractive primary minimum (III). As separation distance between two surfaces increases, the different components of the interaction energies diminish from their corresponding values at near-contact following a unique decay pattern, which affects the shape of the resulting energy plot [4]. For instance, the secondary minimum develops because repulsive electrostatic interactions decay exponentially with separation distance while attractive van der Waals interactions decay somewhat more slowly over a longer range. The magnitudes of different interaction energies depend on particle size and thus, so does the overall interaction energy curve. For example, as particle size decreases, the height of the energy barrier, if present, also decreases; similarly, the secondary minimum becomes more shallow with decreasing particle size. This is particularly evident at separation distances greater than 5 nm. Below this separation distance, shorter-range interactions tend to be less impacted by particle size. Nevertheless, size effects on interaction energies have consequences with regards to the stability of particle dispersions and particle mobility [6].

The components of interaction energy that are accounted for in the DLVO and other similar theories are discussed in more detail in the following sections. These discussions, however, are not meant to be exhaustive reviews of these subjects; the reader is instead referred to other sources [4, 7, 8] for such detailed overviews. The purpose here is to consider the effect of diminishing particle size on those interaction energies as they may be relevant in determining nanoparticle fate and transport in the environment.

Van der waals interactions.   Van der Waals interactions are in most cases attractive and act universally between surfaces in aqueous media. These interactions incorporate three different electrodynamic forces: dispersion, induction, and orientation [4]. The three electrodynamic forces may be summed together to equal the total van der Waals interfacial energy term. Lifshitz-van der Waals interaction energies decay according to L^{-2}, where L represents the distance between two infinitely long flat plates [3]. Using Derjaguin's approximation the LW interaction energy between a flat surface and a spherical particle may be calculated as follows:

$$U_{123}^{LW} = -\left(\frac{Aa_p}{6h}\right)\left(1 + \frac{14h}{\lambda}\right)^{-1} \qquad (2)$$

where A is the Hamaker constant for the interacting surfaces across the medium; λ is the characteristic wavelength of the dielectric, usually taken to be equal to 100 nm; a_p is the radius of the spherical particle; and h is the surface to surface separation distance. Figure 7.2 shows the van der Waals interaction energy for a sphere and an infinite flat surface for different particle sizes. As particle size decreases, the van der Waals attraction similarly decreases in magnitude and acts over shorter separation distances, consistent with Eq. 2 [9]. Van der Waals interactions may be weak for nanoparticles for a variety of reasons. This is particularly significant when considering nanoparticle transport in porous media [10]. We will examine the role of van der Waals interactions as they affect particle deposition in porous media in the sections "Deposition" and "Detachment."

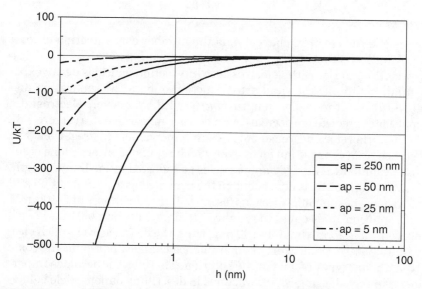

Figure 7.2   van der Waals interaction energy calculated for a sphere-flat plate as a function of separation distance. The interaction energy is normalized to the thermal energy (kT) of the suspension ($T = 20°C$; pH = 7; $A_H = 10^{-20}$ J). The curves were calculated as a function of particle size.

Electrostatic interactions. Hogg et al. [11] derived an expression for the electrostatic interaction energy between two surfaces under the assumption of constant surface potential. When modeling electrostatic interactions between surfaces, a theoretical constraint of constant surface potential or constant surface charge is generally made [7]. In reality, the actual condition lies between these two experimental extremes. The electrostatic interaction energy per unit area between a spherical particle and a flat surface decays with separation distance according to:

$$U_{123}^{El}(h) = \pi\varepsilon_r\varepsilon_0 a_p\left[2\zeta_1\zeta_3\ln\left(\frac{1 + e^{-\kappa h}}{1 - e^{-\kappa h}}\right) + (\zeta_1^2 + \zeta_3^2)\ln(1 - e^{-2\kappa h})\right] \qquad (3)$$

where $\varepsilon_0\varepsilon_r$ is the dielectric permittivity of the suspending fluid; κ is the Debye constant; and ζ_1 and ζ_2 are the surface potentials of the interacting surfaces. The inverse of the Debye constant is a measure of the diffuse ionic double layer thickness that surrounds charged surfaces in aqueous systems. For this reason it is often referred to as the inverse Debye screening length and is determined according to the following equation for z-z electrolytes:

$$\kappa = \sqrt{\frac{e^2\sum n_iz_i^2}{\varepsilon_r\varepsilon_0 kT}} \qquad (4)$$

where e is the electron charge; n_i is the number concentration of ion i in the bulk solution; and z_i is the valence of ion i.

According to Eq. 3, the electrostatic interaction energies decay exponentially with distance and are a function of both the separation distance (h) and the Debye length ($1/\kappa$) [7, 12]. Moreover, electrostatic repulsion is predicted to decrease with decreasing particle size (Figure 7.3), and may therefore decrease barriers to nanoparticle aggregation. The distance from charged surfaces over which repulsive interactions occur is often similar to the size nanoparticles as reflected in the Debye length. The variability of $1/\kappa$ is dependent on the ionic strength of the solution and the valency of the ionic species present [13]. For instance, in a 100 mM NaCl solution $1/\kappa$ is equal to roughly 1.0 nm, while for a 0.01 mM NaCl solution $1/\kappa$ is approximately 100 nm. Thus, the distance over which electrostatic interactions exert their influence changes based on the ionic strength and types of ions in the solution. Therefore, it becomes important when considering EL interactions in describing nanoparticle behavior in water to consider the salt concentration and valence of the ions present. Furthermore, the thickness of the diffuse double layer has specific consequences unique to nanoparticles as they may be present

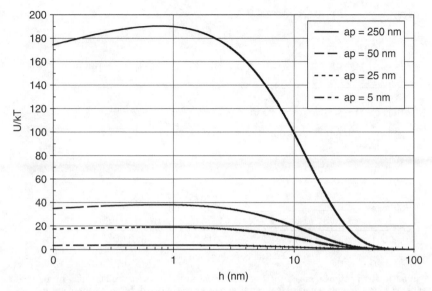

Figure 7.3 Electrostatic repulsion between a spherical particle and a flat surface plotted for several different particle sizes as a function of separation distance ($\zeta 1 = -30$ mV; $\zeta 2 = -25$ mV; pH = 7; T = 20°C; I = 1 mM NaCl).

at similar scales (1 to 100 nm). This is important as many models are based on the assumption that κh is much less than the diameter of the particle.

Nanoparticles have large surface area to volume ratios and potentially high sorption capacities for other aqueous species, such as ionic materials and natural organic matter [14], that would tend to favor complexation processes. For example, the high electron-affinity of fullerenes has been shown to facilitate covalent, charge transfer, and donor-acceptor interactions with other compounds [15]. Additionally, because a significant fraction of atoms are exposed at the nanoparticle surface, rather than contained in the bulk interior, nanoparticle surface chemistry can be significantly altered by surface complexation processes [14, 16]. Such processes can have a dramatic affect on nanoparticle surface charge characteristics and in turn the electrostatic interactions with other surfaces. Adsorption of ionic species can in fact impart a charge to an otherwise uncharged particle. For example, adsorption of water and subsequent deprotonation to form hydroxyl groups have previously been observed for hydrophobic oil droplets in water, which was concluded to be the source of the measured electrophoretic mobility for these particles [17]. Similarly, the adsorption of hydroxyl groups and charge transfer interactions with solvents have been proposed as likely

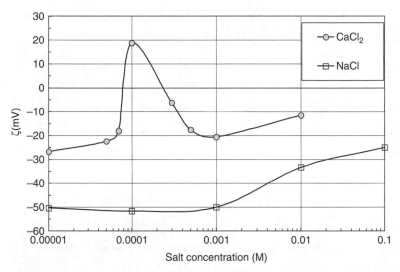

Figure 7.4 Zeta potential as a function of salt concentration measured for nC_{60} nanoclusters. Zeta potential was measured in the presence of two different types of salts and illustrates the charge reversal experienced by the nC_{60} in the presence of calcium chloride (a_p = 168 nm; pH = 7; T = 20°C).

charging mechanisms for fullerene C_{60} nanoclusters referred to as nC_{60} [18, 19]. These processes are then particularly significant for fullerene nanoparticles that might otherwise have little affinity for the aqueous phase and, thus, decreased stability and mobility. Furthermore, the adsorption of ionic species may lead to charge reversal for some nanoparticles (Figure 7.4) [18]. In this example, colloidal aggregates of nC_{60} are observed to have a negative charge at low ionic strengths. The origin of this charge appears to be due to the structuring of water at the fullerene surface, although impurities such as residual solvent used to fabricate some of these colloids, or perhaps even present in stock C_{60} may also play a role. As the concentration of an indifferent electrolyte (assumed to be nonadsorbing) increases, the electrophoretic mobility of the nC_{60} becomes less monotonically negative due to charge shielding. However, when the divalent ion calcium is used as an electrolyte, a reversal of charge occurs at low concentrations, followed by a re-reversal at slightly higher concentrations, suggestive of the formation of a complex on the nC_{60} surface.

Born repulsion. Born repulsion results from the strong repulsive forces between atoms as their electron shells begin to overlap. This is a short-range interaction that acts over a distance of up to several nanometers. Prieve and Ruckenstein [20] derived an expression for the Born repulsion

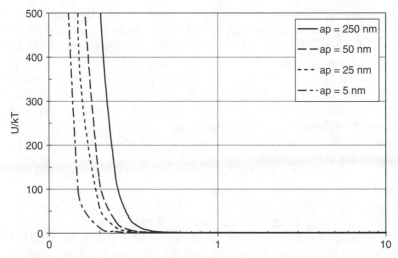

Figure 7.5 Born repulsion as a function of nanoparticle surface- to-surface separation distance plotted for different particle sizes ($A_H = 10^{-20}$ J; collision diameter = 0.5 nm; pH = 7; T = 20°C). The interaction energy was calculated for a sphere-plate geometry.

interaction energy between a spherical particle and a flat surface assuming pairwise additivity of the atomic Lennard-Jones potential:

$$U_{123}^{BO} = \frac{A\sigma_B^6}{7500} \left[\frac{8a_p+h}{(2a_p+h)^7} + \frac{6a_p-h}{h^7} \right] \tag{5}$$

where σ_B is the Born collision diameter. A value for σ_B of 0.5 nm is often assumed [6]. The separation distance at which this force becomes important is predicted to decrease with particle size (Figure 7.5). Because repulsive Born interactions act over such a short distance they may not affect nanoparticle interactions on approach in aqueous media [12]. However, the Born repulsion does significantly affect the depth of the primary minimum and may possibly affect the reversibility of nanoparticle attachment relative to larger particles.

Acid-base interactions and the hydrophobic effect.   Acid-base (AB) interactions characterize the hydrogen bonding properties of a surface or interacting surfaces and thus describe how that surface interacts with water. Water molecules interact with one another and structure themselves through hydrogen bonding [4, 21]. This structuring tendency can result in either attractive hydrophobic or repulsive hydrophilic interactions between particles in water. Functionality on a particle's surface results in the coordination of water molecules on the surface, which subsequently provides a preferential orientation of water through hydrogen bonding. The adsorption of water at functional sites on a surface may be associated with the net release of a proton or hydroxide to the solution. Thus, the surface may act as an acid or a base. Layer(s) of relatively ordered water on surfaces may impede particle attachment or aggregation due to the additional energy required to "squeeze" water out as separation distances become small. In contrast, hydrophobic surfaces do not interact with water and are essentially pushed toward one another as water molecules preferentially bond with one another.

This latter phenomenon can be interpreted as an attractive interaction referred to as the hydrophobic effect [3, 21]. The interactions between particles and the bulk water are a function of the size of the particles or hydrophobic surfaces [21]. Hydrophobic nanoparticles will perturb water structuring to a much smaller degree than will larger particles with a similar chemistry [9]. In the case of small nanoparticles or other hydrophobic molecules, water may form a cage structure around the nanoparticle or molecular core resulting in a relatively stable configuration referred to as a clathrate. Methane clathrates in deep ocean waters are thought to represent an enormous reservoir of hydrocarbons on our planet. Clathrate formation has been suggested as a mechanism behind the hydration of otherwise hydrophobic C_{60} nanoparticles (which have a diameter of a little over 1 nm) and may explain how these fullerenes acquire a charge and interact with water to form stable colloidal suspensions of C_{60} aggregates.

As particle size becomes larger than 2 nm, an energetically unfavorable condition results. To compensate for a larger "disruption" between hydrogen bonds, such as hydrophobic particle, the surrounding water molecules must still restructure themselves to maintain their hydrogen-bonding network with other water molecules. This water depletion or drying between two interacting surfaces results a net attraction between hydrophobic surfaces. It is important to bear in mind that the surfaces themselves do not actually attract each other; in fact, the water simply "likes" itself too much to allow the surface to remain exposed. The balance between particle-solvent and solvent-solvent interactions and hydrogen bonding energies ultimately determines whether cavitation or strict microscopic dewetting will occur. Therefore, for all but the smallest size fraction of nanoparticles (1 nm < d < 2 nm) hydrophobic interactions will be significant and will tend to favor particle aggregation.

In contrast with hydrophobic surfaces, hydrophilic surfaces possess surface groups that may coordinate water molecules through hydration [3, 4]. Subsequent layers of water molecules may hydrogen bond with the hydrated water, resulting in several layers of relatively ordered water extending from the surface. The kinetic energy of the water molecules will tend to break hydrogen bonds. Thus, lower temperatures should favor extension of ordered water further into the bulk. Although there is some controversy regarding the extent to which water may extend into the bulk, it appears that at least two to three layers of ordered water are likely present at most hydrated surfaces. As two hydrated surfaces approach one another, dehydration must occur before the underlying surfaces are in direct contact. The additional free energy required for dehydration represents a repulsive barrier between the two approaching surfaces. Hydration forces act over a shorter range [decay length (λ) = 0.2–1.1 nm] than attractive hydrophobic interactions and decay exponentially with separation distance.

The acid-base interaction energy (U^{AB}) between a sphere and a flat surface is predicted to decay exponentially with separation distance according to the following expression:

$$U_{mlc}^{AB}(h) = 2\pi a_p \lambda \Delta G_{y_0}^{AB} \exp\left[\frac{y_0 - h}{\lambda}\right] \qquad (6)$$

where $\Delta G_{y_0}^{AB}$ is the acid-base free energy of interaction at contact; λ is the characteristic decay length of AB interactions in water, whose value is between 0.2 and 1.0 nm, a value of 0.6 nm is commonly used [3, 4]. Acid-base interactions decrease with decreasing particle size (Figure 7.6) but can be nonetheless comparable to electrostatic interactions even when particle size is below several tens of nanometers.

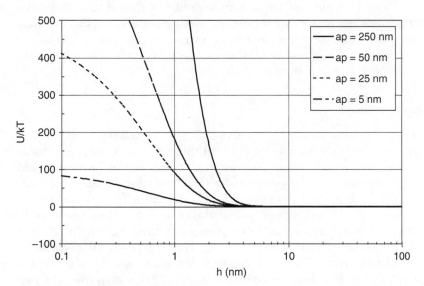

Figure 7.6 Repulsive acid-base interaction energies calculated for a sphere–flat plate system as a function of separation distance and particle size. The interaction energy is normalized to the thermal energy (kT) of the suspension ($\gamma 1^+ = 1$; $\gamma_1^- = 35$; $\gamma_1^+ = 1$; $\gamma_1^- = 38$; T = 20°C; pH = 7).

Implications of the continuum approximation for nanoparticles. The foregoing discussion of the configuration of water on surfaces underscores a key limitation in our ability to describe many interactions at the nanometeric scale. DLVO and other theories describing particle behavior in aqueous media typically treat the intervening fluid (water) as a uniform, structureless medium that is well described in terms of its bulk properties [4] such as density, viscosity, and dielectric constant. As illustrated in the case of ordered water near surfaces, a molecular view of interactions between particles, surfaces, and fluid molecules may be required to adequately describe phenomena that affect nanoparticle aggregation and deposition. A primary challenge in this regard lies in bridging phenomena that apply at the atomic or molecular scale with those observed in the larger scale system. Given the size and complex composition of any real system, it is not possible to simply calculate and sum all of the interactions that occur at the molecular scale. The problem remains computationally intractable. Approaches for bridging this gap in length-scales include averaging across many molecular interactions at a given scale, or using bulk properties as boundary conditions for performing detailed calculations at a given location at the molecular scale.

Limitations on theories that assume that particles and ions exist in a fluid described as a continuum are particularly apparent when separation distances between two surfaces approach 5 nm or less. When considering particles with dimensions similar to that of ions, molecular interactions, such as steric repulsion, become significant. Similar limitations exist in describing particle surfaces. Errors may be introduced in averaging over many functional groups on a surface as is typically done in surface complexation modeling.

Aggregation

Particle dispersions are thermodynamically unstable if the total free energy of the systems may be lowered through a reduction in interfacial area via aggregation. Aggregation involves the formation and growth of clusters and is controlled by both the reaction conditions and interfacial chemical interactions [12, 14, 22]. The propensity of nanoparticles to aggregate, particularly in natural systems, is an important consideration in determining not only their mobility, fate, and persistence, but also their toxicity. Nanoparticles will have negligible settling rates. However, aggregation may result in a growth in mean particle size to the extent that settling rates increase. The persistence of the aggregated nanomaterial in suspension may therefore decrease as these aggregates settle or flow toward collector surfaces, due to favorable attachment conditions. This reduction in persistence (or equivalently, mobility) will have the effect of reducing concentrations in environmental systems. Aggregation may occur between nanoparticles or with other materials (heteroaggregation). In either case, a reduction in stability associated with an increase in aggregation rate has direct consequences for the removal of nanoparticles from air and water in engineered systems (e.g., water and wastewater treatment systems) as well as in the environment.

The role of aggregation in determining toxicity may be less evident. Reductions in active surface area that occur during aggregation and an increased proximity of particle surface area within aggregates may fundamentally reduce reactivity compared with nanoparticles in an unaggregated state. Recent evidence suggests that destabilized nanoparticles experience a decreased ability to produce reactive oxygen species [23] and this may have implications regarding their toxicity [24]. Aggregation may also alter the bioavailability of the nanomaterial if larger materials are less capable of entering cells. Conversely, aggregation may be at the heart of the toxic response as in the case of carbon nanotube agglomeration within lungs, which has been observed to lead to suffocation in laboratory animals where these materials were intentionally introduced [25, 26]. It is possible that heteroaggregation of nanoparticles with organic molecules or other materials may essentially imbed the nanomaterial in another functionality or may alter the availability of the nanomaterial due to changes in size or chemistry. Aggregation will be equally important in environmental engineering applications of nanomaterials. The proposed use of nanoparticles for groundwater remediation is one example where nanoparticle aggregation may decrease the effective surface area of the particles, which reduces their ability to effectively oxidize or react with other compounds. Aggregation may affect the mobility of the nanomaterial and therefore the ability to deliver nanoparticles to a desired location in the subsurface. In summary, the environmental applications, exposures, and effects of nanoparticles are therefore likely to depend on the conditions under which nanoparticles remain as discrete units or aggregate into clusters.

Some nanomaterials have an inherently small affinity for water. For example, the fullerene C_{60} is negligibly soluble in water. Such nanomaterials may be modified to make them compatible with a given end-use or processing requirements in aqueous systems. Nanoparticles may be further modified to maintain a stable suspension. Particle suspensions are commonly stabilized through changes in solution chemistry or modification of particle surface chemistry [27]. However, functionalizing nanoparticles surfaces may compromise the characteristics of the nanoparticles that make them desirable in the first place.

Suspensions of nanoparticles may be stabilized through selection of solvent polarity [28–32] and solution ionic strength [33, 34]. For processes that require working in aqueous or water-based systems (that may also be more representative of natural systems), manipulating solvent polarity may be problematic and eventually impractical. A variety of techniques have been developed for dispersing particles into suspension, including: solvent exchange for buckminsterfullerene [15, 33, 35], extended agitation (and presumably slow reaction) [18, 36], wrapping or grafting molecules, such as surfactants, that impart hydrophilicity to the nanoparticles [37], or direct functionalization of the nanoparticle to produce hydrophilic groups.

Aggregation kinetics and particle stability

The stability of particle dispersions may be evaluated by comparing the aggregation rates of "sticky" and less sticky particles. Here, the stickiness is accounted for in terms of a collision efficiency [38]. In the case of perfectly sticky particles ($\alpha = 1$), aggregation is assumed to be limited only by the transport of particles up to one another. The ratio of the aggregation rates between these two cases is characterized in terms of a stability ratio, W, which may be written as:

$$W = \frac{k_{11\text{fast}}}{k_{11\text{slow}}} \tag{7}$$

where $k_{11\text{fast}}$ and $k_{11\text{slow}}$ are the rate constants for the early stages of aggregation where individual particles collide and form doublets under conditions where transport (often diffusion) is limiting and where attachment is limiting. The rate constants respectively correspond to diffusion-limited (fast) and reaction-limited (slow) aggregation. In diffusion-limited aggregation, every contact between particles is assumed to result in the formation of a particle-particle bond or adhesion. In reaction-limited aggregation only a fraction of the particle-particle contacts that occur result in the particles adhering to one another. The stability ratio may be evaluated experimentally by comparing the aggregation rate observed under favorable ($\alpha = 1$) and unfavorable ($\alpha \to 0$) interfacial interaction conditions. For particles that are stabilized primarily by electrostatic repulsion, there is a critical ionic strength (which varies as a function of electrolyte valence) referred to as the critical coagulation concentration (CCC), above which aggregation is assumed to be transport-limited and the value of W is taken as unity. At ionic strengths below the CCC, repulsive electrostatic interactions become significant and W is greater than one.

Aggregation rates can be measured using static or dynamic light scattering (depending on particle size) of mean particle diameter. In the case of particles stabilized by electrostatic repulsion, the stability ratio is then determined by taking the ratio of rates observed at ionic strengths below and above the CCC. Rates may be determined for either monodisperse or polydisperse suspensions, though the latter case is exceedingly more complex as a range of particle sizes, shapes, and interaction potentials must be considered. As an alternative to extracting initial aggregation rates from measurements of mean aggregate diameter over time, measurements of particle size distribution over time can be compared with those calculated from a particle population model. In this case, W is treated as a fitting coefficient. Such an approach is well suited to polydisperse suspensions and has been used to track nanoparticle aggregation [16].

The classical kinetic approach of Von Smoluchowski can be adapted to a particle population model that may be used to assess the aggregation behavior of nanoparticles. A population balance describing the kinetics of aggregation and the evolution of the particle size distribution over time can be written as a system of differential equations [39]. The assumption that aggregation occurs in sequential steps of transport and attachment is embodied in these equations by multiplying rate constants that describe particle transport by stickiness coefficients that describe attachment. In the case of an irreversible aggregation, the evolution of the number concentration, n_k, of a particle in size class k can be written as follows:

$$\frac{dn_k}{dt} = \frac{1}{2} \alpha \sum_{i+j \to k} \beta(i,j) n_i n_j - \alpha n_k \sum_i \beta(i,k) n_i \tag{8}$$

where α is the attachment efficiency (or stickiness coefficient, equal to $1/W$), and $\beta(i, j)$ are the collision frequencies (transport) between particles or aggregates of classes i and j. The population balance can be modified to account for factors such as particle sedimentation and particle breakup. Approaches to computing the collision frequency kernel $\beta(i, j)$ typically consider three principal collision mechanisms: Brownian diffusion (β_{br}), shear-induced collisions (β_{sh}), and differential settling (β_{ds}).

Two ideal cases have been presented in the literature for calculating particle collisions by each of these mechanisms. In one case, the rectilinear model, particles are assumed to have no effect on the fluid they are suspended in, as though all fluid passes through the aggregate or particle as it diffuses, settles, or is carried by the fluid. In contrast, the curvilinear model [40] considers the effect of creeping flow around particles and compressed streamlines as two particles approach one another. The curvilinear model predicts many fewer collisions between particles than does the rectilinear model since fluid must be displaced before particle contact occurs. The curvilinear model with particle transport defined primarily by Brownian diffusion would appear to be appropriate for describing fullerene transport early during the aggregation process, particularly if the resulting aggregates are compact and highly ordered. However, if aggregation leads to the formation of porous aggregates, neither the rectilinear nor the curvilinear model yields satisfactory results compared with laboratory observations. An intermediate model [41] is required that considers the entire range of cases where the effect of fluid flow around each aggregate as well as the potential for fluid to pass through the aggregate are taken into account. Closed-form approximations have been proposed [42] for calculating the collision frequency kernel as a function of the ratio of drag force on a permeable aggregate to that on an impermeable sphere, Ω, and the fraction of fluid that passes through an isolated aggregate, η:

$$\beta_{Br(i,j)} = \frac{2kT}{3\mu}\left(\frac{1}{\Omega_i r_i} + \frac{1}{\Omega_j r_j}\right)(r_i + r_j) \qquad (9)$$

$$\beta_{sh(i,j)} = \frac{1}{6}\left(\sqrt{\eta_i r_i} + \sqrt{\eta_i r_i}\right)^3 G \qquad (10)$$

$$\beta_{ds(i,j)} = \pi\left(\sqrt{\eta_i r_i} + \sqrt{\eta_i r_i}\right)^2\left|u_i^* - u_j^*\right| \qquad (11)$$

where r is the radius of the aggregate or particle, μ is the dispersing medium viscosity (0.000890 kg/m sec for water), G is the mean shear rate (s^{-1}), and u^* is the settling velocity of the particle or aggregate. Assuming that the transport step of aggregation can be adequately expressed by calculated values for the collision kernel, β_{ij}, then the stability ratio can be determined by adjusting its reciprocal, α, so that calculations of particle size distributions over time correspond with the observed particle size distributions.

For a suspension of nanoparticles all having the same size, assuming that collisions between particles j leads to irreversible attachment with an efficiency of α, the decrease in number concentration of unit particles of size j in suspension can be estimated as:

$$\frac{dn_j}{dt} = -\alpha\beta_{jj}\left(\frac{n_j}{2}\right)^2 \qquad (12)$$

where β_{jj} is the specific particle collision frequency, and n_j is the number concentration of particles j. If all particle contacts result in aggregation, then $\alpha = (1/W) = 1$ and $\frac{dn_j}{dt} = -\beta_{ii}\left(\frac{n_j}{2}\right)^2 = K_{jj,fast}$. Compared to micron-sized particles, nanoparticles will have considerably smaller inertia and approach velocities prior to impact with another particle [16]. This results in extended times in the interaction fields of the other particle, meaning that nanoparticles are more likely to be deflected away as a result of repulsive forces. This would also suggest that nanoparticles do not "collide" in the literal sense, but rather slow down to a velocity of zero as they approach one another. If we describe aggregation as occurring between two groups of identically sized particles, each with an initial concentration equal to one-half of the total number concentration n_j, the solution to equation 12 for the time t_a required to reduce the initial particle concentration to a fraction $(1 - n_j/n_{j0})$ of its initial value through aggregation is:

$$t_c = \frac{4}{\alpha\beta_{jj}n_{j0}}\left(\frac{n_{j0}}{n_j} - 1\right) \tag{13}$$

From Eq. 13 it becomes evident that t_a varies inversely with the collision frequency β_{ij} and with the initial particle concentration n_{j0}.

In a simple system where particle collisions occur via the rectilinear model as a result of Brownian motion and mechanical stirring in the suspension (thus ignoring settling, which should be minimal for nanoparticles), then $\beta_{jj} = \beta_{jj\,Br} + \beta_{jj\,sh}$ can be expressed from Eqs. 9 and 10 as:

$$\beta = \frac{8kT}{3\mu} + \frac{32}{3}r_j^3 G_m \tag{14}$$

The evolution of β_{jj} and n_j for a fixed particle volume concentration is compared in Figure 7.7 as a function of particle radius r_j. For a given particle

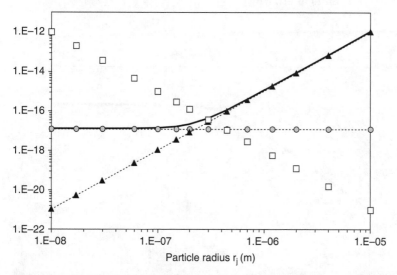

Figure 7.7 Theoretical prediction of nanoparticle aggregation kinetics by simple coagulation mechanisms as a function of size according to Von Smoluchowski model under Brownian motion and mechanical stirring ($G_m = 100 \text{ s}^{-1}$). Particle internal density and total weight are kept constant whatever the size. The calculations shown are n_{j0} (squares), $\beta_{jj\,sh}$ (triangle), $\beta_{jj\,Br}$ (circles), and the total collision frequency β_{jj} (line) in m^3s^{-1}.

volume concentration, the number concentration of particles, n_j decreases inversely proportional to r_j^3. The β_{jj} (nm^3 s^{-1}) coefficient is only dependent on the particle volume or size when other environmental parameters are kept constant. When r_j is sufficiently high so that the Brownian contribution ($\beta_{jj\,Br} = \frac{8kT}{3\mu}$) is negligible with regard to the mechanical shearing contribution ($\beta_{jj\,sh} = \frac{32}{3}r_j^3 G_m$) (eq. 14), then β_{jj} varies as r_j^3. Conversely, as particle size decreases, the contribution from Brownian motion becomes significant and induces a horizontal asymptote at $\beta_{jj\,(rj \to 0)} \approx \frac{8kT}{3\mu}$. In Figure 7.7, the magnitude of the mean shear rate ($G_m = 100$ s^{-1}) is representative of conditions in natural systems. Under these conditions Brownian motion is significant for particles with a radius smaller than 200 nm. Therefore, there are two diffusion mechanisms for particles with radii smaller and larger than 200 nm. For $r_j < 200$ nm, Brownian motion is the dominant contribution for particle mobility: β_{jj} is constant and t_a varies inversely with n_j, and thus decreases with r_j^3 (Figure 7.8). For $r_j > 200$ nm, mechanical stirring becomes dominant and t_a varies with $\frac{1}{\beta_{jj}\,n_{j0}}$, i.e. with r_j^3/r_j^3, and thus reaches a plateau. It is important to bear in mind, however, that changes in environmental parameters such as G, T, or μ will shift this mechanism balance toward lower or larger sizes, according to Eq. 14. Nevertheless, this example demonstrates the important relationship between particle size and aggregation kinetics, especially for nanoparticles, where dispersions are easily formed in the absence of mechanical stirring.

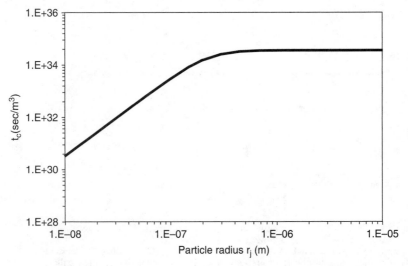

Figure 7.8 The time t_{c90} required for 90% of the initial particle population to form aggregates as a function of particle size.

We can continue our theoretical consideration of aggregation to include calculations of the stickiness coefficient, α or its reciprocal, W. The stability ratio can be calculated from theory as an integral measure of the interaction potential between two particles:

$$W = 2r \int_{2r}^{\infty} \frac{\exp[V(R)/kT]}{R^2 G(R)} \, dR \qquad (15)$$

where R is the center to center separation distance between two particles; $V(R)$ is the interaction potential between two particles at distance R; and $G(R)$ is a dimensionless hydrodynamic resistance function. The coefficient $G(R)$ accounts for the additional resistance caused by the squeezing of the fluid molecules between two approaching particles, thus as particle size decreases the resistance imposed by the fluid molecules also decreases (imagine a basketball versus a golf ball traveling through water). The hydrodynamic resistance function is close to unity for nanoparticles and may therefore generally be neglected in these cases. Unfortunately, in many cases observed stability ratios do not typically compare quantitatively with those calculated from theory, which is attributed to an incomplete assessment of the interfacial energy conditions between interacting surfaces in water and geometrical considerations. Calculations of the stability ratio have typically relied on extended DLVO theory to describe the interaction potential. Although calculated values of the stability ratio provide insight into conditions that favor aggregation and those that do not, there is not quantitative agreement between calculated and experimentally observed values of the stability ratio.

Formation of nanoparticle aggregates

While we tend to think about nanoparticles as unusually small objects, in fact they will often be present in aqueous systems as larger aggregates of the primary nanoparticles, even in the absence of any potential destabilizing agents (e.g., salts, polymers, organic materials). Spontaneous aggregation of nanoparticles can be an important impediment to nanomaterial handling. We are therefore interested in understanding when nanoparticles remain as discrete units or when they aggregate into clusters through attractive interfacial interactions and/or Brownian collisions. For example, upon agitation in water, fullerene C_{60} molecules tend to form clusters (nC_{60}) between 100 to 200 nm in diameter that are relatively stable [43]. The stability of these clusters is interesting in itself since the initial C_{60} is virtually insoluble in water [44–46]. The fact that cluster formation occurs is suggestive of changes in the surface chemistry of the C_{60} with time and exposure to water. This hypothesis is supported by adsorption isotherms for C_{60}, which illustrate the affinity of these materials for water (Figure 7.9). Here, the sample mass is measured as a function of varying the water vapor pressure $[P_{(H_2O)}/P_0]$ from 0 to 1 (adsorption isotherm) and back to 0 (desorption isotherm). The isotherm can be decomposed in two steps: 1) at $P/P_0 < 0.7$, the low slope of the adsorption isotherm indicates a weak affinity for water, and 2) at $P/P_0 = 0.7$, the step in the slope indicates a new surface chemistry resulting in an exponential increase of the isotherm and the multilayer adsorption of water molecules at higher vapor pressures. The C_{60} is thus hydrophilic for $P/P_0 > 0.7$. Moreover, this modification in the C_{60} surface appears to be irreversible since a significant hysteresis persists between the adsorption/desorption steps over the entire vapor pressure range. After desorption, one monolayer of water remains adsorbed on the C_{60} surface, perhaps due to clathrate formation [15, 19] or surface hydroxylation around $P/P_0 = 7$ during the adsorption step. The resulting nC_{60} aggregates have a negative surface charge as measured by zeta potential.

Light scattering and Transmission Electron Microscope (TEM) analyses of several common nanoparticles suggest that the spontaneous formation of nanoparticle aggregates occurs with many nanomaterials dispersed in water (Figure 7.10).

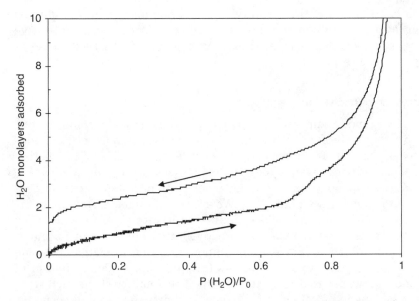

Figure 7.9 Adsorption/desorption isotherm of water vapor onto pristine C_{60} powder obtained from gravimetric analysis and nitrogen adsorption isotherms.

This may be attributed in part to a decreasing electrostatic repulsion between charged particles with decreasing size for those particles that develop a charge in water. This topic was covered earlier in this chapter showing the influence of particle size on the interfacial energy for particles in water (Figure 7.6). The size, structure, and chemical properties of these clusters are dependent on the characteristics of the constituent nanoparticles [47] and the process by which the particles are put into suspension [43, 47]. Unmodified titania nanoparticles form stable clusters with a narrow size distribution. Cluster size measured by light scattering is confirmed by TEM imagery, and the primary TiO_2 particles within the cluster are evident. A striking characteristic of the nC_{60} shown in

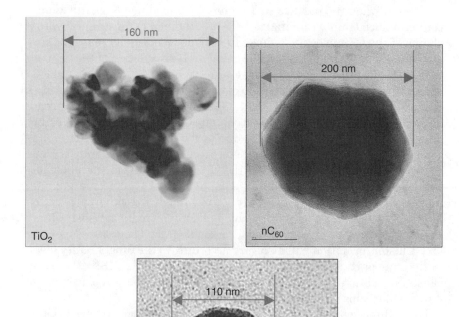

Figure 7.10   TEM images of titanium dioxide, nC_{60}, and fullerol nanoclusters formed after introduction into water.

Figure 7.10 is the presence of well-defined facets and hexagonal shape in 2D projection. In this case, the nC_{60} is formed by solvent exchange, where the C_{60} is initially dissolved in an organic solvent and then mixed with water. The images of clusters produced in this case suggest that aggregate formation may resemble a process of crystal growth rather than undirected particle aggregation. Similar observations have been made for nC_{60} produced through other solvent-exchange techniques for dispersing these materials in water [43]. When nC_{60} is formed by extended mixing in water without the intermediary use of organic solvents, the resulting aggregates are much less organized.

It is also interesting to note that modification of the nanoparticle surface chemistry to enhance the solubility of the nanoparticle may not in fact result in a molecular dispersion [48]. In other words, cluster formation may occur even when the particle surface has been modified (e.g., functionalization or surfactant wrapping) to enhance stability. For instance, hydroxylation of the C_{60} molecule to make fullerol does indeed increase the rate at which stable suspensions can be formed. Hydroxylation of the C_{60} molecule to form fullerol increases the affinity of the C_{60} for the aqueous phase. However, these molecules do not appear to exist as discrete entities (Figure 7.10). Instead, as shown in the TEM of a fullerol cluster, they readily form spherical clusters composed of many $C_{60}OH_{20-24}$ molecules [49]. For fullerol, cluster formation is likely due to the distribution of OH groups across the C_{60} surface resulting in heterogeneous interfacial interactions. The distribution of –OH groups on the C_{60} results in the formation of distributed hydrophobic/hydrophilic regions on the C_{60} molecule [48, 50]. Agglomeration of the $C_{60}OH_{20-24}$ may occur through attractive patch-patch interactions between the hydrophobic regions. Graphite similarly shows a strong tendency to aggregate and form highly fractal aggregates in water [51]. Cluster formation will also vary according to the functionality given to the nanoparticle surface as has been found for C_{60} [48, 49, 52, 53]. Cluster formation will thus vary depending on the type and placement of functional groups on the nanoparticle surface.

The influence of surface modification on nanoparticle aggregation is further illustrated in the case of maghemite (Fe_2O_3) nanoparticles. At a pH of 7, maghemite nanoparticles aggregate to form large clusters (Figure 7.11a). When the pH is reduced to a value of 3, protonation of the maghemite surface promotes nanoparticle dispersion. To reduce or prevent this tendency of nanometric maghemite to aggregate in water at natural pH, the nanoparticle surface may be modified by applying a coating of meso-2,3-dimercaptosuccinic acid (DMSA, $C_4S_2O_4H_6$) (Figure 7.11b) [54]. The coating procedure begins with an acidic solution containing a molar ratio of DMSA to Fe total of 9.2 percent. Then the solution is alkalinized as illustrated in Figure 7.11b. For pH < 8.5, the DMSA surface coating does not produce sufficient surface charge to prevent the

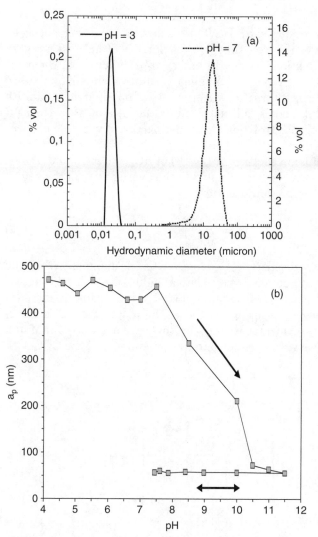

Figure 7.11   The pH-dependence of the nano-Fe_2O_3 size distribution
a) without and b) with a DMSA coating procedure [1].

aggregation of the 6 nm particles, and the measured size is approximately
450 ± 20 nm. For pH > 8.5, the DMSA generates enough negative surface
charge to enhance the stability of the suspension, and the maghemite
nanoparticles are well dispersed as small aggregates with an average
size of 20 ± 1 nm at pH = 11. Once these modified maghemite nanoparticles are exposed to a solution pH > 8.5 they are capable of maintain-
ing a stable suspension and do not aggregate even when the pH is
reduced to the acidic range [1].

Ionic strength effects

Qualitative trends predicted by DLVO-type models have been observed in nanoparticle suspensions—that is, that particles tend to aggregate more quickly at higher ionic strengths and/or at pH values near the point of zero charge [55]. Both of these changes in solution chemistry reduce repulsive electrostatic interactions. Adherence of nanoparticle stability to classical colloidal models is illustrated by the case of nC_{60} aggregation in two different electrolytes, NaCl and $CaCl_2$, of variable concentration (Figure 7.12).

The average diffusion coefficient ($D^{nC_{60}}$) of nC_{60} clusters and higher order aggregates is inversely correlated to the average size of the aggregates via the Stokes-Einstein relation:

$$R_h = \frac{kT}{6\pi\mu D} \tag{16}$$

where R_h is the average cluster (nC_{60}) hydrodynamic diameter. When the salt concentration increases above a critical value, the critical coagulation concentration (CCC), D decreases abruptly, indicating an increase in average size due to cluster aggregation. The respective CCCs measured with $CaCl_2$ and NaCl for the nC_{60} in this case are 2 ± 0.4 mM and 100 ± 0.4 mM. These values conform to the dependence of the CCC with

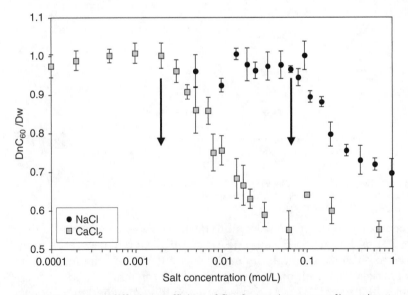

Figure 7.12   Average diffusion coefficient of C_{60} clusters in aqueous dispersion as a function of salt concentration and valence; $CaCl_2$ (squares) and NaCl (solid circles). Data are normalized to the reference diffusion coefficient in absence of salt D_w. The arrows point the respective critical coagulation concentrations, above which the dispersion is destabilized.

the sixth power of the counterion valence, as predicted by the Schulze-Hardy rule for ideal systems [56].

It is interesting to note that these CCCs are on the order of the salinity of water in freshwater and ocean environments. This implies that in the natural aquatic system, the stability of these nanoparticle clusters should be highly sensitive to the water salinity. The influence of ionic strength on the initial formation of nC_{60} suggests that the stability of fullerene dispersions is largely electrostatic in origin and that the magnitude and range of these interactions determine cluster growth rates and size [33, 34, 49].

Electrophoretic mobility measurements (taken as an indicator of surface charge) for a variety of different nanoparticles as a function of pH reveal a classic curve of increasingly negative electrophoretic mobility as solution pH becomes more basic (Figure 7.13), while mobility approaches zero with increasing ionic strength [4, 12, 57]. DLVO calculations suggest that as ionic strength increases, there is a reduction in the energy barrier between nanoparticles due to compression of the electric double layer, which allows for the attractive van der Waals interactions to dominate, leading to the formation of m-scale aggregates. Here m-scale implies agglomerates of initially formed, stable n-scale clusters of nanoparticles.

The reader must bear in mind, however, that aggregation may differ amongst different nanoparticle size fractions. In other words, nanomaterials cannot be treated as a single class of materials where for example

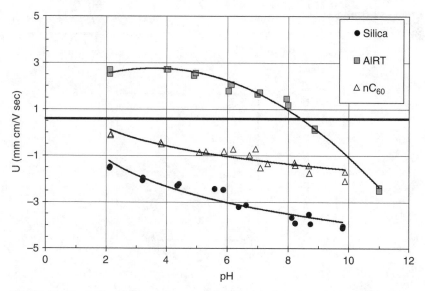

Figure 7.13 Evolution of electrophoretic mobility with changing solution pH for silica (d = 50 nm), alumoxane (AlRT, d = 24 nm), and nC_{60} (d = 168 nm) nanoparticles (I = 10 mM NaCl, T = 25°C).

a 10-nm particle is treated in the same manner as that of a 100-nm particle. In fact, a range of behaviors may be observed for particles that are characterized by sizes on two different ends of the nano-spectrum (1 to 100 nm). For instance, the cases illustrated above were clusters of nanoparticles in the larger nano-size fraction (d ~ 100 nm) as defined here. Indeed, as particle size approaches that of the lower size range (d < 50 nm) a deviation from classical aggregation behavior emerges. This deviation is linked to changes in the interfacial interactions that occur as particle size approaches the nano-domain. To illustrate this point, let us consider the findings of Kobayashi et al. [22], who observed a size effect on the aggregation rate constant for particles of similar surface chemistries. Kobayashi et al. [22] examined three different sizes of silica nanoparticles (80, 50, and 30 nm) and determined that only the aggregation of the 80 nm particles followed predictions based on DLVO theory (Figure 7.14). Smaller 50 nm

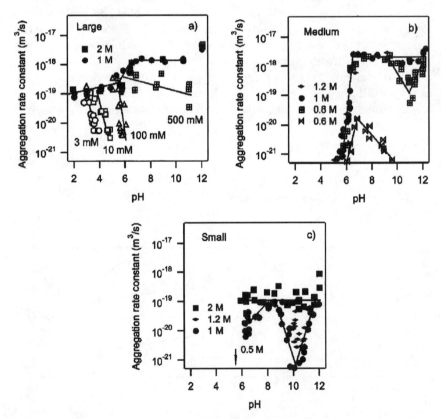

Figure 7.14 Experimental results of the aggregation rate constants for silica particles as a function of pH at different ionic strength: (a) d = 80 nm, (b) d = 50 nm, and (c) d = 30 nm silica particles. The lines serve to guide the eye.

particles aggregated more slowly at lower pH values, while the 30 nm particle dispersions remained stable under these conditions. These differences were attributed to additional interfacial interactions (hydration and the presence of a gel layer) that became more pronounced with decreasing particle size.

Deposition

The deposition of nanoparticles on stationary surfaces will be a key factor in determining their mobility, persistence, and fate in environmental and engineered systems. Similarly, the effective exposure and dose of nanoparticles experienced by organisms will be influenced to a large degree by the potential for deposition on biological surfaces such as gills and lungs. Analogous to particle aggregation, we will consider particle deposition as a process of transport and attachment, focusing on the case of particle deposition from aqueous suspensions in porous media such as aquifers or filters.

Particle deposition in porous media

The mobility of submicron-sized particles in aqueous systems has been explored extensively, particularly for the case of latex suspensions [6]. Analysis of the transport and deposition of particles in porous media typically begins with the convective-diffusion equation, which under steady-state conditions is generally written as [12]:

$$\frac{\partial n_j}{\partial t} + \nabla \cdot (\mathbf{u} n_j) = \nabla \cdot (\mathbf{D} \cdot \nabla n_j) - \nabla \left(\frac{\mathbf{D} \cdot \mathbf{F}}{kT} n_j \right) + \frac{\partial n_j}{\partial t}_{\text{RXN}} \qquad (17)$$

where n_j is the number concentration of particles of size or type j in the suspension; $\mathbf{D}$ is the particle diffusion tensor; $\mathbf{u}$ is the particle velocity vector induced by the fluid flow; $\mathbf{F}$ is the external force vector, and $\frac{\partial n_j}{\partial t_{\text{RXN}}}$ is the reaction of these particles with a surface (deposition) or with other particles (aggregation). The external force vector may be used to account for forces such as those arising from gravity and interfacial chemical interactions typically accounted for in the extended DLVO model (see the section entitled "Physicochemical Interactions").

As in the case of aggregation, particle deposition can be considered as a sequence of particle transport (in this case to an immobile collector) followed by attachment to the surface [58]. In a porous medium such as a filter or groundwater aquifer, fluid flow can be described by the Happel sphere in cell model [59], in which grains within the porous medium are assumed to be spherical collectors of radius a_c, surrounded by an imaginary outer fluid sphere of radius b with a free surface (Figure 7.15). The imaginary fluid envelope contains the same amount of fluid as the relative volume of fluid to the collector volume in the entire medium [i.e., $(a_c/b)(1-\varepsilon)^{1/3}$] where ε is the porosity of the given medium. Particles can be transported to the collector surface through a combination of interception, gravitational settling, and diffusion transport mechanisms. Interception occurs when fluid streamlines pass sufficiently close to the collector so that contact results. Contact through gravitational settling and diffusion results from the particles crossing the fluid streamlines via the respective mechanisms within a critical region around the collector and contacting the surface.

Analytical solutions for particle transport due to Brownian diffusion have been combined with particle trajectory calculations to yield a closed-form solution for the transport of particles to the surface of spherical collectors expressed as the theoretical single collector contact efficiency (η_0) [10]:

$$\eta_0 = \eta_D + \eta_I + \eta_G \tag{18}$$

where η_D is the single collector contact efficiency for transport by diffusion; η_I is the single collector contact efficiency for transport by interception; and

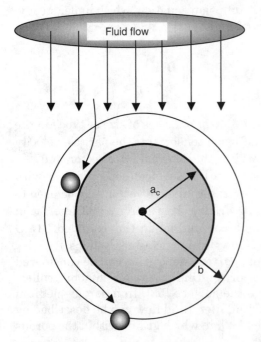

Figure 7.15 Illustration of the Happel sphere model in which the spherical collector having a radius a_c is surrounded by a fluid envelope having a radius b.

η_G is the single collector contact efficiency for transport by gravity. Eq. 18 is based on the additivity assumption where η_0 is determined by summation of the three independently determined transport mechanisms.

It is instructive to look at the components of η_0 plotted as a function of particle size (Figure 7.16). For particles in the micron size range, transport to a collector surface is largely governed by interception and gravitational settling mechanisms. However, nanoparticle transport to a collector will typically be dominated by diffusion [10]. It is evident that as particle size approaches that of the nanoscale (d <100 nm) the interception and gravity terms begin to approach zero. On the other hand, once particle size is smaller than approximately 300 nm, the particle contact efficiency is largely controlled by η_D. The diffusion component η_D is a function of the porosity of the transporting medium, the aspect ratio between the collector and particle sizes, the particle approach velocity to the collector surface, and the Hamaker constant for the interacting surfaces. With the high diffusion component for nanoparticles their mobility may be especially low in porous media characterized by low flow rates or Peclet numbers, such as groundwater aquifers.

We obtain a similar picture of particle mobility when the efficiency of single collectors is integrated across a length of the porous medium that

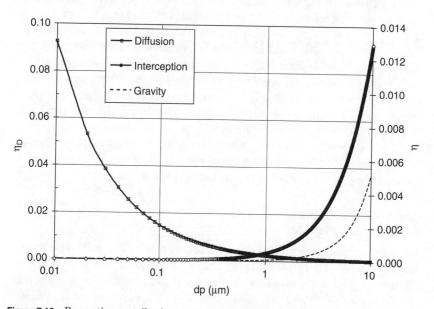

Figure 7.16 Respective contributions to the total single collector contact efficiency from each of three transport mechanisms: diffusion (η_D), interception (η_I), and gravitational settling (η_G). The magnitude of each mechanism is plotted as a function of particle size while holding other particle characteristics constant.

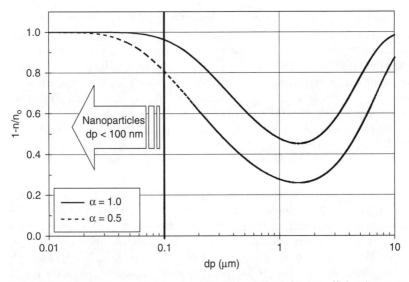

Figure 7.17 Particle removal assuming two different attachment efficiencies as a function of particle size calculated using equation 4.4 ($v = 0.14$ cm/sec, $\varepsilon = 0.43$, $a_c = 355 \mu$m, L = 9.25 cm)

they comprise. In Figure 7.17, particle removal, represented as the mass concentration of particles removed, $1-n/n_0$, is plotted as a function of particle size on a semi-log scale. Increasing the fluid flow rate or flow velocity serves to increase particle mobility. Particles in the nano-size range ($d_p = 1$ to 100 nm) are relatively immobile, which is indicated by their high removal values. As particle size increases, mobility increases until a relative maximum in mobility is reached at around 1.5 μm. For particles larger than approximately 1.5 μm mobility begins to decrease as the collector efficiency increases due to a larger contribution by interception and gravity. Holding all other variables constant, nanoparticles should be relatively immobile under a given set of conditions in comparison to other micron-sized particles of similar surface chemistry.

Quantifying nanoparticle deposition and mobility

It is somewhat nonintuitive that the very small size of nanoparticles may actually lead to less mobility (a higher deposition rate) in a porous medium such as a groundwater aquifer or a filter in comparison with larger particles. Observations of particle deposition in porous media over a wide range of particle sizes suggest that the current description of particle transport to collectors is reasonably complete [59–61]. However, when particle attachment is not favorable only a fraction of the particles transported to the collector surface will attach; therefore, the single collector efficiency must be modified.

Indeed, transport is only part of the story. The surface chemistry of nanoparticles rather than their size is likely to be the factor that determines mobility. Just as aggregation was described mathematically to be the product of factors transport (β) and attachment (α), the probability that a particle approaching a single collector deposits on the collector can be written[38] as the product of the collector contact efficiency and the attachment efficiency:

$$\eta = \eta_0 \alpha \qquad (19)$$

When particle removal is calculated for a lower value of the attachment or stickiness coefficient, particle removal decreases, corresponding to an increase in mobility (Figure 7.17).

The ratio of the rate of particle deposition on a collector to the rate of collisions with that collector is the attachment efficiency factor, α, and is analogous to the stickiness coefficient or $1/W$. Theoretical predictions of the attachment efficiency in this case are identical to those for particle aggregation, typically based on DLVO-type calculations that consider the balance of forces arising from interactions at very small separation distances between particles and the collector surface. These phenomena may be important over length scales that are large by comparison with nanoparticle dimensions. Similar to the case where the stickiness coefficient is treated as a fitting variable in the particle population models, the attachment efficiency can be treated as an empirical parameter that captures all aspects of particle deposition not described by the more extensively validated particle transport models. The empirically determined attachment efficiency should vary with changes in surface and solution chemistry.

Measurements of particle removal across a length (L) of a homogeneous porous medium composed of spherical grains of radius (r_c) and porosity [38] can be combined with calculations of particle transport to yield estimates of the attachment efficiency factor [38]:

$$\alpha = \frac{4a_c}{3(1 - \varepsilon)\eta_0 L} \ln\left(\frac{n_j}{n_0}\right) \qquad (20)$$

where n_j and n_0 are respectively the particle number concentrations present at distance L in the column effluent and influent to the column; η_0 is the clean bed single collector contact efficiency, which describes the particle transport to an individual collector and can be calculated as a function of the Darcy velocity, porous medium grain size, porosity, and temperature among other variables using Eq. 18. Using experimentally measured n_j/n_0 values (fraction of influent particles remaining after passing through the porous medium) and theoretical η_0 values, values of α can be calculated for a given particle suspension.

Particle detachment from a collector is dependent on the nature of the interaction between the two surfaces. In the absence of an energy barrier, the rate of particle detachment from the collector surface will be controlled by the ability of the particle to diffuse across the diffusion boundary layer [62]. When an energy barrier is present, however, the deposited particle must overcome an energy (ϕ_T) that is equivalent to the depth of the primary minimum (ϕ_1) plus the height of the energy barrier (ϕ_2) (Figure 7.18) in order to go back into the bulk suspension. Similarly, nanoparticles deposited in a secondary minimum must overcome an energy that is equivalent to its depth (ϕ_{min}). As noted in the section "Physicochemical Interactions," the energy barrier height of primary and secondary minima depths decrease with decreasing particle size. This being the case, nanoparticle remobilization should be more sensitive to changes in solution chemistry than larger particles.

Particle deposition or attachment may also take place in the attractive secondary minimum if it is present [6]. The ability of particles to deposit in a secondary minimum is dependent on their size and on the

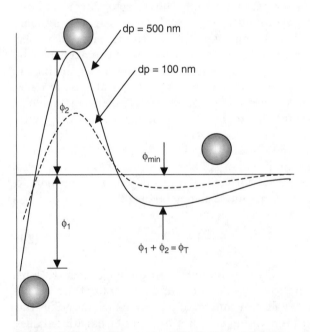

dp = 500 nm

dp = 100 nm

ϕ_2

ϕ_{min}

ϕ_1

$\phi_1 + \phi_2 = \phi_T$

Figure 7.18 Interaction energy profile for two particle sizes illustrating the total interfacial energy (ϕ_T) that must be overcome in order for a particle to detach from a collector surface and become resuspended in the bulk suspension.

kinetic energy imparted to them by the fluid flow. As particle size decreases, the secondary minimum becomes shallower, while the energy barrier decreases in height. In this way a transition between deposition in the secondary and primary minima will exist according to particle size. For example, Petit et al. [1973] found that for selenium sols this transition from primary to secondary minima deposition occurred at a particle size of around 55 nm. In other words, particles larger than 55 nm deposited in the secondary minimum while those smaller than 55 nm deposited in the primary minimum. This value of course will vary as a function of solution chemistry and particle-surface chemistry.

In most cases relatively good agreement exists between model predictions and experimental results when favorable deposition conditions exist, while there is more disagreement when unfavorable conditions are present. Deposition in the secondary minimum may resolve some of the discrepancy between theoretical and observed values of the attachment efficiency.

When electrostatic repulsion is a primary source of particle stability, the value of α, may be manipulated through changes in solution chemistry (ionic strength and pH). In Figure 7.19 the experimentally determined α for fullerene nanoclusters passing through a packed column of silicate glass beads is plotted as a function of changing solution chemistry, in this case ionic composition and strength. As the ionic strength and valency increase, the value of α increases. The increase in α in this case indicates that repulsive interactions that previously prevented surface

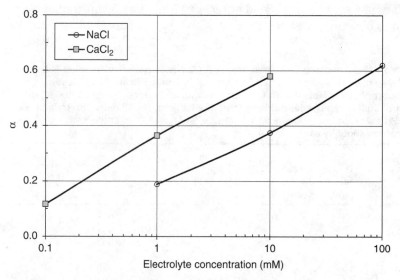

Figure 7.19 Experimentally determined attachment efficiency for nC_{60} clusters as a function of salt concentration (d_p = 168 nm; pH = 6).

attachment are reduced. This is best explained by a reduction in repulsive electrostatic interactions, which are most sensitive to changes in ionic strength and valency.

Theory suggests that as particle size decreases, the attachment efficiency should increase [63] due to a reduction in the height of the energy barrier. However, experiments with model systems looking at both particle stability and mobility have found that the attachment efficiency is largely insensitive to particle size. This discrepancy between theory and reality is attributed to an incomplete assessment of interfacial interaction energies and to changes in the nature of interactions with decreasing particle size. It is therefore unclear whether nanoparticles should be inherently less stable or more prone to depositing onto surfaces under favorable conditions than are larger particles.

Attachment efficiencies as calculated from experiments of particle deposition in a well-characterized medium [64] show wide variability as a function of composition and size (Table 7.1).

Information on nanoparticle transport and attachment can be combined to calculate an index of particle mobility in porous media based on the characteristic distance for removal from an initial source. The reciprocal of this distance, commonly referred to as the filter coefficient [38], is calculated according to the following expression:

$$\lambda = \frac{3}{2} \frac{(1 - \varepsilon)}{d_c} \alpha \eta_0 \qquad (21)$$

where λ has units of inverse distance. (When $(1/\lambda)$ is multiplied by ln n/n_0 one obtains an estimate of the distance required to reduce the

TABLE 7.1 Measured and Calculated Transport Characteristics of Selected Fullerene and Metal Oxide Nanoparticles in a Column Packed with Silicate Glass Beads. Here U is Electrophoretic Mobility, Which Is an Indicator of Surface Charge, and ϕ_{max} Is the Height of the Energy Barrier Between the Two Surfaces. The Distance to Reduce the Initial Particle Concentration by 99.9% Can Be Taken as a Qualitative Measure of Nanoparticle Mobility

	dp (nm)	U (μm cm/ s^{-1}V^{-1})	C/C$_0$	ϕ_{max} (kT)	α	Distance to 99.9% Deposition (m)
TTA/nC$_{60}$	44	−0.86	0.56	10	0.0001	14
C$_{60}$OH$_{20-24}$	106	−1.50	0.99	75	0.001	9.8
SWNT	16	−3.88	0.94	37	0.008	2.4
Silica ZL	106	−3.20	0.68	297	0.039	0.6
Silica OL	57	−1.95	0.97	135	0.169	0.2
Anatase	142	−0.60	0.56	50	0.298	0.1
FeRT	51	1.23	0.30	0	0.336	0.1
AlRT	24	1.72	0.85	0	0.895	0.06

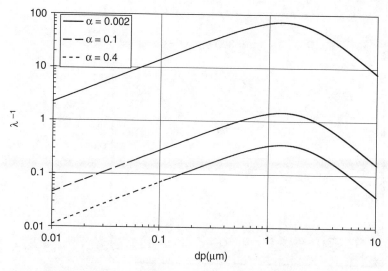

Figure 7.20 Travel distance (meters) in a homogeneous porous medium as a function of particle size ($v = 0.02$ cm sec^{-1}). Three different attachment efficiencies are assumed accounting for unfavorable and favorable deposition conditions.

initial particle concentration n_o to a concentration of n.) Plotting reciprocal λ as a function of particle size (Figure 7.20), it is evident that even for nanoparticles with low attachment efficiencies, the characteristic distance to which they will travel in a homogeneous media is low. Real systems present numerous complexities that may increase or decrease the true mobility of nanoparticles. For example, physical inhomogeneities such as fissures may increase mobility by providing preferential flow paths. Heterogeneities in media size may decrease porosity, thereby increasing deposition. Heterogeneities in surface chemistry may decrease the effective surface area available for deposition.

However, particle size also interacts with heterogeneities to determine the effective surface of the collectors that may be "visible" to a particle. While larger particles may see an average surface that is unfavorable to deposition, nanoparticles may be able to sample the surface at a finer scale, seeking out more favorable attachment sites. The ability to access more of the surface may combine with a higher rate of transport to the surface due to Brownian diffusion to reduce nanoparticle mobility. In one study by Schrick et al. [2], it was observed that the mobility of iron nanoparticles in soil columns was less than that of larger iron particles (Figure 7.21).

This observation has direct implications for the use of nanoparticles in groundwater remediation applications and removal by granular media filters. Highly reactive nanoparticles, such as nano-iron, have been proposed as a possible remediation tool for contaminants susceptible to reduction by Fe$^\circ$. However, both theory and experiments suggest that

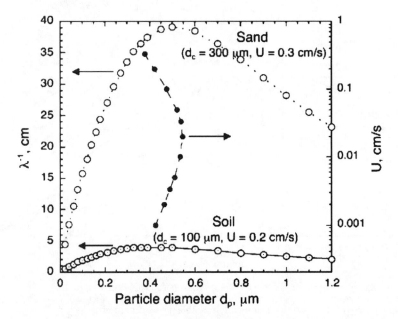

Figure 7.21   Open circles: Calculated filtration length (λ^{-1}) as a function of iron particle diameter, dp. Upper curve shows a calculation for Ottawa sand (ε = 0.42, dc = 300 m, Darcy flow velocity U = 0.3 cm/s). Lower curve shows a calculation for an average soil (θ = 0.42, dc = 100 m, U = 0.2 cm/s). Filled circles: Particle diameter with maximum calculated value of −1 as a function of Darcy flow velocity for soil (θ = 0.42, dc = 100 m). The shaded region highlights particle diameters associated with the greatest predicted mobilities. (Adapted from reference [2].)

these materials should have relatively low mobilities in such applications. Adjustments in particle size and surface chemistry to balance reactivity with mobility may resolve this issue. Low nanoparticle mobilities in porous media also suggest that it may be possible to remove many of these materials from water using granular media filters.

Detachment.   Once a particle attaches to a collector surface, it is subjected to a number of forces that simultaneously act to retain or displace it (Figure 7.22) [62]. The balance of forces acting on the particle determines whether the deposited particles may become remobilized through changes in solution chemistry and/or hydrodynamic conditions. The three principal forces acting on a deposited particle in a porous medium are typically taken to be the fluid drag force (F_D), the lift force (F_L), and the adhesive force (F_A). The adhesive forces act to retain the particle on the collector, while the hydrodynamic forces (F_D and F_L) act to favor particle detachment. The fluid drag force acting on a retained particle on a collector surface is calculated according to [62, 65]:

$$F_D = (1.7005)6\pi\mu v_p a_p \qquad (22)$$

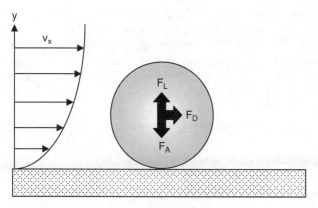

Figure 7.22 Illustration showing the fluid velocity gradient and forces acting on a particle once it has deposited onto a surface.

where the leading coefficient (1.7005) accounts for wall effects near the collector surface; a_p is the radius of the retained particle; μ is the viscosity of the fluid; and v_p is the fluid velocity at the center of the retained particle. The fluid velocity at the center point of the retained particle is calculated using the following relationship, which is derived using a representative pore structure such as the constricted tube model [65].

$$v_p = \frac{Q/N_{\text{pore}}}{(\pi/4)d_z^2} \frac{4(d_z/2 - a_p)}{(d_z/2)^2} \tag{23}$$

where Q is the volumetric flow rate through the porous medium; N_{pore} is the number of pores in a cross section of the packed column; and d_z is the diameter of the pore space in between the collectors. In this case, the pore space is comprised of a series of parabolic constrictions with the diameter being a function of the distance along the pore (z).

$$d_z = 2\left\{ \frac{d_{\max}}{2} + \left[4\left(\frac{d_c}{2} - \frac{d_{\max}}{2} \right)\left(0.5 - \frac{z}{h} \right)^2 \right] \right\} \tag{24}$$

where d_z is the constriction diameter at a distance z along the pore; d_c is the equivalent diameter of the constriction; $d_{\max}$ is the maximum pore diameter, and h is the pore length. For less well-defined and more complex flow geometries, accurately modeling the hydrodynamic torque will be difficult. The lift force on a spherical particle attached to a collector surface may be approximated as follows:

$$F_L = \frac{81.2 a_p^2 \mu \omega^{0.5} v_p}{v^{0.5}} \tag{25}$$

where ω is the velocity gradient at the collector surface and v is the kinematic viscosity of the fluid. The lift force arises from the different pressure forces acting on the top and bottom of the particle due to the velocity gradient. For nanoparticles, F_L is negligible in comparison to the adhesive force and may likely be omitted [62]. The result is a torque exerted on the particles due to the hydrodynamic drag given by:

$$T_D = 1.399a_p F_D \qquad (26)$$

where the leading coefficient (1.399) indicates that the drag force acts on the deposited particle at an effective distance of $1.399a_p$ from the surface [62]. In other words, the hydrodynamic torque acts over a lever arm of $l_y = 0.339a_p$.

The adhesive forces are a result of physicochemical interactions occurring between the two interacting bodies. The favorability or adhesion strength may be quantified in terms of the free energy of adhesion (W_A) using an XDLVO type approach for determining the free energy at contact. The adhesion force (F_A) may then be determined using different scaling models, such as the Johnson-Kendell-Roberts (JKR) [66] and Derjaguin-Mullen-Toporov (DMT) [67] models. This adhesive force resists the torque exerted by the drag force, and is manifested as a torque acting over a specified lever arm of l_x:

$$T_A = F_A l_x \qquad (27)$$

The value of l_x is taken to be equal to the radius of the contact area between the collector and the retained nanoparticle. The size of the contact area is determined by a number of different parameters, the most significant of which are the elasticity of the interacting bodies, particle size, and surface roughness.

Elasticity describes the stiffness or malleability of a material. As elasticity increases, the contact area will also increase in size as the surfaces can deform and wrap around each other [68]. As is illustrated in Figure 7.23, the contact area also decreases both on approach and at contact with decreasing particle size. Nanoparticles interact with a smaller region of the collector surface as they approach and ultimately make contact with the collector surface. Surface roughness may result in either an increase or a decrease in the contact area depending on particle size and the density of asperities on the collector surface. Roughness is likely to result in an increased contact area between the collector and the nanoparticle, resulting in an increase in the adhesive force.

Typically, fluid drag or hydrodynamic shearing forces are significant only for particles larger than several hundred nanometers when deposition occurs in the primary minimum [69]. This is due to several factors,

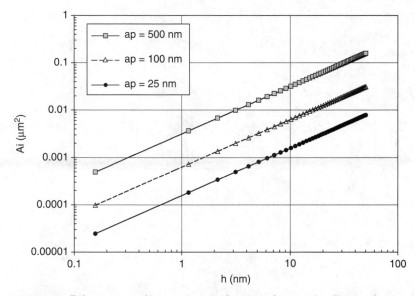

Figure 7.23 Relevant area of interaction as a function of separation distance for various particle sizes on approach to a flat surface. The interaction area decreases with decreasing particle size, illustrating the importance of surface heterogeneities on surface interactions.

such as surface roughness effects and the relative insignificance of hydrodynamic interactions in relation to thermodynamic ones at the collector interface. However, drag forces play a larger role in nanoparticle detachment when deposition occurs in the secondary minimum where the thermodynamic interactions are much weaker (see "Physicochemical Interactions"). The importance of hydrodynamic interactions relative to thermodynamic ones for nanoparticles can be understood by comparing the relative magnitudes of each. The mean kinetic energy (U_{KE}) imparted to a particle in a fluid flowing through a packed bed is determined according to [69]:

$$U_{KE} = \frac{1}{2}\, m_p \left(\frac{U}{\varepsilon}\right)^2 \qquad (28)$$

where U is the superficial pore velocity and m_p is the particle mass. This assumes that the particle velocity does not lag the fluid velocity, an assumption that should be valid for nanoparticles. The relationship between particle size and kinetic energy in a porous media is illustrated in Figure 7.24. Calculations of kinetic energy are plotted for two different fluid velocities that are representative of typical groundwater Darcy velocities. Particles smaller than 100 nm are the least impacted by changes in fluid velocity and have a relatively weak kinetic energy

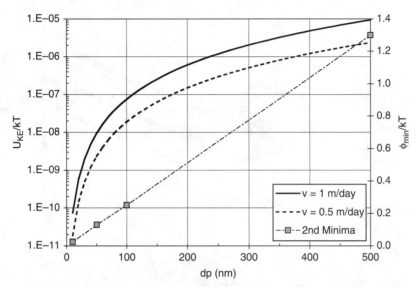

Figure 7.24   Kinetic energy and the depth of the secondary interaction energy minima as a function of particle size ($\zeta_p = -30$ mV; $\zeta_c = -20$ mV; $H_A = 10^{-20}$ J; $\varepsilon = 0.4$).

associated with them. The low kinetic energy associated with nanoparticles suggests that deposition into even shallow secondary minima may be possible. This may explain a lack of dependence of particle mobility on fluid flow velocity if the kinetic energy does not exceed the depth of a secondary minima or the height of the energy barrier.

Effect of surface roughness

Surface heterogeneities are commonly cited as the principal reasons for discrepancies between theoretical predictions and experimental results for surface controlled processes and have received considerable attention in the research literature [70–75]. These heterogeneities may be physical (e.g., roughness) or chemical (e.g., charge distribution) [4, 12]. Heterogeneity tends to become more apparent at smaller length-scales. Because the interaction area decreases with decreasing particle size (Figure 7.23), nanoparticles will be more affected by surface heterogeneities than larger ones [76].

One form of chemical heterogeneity is that of an uneven distribution of charge resulting from the uneven distribution of surface functional groups and crystalline structure defects, and the presence of surface impurities or contaminants, such as ferric, aluminum, and manganese oxides. The distribution of these heterogeneities may be thought of in terms of patches having different charge properties [77]. These patch-patch interactions appear to explain some of the observed variability in particle transport in chemically heterogeneous systems [73].

Greater surface roughness has been observed to increase deposition rate coefficients observed for some nanoparticles [65]. Similar observations have also been made for colloid deposition onto membrane surfaces [78, 79]. Enhanced particle deposition rates on rough surfaces are attributed to a combination of a reduced energy barrier on approach and physical trapping of the nanoparticles by surface features. For ideally smooth surfaces, interactions occur normal to the interacting surfaces; however, for a particle approaching a rough surface, it experiences both normal and tangential forces that can capture it in surface depressions. Furthermore, the variable height of the surface topography means that the approaching particle is experiencing a range of interactions as some are more prominent than others at different separation distances. For this reason it is difficult to fully describe a true surface interaction between heterogeneous surfaces in terms of a single energy curve. Surface roughness can also alter the magnitude and type (repulsive or attractive) of interaction between two surfaces [78]. This results from a reduction in the relative interaction area between the particle and the collector surface. One approach is to model surface asperities as a series of small hemispheres instead of a flat surface. The resulting calculations yield smaller interaction energies as the interaction energy is determined by the radii of curvature of the protrusions rather than that of the interacting bodies. Consequently, roughness reduces the height of the energy barrier between two surfaces in aqueous media, thus making particle deposition more favorable [74, 75]. For instance, Suresh and Walz [74] found that the van der Waals interaction energy significantly increased when the separation distance between two surfaces approached the height of the surface asperities. The number of asperities per unit area on the surface had a smaller influence on the magnitude of the interaction energy than did asperity size. A similar conclusion was reached by these authors regarding the electrostatic interactions—electrostatic repulsion occurs at a larger separation distance than would be expected for a smooth surface. At small separation distances, surface roughness has a more profound impact on van der Waals interactions (i.e., makes the attraction stronger). Hence, at shorter separations the height of the energy barrier is substantially reduced. In summary, surface roughness tends to reduce the depth of a secondary minimum and shifts it to longer separation distances while the repulsive energy barrier is decreased and the primary minimum is moved to larger separation distances.

Roughness features may also physically trap nanoparticles on the surface once contact has been made. This occurs through a combination of enhanced surface adhesion and modification of the surface fluid velocity pattern. Upon deposition, particles are principally subject to shearing forces that would cause them to "roll" across a surface. This rolling motion is resisted by the adhesion force acting between the surfaces, as discussed earlier in this section. Surface asperities act to increase the particle's adhesion lever arm while also reducing the shearing forces acting on the deposited nanoparticle. The impact of surface roughness on the adhesion between two interacting bodies has been explored at length [74–76, 78–80]. Even small asperities (< 1 nm) on a collector surface may dramatically increase the contact area between the nanoparticle and collector surfaces and, in turn, the adhesive lever arm for the nanoparticle (Figure 7.25) [65]. This is in contrast with the case of larger particles where the adhesion is reduced through a reduction in the contact area due to the asperities [76, 81, 82]. For larger colloids, roughness may decrease

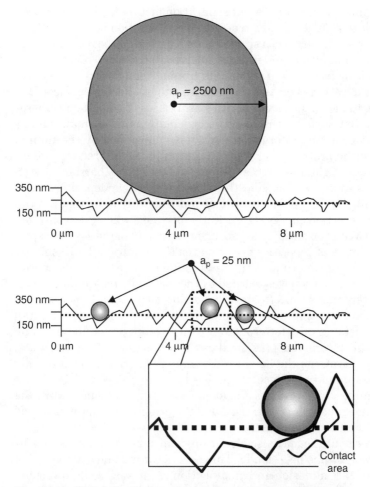

Figure 7.25 Scaled illustration of a large and small particle(s) on an AFM-generated height profile of a rough surface. The surface had an average roughness of 65 nm. The contact area between one of the small (a_p = 25 nm) particles with the rough surface is highlighted.

the actual contact area leading to higher mobilization rates. However, the opposite scenario is likely for nanoparticles as their size will allow them to fit in between even tightly spaced surface asperities. In other words, if the particle diameter is greater than the distance between any two asperities, adhesion will tend to be low, as the contact area is low [80]. However, if the particle diameter is less than this distance, the adhesion will be higher, due to a higher contact area. Thus, roughness should tend to increase the contact area between the collector and the depositing nanoparticle [78] and resuspension should be reduced.

Nanoparticle Behavior in Heterogeneous Systems

Environmental systems are characterized by a number of naturally occurring colloidal materials, such as clays, products from mineral weathering, bacterial polysaccharides, and other organic macromolecules. Natural organic matter (NOM) [83] is a broad class of organic macromolecules and is particularly relevant as it is ubiquitous to nearly all soil and water systems. Here we will consider cases where NOM and inorganic materials may interact with nanomaterials and alter their subsequent behavior in aqueous systems.

In laboratory analyses, particle properties are often characterized in relatively simple model systems such as in solutions of one or two background electrolytes added to purified and deionized water. The large number of components present in natural waters may include mono- and multi-valent salts, NOM, clays, microorganisms, and other colloidal materials. Each of these materials, either individually or in concert, can alter the characteristics of particles [84].

For example, the size and highly reactive nature of NOM confer on these molecules a high interfacial reactivity that influences physico-chemical interactions in rhizospheric and aquatic environments [85]. Nanoparticles may have similar properties. The diversity of nanoscale materials and the complex composition of soil suggests a nearly infinite number of interactions. Two likely possibilities that merit consideration in assessing the role of organics in determining the state of nanoparticles in aqueous systems and exposures to living organisms are: 1) nanoparticles adsorbed into organo-mineral complexes, and 2) nanoparticles diffusing through (or being entrapped within) extracellular materials and biofilms.

Particles in natural waters, regardless of size, may associate with organic macromolecules that subsequently alter the particle's interfacial and physical characteristics (e.g., charge, reactivity, size) [86]. For example, association of NOM with particles can alter their effective charge, reactivity, and hydrodynamic radius [84]. Nanoparticles in

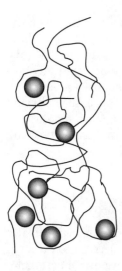

Figure 7.26 Illustration of nanoparticles entangled in a collection of organic macromolecules.

particular may be susceptible to the adsorption of or on other materials given their high reactivity and large surface areas. If nanoparticles are present as individual entities, adsorption of materials on these particles may not be an appropriate model since the material interacting with the nanoparticle may be similar or even larger in size. For nanoparticles with dimensions approaching several tens of nanometers or less, it is plausible that they may become incorporated into organic macromolecule structures (Figure 7.26). In this scenario the organic material acts as the adsorbing surface and thus, mobility may be dominated by the properties of the macromolecule rather than the nanoparticle.

However, in many cases nanoparticles may be present as larger colloidal aggregates, presenting a clear interface for adsorption. NOM or ions may form direct chemical bonds on the particle surface (inner sphere adsorption). Conversely, in outer sphere adsorption, no direct chemical bonds are formed and instead the adsorbate is held at the particle surface through electrostatic and/or hydrogen bonding forces. Adsorption of NOM to particle surfaces is a function of the solution chemistry (pH and ionic strength) and the characteristics of the organic molecule, such as molecular size and charge [84]. NOM has been shown to adsorb to negatively charged particles in aqueous media (e.g., iron oxide colloids). Under environmentally relevant conditions, humic acids readily adsorb to other negatively charged surfaces and in turn modify the physicochemical characteristics (e.g., charge) of the solid-liquid interface. The thickness of the adsorbed layer generally increases with increasing ionic strength [84, 86] and with increasing concentration of the organic macromolecule [86]. Adsorption of NOM on colloidal iron particles appears to be favored by neutral pH values (pH in the range of 4 to 6) and ionic strengths of 1 to 10 mM NaCl. Similar conditions may favor association with other nanoscale particles.

Naturally occurring organic matter and particle charge

Although the majority of surfaces in aqueous environments carry a net negative charge, the magnitude of this charge can vary considerably with differences in functionality and charge density. Adsorption of NOM to a surface will alter the charge properties of that surface, in most cases making it more negatively charged [86]. Modification of the particle surface charge will depend on a variety of factors, including the charge characteristics of the NOM (e.g., charge density and functionality), the mode of adsorption (inner versus outer sphere) [87], particle surface chemistry, and the solution chemistry.

The impact of humic acid on the surface charge of two types of fullerene nanoclusters, SON/nC_{60} and fullerol, is shown in Figure 7.27. Recalling that these clusters are very similar with the exception of their surface chemistries, a clear difference exists in their interaction with the NOM. Clusters of the fullerol (hydroxylated C_{60}) readily adsorb the tannic acid as indicated by the increasingly negative surface potential. On the other hand, very little change in surface charge is observed for

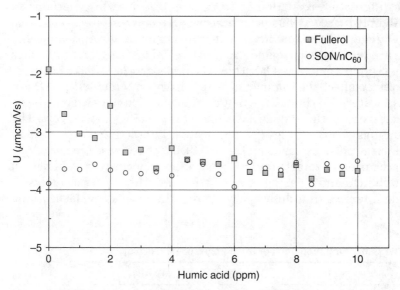

Figure 7.27   Change in nanoparticle electrophoretic mobility with increasing humic acid solution concentration in water (pH = 7.2; T = 25°C). Measurements were performed for fullerol clusters and C_{60} clusters formed through toluene exchange into water using sonication.

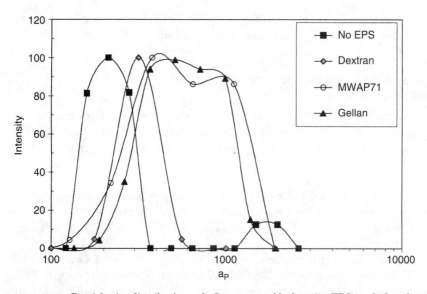

Figure 7.28   Particle size distributions of nC_{60} measured before (no EPS) and after the addition of three polysaccharides. In the presence of polysaccharide, these fullerene aggregates are destabilized, leading to an increase in aggregate size.

the SON/nC_{60}, suggesting that there is less adsorption of the NOM. These differences are attributed to the presence of hydroxyl groups on the fullerol surface as has been observed for magnetite nanoparticles [88]. For both fullerol and magnetite, the dominant adsorption mechanisms are thought to be ligand-exchange reactions between the surface hydroxyl groups and the NOM. Thus, the nanoparticle surface chemistry will greatly affect the manner in which it will interact with NOM in natural waters. Of course, the degree of adsorption may vary with the characteristics of the organic molecule. This hypothesis is supported by the findings shown in Figure 7.28 for nC_{60} interacting with a variety of bacterial polysaccharides (Table 7.2). In these cases, each of the polysaccharides interacts favorably with the nanoclusters as indicated by the change in the particle size distribution. However, each polysaccharide interacts in a different way with the nC_{60} as evidenced by a

TABLE 7.2   Structural Properties of Three Polysaccharides [89]

	Dextran	MWAP71	Gellan
Molecular Weight (Da)	2×10^6	5.7×10^5	3×10^6
Radius of Gyration (nm)	44	54.9	300
Charge Density	0	0.49 (anionic)	0.7 (anionic)

different evolution in the size distribution produced in each instance (Figure 7.28).

The unique aggregation results found observed for the various polysaccharides in Figure 7.28 may be traced to their different molecular chemistries and do not demonstrate a clear size effect. In this example, the acetate groups (CH_3COO^-) that are characteristic of the MWAP71 and gellan seem to promote aggregation. Conversely, no trend was evidenced with regards to molecular size and the degree of aggregation. Indeed, the smaller MWAP71 produced a similar degree of aggregation as the larger gellan, while less severe aggregation was seen with the dextran. While not considered as conclusive evidence, these results suggest that chemical interactions and not purely size must be considered when evaluating nanoparticle interactions with complex molecules like polysacharrides. In other words, we see that chemistry, and not size alone, continues to be important.

Beyond merely altering particle surface charge, adsorption of NOM may affect the type of interactions that occur between particles in water. For example, the thickness of the adsorbed organic layer may, under specific conditions (high salinity), exceed the thickness of the diffuse double layer, ultimately altering the nature of the interfacial interactions (Figure 7.29) [86]. Under conditions of high ionic strength, as two nanoparticles approach one another the interfacial interaction would more likely be determined by chemical interactions between the

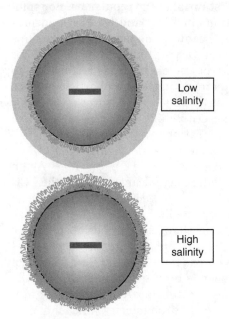

Low salinity

High salinity

Figure 7.29 Illustration showing the diffuse double layer and adsorbed organic layer thickness at low and high salt concentrations.

adsorbed organic layers. As ionic strength decreases and the thickness of the diffuse double layer increases, the interfacial interactions will be due primarily to EL interactions, in which the presence of the adsorbed organic molecules will likely only increase the EL repulsion. Steric interactions are also more likely when the ionic strength is low as the organic macromolecules may expand under such conditions as a result of increased electrostatic repulsion between the molecular branches and with other adsorbed organics.

NOM adsorption and nanoparticle aggregation

In changing nanoparticle surface chemistry, NOM changes the stability and aggregation behavior of the nanoparticle suspension [37, 84, 86, 90]. NOM adsorption affects nanoparticle stability in two principal ways. First, it modifies the effective interfacial charge and thus the magnitude and possibly sign of the electrostatic interactions between surfaces. Second, short-range steric barriers and hydration forces are produced that prevent interparticle approach and diminish the impact of attractive van der Waals interactions [87]. The degree to which NOM adsorption alters nanoparticle stability is dependent on the solution chemistry and the characteristics of the nanoparticle and the organic macromolecules. As an example, steric interactions are a strong function of the molecular weight (an indicator of size) of the NOM; larger macromolecules result in more substantial steric repulsions. For solutions composed of monovalent cations like Na^+ and K^+, NOM adsorption will generally result in more stable nanoparticle suspensions (Figure 7.30). For example, Mylon et al. [84] found that the CCC for NOM-coated hematite nanoscale particles increased from 30 to 107 mM NaCl. This is due to an increase in the nanoparticle surface charge, as was previously discussed. In this case, adsorption of NOM to the nanoparticle surface enhances its stability in the presence of increasing monovalent cation concentrations.

A different interaction behavior sometimes occurs for NOM-coated particles in the presence of divalent cations such as Ca^{2+} and Mg^{2+}. Again referring to the work of Mylon et al. [84], no significant change in the CCC (decreased from 2.4 to 1.8) was observed when the NOM-coated nanoparticles were dispersed in a calcium chloride solution. The lack of a change here may be attributed to conformational changes in the adsorbed organic layer due to interactions between the Ca^{2+} and the organic macromolecules. This type of interaction however is not universal for all types of NOM-coated nanoparticles and depends on the characteristics of the organics. Characteristics and properties that are particularly significant in this respect include charge, hydrophobicity, and chemical

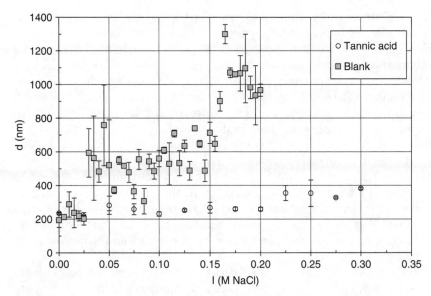

Figure 7.30 Aggregation of unmodified and modified fullerol clusters using sodium chloride (pH = 7.2). The fullerol clusters were modified by first conditioning them in a 2-ppm tannic acid solution.

composition as per the previous discussion on polysaccharides interacting with nC_{60}.

In addition to influencing nanoparticle aggregation, NOM adsorption can alter aggregate structure [84]. This is because aggregate structure is generally determined by both aggregation kinetics and short-range chemical interactions between the agglomerating particles. As NOM adsorption alters these interactions, it will ultimately influence aggregate structure. For instance, attractive van der Waals interactions between adsorbed NOM may result in the formation of a tighter aggregate structure.

NOM adsorption and particle transport

Adsorption of organic macromolecules may either inhibit or enhance fullerene aggregation and deposition [86, 91, 92] in environmental systems by altering their surface chemistry, and in turn the respective interfacial interactions between other surfaces. By associating with nanoparticles, organic compounds alter the particle surface charge (making them more negatively charged), and possibly other interfacial properties, and inhibit particle deposition by generating short-range barriers (e.g., hydration forces and steric interactions) to surface contact [92]. In column experiments with colloidal aggregates of fullerenes (fullerol aggregates, THF/nC_{60}, and TTA nC_{60}) the initially hydroxy-lated fullerols appear to be relatively "immune" to changes in the solution chemistry, while the two varieties of nC_{60} exhibited differential changes in deposition within the column. The observation that fullerol aggregates were relatively unaffected by the presence of the tannic acid contrasts with the observation of reduced aggregation of fullerol in tannic acid solutions (Figure 7.30). However, similar to the aggregation experiments, tannic acid (TA), tended to stabilize the nC_{60}, resulting in less removal in the porous medium. Also, just as natural organic matter in the form of polysaccharides tended to *promote* aggregation of nC_{60} (Figure 7.28), algenate polysaccharides in solution also appear to increase the fraction of nC_{60} deposited in the porous medium, thereby reducing fullerene mobility. Thus, adsorbing organic species in environmental or physiological solutions may be expected to either increase or decrease exposure and bioavailability, depending on the nature of the organic matter.

Indeed, we have also observed that depending on the cell culture media used, very different results are obtained regarding the propensity of fullerenes to aggregate. For example, when placed in Dulbecco's Modified Eagle Medium (primarily a solution of glucose, amino and other organic acids with inorganic salts with an ionic strength on the order of 10^{-1} M) nC_{60} and fullerol colloids rapidly aggregate. In contrast, when placed in a solution of mesenchymal stem cell medium (MSC), these materials are stabilized. Such differences in stability of nanomaterials in these two growth media underscore the importance of considering solute-nanoparticles interactions in performing toxicity studies and in assessing environmental risks posed by fullerenes or other nanomaterials.

Finally, we note that organic matter may also act to mobilize otherwise immobile materials, as has been observed for organic acids in promoting colloidal transport [90, 92]. This process of organic molecules "piggybacking" on particles is referred to as facilitated transport, and has also been observed for mercury and radionucleotides transported by colloids [92].

Nanoparticle transport in matrices such as gels and biofilms

Nanoparticles may be present in complex structures such as porous flocs and biofilms. For example, as a result of heteroaggregation, highly random structures may be formed that include nanoparticles within the structure. In many environmental systems, surfaces exposed to nonsterile and humid conditions can be colonized by microorganisms that encage themselves in hydrated exopolymeric substances (EPSs), which contain mixtures of polysaccharides, proteins, and nucleic acids. This process is collectively called biofilm formation [93]. The bioavailability of nanoparticles to aquatic organisms, and in turn their access to the chain food, is likely to depend on the filtering properties of the biofilms that surround various microbes. Attractive interactions between the nanoparticles and biofilms are expected to favor retention [94–96], while repulsive interactions will prevent nanoparticle diffusion in the organic network.

Diffusion in biofilms and gel-like structures. In porous and disordered media such as soil, biofilms, or bacterial flocs, the random movement of diffusing particles is constrained. Diffusion in these disordered gel-like systems is often modeled as transport through a fractal structure [97]. In uniform Euclidean systems, for every number of dimensions, the random diffusion motion of a nanoparticle denoted by superscript A is described as [98]:

$$\overline{x^2(t)}^A = 2dD^A t \tag{29}$$

where $\overline{x^2(t)}^A$ is the mean square displacement of the diffusing solute at time t; D is the nanoparticle diffusion coefficient; and d is the dimensionality of space. The diffusion properties of small particles in gels has been related to three distinct parameters [99–101] that affect the distribution of particles in a gel saturated with water described by a global partition coefficient Φ [99]:

$$\Phi = \vartheta\alpha\pi = \frac{[A]_g}{[A]_w} \tag{30}$$

where $[A]_g$ and $[A]_w$ are the nanoparticle concentrations in the gel and in the bulk solution, respectively; and ϑ, α, and π are the respective contributions of purely steric, chemical, and electrostatic effects.

Steric obstruction. Steric interactions arise in disordered systems when the nanoparticle approaches the size of the interconnected pores in the media. (This is different from the steric interactions affecting particle stability discussed earlier.) The potential for collisions with the pore walls in the media reduces the degrees of freedom for particle movement, with a resultant reduction in the macroscopic diffusion coefficient of the particle in the disordered media compared with unconstrained diffusion in the bulk fluid. These obstructions may be characterized in terms of a steric obstruction coefficient [102]:

$$\vartheta = (1 - \phi) \times \left(1 - \frac{R_A}{R_p}\right)^2 \tag{31}$$

where R_A and R_p are the respective diameters of the particles and the pores; ϕ is the gel volume fraction. This equation is based on a number of limiting assumptions such as the particles being perfect spheres and the particle and pore sizes being described each by a single value. Although the theoretical constraints of this equation will rarely be met in natural gels and biofilms, it provides some insight into the role of particle and pore size as they affect the steric contribution.

Steric effects on diffusion can also be viewed in terms of decrease in mean squared displacement, $\overline{x^2(t)}^A$, that occurs as the particle moves through a fractal structure [97, 103–107]:

$$\overline{x^2(t)}^A = \Gamma t^{2/d_w} \tag{32}$$

where d_w is the fractal dimension of diffusion and Γ is proportional to the constrained diffusion coefficient. In normal random diffusion, d_w is equal to 2 (Eq. 29). In anomalous diffusion, d_w greater than 2 corresponds to the slowing down of the transport caused by the delay of the diffusing particles in the disordered structure. By measuring the respective characteristic times $t_c(x)^A$ required for the nanoparticle to cover several distances x in the disordered structure shown in Figure 7.31, one

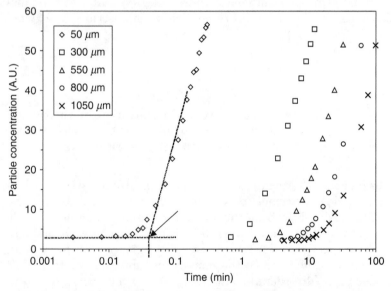

Figure 7.31 Concentration profile according to time and distance in an agarose gel (1.5 wt%) of a nanoscale particle (protein R-phycoerythrin, Rh ~ 4.5 nm) incorporated at time 0 and distance 0. The t_c value obtained from the intersection between the horizontal background concentration and the linear increase in concentration in semi-log plot, points the characteristic time at which the nanoparticle reaches the corresponding distance. Fluorescence correlation spectroscopy data from [108].

can calculate d_w by plotting Eq. 32 and solving for the inverse slope of a log/log plot of $t_c(x)$ versus x [e.g., $\log(x) \sim \frac{1}{d_w}\log(t)$].

Due to this steric contribution, as particle size approaches that of the pore size particles diffuse within a disordered structure in a constrained fashion. In bacterial biofilms, Lacroix-Gueu et al. [109] showed the enabled anomalous diffusion of latexes and bacteriophages with 55 nm radius. In 1.5 to 2 weight percent agarose gels, the steric obstructions appear with solute sizes above 10 nm, and a critical size is measured around 70 nm [100, 110]. However, the critical size is logically inversely correlated to the gel density. Labille et al. [108] showed with the same agarose gel that in a gel/solution interface, the fiber density is locally increased from 1.5 to 5 percent by weight in a 100-μm thickness interphase layer, through which the maximum size of diffusing particles is decreased to around 50 nm radius. The diffusion motion of macromolecules has also been measured in intracellular cytoplasm, where dextran with molecular weight up to 2 106 Da (~ 45 nm radius) undergoes impaired diffusion [111].

Electrostatic interactions.   Most gels, flocs, and biofilms encountered in aquatic and soil environments are mainly composed of polysaccharides and humic substances. Both of these highly reactive components are negatively charged in a large pH range due to the presence of acidic groups in their chemical structure [89, 112]. This induces a negative surface charge to the global diffusing media, which is characterized as follows by the Donnan potential Ψ, which is the average difference in potential between gel and bulk water:

$$\psi = \frac{kT}{zF}\sinh^{-1}\frac{\rho}{2zFc} \tag{33}$$

where ρ is the charge density of the gel; c and z are respectively the molar concentration and charge of the electrolyte in the external bulk solution; and F is the Faraday constant. The diffusion motion in the gel of charged nanoparticles is thus affected by an electrostatic contribution π that can be described by a Boltzmann distribution:

$$\pi = \frac{[A]_P}{[A]_w} = \exp\left(-\frac{Z_A F\psi}{kT}\right) \tag{34}$$

where $[A]_P$ is the concentration of particles A in gel pores exclusively controlled by electrostatic interactions and Z_A is electrical charge of the particles. Figure 7.32 exemplifies the effect of a negatively charged agarose gel on the diffusion motion of a positively charged solute. The anionic gel charge is neutralized by protonation at pH < 3.5, and is screened at pH > 9 by increased ionic strength required for pH adjustment. The

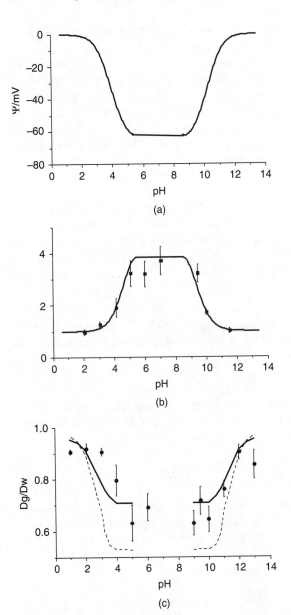

Figure 7.32   Effect of pH on the diffusion motion of a positively charged nanometric solute Rhodamine R6G^{2+} in a negatively charged agarose gel. The partition coefficient Φ (b) controlled by electrostatic attractions is inversely correlated to the negative Donnan potential Ψ (a), and to the diffusion coefficient of the solute in the gel normalized to that in water D_g/D_w (c). Data from [99].

influence of gel charge, i.e. the Donnan potential (Figure 7.32a), is thus maximized in the pH range of 3.5 < pH < 9, where Φ is maximized (Figure 7.32b) and the average diffusion coefficient D_g is minimized (Figure 7.32c) due to electrostatic attractions with oppositely charged particles. The electrostatic contribution π can thus be easily cancelled by neutralizing the sum of the electrostatic interactions by changes in pH or ionic strength. A minimum salt concentration of approximately 10^{-3} M NaCl has been found to be required for π cancellation. This salinity is largely below that of natural sea or river waters, implying the bioavailability of nanoparticles via diffusion through charged or uncharged organic gels.

Specific chemical bonds. When the particle surface presents a specific chemical affinity for sites S on the gel fibers, an adsorption reaction occurs to form a complex SA according to the following intrinsic equilibrium constant K_A^{int}:

$$K_A^{int} = \frac{[SA]}{[A]_P[S]} \tag{35}$$

where [SA] and [S] are the respective concentrations of complexes SA and free sites S. Then, when $\Psi \neq 0$, by combining Eqs. 30, 34, and 35, one obtains the following expression for Φ:

$$\Phi = \frac{[SA] + [A]_P}{[A]_W} = \exp\left(-\frac{Z_A F \Psi}{kT}\right)[1 + K_A^{int}(S)] = \pi\alpha \tag{36}$$

The existence of specific chemical interactions has been put in evidence for example in the case of silica nanoparticles (ludox HS_{30}) and amine- or carboxylate-terminated dendrimers, diffusing in an agarose gel [99, 100], where the diffusion coefficients measured by FCS showed more reduced diffusivity, compared to the pure steric effect expected.

Airborne Nanoparticles

A detailed consideration of the origins, transport, and characteristics of airborne nanoparticles is beyond the scope of this chapter. However, these materials are of major concern for human health and occupational safety [113]. In fact, nanoscale particles are perhaps more ubiquitous in the atmosphere than in any other environment. Clinical studies have suggested a strong link between particulate air pollution and respiratory disease. Many of the principles governing nanoparticle transport in aqueous systems apply to atmospheric systems as well. In particular, the framework for analyzing the kinetics of particle aggregation and deposition are directly applicable to atmospheric particles. However, the origins and characteristics of particles in air and water may differ. Due to the importance of combustion as a source of exposure to nanoscale particles, and the potential similarity between soot and fullerenes, we briefly address this aspect.

Atmospheric nanoparticles may be detected and characterized by simultaneously measuring the light scattering, the photoelectric charging, and the diffusion charging [114]. These methods may also be used to determine the likely source of the nanoparticles and differentiate between those from combustion processes and background nanoparticles. Particles carrying polycyclic aromatic hydrocarbons are detected by their large photoelectric charging signature, whereas particles from other sources only exhibit light scattering and diffusion charging. Although carbon nanotubes have not yet been detected as unintentional combustion products, C_{60} maybe present in some cases.

Nanoscale particles may be produced and dispersed in the atmosphere through both natural processes and human activities. A significant source of manmade nanoscale particles is the burning of hydrocarbon fuels such as diesel, gasoline, and propane [113, 115]. Industrial processes are another source of atmospheric nanoparticles, though their contribution is far less than that of vehicle exhaust. Thus, relatively high concentrations of these nanoparticles occur along roadsides and other areas immediately surrounding combustion sources [113]. Nanoscale particles formed from the burning of these fuels generally fall in a size range of 10 to 60 nm [113, 116] and are primarily composed of unburned oils, polycyclic aromatic hydrocarbons, inorganic compounds, and sulfates [115, 116]. The composition and properties of the airborne nanoparticles will vary according to the type of fuel and engine that is used (Table 7.3) [115, 116]. For instance, gasoline engines tend to produce smaller sized nanoparticles while diesel engines emit larger agglomerates in a size range of 50 to 1000 nm. Similar to nanoparticles in aqueous media, the behavior and properties of atmospheric nanoparticles are dynamic [113, 116].

TABLE 7.3 **General Composition of Nanoparticles Emitted from Diesel and Gasoline Engines [116]. Particles in Diesel and Gasoline Exhaust at the Point of Emission Are Generally Smaller than 50 nm, with Diesel Exhaust also Containing Some Larger Particles up to 1000 nm in Diameter.**

	Diesel exhaust	Gasoline exhaust
Carbon Black	68%	32%
Organic Carbon	31%	61%
Other Materials	1%	7%

Conventional diesel exhaust (in the absence of after-treatment) emits 10–100 times more particle mass and up to 10^5 times more particle numbers than gas engines. For instance, soot produced from the burning of diesel fuel has been characterized as consisting of agglomerated spherical particles with a mean diameter between 20 nm to several microns. The primary particles here are homogeneous small particles with a mean size distribution of 25 nm. Soot had a surface area of 175 m^2/g, compared to 11 m^2/g for commercial carbon black. The core of the soot was characterized as containing disordered polycyclic aromatic hydrocarbons, which have been reported to act as nuclei for soot formation [117]. The small primary particle size and high surface area, and presence of adsorbed hydrocarbons all contribute to the high reactivity of these particles [115]. Furthermore, absorbed material in diesel exhaust particulate matter is specifically responsible for adverse health effects; particles in the smaller size fraction of the particulate matter may have a larger fraction of absorbed material.

The size distribution of combustion-derived atmospheric nanoparticles evolves as they are dispersed from the point source and is a function of the characteristics of the system and the particles [116]. System properties that have been found to be particularly significant include the meteorological conditions (wind speed, wind direction, atmospheric temperature, and relative humidity), particle concentration, presence and concentration of trace gases (e.g., NO_x), and the concentration of materials that may induce coagulation, condensation, and evaporation processes [113]. The variation in particle size depends on the balance between growth through coagulation and condensation and shrinkage by evaporation. Nanoparticles may aggregate either with each other (self-coagulation) or with other larger background particles, a process known as heteroaggregation [116]. Aggregation of atmospheric nanoparticles is typically a rapid process [116] and accounts for the more substantial concentrations of atmospheric nanoparticles in the immediate vicinity of combustion or other generation sources.

Given the dependence of particle size distribution on meteorological conditions, or actually the "solution chemistry," it is reasonable to expect that the characteristics of the nanoparticles will have seasonal variations. Minoura et al. [113] observed that nanoparticles had on average a larger peak diameter in the summer months compared to those in the winter months. This difference may be attributed to a variety of factors as has been discussed here. Additionally, it was found that nanoparticles in the winter were determined to come from a point source, while those in the summer were thought to come from both a point source and through photochemical nucleation, which takes place primarily in summer. The more favorable formation and growth conditions (e.g., higher humidity and particle concentrations) in summer months also result in higher particle aggregation or growth rates [113]. For example, higher particle concentrations result in higher particle growth rates (60 nm/day in the winter and 103 nm/day in the summer) as a result of higher collision frequencies.

Summary

For many nanoparticles larger than several tens of nanometers, many traditional relationships and models used for colloidal systems may be used for describing nanoparticle behavior in aqueous systems. However, when particles are smaller than approximately 20 nm, particle behavior increasingly resembles that of a molecular solute and intermolecular forces play a greater role in determining the transport, aggregation, and deposition of these materials. Heterogeneities of the surfaces with which nanoscale particles may interact will also play an increasingly important role, and the characterization of these surfaces is therefore important in predicting nanoparticle behavior.

Nanoparticle transport at the mesoscopic scale in aqueous systems is dominated by their characteristically high diffusion coefficients as a result of their small size. While this confers a high mobility to nanoparticles in a liquid or gas, it also results in them having high contact efficiencies with potential collector surfaces, making them relatively immobile even when they possess low attachment efficiencies ($\alpha < 0.1$). The stability of nanoparticle dispersions will largely determine how long nanoparticles are likely to remain in the nano-domain. Nanoparticles often aggregate to form clusters both with and without the presence of destabilizing agents or changing chemical conditions. It is therefore necessary to consider the transport of nanomaterials both as nanoparticles and as materials that may transition into the colloidal domain or larger, where they may be subject to transport mechanisms such as gravitational settling. Aggregation will therefore likely reduce the persistence of these materials in the environment and possibly their bioavailability.

The deposition of nanoparticles on surfaces ranging from sand grains in aquifers to leaves on trees will also reduce their persistence. Nanoparticles can be entrapped by myriad media ranging from mineral formations to biofilms. The interaction between nanoparticles and surfaces may be assessed within the context of various established interfacial models, such as the DLVO model discussed here.

According to the DLVO theory and its extensions, the stability of particle dispersions results from the sum of repulsive and attractive forces at their interface with neighboring solid interfaces. Generally, repulsive electrostatic interactions between like charged particles controls dispersion stability, though the nature of these and other interactions changes with particle size. It is therefore imperative that size effects on interfacial interaction energies be considered, particularly given that some interactions may become significant for nanoparticles that may otherwise not be so for colloidal-sized materials of the same composition (e.g., hydration interactions and silica particles).

The surface reactivity of suspended particles is a key parameter that controls their interaction with other materials that may be present in suspension, and thus their behavior in the given dispersion; even more so for nanoparticles, for which their large surface area to volume ratio induces a high sorption capacity for foreign species. The nature of these interactions is always specific to the chemical characteristics of the nanoparticles, and their affinity for the foreign species present in the system. In many cases, adsorption occurs through charge-mediated interactions, or via the formation of specific chemical bonds. In most cases, bond formation between nanoparticles and other materials results in a modification of the surface reactivity of both, tending to lower the system free energy. This favors intraparticle attractions and aggregation. One obvious example of this phenomenon is the absorption and embedding of nanoparticles into organo-mineral flocs, which cloaks the properties of the nanoparticles producing transport and fate behavior that is determined by the floc. Similarly, adsorption or reaction with other materials will affect nanoparticles' fate and transport.

References

1. Auffan, M., et al., *In vitro interactions between DMSA-coated maghemite nanoparticles and human fibroblasts: a physicochemical and cyto-genotoxical study.* Environmental Science & Technology, 2006.
2. Schrick, B., et al., *Delivery vehicles for zerovalent metal nanoparticles in soil and groundwater.* Chemistry of Materials, 2004. 16(11): p. 2187–2193.
3. van Oss, C.J., *Acid-base interfacial interactions in aqueous media.* Colloids and Surfaces A, 1993. 78: p. 1.
4. Israelachvili, J.N., *Intermolecular and Surface Forces.* 2nd ed. 1992, London: Academic Press Harcourt Brace Jovanovich. 450.
5. Brant, J.A., and A.E. Childress, *Assessing short-range membrane-colloid interactions using surface energetics.* Journal of Membrane Science, 2002. 203: p. 257–273.
6. Hahn, M.W., D. Abadzic, and C.R. O'Melia, *Aquasols: On the role of the secondary minima.* Environmental Science & Technology, 2004. 38(22): p. 5915–5924.
7. Hunter, R.J., *Foundations of Colloid Science.* Vol. 2. 1989, Oxford: Clarendon Press.
8. van Oss, C.J., *Interfacial Forces in Aqueous Media.* 1994, New York: Marcel Dekker. 179.
9. Choudhury, N. and B.M. Pettitt, *On the mechanism of hydrophobic association of nanoscopic solutes.* Journal of the American Chemical Society, 2005.
10. Tufenkji, N., and M. Elimelech, *Correlation equation for predicting single-collector efficiency in physiochemical filtration in saturated porous media.* Environmental Science & Technology, 2004. 38: p. 529–536.
11. Hogg, R.I., T.W. Healy, and D.W. Fuerstenau, *Mutual coagulation of colloidal dispersions.* Transactions of the Faraday Society, 1966. 62: p. 1638.
12. Elimelech, M., et al., *Particle Deposition and Aggregation: Measurement, Modelling, and Simulation.* Colloid and Surface Engineering Series. 1995, Oxford: Butterworth-Heinemann. 441.

13. van Oss, C.J., *Hydrophobic, hydrophilic and other interactions in epitope-paratope binding.* Molecular Immunology, 1995. 32(3): p. 199–211.

14. Fukushi, K., and T. Sato, *Using a surface complexation model to predict the nature and stability of nanoparticles.* Environmental Science & Technology, 2005.

15. Andrievsky, G.V., et al., *On the production of an aqueous colloidal solution of fullerenes.* Chemical Communications, 1995: p. 1281–1282.

16. Schwarzer, H.-C., and W. Peukert, *Prediction of aggregation kinetics based on surface properties of nanoparticles.* Chemical Engineering Science, 2005. 60(1): p. 11–25.

17. Marinova, K.G., et al., *Charging of oil-water interfaces due to spontaneous adsorption of hydroxyl ions.* Langmuir, 1996. 12: p. 2045–2051.

18. Brant, J.A., et al., *Comparison of electrokinetic properties of colloidal fullerenes (n-C_{60}) formed using two procedures.* Environmental Science & Technology, 2005, 39(17):6343–6351.

19. Andrievsky, G.V., et al., *Comparative analysis of two aqueous-colloidal solutions of C_{60} fullerene with help of FTIR reflectance and UV-Vis spectroscopy.* Chem. Phys. Lett., 2002. 364: p. 8–17.

20. Prieve, D.C., and E. Ruckenstein, *Rates of deposition of brownian particles calculated by lumping interaction forces into a boundary condition.* J. Colloid Interface Sci., 1976. 57(3): p. 547–550.

21. Yaminsky, V.V.V., E.A., *Hydrophobic hydration.* Current Opinion in Colloid & Interface Science, 2001. 6: p. 342–349.

22. Kobayashi, M., et al., *Aggregation and charging of colloidal silica particles: effect of particle size.* Langmuir, 2005.

23. Guldi, D.M., and M. Prato, *Excited-state properties of C_{60} fullerene derivatives.* Accounts of Chemical Research, 2000. 33(10): p. 695–703.

24. Brunner, T.J., et al., *In vitro cytotoxicity of oxide nanoparticles: comparison to asbestos, silica, and the effect of particle solubility.* Environmental Science & Technology, 2006.

25. Lam, C.W., et al., *Pulmonary toxicity of single-wall carbon nanotubes in mice 7 and 90 days after intratracheal installation.* Toxicological Sciences, 2004. 77: p. 126–134.

26. Warheit, D.B., et al., *Comparative pulmonary toxicity assessment of single-wall carbon nanotubes in rates.* Toxicological Sciences, 2004. 77: p. 117–125.

27. Hyning, D.L.V., W.G. Klemperer, and C.F. Zukoski, *Characterization of colloidal stability during precipitation reactions.* Langmuir, 2001. 17: p. 3120–3127.

28. Bokare, A., and A. Patnaik, *C_{60} aggregate structure and geometry in nonpolar oxylene.* Journal of Physical Chemistry B, 2005. 109(1): p. 87–92.

29. Bulavin, L.A., et al., *Self-organization C_{60} nanoparticles in toluene solution.* Journal of Molecular Liquids, 2001. 93: p. 187–191.

30. Nath, S., H. Pal, and A.V. Sapre, *Effect of solvent polarity on the aggregation of C_{60}.* Chemical Physics Letters, 2000. 327: p. 143–148.

31. Nath, S., H. Pal, and A.V. Sapre, *Effect of solvent polarity on the aggregation of fullerenes: a comparison between C_{60} and C_{70}.* Chemical Physics Letters, 2002. 360: p. 422–428.

32. Ying, Q., J. Marecek, and B. Chu, *Slow aggregation of buckminsterfullerene (C_{60}) in benzene solution.* Chemical Physics Letters, 1994. 219: p. 214–218.

33. Deguchi, S., G.A. Rossitza, and K. Tsujii, *Stable dispersions of fullerenes, C_{60} and C_{70} in water. Preparation and characterization.* Langmuir, 2001. 17: p. 6013–6017.

34. Mchedlov-Petrossyan, N.O., V.K. Klochkov, and G.V. Andrievsky, *Colloidal dispersions of fullerene C_{60} in water: some properties and regularities of coagulation by electrolytes.* Journal of the American Chemical Society Faraday Transactions, 1997. 93(24): p. 4343–4346.

35. Scrivens, W.A. and J.M. Tour, *Synthesis of ^{14}C-labeled C_{60}, its suspension in water, and its uptake by human keratinocytes.* Journal of the American Chemical Society, 1994. 116: p. 4517–4518.

36. Cheng, X., A.T. Kan, and M.B. Tomson, *Naphthalene adsorption and desorption from aqueous C_{60} fullerene.* Journal of Chemical and Engineering Data, 2004. 49: p. 675–683.

37. Saleh, N., et al., *Absorbed triblock copolymers deliver reactive iron nanoparticles to the oil/water interface.* Nano Letters, 2005. 5(12): p. 2489–2494.

38. Newbury, R.W.e.P., *Hydrologic determinants of aquatic insect habitat. Chapter 11*, in *The Ecology of Aquatic Insects*, V.H. Resh and D.M. Rosenberg, editors. 1984, Praeger: New York. p. 323–357.

39. Smoluchowski, M., *Versuch einer Mathematischen Theorie der Koagulations- Kinetik Kolloider Losungen.* 1917. 92: p. 129.
40. Han, M., and D.F. Lawler, *Interactions of two settling spheres: settling rates and collision efficiency.* J. Hydraulic Engrg.-ASCE, 1991. 117(10): p. 1269–1289.
41. Veerapaneni, S., and M.R. Wiesner, *Particle deposition to an infinitely permeable surface: dependence of deposit morphology on particle size.* Journal of Colloid and Interface Science, 1996. 162(1): p. 110–122.
42. Veerapaneni, S., and M.R. Wiesner, *Hydrodynamics of fractal aggregates with radially varying permeability.* Journal of Colloid and Interface Science, 1996. 177: p. 45–57.
43. Brant, J.A., et al., *Characterizing the impact of preparation method on fullerene cluster structure and chemistry.* Langmuir, 2006. 22: p. 3878–3885.
44. Ruoff, R.S., et al., *Solubility of C_{60} in a variety of solvents.* Journal of Physical Chemistry, 1993. 97: p. 3379–3383.
45. Marcus, Y., et al., *Solubility of C_{60} fullerene.* Journal of Physical Chemistry B, 2001. 105: p. 2499–2506.
46. Nakamura, E., and H. Isobe, *Functionalized fullerenes in water. The first 10 years of their chemistry, biology, and nanoscience.* Accounts of Chemical Research, 2003. 36(11): p. 807–815.
47. Barnard, A.S., and L.A. Curtiss, *Prediction of TiO_2 nanoparticle phase and shape transitions controlled by surface chemistry.* Nano Letters, 2005. 5(7): p. 1261–1266.
48. Georgakilas, V., et al., *Supramolecular self-assembled fullerene nanostructures.* Proceedings of the National Academy of Sciences, 2002. 99(8): p. 5075–5080.
49. Guo, Z.-X., et al., *Nanoscale aggregation of fullerene in nafion membrane.* Langmuir, 2002. 18: p. 9017–9021.
50. Natalini, B., et al., *Chromatographic separation and evaluation of the lipophilicity by reversed-phase high performance liquid chromotography of fullerene-C_{60} derivatives.* Journal of Chromatography A, 1999. 847: p. 339–343.
51. Moraru, V., N. Lebovka, and D. Shevchenko, *Structural transitions in aqueous suspensions of natural graphite.* Colloids and Surfaces A: Physicochemical and Engineering Aspects, 2004. 242(1–3): p. 181–187.
52. Angelini, G., et al., *Study of the aggregation properties of a novel amphilic C_{60} fullerene derivative.* Langmuir, 2001. 17: p. 6404–6407.
53. Guldi, D.M., et al., *Ordering fullerene materials at nanometer dimensions.* Accounts of Chemical Research, 2005. 38(1): p. 38–43.
54. Fauconnier, N., et al., *Thiolation of maghemite nanoparticles by dimercaptosuccinic acid.* Journal of Colloid and Interface Science, 1997. 194(2): p. 427–433.
55. Bellona, C., et al., *Factors affecting the rejection of organic solutes during NF/RO treatment—a literature review.* Water Research, 2004. 38: p. 2795–2809.
56. Hsu, J.-P. and Y.-C. Kuo, *An algorithm for the calculation of the electrostatic potential distribution of ion-penetrable membranes carrying fixed charges.* J. Colloid Interface Sci., 1995. 171: p. 483–489.
57. Hunter, R.J., *Zeta Potential in Colloid Science: Principles and Applications.* 1981, London: Academic Press.
58. O'Melia, C.R., *Particle-particle interactions in aquatic systems.* Colloids and Surfaces, 1989. 39: p. 255.
59. Happel, J., *Viscous flow in multiparticle systems: slow motion of fluids relative to beds of spherical particles.* A.I.Ch.E.,J, 1958. 4(2): p. 197–201.
60. Derjaguin, B.V., and L.D. Landau, *Theory of the stability of strongly charged lyophobic sols and of the adhesion of strongly charged particles in solutions of electrolytes.* Acta Physicochim. URSS, 1941. 14: p. 733–762.
61. Elimelech, M., and C.R. O'Melia, *Effect of particle size on collision efficiency in the deposition of Brownian particles with electrostatic energy barriers.* Langmuir, 1990. 6: p. 1153–1163.
62. Ryan, J.N., and M. Elimelech, *Colloid mobilization and transport in groundwater.* Colloids and Surfaces A: Physicochemical and Engineering Aspects, 1996. 107: p. 1–56.
63. O'Melia, C.R., ed. *Kinetics of Colloid Chemical Processes in Aquatic Systems.* ed. W. Stumm. 1990, John Wiley & Sons, Inc.: New York.

64. Lecoanet, H.F., and M.R. Wiesner, *Velocity effects on fullerene and oxide nanoparticle deposition in porous media.* Environmental Science & Technology, 2004. 38(16): p. 4377–4382.
65. Li, X., et al., *Role of hydrodynamic drag on microsphere deposition and re-entrainment in porous media under unfavorable conditions.* Environmental Science & Technology, 2006. 39(11): p. 4012–4020.
66. Johnson, K.L., K. Kendall, and A.D. Roberts, *Surface energy and the contact of elastic solids.* Proceedings of the Royal Society of London. Series A, Mathematical and Physical Sciences, 1971. 324(1558): p. 301–313.
67. Derjaguin, B.V., J.J. Muller, and Y.P. Toporov, *Effect of contact deformations on the adhesion of particles.* Journal of Colloid and Interface Science, 1975. 53: p. 314–326.
68. Johnson, K.L., *Mechanics of adhesion.* Tribology International, 1998. 31(8): p. 413–418.
69. Hahn, M.W., and C.R. O'Melia, *Deposition and reentrainment of Brownian particles in porous media under unfavorable chemical conditions: some concepts and applications.* Environmental Science & Technology, 2004. 38: p. 210–220.
70. Brant, J.A., K.M. Johnson, and A.E. Childress, *Characterizing NF and RO membrane surface heterogeneity using chemical force microscopy.* Colloids and Surfaces A: Physicochemical and Engineering Aspects, 2006.
71. Elimelech, M., et al., *Relative insignificance of mineral grain zeta potential to colloid transport in geochemically heterogeneous porous media.* Environmental Science & Technology, 2000. 34: p. 2143–2148.
72. Johnson, P.R., N. Sun, and M. Elimelech, *Colloid transport in geochemically heterogeneous porous media: modeling and measurements.* Environmental Science & Technology, 1996. 30: p. 3284–3293.
73. Song, L., and M. Elimelech, *Transient deposition of colloidal particles in heterogeneous porous media.* J. Colloid Interface Sci., 1994. 167: p. 301–313.
74. Suresh, L.W., and J.Y. Walz, *Effect of surface roughness on the interaction energy between a colloidal sphere and a flat plate.* Journal of Colloid and Interface Science, 1996. 183: p. 199–213.
75. Suresh, L.W., and J.Y. Walz, *Direct measurement of the effect of surface roughness on the colloidal forces between a particle and flat plate.* Journal of Colloid and Interface Science, 1997. 196: p. 177–190.
76. Brant, J.A., and A.E. Childress, *Colloidal adhesion to hydrophilic membrane surfaces.* Journal of Membrane Science, 2004. 241(2): p. 235–248.
77. Vaidyanathan, R., *Double layer calculations for the attachment of a colloidal particle with a charged surface patch onto a substrate.* Sep. Technol., 1992. 2: p. 98–103.
78. Elimelech, M., et al., *Role of membrane surface morphology in colloidal fouling of cellulose acetate and composite aromatic polyamide reverse osmosis membranes.* Journal of Membrane Science, 1997. 127(1): p. 101–109.
79. Vrijenhoek, E.M., S. Hong, and M. Elimelech, *Influence of membrane surface properties on initial rate of colloidal fouling of reverse osmosis and nanofiltration membranes.* Journal of Membrane Science, 2001. 188: p. 115–128.
80. Cappella, B., and G. Dietler, *Force-distance curves by atomic force microscopy.* Surface Science Reports, 1999. 34: p. 1–104.
81. Bowen, W.R., et al., *The effects of electrostatic interactions on the rejection of colloids by membrane pores-visualisation and quantification.* Chemical Engineering Science, 1999. 54: p. 369–375.
82. Bowen, W.R., and T.A. Doneva, *Atomic force microscopy studies of membranes: effect of surface roughness on double-layer interactions and particle adhesion.* Journal of Colloid and Interface Science, 2000. 229: p. 544–549.
83. Yamago, S., et al., *In-vivo biological behavior of a water-miscible fullerene—C-14 labeling, absorption, distribution, excretion and acute toxicity.* Chemistry & Biology, 1995. 2(6): p. 385–389.
84. Mylon, S.E., K.L. Chen, and M. Elimelech, *Influence of natural organic matter and ionic composition on the kinetics and structure of hematite colloid aggregation: implications to iron depletion in estuaries.* Langmuir, 2004. 20(21): p. 9000–9006.
85. Labille, J., et al., *Destabilisation of montmorillonite suspension by polysaccharide and Ca^{2+}.* Clay Minerals, 2002. 38: p. 173.

86. Franchi, A., and C.R. O'Melia, *Effects of natural organic matter and solution chemistry on the deposition and reentrainment of colloids in porous media.* Environmental Science & Technology, 2003. 37: p. 1122–1129.
87. Johnson, S.B., et al., *Adsorption of organic matter at mineral/water interfaces. 6. effect of inner-sphere versus outer-sphere adsorption on colloid stability.* Langmuir, 2005. 21(14): p. 6356–6365.
88. Illes, E., and E. Tombacz, *The effect of humic acid adsorption on pH-dependent surface charging and aggregation of magnetite nanoparticles.* Journal of Colloid and Interface Science, 2006. 295: p. 115–123.
89. Labille, J., et al., *Flocculation of colloidal clay by bacterial polysaccharides: effect of macromolecule charge and structure.* Journal of Colloid and Interface Science, 2005. 284(1): p. 149–156.
90. Lowry, G.V., et al., *Macroscopic and microscopic observations of particle-facilitated mercury transport from New Idria and sulfur bank mercury mine tailings.* Environmental Science & Technology, 2004. 38: p. 5101–5111.
91. Thoral, S., et al., *XAS study of iron and arsenic speciation during Fe(II) oxidation in the presence of As(III).* Environmental Science & Technology, 2005. 39(24): p. 9478–9485.
92. Slowey, A.J., et al., *Role of organic acids in promoting colloidal transport or mercury from mine tailings.* Environmental Science & Technology, 2005. 39: p. 7869–7874.
93. Costerton, J.W., et al., *Bacterial biofilms in nature and disease.* Annual Review of Microbiology, 1987. 41: p. 435–464.
94. Ballance, S., et al., *Influence of sediment biofilm on the behaviour of aluminum and its bioavailability to the snail Lymnaea stagnalis in neutral freshwater.* Canadian Journal of Fisheries and Aquatic Sciences, 2001. 58(9): p. 1708–1715.
95. Battin, T.J., et al., *Contributions of microbial biofilms to ecosystem processes in stream mesocosms.* Nature, 2003. 426(6965): p. 439–442.
96. Jordan, R.N., *Evaluating biofilm activity in response to mass transfer-limited bioavailability of sorbed nutrients.* Biofilms, 1999. 310: p. 393–402.
97. Havlin, S., and D. Ben-Avraham, *Diffusion in disordered media.* Advances in Physics, 2002. 51(1): p. 187–292.
98. Havlin, S., and D. Ben-Avraham, *Diffusion in disordered media.* Advances in Physics, 1987. 36(695–798).
99. Fatin-Rouge, N., et al., *Diffusion and partitioning of solutes in agarose hydrogels: the relative influence of electrostatic and specific interactions.* J. Phys. Chem. B, 2003. 107(44): p. 12126–12137.
100. Fatin-Rouge, N., K. Starchev, and J. Buffle, *Size effects on diffusion processes within agarose gels.* Biophysical Journal, 2004. 86: p. 2710–2719.
101. Fatin-Rouge, N., et al., *A global approach of diffusion in agarose gels from SANS and FCS experiments,* to be submitted.
102. Giddings, J.C., et al., *Statistical theory for the equilibrium distribution of rigid molecules in inert porous networks. Exclusion chromatography.* Journal of Physical Chemistry, 1968. 72(13): p. 4397–4408.
103. Alexander, S., and R. Orbach, *Density of States on Fractals-Fractons.* Journal De Physique Lettres, 1982. 43(17): p. L625–L631.
104. Webman, I., *Effective-medium approximation for diffusion on a random lattice.* Physical Review Letters, 1981. 47(21): p. 1496–1499.
105. Benavraham, D., and S. Havlin, *Diffusion on percolation clusters at criticality.* Journal of Physics a-Mathematical and General, 1982. 15(12): p. L691–L697.
106. Gefen, Y., A. Aharony, and S. Alexander, *Anomalous diffusion on percolating clusters.* Physical Review Letters, 1983. 50(1): p. 77–80.
107. Rammal, R., and G. Toulouse, *Random-walks on fractal structures and percolation clusters.* Journal De Physique Lettres, 1983. 44(1): p. L13–L22.
108. Labille, J., N. Fatin-Rouge, and J. Buffle, *Solutes diffusion in agarose gel. Effects of gel relaxation and local structure at gel/solution interface.* Environmental Chemistry, submitted.
109. Lacroix-Gueu, P., et al., *In situ measurements of viral particles diffusion inside mucoid biofilms.* Comptes Rendus Biologies, 2005. 328(12): p. 1065–1072.
110. Pluen, A., et al., *Diffusion of macromolecules in agarose gels: comparison of linear and globular configurations.* Biophysical Journal, 1999. 77(1): p. 542–552.

111. Verkman, A.S., *Solute and macromolecule diffusion in cellular aqueous compartments.* Trends in Biochemical Sciences, 2002. 27(1): p. 27–33.
112. Rinaudo, M., et al., *Electrostatic interactions in aqueous solutions of ionic polysaccharides.* International Journal of Polymer Analysis and Characterization, 1997. 4(1): p. 57–69.
113. Minoura, H.. and H. Takekawa, *Observation of number concentrations of atmospheric aerosols and analysis of nanoparticle behavior at an urban background area in Japan.* Atmospheric Environment, 2005. 39: p. 5806–5816.
114. Siegmann, K., and H.C. Siegmann, *Fast and reliable "in situ" evaluation of particles and their surfaces with special reference to diesel exhaust.* 2000.
115. Muller, J.-O., et al., *Diesel engine exhaust emission: oxidative behavior and microstructure of black smoke soot particulate.* Environmental Science & Technology, 2006.
116. Jacobson, M.Z., and J.H. Seinfield, *Evolution of nanoparticle size and mixing state near the point of emission.* Atmospheric Environment, 2004. 38: p. 1839–1850.
117. Ishiguro, I., Y. Takitori, and K. Akihara, *Microstructure of diesel soot particles probed by electron microscopy: first observation of inner core and outer shell.* Combust Flame, 1997. 108: p. 231–234.

Environmental Applications of Nanomaterials

8

Nanomaterials for Groundwater Remediation

Gregory V. Lowry *Carnegie Mellon University, Pittsburgh, Pennsylvania, USA*

Introduction

Contamination of subsurface soil and groundwater by organic and inorganic contaminants is an extensive and vexing environmental problem that stands to benefit from nanotechnology. The EPA reports that contamination by chlorinated organic pollutants such as trichloroethylene (TCE), and heavy metals such as lead and hexavalent chromium are primary concerns at Superfund National Priorities List sites, which are the most contaminated sites in the United States. Associated health risks have led to an extensive remediation effort for the past 30 years. Remediation is costly and poses significant technical challenges. For example, life-cycle treatment costs are estimated to exceed $2 billion for approximately 3,000 contaminated Department of Defense sites (Stroo et al. 2003).

Remediation efforts aimed at removing deep subsurface contamination (e.g., pump and treat) have had limited success because most pollutants are not highly mobile in the subsurface. For example, many organic contaminants are only weakly water-soluble and tend to remain as a separate nonaqueous phase liquid (NAPL) in the subsurface. Many organic contaminants are denser than water (DNAPL) and migrate downward in the aquifer. As sketched in Figure 8.1, DNAPL residual ganglia and pooled DNAPL are trapped in the porous soil. Heavy metals such as Pb(II) or Cr(VI) also tend to be concentrated in areas near their release point. The residual saturation acts as a long-term source for contaminant

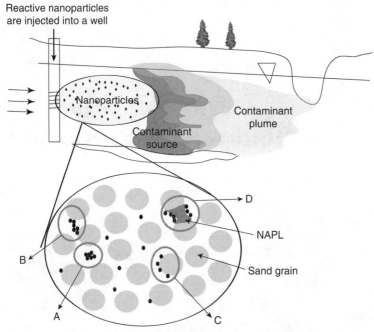

Reactive nanoparticles
are injected into a well

Figure 8.1   The macroscale problem. Illustration of DNAPL distribution as residual saturation (sources) and a plume of dissolved contaminants in an aquifer. Nanotechnology offers the potential to effectively target the chemical treatment to the residual saturation zone *in situ*. Reactive nanoparticles are injected into the aquifer using a well. The particles are transported to the contaminant source where they can degrade the contaminant. Nanoparticles can aggregate (A) and be filtered from solution via straining (B) or attachment to aquifer grains (C). Methods to target the nanoparticles to the contaminant (D) could improve the efficiency of the technology. Factors affecting the mobility of nanoparticles in the subsurface and the ability to target contaminants are discussed later in this chapter.

leaching to the groundwater, resulting in large plumes of dissolved contaminants and decades-long remediation times. Often the contaminant source may be difficult to locate or is too deep for excavation to be a cost-effective solution, and the prevailing "pump-and-treat" technologies cannot meet cleanup targets in a reasonable amount of time in most cases (Water Science Technology Board 2004). "Pump-and-treat" attempts to remove the contamination by pumping groundwater from the contaminated plume to the surface to remove the contaminants. "Pump-and-treat" and other technologies that address primarily the plume tend to fail because they do not address or remove the source. Accordingly, the Department of Energy recently advocated the development of novel *in situ* technologies to accelerate the rate at which contaminated sites are restored back to an acceptable condition (US DOE 2005).

Nanotechnology has the potential to create novel and effective *in situ* treatment technologies for groundwater contaminant source zones. Rapid advances in nanotechnology have led to the creation of novel nanoparticles with unique and tunable physical and chemical properties. Their properties can be adjusted to make them highly reactive with common organic pollutants, and to minimize the formation of unwanted toxic by-products. Highly reactive nanoparticles such as nanoscale zerovalent iron ("nanoiron" or NZVI) (Liu et al. 2005a; Liu et al. 2005b), nanocatalysts (e.g., Au/Pd bimetallic nanoparticles [Nutt et al. 2005]), or nanosized sorbents (Tungittiplakorn et al. 2005; Tungittiplakorn et al. 2004) have been developed specifically to remediate contamination by organic and inorganic contaminants. In principle, their small size (10–500 nm) also provides an opportunity to deliver these remedial agents to subsurface contaminants *in situ*, and provides access to contamination trapped in the smallest pores in an aquifer matrix. Highly mobile nanoparticles are needed that can transport in the subsurface to where contaminants are located, which increases the potential for off-site migration and thus the potential for any unwanted ecotoxicity or human health risks. The high reactivity, and the potential for facile delivery directly to the contaminant source, suggests that nanoparticles can accelerate the degradation rate of contaminants in the source zone, and decrease the time and cost of remediation relative to traditional treatment technologies that address the plume.

Here, we focus on recent advances in the use of nanoparticles for *in situ* remediation of contaminated groundwater. In particular, the focus is on the use of nanoscale zerovalent iron and bimetallics (e.g., Fe^0/Pd) for the rapid *in situ* degradation of chlorinated organic compounds and reduction of heavy metals in contaminant source zones. *In situ* source zone treatment using nanoiron is one of the early adopters of environmental nanotechnology. There are already many documented applications of nanoiron ranging from pilot to full-scale, and early data suggest that nanoiron can be effective and can lower the costs of contaminant source zone remediation in groundwater (Gavaskar et al. 2005). Despite this rush to market, there are critical aspects of nanoparticle-based remediation strategies that have not yet been fully addressed. The delivery of nanomaterials in the subsurface is analogous to filtration in porous media. Particle slurries injected into the subsurface must migrate from the point of injection through the subsurface to the source of the contamination. Natural geochemical conditions (e.g., pH and ionic strength) can destabilize the nanoparticles and allows aggregation (labeled A in Figure 8.1). Particle aggregates may then be removed by straining and pore plugging (labeled B in Figure 8.1). These same conditions will also increase nanoparticle deposition onto media grains (labeled C in Figure 8.1). High deposition rates onto media grains or straining and pore plugging will limit nanoparticle transport. While maximizing transportability, ideally the nanomaterials will have an affinity for the target contaminant. For example, nanomaterials with hydrophobic character may partition to the DNAPL/water interface thereby delivering the particles to where they are most needed (labeled D in Figure 8.1). A fundamental understanding of the particle properties controlling the reactivity (rates and products/intermediates) and lifetime of iron nanoparticles, optimal methods of nanoiron injection/transport to the contaminant source zone, and methods for targeting specific sub-surface contaminants or source areas to increase their effectiveness are needed.

Reactivity, Fate, and Lifetime

Nanoparticles can sequester groundwater contaminants (via adsorption or complexation), making them immobile, or can degrade or transform them to innocuous compounds. Contaminant transformations by nano-iron, which is a strong reductant, are typically redox reactions. When the oxidant or reductant is the nanoparticle itself, it is considered a reactive nanoparticle (Figure 8.2). A good example of a reactive nanoparticle is TCE dechlorination by nanoiron. Oxidation of Fe^0 in the particle provides electrons for the reduction of TCE, which acts as the oxidant. The Fe^0 core shrinks and is used up in the reaction. The particles are no longer active once all of the Fe^0 is oxidized (Figure 8.2a). Nanoparticles that *catalyze* redox reactions but are not themselves transformed are cat-alytic nanoparticles, which requires an additional reagent that serves as the reductant or oxidant (e.g., Pd nanoparticles require H_2 as a reduc-tant [Lowry and Reinhard 1999]). H_2 (the reductant) is activated by the

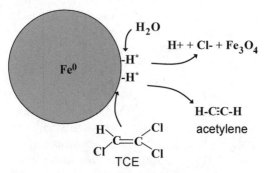

Figure 8.2a   Reactive nanoparticles. TCE dechlori-nation by nanoiron. Fe^0 oxidation provides electrons for the reduction of TCE. The Fe^0 core shrinks while the Fe_3O_4 oxide shell grows. The particles are no longer active once all of the Fe^0 is oxidized.

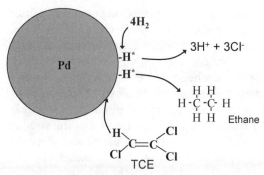

Figure 8.2b Catalytic nanoparticles. TCE dechlorination by catalytic Pd particles. H_2 is supplied as the reductant for TCE dechlorination. In principle, the Pd catalyst is not altered by the reaction and can remain active as long as H_2 is supplied. In practice, catalyst deactivation occurs and the particle lifetime is finite. Catalyst regeneration may extend the life of the particle.

Pd to form adsorbed reactive H species. TCE adsorbs to the Pd surface where it is reduced by the reactive H species on the Pd surface (Figure 8.2b). In principle, the catalyst can repeat this indefinitely so long as reductant is continually supplied. In practice, precipitation of minerals or natural organic matter on the Pd surface or adsorption of reduced sulfur species deactivate the catalyst and it has a finite lifetime (Lowry and Reinhard 2000). Some nanomaterials are engineered to strongly sequester contaminants (Figure 8.2c). The high affinity for the contaminant allows the nanoparticle to significantly lower the aqueous phase concentrations, to out-compete natural geosorbents such as organic carbon, and serves to concentrate the contaminants onto the particles. Once concentrated onto the nanoparticles, the contaminants can be removed along with the nanoparticles. This can be highly effective for hydrophobic organic contaminants such as PCBs and PAHs and for heavy metals. For *in situ*

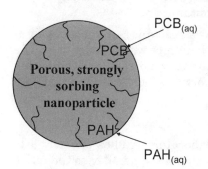

Figure 8.2c Adsorbent nanoparticles. Nanoparticles engineered as very strong sorbents can be used to strongly sequester organic or inorganic contaminants. Once adsorbed, the contaminants are no longer bioavailable.

remediation with any type of nanoparticle, it is important to know which groundwater contaminants will respond to the treatment and which will not. It is also important to know how long the reactive or catalytic particles will remain active as this will determine important operation decisions such as how much to inject and when reinjection may be necessary.

Degradation of halogenated hydrocarbons, particularly chlorinated solvents, occurs via a reductive process. The Fe^0 in the nanoiron is oxidized by the chlorinated solvent, which is subsequently reduced. For chlorinated hydrocarbons, the reduction typically results in the replacement of a chlorine atom with a hydrogen atom. For heavy metals, the metal, such as Pb(II) or Cr(VI), is reduced to its zerovalent form on the nanoiron surface, or forms mixed (Fe-Metal) precipitates that are highly insoluble (Ponder et al. 2000). The general half-reactions for the oxidation of iron and the reduction of chlorinated organic compounds (COC) or heavy metals are given in Eqs. 1 to 3, where Me is a metal ion of charge a.

$$Fe^0 \rightarrow Fe^{2+} + 2 \cdot e^- \qquad (1)$$

$$COC + n \cdot e^- + m \cdot H^+ \rightarrow products + 3Cl^- \qquad (2)$$

$$Me^{a+} + b \cdot e^- \rightarrow Me^{a-b} \qquad (3)$$

In the case of nanoiron, or Fe^0-based bimetallics, the reduction of the contaminant is surface-mediated, and the particle itself is the reductant. The attractiveness of nanoiron is that the particles have a high surface-to-volume ratio and therefore have high reactivity with the target contaminants. The following generalizations can be made about the reactivity and lifetime of all nanoparticulate remedial agents that are themselves the reactive material—that is, not true catalysts according to the formal definition of a catalyst:

- Any process that affects the surface properties of the particles (e.g., formation of an Fe-oxide on the surface) can affect their *reactivity*.

- Any oxidant (e.g., O_2 or NO_3^-) competing with the target contaminant will utilize electrons and may lower the rate and efficiency of the nanoiron treatment for the target contaminants.

- Reactive nanoparticles that serve as a reactant rather than a catalyst will have a finite lifetime, the length of which depends on the concentration of the target contaminant, the presence of competing oxidants, and the selectivity of the particles for the desired reaction.

Nanoiron reactivity

Since 1997, there have been many laboratory studies conducted to determine the range of contaminants that are amenable to reductive dechlorination by nanoiron or bimetallics such as Fe^0/Pd nanoparticles. A summary of the reactive surface area normalized pseudo-first-order reaction rate constants for each type of contaminant and iron is given in Table 8.1. For some forms of nanoiron, the reported surface area-normalized reaction rate constants are similar to those reported for iron filings, which have been used successfully as *in situ* reactive barriers for groundwater remediation in shallow aquifers (Wilkin et al. 2005), suggesting that there is not a quantum nanosize effect for these particles (Liu et al. 2005b; Nurmi et al. 2005). Rather, the high surface area of nanoiron (10–50 m^2/g) compared to iron filings (0.01–1 m^2/g) makes them attractive for *in situ* remediation of groundwater because of the higher reaction rates that they can provide. This is exemplified in the reaction with perchlorate (Cao et al. 2005); perchlorate reduction by nanoiron was rapid enough to make it useful for *in situ* perchlorate reduction, whereas iron filings reduced perchlorate far too slowly to be useful (Moore et al. 2003). There are instances where nanosize effects may be evident. Nanoiron made from the borohydride reduction of dissolved Fe has exhibited behavior that may be attributable to small particle size (i.e., a nanosize effect). First, nanoiron made from reduction of dissolved Fe by sodium borohydride has the ability to activate H_2 gas and use it in the reduction of chlorinated ethenes (Liu et al. 2005a; Liu et al. 2005b). This ability was attributed to the amorphous structure and the nanosized Fe^0 crystallites (~1 nm) in these particles. Micron-scale iron filings and other forms of nanoiron with larger crystallites cannot active H_2. Second, this nanoiron that does not have any noble metal catalyst can dechlorinate polychlorinated biphenyl compounds (PCBs) at ambient temperature and pressure (Lowry and Johnson 2004; Wang and Zhang 1997). Iron filings and other types of nanoiron without an added catalyst cannot dechlorinate PCBs under these conditions. Novel nanomaterials that leverage the nanosize effects may degrade highly recalcitrant compounds that are not amenable to degradation by currently available materials.

Laboratory investigations (Table 8.1) indicate that the reactivity afforded by nanoiron and Fe^0/Pd bimetallic nanoparticles is sufficient to make it useful for *in situ* remediation of groundwater. It is important to note, however, that most of these studies use fresh nanoiron particles and do not observe reactivity over the lifetime of the particles so they should be considered as initial rates. There have been few investigations on the effect of aging (oxidation) or precipitation of Fe-containing minerals (e.g., $FeCO_3$) on nanoiron reactivity throughout its lifetime. It is likely that many of the conclusions regarding geochemical effects on the reactivity of iron filings will be applicable to nanoiron systems, but this has not yet been demonstrated. If reactivity of nanoiron is found to be

TABLE 8.1   Organic Contaminants and Their Reaction Rates in Reactions with Nanoscale Iron or Bimetallic Iron Particles

Compound	Iron type	Pseudo-first-order k_{rxn} $(\times 10^{-3}$ L/hr $\cdot$ m^2)	Ref.
Chlorinated Methanes			
CCl$_4$	Fe^0	0.5–240	(Lien and Zhang 1999; Nurmi et al. 2005)
CHCl$_3$	Pd/Fe	0.08	(Lien and Zhang 1999)
Chlorinated Ethanes			
HCA	Fe^0	770±70	(Song and Carraway 2005)
PCA		796±63	
1,1,2,2-TeCA		538±56	
1,1,1,2-TeCA		30.3±4.2	
1,1,2-TCA		151±13	
1,1,1-TCA		2.31±0.34	
1,2-DCA		0	
1,1-DCA		0.02±0.004	
HCA			(Lien and Zhang 2005)
PCA	Fe/Pd	20	
1,1,1,2-TeCA		27	
1,1,2,2-TeCA		21	
1,1,1-TCA		9	
Chlorinated Ethenes		6	
TCE	Fe^0	1–14	(Liu et al. 2005a) (Liu et al. 2005b) (He and Zhao 2005)
	Fe^0-starch stabilized	20	(He and Zhao 2005)
	Fe/Pd	18	(Wang and Zhang 1997)
	Fe/Pd-starch stabilized	67	(He and Zhao 2005)
PCBs	Fe^0	$10^{-6}-5 \times 10^{-5}$ L/yr $\cdot$ m^2	(Lowry and Johnson 2004)
	Fe^0	0	(He and Zhao 2005)
	Fe/Pd	0.053	
	Fe/Pd-starch stabilized	0.31	
Inorganics/metals			
Perchlorate	Fe^0	0.013 mg Cl/g Fe/hr	(Cao et al. 2005)
Nitrate	Fe^0	70	(Caoe et al. 2000)
Selenate	Fe^0	0.9	(Mondal et al. 2004)
	Fe/Ni	2.5	
Arsenic(III)	Fe^0	200–600	(Kanel et al. 2005)
Cr(VI)	Fe^0	110–230	(Ponder et al. 2000)
Pb(II)	Fe^0	130	

insufficient, addition of Pd or some other noble metal catalyst increases the reactivity, but this also increases the cost, which currently ranges from \$50–110/kg for nanoiron without added catalyst.

Reaction products, intermediates, and efficiency

There are two reasons why *in situ* remediation using reactive nanoparticles must consider the reaction products formed. First, a reaction product may potentially be more toxic or mobile than the parent compound, and thus increases the risks posed by the site rather than decreasing them. Second, the products formed can strongly influence the effectiveness and costs of remediation. Developers of reactive nanoparticles for remediation need to consider the production and potential negative consequences of reaction intermediates and products. TCE offers a good example of how *in situ* remediation may increase the risks at a site rather than decrease them. Under reducing conditions afforded by nanoiron, TCE can be sequentially dechlorinated from TCE to dichloroethene (DCE), and to vinyl chloride (VC), and then to ethene, and under some conditions to ethane (Figure 8.3). VC is classified as a known human carcinogen, while TCE is classified as a suspected human carcinogen. Thus, conversion of TCE to VC may in fact increase the risk at a site due to VC's higher toxicity. In the case of nanoiron, most laboratory investigations have shown negligible production of chlorinated intermediates

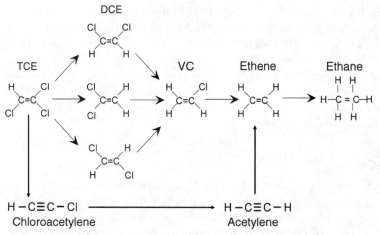

Figure 8.3   Dechlorination pathways of TCE. Partial dechlorination may lead to production of dichloroethene isomers and vinyl chloride, which is more toxic (known carcinogen) than the parent compound TCE (suspected carcinogen).

for most chlorinated solvents, which is optimal. The exception is carbon tetrachloride (CCl_4), where reduction by nanoiron results in chloroform ($CHCl_3$) as a long-lived reactive intermediate or product. Chloroform is more mobile and toxic than carbon tetrachloride, and reactive particles that can promote the complete conversion of carbon tetrachloride to methane (CH_4) or to carbon dioxide (CO_2) is most desirable.

For reactive nanoparticles, where the particle itself is the reductant and therefore has a finite reducing power, it is important to understand the factors influencing the Fe^0 utilization efficiency. The iron utilization efficiency is defined as the mass of target contaminant degraded per unit mass of nanoiron added. The degradation products formed can strongly influence the nanoiron mass required to degrade a given mass of contaminant to innocuous products. For example, using Fe^0 as the reductant TCE can dechlorinate to partially or fully saturated dechlorination products (Eq. 4), where the values of the coefficients a and b are determined by the products formed (e.g., $a = 4$, $b = 3$ for ethane, and $a = 2$, $b = 3$ for acetylene). Eq. 4 is the net reaction derived from the two half-reactions shown in Eqs. 1 and 2.

$$a Fe^0 + TCE + (2a - b)H^+ \rightarrow prod + a Fe^{2+} + b Cl^- \qquad (4)$$

For TCE, removing all of the chlorines to form the corresponding hydrocarbon makes it nontoxic. For nanoiron made from borohydride reduction of dissolved iron, the dominant reaction product is ethane (C_2H_6) (Liu et al. 2005a). For another type of nanoiron, the dominant reaction product is acetylene (C_2H_2) (Liu et al. 2005b). Both are nonchlorinated and are equally nontoxic, however, the carbon in ethane is highly reduced (C average oxidation state is –III), while the carbon in acetylene is not as reduced (C average oxidation state in acetylene is –I). The average oxidation state of C in the parent compound (TCE) is +I, so reduction of TCE to ethane requires twice as much Fe^0 as for reduction of TCE to acetylene. This is also evident in the value of a for ethane ($a = 4$) versus acetylene ($a = 2$). Reactive nanoparticles can potentially be designed for optimal efficiency by selecting for products that are nontoxic but require the smallest possible redox swing.

Fate and lifetime

Nanoparticles have great potential to benefit the environment by improving groundwater remediation, but before their widespread release it is prudent to evaluate the potential risks associated with their use. This requires a fundamental understanding of their long-term fate and lifetime. Most nanoiron is comprised of primary particles ranging in size from ~40 to 100nm that have an Fe^0 core and Fe-oxide shell that is

RNIP (FeOOH → Fe⁰ → Fe⁰/Fe₃O₄)

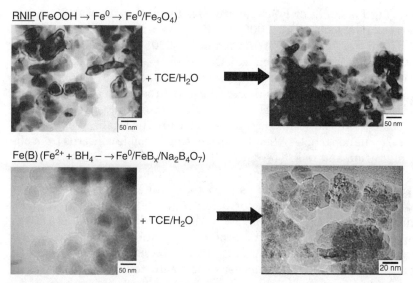

Fe(B) (Fe²⁺ + BH₄⁻ → →Fe⁰/FeBₓ/Na₂B₄O₇)

Figure 8.4   TEM images of fresh nanoiron samples showing before reaction and after reaction with TCE in water. Both particle types have an apparent core-shell morphology. RNIP, made from gas phase reduction of Fe-oxides in H_2, appear to have a shrinking core during reaction in water. Fe(B), made from reduction of Fe^{2+} in a water/methanol solution using sodium borohydride appear to undergo oxidative dissolution followed by precipitation of the dissolved Fe to hematite (Liu et al. 2005b). The shell on Fe(B) is predominantly borate (Nurmi et al. 2005).

primarily magnetite and maghemite (Figure 8.4). Magnetite (Fe_3O_4) is the dominant Fe-oxide at the Fe^0/Fe-oxide interface, while maghemite (Fe_2O_3) is the dominant Fe-oxide at the Fe-oxide/water interface. As the particles oxidize (and the contaminants reduce), the Fe^0 core shrinks and ultimately the particles become Fe-oxide. For some nanoiron, the particles undergo reductive dissolution, followed by precipitation of the dissolved Fe as another Fe-oxide form (Liu et al. 2005a).

Nanoiron is a unique nanoparticle because its corrosion by water produces H_2 gas (Eq. 5) (Liu et al. 2005a; Liu et al. 2005b), which can be used as an energy source by hydrogenotrophic bacteria (Sorel et al. 2001; Vikesland et al. 2003) that also remove surficial H_2 (i.e., cathodic depolarization) (Hamilton 1999) to enhance or sustain iron surface reactivity.

$$\text{Fe} + \text{H}^+ + e^- \xrightarrow{k_1} \text{Fe} - \text{H}^{\cdot} \xrightarrow{k_2} \frac{1}{2}\text{H}_2 \qquad (5)$$

Further, dissimilatory iron reducing bacteria (DIRB) such as *Geobacter* or *Shewanella* spp. could maintain active Fe^0 surfaces by reductively dissolving Fe-oxides or Fe-oxyhydroxides formed at the water/particle interface (Gerlach et al. 2000), or by generating reactive surface-associated Fe(II) species (Williams et al. 2005). Homoacetogens are another group that could enhance TCE removal—either directly through cometabolism (using H_2 as primary substrate), or indirectly by stimulating heterotrophic activity through acetate production ($4Fe^0 + 2CO_2 + 5H_2O \longrightarrow CH_3COO^- + 4Fe^{+2} + 7OH^-$) (Oh and Alvarez 2002). Microbially induced corrosion could also ensure the localized dissolution of the iron nanoparticles, thereby eliminating possible concerns from off-site migration and risk. The data collected thus far on the fate of nano-iron have been largely collected in the laboratory in deionized water. Porewater constituents such as carbonate, sulfate, and chloride may inhibit or promote nanoiron corrosion (Agrawal and Tratnyek 1996; Bonin et al. 1998; Phillips et al. 2000; Vogan et al. 1999). It is critical to know what the end product of nanoiron oxidation is under real environmental conditions, but a well controlled *in situ* study of the fate of nanoiron in the subsurface has yet to be conducted.

Effect of pH. The role of pH is an important one for nanoiron lifetime. As seen in Eq. 5, decreasing the pH should increase the rate of H_2 evolution from nanoiron if the reduction of H^+ at the iron surface is the rate controlling step. According to Eq. 5, in the absence of any other oxidants and assuming that the reaction is first order with respect to the specific surface area of the iron, the rate of H_2 evolution is given by Eq. 6.

$$\frac{dH_2}{dt} = k_{H_2} \cdot [SA] \cdot [H^+]^b = k_{H_2obs}[H^+]^b \qquad (6)$$

k_{H_2} is the first-order reaction rate constant for H_2 evolution and [SA] is the specific surface area for the reaction, k_{H_2obs}, is the observed pseudo-first-order reaction rate constant. The H_2 evolution from nanoiron in solution at pH ranging from 6.5 to 8.9, the equilibrium pH for an iron/water system (Pourbaix 1966), is shown in Figure 8.5a. Indeed, as the pH decreases, the rate of H_2 evolution increases. For pH ranging from 6.5 to 8.0, there is a log linear relationship between the rates, indicating that b in Eq. 6 is ≈ 1 (Figure 8.5b). Thus H_2 evolution from RNIP is pseudo-first-order with respect to H^+. At pH greater than 8, the mechanism for Fe^0 oxidation by H^+ appears to change, although some H_2 evolution is occurring at pH > 8. This agrees with published reports for hydrogen evolution from zerovalent metals at near neutral pH where k_1 (Eq. 5) has been shown to be the rate controlling step (Wang and Farrell, 2003; Bockris et al., 1987). Most importantly is the effect of pH on the lifetime of RNIP. Using the data in Figure 8.5a, at pH > 8, the

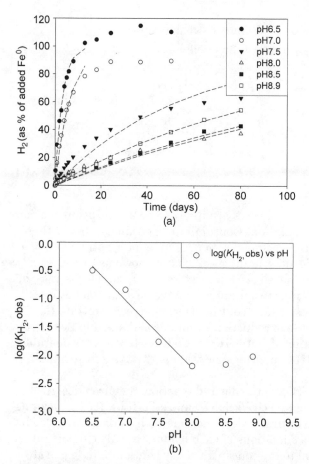

Figure 8.5 H_2 production from RNIP. (a) H_2 production from 500 mg/L RNIP with an Fe^0 content of 27 wt% at different pHs.pH was buffered using 50 mM HEPES buffer. Data are from duplicate reactors. The curve represents the first-order data fit (kobs, H_2 = 0.0067 1/day, r^2 = 0.996). (b) k_{obs}, H_2 vs. pH production from 500 mg/L RNIP with an Fe^0 content of 27 wt%. Lines are interpolated (not fit) and only meant to guide the eye.

lifetime of RNIP is expected to be on the order of 9 to 12 months, whereas at pH near neutral (typical for soils) the lifetime of RNIP will be on the order of 2 weeks. This only considers the unintended competing reaction of RNIP with water/H+. In the presence of additional oxidants such as the target contaminant (e.g., TCE), the lifetime could be even shorter because this adds another oxidant to the system that will utilize the Fe^0 in the particles.

Effect of competing oxidants. Any *in situ* groundwater remediation technology that employs nanomaterials will have to consider the cost of

$$+ \quad \text{TCE} \quad \xrightarrow{k_2} \quad \text{Nontoxic dechlorination products}$$

$$\text{Fe}^0 \quad + \quad \text{NO}_3^- \quad \xrightarrow{k_3} \quad \text{NO}_2^-, \text{NH}_4^+, \text{N}_2$$

$$+ \quad \text{H}^+ \quad \xrightarrow{k_1} \quad \text{H}_2, \text{OH}^-$$

Figure 8.6 Parallel reaction pathways for competing oxidants in a contaminated aquifer.

those materials, as material costs are likely to be a significant portion of the total cost of remediation. Contaminant specificity should therefore be considered in the design of novel reactive (or sorptive) nanomaterials. The concept of contaminant specificity is known as the selectivity ratio. At many contaminated sites a mixture of contaminants may be present and nanoiron injected there may have several competing oxidants. For example, a site may contain TCE along with nitrate (NO_3^-), which are both amenable to reduction by nanoiron (Yang and Lee 2005). In addition, the oxidation of water/H^+ to produce H_2 is always operable, which provides parallel competing oxidation pathways for the nanoiron (Figure 8.6).

If the reaction with TCE is the desired reaction, nanoiron designed to give selectivity ratios of $k_2[\text{TCE}]/k_3[\text{NO}_3^-]$ and $k_2[\text{TCE}]/k_1[\text{H}^+]$ as large as possible is desirable. For first-order reactions such as those provided by RNIP, the relative concentrations of each oxidant are also important as the rate is proportional to the concentration of contaminant (e.g., rate = $k_2[\text{TCE}]$). Any application of reactive nanomaterials for groundwater remediation will require extensive site characterization to determine the suitability of the materials for that particular site. Reactive or sorptive nanomaterials may not be suitable at a site where the concentrations of competing oxidants at a given site are high, especially if the nanomaterial cannot be designed with a selectivity ratio that minimizes the effect of competing oxidants. There is considerable room for improvement regarding the selectivity ratio of nanomaterials, and controlling the reactivity of these materials at the nanoscale is an exciting research avenue that will continue to be pursued.

The economic feasibility of reactive nanoiron for TCE DNAPL source zone remediation highlights the need to understand the products formed and the potential competing reactions. The goal is to reductively dechlorinate the contaminant (TCE) to nontoxic products using electrons derived from the nanoiron. General equations describing the transformation of trichloroethylene (TCE) were given in Eq. 4 (Liu et al. 2005a).

The two electrons from Fe^0 oxidation can be used to dechlorinate TCE (Eq. 4), or can be used to produce H_2 (Eq. 5). The mass of TCE dechlorinated per mass of Fe^0 will depend on the relative rates of each of these reactions, and on the value of n in Eq. 4, which is a function of the dechlorination products formed. The products of TCE degradation using nanoiron vary from acetylene (using RNIP) to ethane (using iron made from the borohydride reduction of dissolved iron). Acetylene is the best-case scenario as this pathway only requires 4 moles of electrons per mole of TCE ($n = 4$). It requires twice as many to transform TCE to ethane ($n = 8$). Assuming nanoiron costs $50/kg and that all injected zerovalent iron is used to transform TCE, the cost of treatment ranges from $44/kg TCE (for acetylene) to $88/kg TCE (for ethane). If oxidants other than TCE were present (e.g., NO_3^- or dissolved oxygen), the process efficiency may decrease and the cost for iron would increase. For example, if only 10 percent of the electron from iron oxidation were used to transform TCE, the cost would increase by a factor of 10. Careful consideration of the site geochemistry and appropriate bench scale feasibility testing are necessary to assess the economic feasibility of reactive nanoparticles at a particular site.

Delivery and Transport Issues

The use of reactive nanomaterials for groundwater remediation is promising considering the availability and effectiveness of many types of nanomaterials for degrading or sequestering environmental contaminants of concern. The attractiveness of nanomaterials is their potential to be used *in situ*, directly degrading the contaminants in the subsurface without the need to excavate them or pump them out of the ground. Realizing the potential of nanomaterials will require the ability to inject them into the subsurface and transport them to the contaminant source zone where they rapidly degrade the contaminants. If particles cannot be delivered to and remain at the source of contamination, they will not be utilized efficiently.

Injection methods and delivery vehicles

Reactive nanoparticles can be injected into a well and allowed to transport down gradient from the injection site to the contaminated area. Typically this is done in existing wells, or in wells drilled specifically for the remediation process. Drilling and packing a well, even a small diameter (3-inch) well is quite expensive. Direct push wells (Figure 8.7) are a lower cost alternative to drilled wells, and are the most often used delivery tool for remediation with nanoiron. These are hydraulic devices mounted on a truck or tractor. They can create wells that are 2 to 3 inches

Figure 8.7 Truck-mounted Geo-Probe® for creating direct push wells for nanoiron injection. Photo is courtesy of GeoProbe® Systems, Inc., Salina, Kansas.

in diameter that can then be used for injecting reactive nanoparticles. A nanoparticle slurry can be injected along part of all of the vertical range of the probe (Figure 8.8) to provide treatment to specific regions in the aquifer. This helps to target the reactive nanoparticles to the contaminated regions of the subsurface where they are needed.

Transport

The effectiveness of remediation, as well as the potential for unwanted exposure of humans and other biota to these reactive nanoparticles, depends on how easily these materials transport in porous media. Uncertainties regarding the fate, transport, and potential toxicity of engineered nanomaterials have prompted investigations on the fate and transport of various nanoparticles (very fine colloids) ranging in size from 1.2 nm to 300 nm (Lecoanet et al. 2004; Lecoanet and Wiesner 2004; Royal Academy of Engineering 2004; Saleh et al. 2007; Schrick et al. 2004). These studies have demonstrated that nanoiron and many other types of engineered nanomaterials do not transport easily in saturated porous media. Transport distances range from a few centimeters for

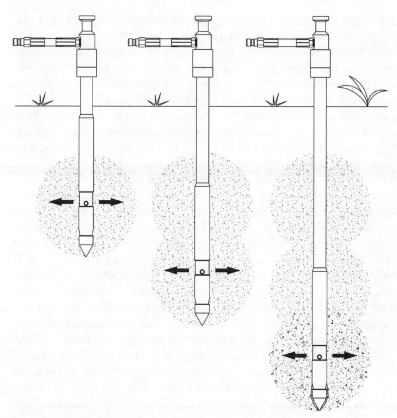

Figure 8.8 GeoProbe® injection of nanoparticles into the subsurface. Multiple injection points along the probe can provide a zone of influence over a portion of or the entire vertical section of the contaminated area. Photo is courtesy of GeoProbe® Systems, Inc., Salina, Kansas.

nanoiron (Saleh et al. 2007; Schrick et al. 2004) to a few meters for carbon nanomaterials under typical groundwater conditions (Lecoanet et al. 2004; Lecoanet and Wiesner 2004). This has prompted the use of supports such as hydrophilic carbon or polyacrylic acid to enhance the transport of nanoiron (Schrick et al. 2004), the use of surfactant micelles to deliver nanoiron directly to entrapped NAPL (Quinn et al. 2005), or adsorbed polymers to improve transport distances and target entrapped DNAPL (Saleh et al. 2007). The reasons for limited transport and the need for advanced delivery vehicles are addressed here.

Investigations of the transport and fate of colloids (micron and submicron particles) in the environment has been extensively studied for decades, as these processes are critical to understanding deep bed filtration (Yao et al. 1971) and the movement of colloids in natural systems (Elimelech et al. 1995; Ryan and Elimelech 1996). Typically this has focused on particles on the order of 1 micrometer, but particles that are a few hundred nanometers have also been studied. As previously discussed, the fate and transport of nanoparticles in porous media can be considered a filtration problem (Figure 8.1). Filtration theory indicates that the magnitude and rate of aggregation or deposition will depend on *physical* properties, including the nanoparticle size, pore size distribution of the media, and the flow velocity, and by the *chemical* properties such as the pH, ionic strength, and ionic composition, which control the magnitude and polarity of the attractive and repulsive forces between the nanoparticles and between the nanoparticles and mineral grains. Thus, the *hydrogeochemistry* of the system, along with the properties of the nanoparticles, will determine the transportability of nanoparticles at a specific site. Increasing ionic strength or the presence of divalent cations such as Ca^{2+} and Mg^{2+} can destabilize the particles by decreasing electrostatic double layer (EDL) repulsions between particles and allow aggregation (labeled A in Figure 8.1). This will also increase nanoparticle deposition onto media grains (labeled C in Figure 8.1). Greater attachment efficiency to media grains (or membrane surfaces in a membrane filter) will limit nanoparticle transport. Alternatively, aggregation of nanoparticles to larger-sized aggregates may potentially serve to increase particle mobility. The diffusion rate of the larger aggregates will be slower than for the larger particles, which may decrease the rate of particle-media interactions. Nanoparticle aggregation or the presence of high concentrations of particles could also lead to straining (labeled B in Figure 8.1), which will limit or retard transport. The flow velocity also plays a significant role. At a high porewater velocity, the residence time of the nanoparticles at the collector surface may be too short to allow for attachment to occur. Low attachment efficiency will result in longer transport distances. For nanoparticles (d_p < 100 nm) in environmental media (e.g., soil, sediments), it is difficult to predict *a priori* how the various interactions among the particles (aggregation) and media grains will affect their transport.

There are limitations to applying standard deep-bed (or clean-bed) filtration models in natural systems, and especially for remediation applications where concentrated suspensions of particles will be injected. First, typical filtration models have to make many simplifying assumptions, such as homogeneous media, monodisperse particle size distribution of nanomaterials, constant porewater velocity, and uniform surface properties of nanoparticles and filter media. These are typically not applicable in real environmental systems, which are physically and chemically heterogeneous. Another assumption of filtration models is that the bed is clean—that is, only particle-media grain inter-actions are explicitly accounted for. This may not be the case for groundwater remediation using engineered nanoparticles where high concentration slurries of polydisperse particles will be injected to achieve a 0.1 wt% to 0.5 wt% concentration of particles in the aquifer. Under these conditions, particle-particle interactions are important, and filter-ripening models may be more appropriate than clean bed fil-tration models. Regardless of the model type (clean bed or filter ripen-ing), model inputs will include particle size, attachment coefficients, flow velocity, ionic strength, and ionic composition as discussed below. Aggregation and surface modification, and how these affect the mobil-ity or engineered nanomaterials and the ability to target specific regions in the subsurface, are discussed.

Effect of aggregation. If aggregation is rapid, and creates particles that are greater than a few microns in diameter, there is potential for this aggregation to limit their transportability. For many nanoparticles, aggre-gation will likely be a mechanism that significantly limits their transport in the environment, and makes it possible to remove them using standard water-treatment practices such as flocculation/sedimentation or mem-brane filtration. The rate of particle aggregation and the size and mor-phology of the aggregates formed depends on both the collision frequency (transport) and the collision/attachment efficiency (i.e., the magnitude of the attractive and repulsive forces between the particles).

Experience shows that many nanoparticle suspensions are typically highly unstable and rapidly aggregate (Saleh et al. 2005a), making them difficult to handle. The reason for the rapid aggregation is that their small size (~100 nm) gives them high diffusion coefficients (trans-port rates) and an exceptionally high number of collisions, so even dilute suspensions of nanoparticles with relatively low sticking coef-ficients would rapidly aggregate and not remain as individual nanopar-ticles in suspension under normal environmental conditions. Moreover, the presence of cations and anions, particularly divalent cations such as Mg^{2+} and Ca^{2+}, further destabilizes nanoparticles and increases their rates of aggregation. For nanoiron, 100-nm particles present at a volume fraction (Φ) as low as $\Phi = 10^{-5}$ rapidly aggregate into 5-micron particles in just a few minutes (Figure 8.9). This problem is exacer-bated for nanoparticles such as nanoiron that are magnetic and there-fore subject to non-DLVO magnetic attractive forces. The time scale of colloid dispersion stability is determined by the magnitude of the energy barrier between particles. According to classical DLVO theory, the major attractive energy is van der Waals energy (V_{vdW}) while the major repulsive energy is electrostatic interaction energy (V_{ES}) (de Vicente 2000; Elimelech et al. 1995; Evans 1999; Heimenz 1997;

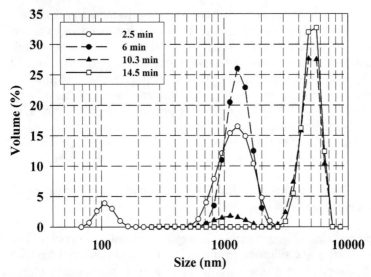

Figure 8.9 DLS indicates that nanoiron particles (100 nm) present at 80 mg/L (volume fraction ~10-5) rapidly flocculate to form 5-micron size aggregates.

Plaza 2001; Strenge 1993). The V_{vdW} attractive force between spherical particles can be expressed as (de Vicente 2000)

$$V_{vdW} = \frac{-A}{6}\left[\frac{2r^2}{s(4r+s)}+\frac{2r^2}{(2r+s)^2}+\ln s\,\frac{(4r+s)}{(2r+s)^2}\right] \qquad (7)$$

where A is the Hamaker constant, which is 10^{-19} N · m for Fe, γ-Fe$_2$O$_3$ and Fe$_3$O$_4$ (Rosensweig 1985). r(m) is the radius of particles, and s(m) is distance between surfaces of two interacting particles. Electrostatic repulsion between two identical particles, V_{ES} can be expressed as (de Vicente 2000)

$$V_{ES} = 2\pi\varepsilon_r\varepsilon_0 r\zeta^2\ln[1 + e^{-\kappa s}] \qquad (8)$$

where ε_r is the relative dielectric constant of the liquid, ε_0 is the permittivity of the vacuum, ζ is electrokinetic or zeta potential of diffuse layer of charged particles, and κ is the reciprocal Debye length. Applying classical DLVO theory, an energy barrier of RNIP is predicted to be 7.0 k_BT (Figure 8.10). This energy barrier is sufficient to prevent rapid aggregation, suggesting that dispersions of these particles should be stable. This contrasts the observed behavior.

Iron nanoparticles that behave as a single domain magnetic particle have an intrinsic permanent magnetic dipole moment $\mu = (4\pi/3)r^3M_s$ even in the absence of an applied magnetic field (Butter 2003a;

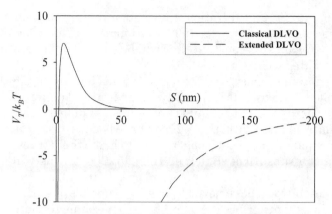

Figure 8.10 Classical DLVO simulations including EDL repulsions and van der Waals attractive forces predict an energy barrier to aggregation of ~7.0 k_BT, which should limit aggregation. Including magnetic attractive forces (dashed line) indicates no energy barrier, and that attractive forces may be as long range as a few hundred nanometers. S is the separation distance between the particles in nanometers. V_T is the sum of the attractive and repulsive forces acting on the particles.

Butter 2003b; de Gennes 1970; McCurrie 1994; Neto 2005; Rosensweig 1985). When particle dipoles are oriented in head-to-tail configuration, the maximum magnetic attraction energy (V_M) can be expressed as (de Vicente 2000)

$$V_M = \frac{-8\pi\mu_0 M_s^2 r^3}{9\left(\frac{s}{r} + 2\right)^3}$$ (9)

where μ_0 is the permeability of the vacuum. The potential energy of interaction for RNIP that includes this magnetic attraction is also shown in Figure 8.10. For magnetic nanoparticles like RNIP and magnetite, the magnetic attraction dominates the interaction energy and there is no longer a predicted energy barrier to aggregation. In fact, extended DLVO suggests that there are relatively long-range attractive forces (~250 nm) for nanoiron particles. This is in agreement with the rapid aggregation observed for RNIP.

Rapid aggregation makes it difficult to predict their transport in porous media since the rates of these transport processes are influenced by particle size. It also makes it difficult to predict the potential toxicity associated with these particles—it is not known if 5-micron-sized aggregates of nanometer-sized particles illicit a toxic response that is the same as or different from a concentrated suspension of the 100-nm-sized particles of equal mass concentration. Rapid aggregation, however, may have advantages. Because flocculation/aggregation of fine particles followed by sedimentation and filtration are commonly used to treat drinking water (Viessman and Hammer 1998), the tendency of nanoparticles to rapidly aggregate suggests that they will be easily removed in traditionally used water-treatment systems.

The rapid aggregation of most types of nanoparticles has led to the use of coatings to modify their surface chemistry in a way that minimizes aggregation and increases the stability of aqueous dispersions. Surfactants and natural and synthetic polymers have been proposed as a means to stabilize nanoiron suspensions in order to make them transport effectively in the subsurface (Saleh et al, 2005, 2007). This will affect their fate and transport characteristics in the environment, and may also affect their ability to be removed from water supplies or during treatment, so the impact of surface coatings on the environmental fate and transport should be considered in the design process for the coatings. For example, coatings could be used to deliver particles in the subsurface, but then biodegrade, transform, or desorb after some time so that the particles are no longer mobile and the potential for exposure is minimized.

Surface modification.   Effective use of many of the new nanoparticles being developed will require that particle aggregation be limited. For example, metal-containing nanoparticles being considered as MRI contrast agents will have to be designed to be dispersible in aqueous environments with high osmolarity such as blood (Sitharaman et al. 2004; Veiseh et al. 2005). Surface functionalization is a common tool for minimizing or controlling nanoparticle aggregation. This niche of nanotechnology has the potential to create nanoparticles that could be widely distributed and mobile in the environment, easily assimilated into people or other biological organisms, and difficult to remove by remediation and treatment if needed. The uncertainties surrounding the effect of surface coatings on the fate, transport, and potential toxicity of engineered nanomaterials makes this a rich area for current and future research on nanotechnology. Nanoparticles used for *in situ* groundwater remediation will require surface coatings for their intended application (Saleh et al. 2005a; Saleh et al. 2007). There are three classes of typical surface coatings used, including polymers, polyelectrolytes, and surfactants. These coatings can impart charge to the particles (positive or negative) and can provide three modes of colloidal stabilization that make them mobile in the environment: electrostatic, steric, or electrosteric (Figure 8.11). Both natural and synthetic varieties of each type of modifier are widely available and used (Table 8.2).

In general, high molecular weight polymers (synthetic or natural) provide steric repulsions that may limit nanoparticle-bacteria interactions (Figure 8.11). Polyelectrolytes are large polymers containing

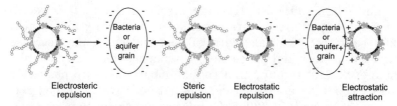

Figure 8.11 Surface coatings may prevent interactions between nanoparticles and bacteria or mineral grains. Polyelectrolytes provide electrosteric repulsive forces, while polymers provide steric repulsive forces. Short chain anionic or cationic surfactants can provide electrostatic repulsive or attractive forces.

charged functional groups (anionic or cationic) and provide electrosteric repulsions. Surfactants can provide electrostatic repulsive or attractive forces (depending on their charge), but are less effective than polymers or polyelectrolytes because their small size is unsuitable for steric repulsions. Polymers adsorb strongly (effectively irreversible) to nanoparticles, while surfactant adsorption is more easily reversible (Braem et al. 2003; Fleer et al. 1993; Holmberg et al. 2003;

TABLE 8.2   Coatings from Each Class to Be Evaluated

Coating	Charge	Stabilization type	Relevance
Polymers Polyethylene glycol (PEG)	Nonionic	Steric	Common nontoxic polymer used to stabilize NP dispersions. Adsorbed PVA reduces bacterial adsorption to surfaces (Koziarz and Yamazaki 1999). Cellulose is used to prepare stabilized nanoiron dispersions (He and Zhao 2005). Inexpensive and biodegradable.
Polyvinyl alcohol (PVA)	Nonionic	Steric	
Carboxymethyl cellulose	Nonionic	Steric	
Guargum	Nonionic	Steric	
Polyelectrolytes Triblock copolymers (PMAA$_x$-PMMA$_y$-PSS$_z$)	Anionic	Electrosteric	Used to prepare stabile nanoiron dispersions that partition to the TCE/water interface (Saleh et al. 2005a). PSS homopolymer is a common inexpensive alternative used to stabilize NPdispersions. Polylysine is a biodegradable polypeptide and bactericidal (Roddick-Lanzilotta and McQuillan 1999). Poly(aspartic acid) should be biodegradable and less bactericidal.
Polystyrene sulfonate	Anionic	Electrosteric	
Poly(aspartic acid)	Anioinc	Electrostatic	
Surfactants SDBS	Anionic	Electrostatic	SDBS is a common anionic surfactant shown to enhance nanoiron mobility (Saleh et al. 2007) but resists biodegradation. Alkyl polyglucosides are biodegradable (Matsson et al. 2004).
Alkyl polyglucosides	Nonionic	Steric/Hydration	

Velegol and Tilton 2001). Electrostatic, steric, or electrosteric repulsions decrease nanoparticle interactions with mineral grains (Saleh et al. 2007) and potentially also with soil bacteria, which may decrease the observed bactericidal properties of nanoparticles. For example, Goodman et al. (2004) found that gold nanoparticles with positively charged side chains were toxic to *E. coli*, but negatively charged particles were not. Rose et al. (2005) found that positively charged CeO_2 nanoparticles adsorbed to *E. coli* and were bactericidal, showing a clear dose-response.

Poly(methacrylic acid)-b-poly(methylmethacrylate)-b-poly(styrene sulfonate) triblock copolymers (PMAA-PMMA-PSS), PSS homopolymer polyelectrolyte, sodium dodecylbenzene sulfonate (SDBS), polyethylene glycol (PEG), carboxymethyl cellulose (CMC), and guar gum have been shown to adsorb to and stabilize dispersions of nanoiron and Fe-oxide nanoparticles to improve particle mobility in the environment due to steric, electrosteric, or electrostatic repulsions between the particles and between the particles and soil grains (He and Zhao 2005; Saleh et al. 2007). Poly(aspartic acid), an anionic polypeptide, also adsorbs to iron oxide surfaces via acid-base interactions through the carboxyl groups (Chibowski and Wisniewska 2002; Drzymala and Fuerstenau 1987; Nakamae et al. 1989) and can be an effective stabilizer for metal-oxide nanoparticles. Alkyl polyglucosides are an emerging class of surfactant synthesized from renewable raw materials and are nontoxic, biocompatible, and biodegradable. They adsorb to metal oxide surfaces (Matsson et al. 2004). They are desirable due to their low cost and "green" nature and widely used as detergents in manufacturing. A systematic investigation of the ability of these types of surface modifiers to enhance nanoiron mobility by minimizing particle-particle interactions as well as particle-media grain interactions is underway.

Polyelectrolytes have been used to provide functionality to nanoiron (e.g., affinity for the DNAPL/water interface [Saleh et al. 2005a]) as well as to enhance their transport in laboratory columns. For example, transport of triblock copolymer-modified nanoiron is significantly enhanced in a 10-cm sand column relative to the unmodified nanoiron, especially at the high nanoiron concentration that is needed to be cost-effective (Figure 8.12) (Saleh et al. 2007). In this case, enhanced transport is due to an adsorbed layer of a strong polyelectrolyte that provides electrosteric repulsions, which limit the particle-particle interactions as well as the particle sand-grain interactions. Enhanced mobility comes at a price, however, as modified particles were four to nine times less reactive with TCE than for unmodified particles (Saleh et al. 2007). Despite the lower reactivity, the particles are sufficiently reactive with TCE to be effective *in situ* groundwater remediation agents.

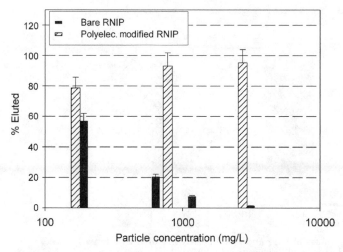

Figure 8.12 Surface modification by strong polyelectrolyte polymers enhances transport of nanoiron (RNIP) transport through saturated porous media. Polymer-modified iron (striped bars) is highly transportable through water-saturated sand columns, while unmodified nanoiron (black bars) does not transport well through the column, particularly at high concentrations (3 g/L).

Effect of modifier type on transport.   Different types of modifiers will provide different modes of stabilizing nanoparticles against aggregation and attachment to aquifer media grains. These differences lead to different elutability and transport distances. For example, the eluted masses of bare MRNIP (a polyaspartate-modified nanoiron), a triblock copolymer-modified nanoiron, and a surfactant-modified (SDBS) nanoiron through a 12.5-cm sand-filled column are shown in Figure 8.13. At 3 g/L, bare RNIP has very low transportability (1.4 ± 3% mass elution) through a saturated sand column at low ionic strength. Retrieving the sand from the column and analyzing for iron revealed that most particles were trapped within the first 1 to 2 cm of the column. MRNIP-, polymer-, and SDBS-modified RNIP elution was much higher, with the triblock copolymer and MRNIP elution at 95 percent and 98 percent, respectively. SDBS was not as effective as the polymer but still improved RNIP elution to approximately 50 percent. These results indicate that surface modification is essential for reasonable transport, even at low ionic strength. These differences could be used to synthesize particles with specific transport distances that can then be used for controlled delivery of nanoparticles to specific regions in the subsurface.

Geochemical effects on transport (pH, ionic strength, and ionic composition).
Each of the modified particles also responds differently to changes in ionic strength and to ionic composition. Both Na^+ and K^+ cations and

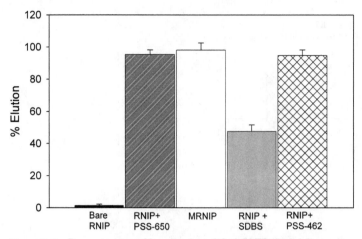

Figure 8.13 Percent mass of bare and modified RNIP eluted through a 12.5-cm silica sand column with porosity of 0.33. Modifying agents, i.e., PMAA-PMMA-PSS polymer or SDBS, were added at 2g/L concentration in each case. Polyaspartic acid (MRNIP) was added at a 6:1 mass ratio. The approach velocity was 93 m/d.

Ca^{2+} and Mg^{2+} cations are prevalent in natural water systems. The presence of these monovalent and divalent cations tends to shield EDL repulsions between particles and between particles and aquifer grains. Shielding these repulsive forces increases the attachment efficiency of the particles to the sand grains and decreases the distance that they can travel through saturated porous media before being filtered from solution. Divalent cations are much more efficient at shielding EDL repulsions than are monovalent cations due to their higher charge density. Surface modifications that provide strong EDL repulsions as well as steric hindrances, termed electrosteric repulsions, should in principle do the best job of minimizing attachment of particles to sand grains. Saleh et al. (2007) measured the elution of each surface modified particle under varying ionic strength conditions and determined that large molecular weight polyelectrolytes indeed provided the best elutability. This was attributed to the ability of the large molecular weight polymers to provide strong electrosteric repulsions compared to short chain polyelectrolytes such as polyaspartic acid, or short chain surfactants such as SDBS. These results demonstrate that the site-specific geochemistry must be taken into account when developing dispersants to enhance nanoiron delivery.

Conditions such as high ionic strength and the presence of divalent cations in even small quantities tend to increase retention of nanoparticles by porous media. Since groundwater aquifers and surface waters typically have ionic strengths in excess of 10^{-4} M and frequently have significant concentrations of calcium or magnesium, conditions should tend to favor nanomaterial deposition. Even the electrosteric repulsions provided by surface modifiers such as polymers or polyelectrolytes can be overcome at high ionic strength and in the presence of divalent cations such as Ca^{2+} and Mg^{2+}. For example, the approximately 100 percent transport of PMAA-PMMA-PSS modified RNIP through a 21-cm sand column up to 10 mM of a monovalent cations (NaCl) (Figure 14a), decreases at ionic strengths of greater than 100 mM. In the presence of divalent cations, such as Ca^{2+} which are much more effective at screening the charge on particles, transport is reduced dramatically at concentrations of Ca^{2+} greater than 5 mM (Figure 8.14b). The electrosteric

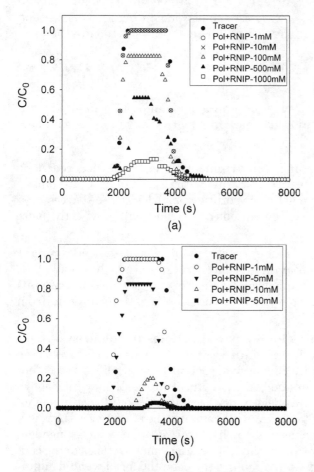

Figure 8.14   (a) Effect of ionic strength (NaCl) and (b) cation type (Ca^{2+}) on transport of polyelectrolyte-modified nano-iron through a 21-cm water-saturated sand column.

repulsions provided by the polyelectrolyte used here are much more effective than for surfactants that rely solely on electrosteric repulsions, or natural polymers or surfactants (e.g., alginate), and should be considered a best-case scenario (i.e., most transportable). Even under these conditions, only transport distances of a few 10s of meters are expected since these cations are present in most surface and subsurface waters. Thus even nanomaterials that are engineered to be highly mobile in the subsurface are not expected to be exceptionally mobile in the subsurface. This poses challenges from a standpoint of delivering the particles *in situ*, but comes as a relief from the standpoint of potential risks posed by releasing these particles into the environment. Further, this implies that the incidental release of engineered nanomaterials that inadvertently contaminant our current water-treatment infrastructure should be easily removed as these processes employ ionic strength increases and addition of divalent cations to destabilize particles to remove them.

Targeting

The ability to target specific contaminants is important in any *in situ* groundwater remediation scheme. Specificity is advantageous for several reasons:

- **Higher contaminant transformation rates.** Since most reactions and sorption processes are first order with respect to the contaminant of interest, placing reactive or sorptive particles in the sources area where contaminant concentrations are highest will provide the most rapid transformations

- **Higher efficiency.** Placing reactive particles at the contaminant source where the contaminant concentrations are highest will allow the preferred process (degradation or sorption of the contaminant) to out-compete the competitive processes (e.g., H_2 evolution from nanoiron).

- **Ability to maintain the reactive particles in the area of contamination and minimize unwanted exposures.** If the reactive particles migrate with the natural groundwater gradient and do not have any affinity for the contaminant of interest, they may move through the source area before they are fully utilized. Targeting nanomaterials to the contaminant will maximize the efficiency of the remediation process and minimize the mass of nanoparticles needed. Further, the ability to maintain the reactive nanoparticles in the contaminant source zone will minimize the potential for unwanted migration and exposure of sensitive biological targets in wetlands, lakes, or streams where groundwater may be discharging.

Targeting approaches

There are several potential approaches to achieve contaminant targeting. The first approach is to impart to the particles some affinity for the specific contaminant of interest. This targeted delivery approach is inspired by targeted drug delivery technology—localizing the remediation agents (the "drug") at the contaminant source (the "diseased tissue") via "smart" thermodynamic interactions between the particles and the contaminant. This approach has been recently evaluated in the laboratory (Saleh et al. 2005a; Saleh et al. 2005b; Saleh et al. 2007). Saleh et al. (2005) used multifunctional triblock copolymers (Figure 8.15) adsorbed to the nanoiron particles to disperse nanoiron into water for good aquifer transportability, minimize undesirable adhesion to mineral and natural organic matter (NOM) surfaces, and preferentially anchor nanoiron at the DNAPL/water interface for source-zone accumulation (Figure 8.16) (Saleh et al. 2005a). The multifunctional polymer assemblies adsorbed on the nanoiron surfaces present hydrophilic blocks that form stable nanoparticle suspensions in water, a requirement for eventual subsurface delivery. The hydrophilic blocks are strong polyanions, so they repel the predominantly negatively charged surfaces of minerals, NOM, and NOM-coated minerals that would be encountered in soils (Beckett and Le 1990; Day et al. 1994). When in contact with DNAPL, hydrophobic polymer blocks respond by anchoring the nanoparticle at the DNAPL/aqueous interface (Figure 8.16).

The affinity of the triblock copolymer modified nanoiron has been examined in the laboratory. RNIP interfacial anchorage by the same polymers that promoted their transport through sand columns was demonstrated by the ability of modified nanoiron to stabilize TCE-in-water

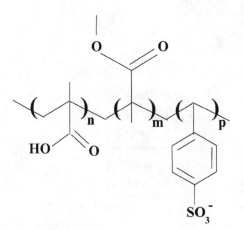

Figure 8.15 Hydrophobic-hydrophilic triblock copolymers containing a short PMAA anchoring group (left), PMMA hydrophobic block (middle), and a hydrophilic sulfonated polystyrene block (right).

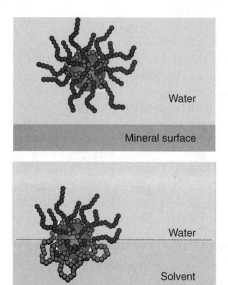

Figure 8.16 The targeting mechanism. Adsorbed block copolymers contain a polyelectrolyte (dark gray) block that has an affinity for water and a hydrophobic (light gray) block that has an affinity for DNAPL. In water, the polyelectrolyte block swells and the hydrophobic block collapses. The reverse happens in the DNAPL phase. This amphiphilicity anchors the particle at the DNAPL/water interface. The polyelectrolyte block is large enough to stably suspend particles in water without aggregating. The strong negative charge in the polyelectrolyte block minimizes particle adhesion to negatively charged mineral or natural organic matter surfaces in the soil before reaching the DNAPL.

emulsions (Saleh et al. 2005a). Emulsification of immiscible fluids is proof of adsorption at the fluid interface, as bare interfaces are unstable to droplet coalescence and macroscopic phase separation. Figure 8.17 shows a micrograph of a TCE-in-water emulsion that was stable for over

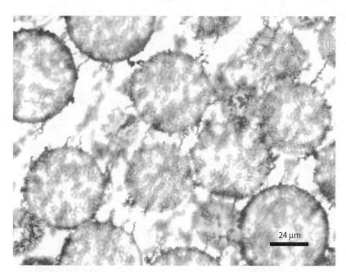

Figure 8.17 Optical micrograph of TCE-in-water emulsion stabilized by PMAA-PMMA-PSS-modified RNIP particles, proving that the modified particles adsorb at the TCE/water interface. Nanoiron did not partition into the TCE droplets.

three months. It was stabilized by RNIP particles that had been modified by irreversible adsorption of $PMAA_{42}$-$PMMA_{26}$-PSS_{466} followed by several cycles of centrifugation and washing. For comparison, TCE and water could not be emulsified by unmodified RNIP particles, proving that the polymer was essential for anchorage at the interface. Likewise, no emulsion was formed using the supernatant of a centrifuged polymer-modified RNIP sample, proving that the interface was stabilized by adsorption of polymer-modified RNIP, and not by adsorption of free polymer that may have been present in solution with the nanoiron. Furthermore, RNIP modified by PSS homopolymers was unable to emulsify TCE and water. This demonstrates that the type and composition of the surface modifiers are critical design variables for interfacial targeting.

There are many other approaches that could be used to achieve significant targeting of contaminants. For example, enzymes-based approaches may be possible. Enzymes are proteins or conjugated proteins produced by living organisms and functioning as biochemical catalysts. They typically have complex active sites that are highly selective for specific compounds. This selectivity may be leveraged to tailor reactive nanoparticles for specific compounds. Enzyme-coated carbon nanotubes have been used as single molecule biosensors (Besteman et al. 2003), and nanoparticles containing a single enzyme protected by a porous inorganic/organic network have been created (Kim and Grate 2003). These approaches may eventually be used to develop groundwater remediation agents with the highest specificity for target compounds possible. Less specific approaches include ethylenediaminetetraacetic acid (EDTA) coatings on TiO_2 nanoparticles designed to sequester radionuclides (Mattigod et al. 2005). EDTA is a strong chelating agent that forms coordination compounds with most divalent (or trivalent) metal ions, such as calcium (Ca^{2+}) and magnesium (Mg^{2+}) or copper (Cu^{2+}), and thus can be used to sequester cationic groundwater contaminants such as Cu^{2+} or Pb^{2+}. Once strongly sequestered, these toxic metals are no longer bioavailable and pose less or no risk to biota. Anatase (TiO_2) nanoparticles coated with EDTA and treated with Cu^{2+} to form an EDTA/Cu(II) complex have been used to then sequester anionic groundwater contaminants such as pertechtinate (TcO_4^-). It is proposed that the pertechtinate forms a strong complex with the bound EDTA/Cu(II) on the TiO_2 surface. Another approach is to use hydrophobic nanoparticles designed to strongly sequester hydrophobic contaminants such as polyaromatic hydrocarbons (PAHs) or polychlorinated biphenyls (PCBs). These contaminants are typically very hydrophobic (log $K_{OW} > 4$) and strongly adsorb to soil and sediment. These hydrophobic nanoparticles are added to the PCB- or PAH-contaminated soil or sediment, where the hydrophobic contaminants can strongly adsorb to them, after which, the nanoparticles are removed, thereby removing the contaminants (Tungittiplakorn et al. 2004). The potential to remove the contaminants from the nanoparticles and reuse them was also demonstrated.

Environmental systems are complex and contain a wide variety of organic and inorganic constituents that can compete for reactive sites regardless of the contaminant of choice. For example, polymeric hydrophobic nanoparticles designed to sequester PAHs or PCBs might also sequester hydrophobic soil organic matter and block access to the contaminants. In the case of DNAPL targeting, the available surface area of DNAPL could be small in comparison to the total mass (e.g., pools), thereby limiting the mass of nanoiron that could be delivered to the DNAPL/water interface. It may be sufficient to use less specific methods to control the travel distance such that reactive groundwater reagents can simply be placed near the vicinity of the source and remain there. This avoids the need for highly complex surface chemistry to provide selectivity, which will likely be expensive. Most metal-oxide nanoparticles will be relatively immobile in saturated porous media without sufficient coatings to make them mobile. As discussed previously, the transport distance for a specific coating type will vary depending on the modifier type and on the geochemical conditions at the site (pH, ionic strength, and ionic composition). This implies the potential to match the modifier type to the geochemical conditions at the site to achieve a specified transport distance. With this approach, remedial agents could be delivered to specific regions within or near the source zone. This could also be done by using surface coatings that desorb at a specific rate such that the nanoparticles travel a specific distance and then stop. Nanoparticle surface modifications, coupled with innovative engineered delivery schemes, will undoubtedly be necessary to achieve adequate targeting of nanoparticulate reagents for groundwater remediation.

Challenges

Subsurface heterogeneity and complex NAPL architecture make remediation difficult (Dai et al. 2001; Daus et al. 2001; Illangasekare et al. 1995). These issues also pose significant challenges to accurately delivering nanoparticulate remedial agents to subsurface contaminants. Nanoparticles will tend to travel along zones of high hydraulic conductivity, which may or may not be the desired placement area. For nanoparticles to reach the NAPL/water interface, particles may be required to diffuse across flow lines as they travel through the porous media. Even 10-nm particles may have prohibitively low diffusion coefficients. For example, a micromodel study by Baumann et al. (2005) investigated the targeting ability of a triblock copolymer-modified nanoiron that had been shown to partition to the NAPL/water interface ex situ (Saleh et al. 2007). At approach velocities of 2 to 13 m/d (residence

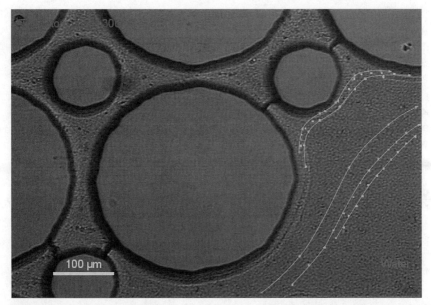

Figure 8.18 Trajectories of polymer-coated nanoiron in an etched silica micromodel containing water and partial TCE saturation. Particles tended to migrate toward the TCE/water interface as the approach velocity decreased. Trajectory points are 300 ms apart (Baumann et al. 2005).

time of 1 to 10 seconds), particles tended to flow past entrapped TCE rather than migrate to the interface (Figure 8.18).

Even though emulsification occurs under high shear conditions that of course are far removed from aquifer conditions, preliminary experiments in sand-packed columns and dodecane-coated sand-packed columns indicate some potential for *in situ* targeting as long as adequate time is available for nanoparticles to diffuse to the NAPL/water interface. Sand column transport studies conducted with NAPL-coated sand, under flow conditions similar to the clean sand experiments described above, indicated a 10 percent reduction in elution for $PMAA_{42}$-$PMMA_{26}$-PSS_{466}-coated RNIP compared to the clean sand column (Saleh et al. 2007). To achieve this, however, the flow had to be stopped for 24 hours to allow time for the particles to transport to the interface. Surface modifications that impart more hydophobicity to the particle should provide better NAPL targeting. Using a higher hydrophobe/hydrophile ratio, or changing the middle hydrophobic block from methyl-methacrylate to butyl-methacrylate to lower the glass transition temperature and thus promote swelling of the hydrophobe in contact with NAPL may further enhance targeting. These targeting experiments indicate potential for *in situ* targeting, but additional research to optimize the block size and type and hydrophile/hydrophobe ratio are needed.

In principle, nanoparticles that have a strong affinity and selectivity for the target contaminant could be injected into the subsurface and allowed to migrate nonspecifically toward the source area where they will locate and degrade or sequester the target contaminant. This is analogous to targeted drug delivery where drugs are introduced into the body and migrate toward the target (diseased tissue). If this is not achievable, delivering nanoparticles near the source area may be acceptable. This will, however, require that the source area be very well characterized in terms of the contaminant masses and locations, and that the hydrogeology at the site be well known. Without this characterization, it will be difficult to engineer and implement an effective delivery system. Source zone characterization is difficult, however, so techniques to effectively characterize these areas are needed. Partitioning tracers (e.g., alcohols) and interfacial tracers (e.g., surfactants) have shown some promise for characterizing DNAPL source zones (Annable et al. 1998). Methods to characterize sources of heavy metals such as Cr(VI) are less well developed.

Summary and Research Needs

Unique and inexpensive reactive or highly sorptive engineered nanomaterials are becoming commercially available. The use of these novel engineered nanomaterials for improved *in situ* remediation of groundwater contaminants is promising, but engineering nanomaterials that are highly selective for the reactions of interest and that have long lifetimes such that they are cost-effective for *in situ* remediation of groundwater contaminants remain active areas of research. Potential concerns about the exposure and human health effects of these engineered nanomaterials must also be addressed to ensure the safe deployment of these materials. Future research developing novel nanomaterials for groundwater remediation will have to consider the reactivity, selectivity, and longevity of materials as well and the potential exposure and human health effects of the materials.

Even though relatively inexpensive commercially produced reactive nanomaterials for *in situ* groundwater remediation (e.g., nanoiron) are available, methods to deliver these materials to subsurface contaminants remain a challenge. Intuitively, the assumption is often made that nanoparticles will be more mobile in porous media due to their small size. However, all other factors being equal, smaller particles are less mobile due to their relatively large diffusivity that produces more frequent contacts with the surfaces of aquifer porous media. The use of surface coatings to enhance subsurface transport and to encourage selectivity toward specific groundwater contaminants will be needed for effective deployment of engineered nanomaterials in groundwater remediation. These coatings can be used to control aggregation (particle-particle interactions) and attachment to aquifer media grains (particle-media grain interactions). The geochemistry at each site (e.g., pH, ionic strength, and ionic composition) will dramatically affect the mobility of engineered nanomaterials in the subsurface and must be considered in their design and application. The ability to tailor the surface coatings to site-specific geochemical conditions for well-controlled placement of engineered nanomaterials in the subsurface appears to be obtainable in the near term and will enhance the cost-effectiveness of this remedial approach. A higher mobility of nanomaterials in the environment due to the use of surface coatings, however, implies a greater potential for exposure as nanomaterials are dispersed over greater distances and their effective persistence in the environment increases. The trade-off between enhanced mobility and effectiveness and potential exposure and risks must be considered on a case-by-case basis.

The ability to target specific contaminants or to concentrate the reactive nanomaterials in the contaminant source zone will be required to make *in situ* groundwater remediation cost-effective. The use of surface coatings to enhance target specificity is very promising and remains an active area of research. Despite the ability to develop nanomaterials with target specificity, there remain significant challenges for *in situ* delivery of nanomaterials in the subsurface due to unfavorable hydrodynamics in many cases. Even at low approach velocities typical of groundwater flow, the diffusion rates of particles that are 10 to 100 nm in diameter across flow lines to adsorbed contaminants may be prohibitively slow to allow them to diffuse across flow lines to reach their targets. Methods to tailor the surface chemistry of nanomaterials to concentrate reactive materials in specific regions in the subsurface appear promising, and may be a more obtainable near-term goal to achieve well-controlled placement of nanomaterials in contaminant source zones.

List of Acronyms and Symbols

aq: aqueous

CMC: carboxy methylcellulose

COC: chlorinated organic compounds

DCA: dichloroethane

DCE: dichloroethene

DIRB: dissimilatory iron reducing bacteria

DLVO: Deraguin-Landau-Verwey-Overbeek

DNAPL: dense nonaqueous phase liquid

d_p: particle diameter (L)

e^-:   electron

EDL: electrostatic double layer

EDTA: ethylenediaminetetraacetic acid

Fe(B):   nanoiron synthesized from reduction of dissolved iron using sodium borohydride

H˙:   adsorbed hydrogen

HCA: hexachloroethane

k_B: Boltzman's constant (J/K)

k_{H_2obs}:   observed rate constant for H_2 evolution from nanoiron corrosion in water (t^{-1})

K_{OW}:   octanol-water partition coefficient, $M_{i,o}/M_{i,w}$ (–)

k_{rxn}:   surface are normalize pseudo-first-order reaction rate constant (L hr^{-1} m^{-2})

Me^{a+}:   metal ion of charge a

MRNIP: modified RNIP (modified with polyaspartic acid)

NAPL: nonaqueous phase liquid

NOM: natural organic matter

PAH: polyaromatic hydrocarbon

PCA: pentachloroethane

PCB: polychlorinated biphenyl

PEG: polyethylene glycol

$PMAA_x$:   poly(methacrylic) acid, subscript represents the degree of polymerization

$PMMA_y$:   poly(methylmethacrylate), subscript represents the degree of polymerization

PSS_z: poly(styrene sulfonate), subscript represents the degree of polymerization

PVA: polyvinyl alcohol

RNIP: reactive nanoscale iron particles

SA: surface area

SDBS: sodium dodecylbenzene sulfonate

T: temperature (K)

TCA: trichloroethane

TCE: trichloroethylene

TeCA: tetrachloroethane

VC: vinyl chloride

V_T: potential (J)

Φ: volume fraction, volume of particles/total volume (–)

References

Agrawal, A., and Tratnyek, P. G. (1996). "Reduction of Nitro Aromatic Compounds by Zero-Valent Iron Metal." *Environmental Science & Technology*, 30(1), 153–160.

Annable, M. D., Jawitz, J. W., Rao, P. S. C., Dai, D. P., Kim, H., and Wood, A. L. (1998). "Field Evaluation of Interfacial and Partitioning Tracers for Characterization of Effective NAPL-Water Contact Areas." *Ground Water*, 36, 495–503.

Baumann, T., Keller, A. A., Auset-Vallejo, M., and Lowry, G. V. "Micromodel Study of Transport Issues during TCE Dechlorination by ZVI Colloids." *American Geophysical Union Fall Meeting*, San Francisco, CA.

Beckett, R., and Le, N. P. (1990). "The Role of Organic Matter and Ionic Composition in Determining the Surface Charge of Suspended Particles in Natural Waters." *Colloids and Surfaces*, 44, 35–49.

Besteman, K., Lee, J.-O., Wiertz, F. G. M., Heering, H. A., and Dekker, C. (2003). "Enzyme-Coated Carbon Nanotubes as Single-Molecule Biosensors." *Nano Letters*, 3(6), 727–730.

Bonin, P. M., Odziemkowski, M. S., and Gilham, R. W. (1998). "Influence of Chlorinated Solvents on Polarization and Corrosion Behaviour of Iron in Borate Buffer." *Corrosion Science*, 40, 1391–1409.

Braem, A. D., Biggs, S., Prieve, D. C., and Tilton, R. D. (2003). "Control of Persistent Nonequilibrium Adsorbed Polymer Layer Structure by Transient Exposure to Surfactants." *Langmuir*, 19, 2736–2744.

Butter, K. B., P.H.; Frederik, P.M.; Vroege, G.J.; Philipse, A.P. (2003a). "Direct Observation of Dipolar Chains in Iron Ferrofluids by Cryogenic Electron Microscopy." *Nature Materials*, 2, 88–91.

Butter, K. B., P.H.H., Frederik, P.M., Vroege, G.J., and Philipse, A.P. (2003b). "Direct Observation of Dipolar Chains in Iron Ferrofluids in Zero Field Using Cryogenic Electron Microscopy." *Journal of Physics: Condensed Matter*, 15, 1415–1470.

Cao, J., Elliott, D., and Zhang, W.-X. (2005). "Perchlorate Reduction by Nanoscale Iron Particles." *Journal of Nanoparticle Research*, 7, 499–506.

Chibowski, S., and Wisniewska, M. (2002). "Study of Electrokinetic Properties and Structure of Adsorbed Layers of Polyacrylic Acid and Polyacrylamide at Fe2O3-Polymer Solution Interface." *Colloids and Surfaces A*, 208, 131–145.

Choe, S., Chang, Y.-Y., Hwang, K.-Y., and Khim, J. (2000). "Kinetics of Reductive Denitrification by Nanoscale Zero-Valent Iron." *Chemosphere*, 41, 1307–1311.

Dai, D., Barranco, F. T. J., and Illangasekare, T. H. (2001). "Partitioning and Interfacial Tracers for Differentiating NAPL Entrapment Configuration: Column-Scale Investigation." *Environmental Science & Technology*, 35, 4894–4899.

Daus, D. A., Kent, B., and Mosquera, G. C. B. (2001). "A Case Study of DNAPL Remediation in Northwestern Brazil." *Journal of Environmental Science & Health*, 36, 1505–1513.

Day, G. M., Hart, B. T., McKelvie, I. D., and Beckett, R. (1994). "Adsorption of Natural Organic Matter onto Goethite." *Colloids and Surfaces A*, 89, 1–13.

de Gennes, P.-G. P., P.A. (1970). "Pair Correlations in a Ferromagnetic Colloid." *Phys. Kondens. Mater.*, 11, 189–198.

de Vicente, J., Delgado, A. V., Plaza, R. C., Durán, J. D. G., and González-Caballero, F. (2000). "Stability of Cobalt Ferrite Colloidal Particles: Effect of pH and Applied Magnetic Fields." *Langmuir*, 212, 14–23.

Drzymala, J., and Fuerstenau, D. W. (1987). "Adsorption of Polyacrylamide, Partially Hydrolyzed Polyacrylamide and Polyacrylic Acid on Ferric Oxide and Silica." *Polymer Process Technology*, 4, 45–60.

Elimelech, M., Gregory, J., Jia, X., and Williams, R. (1995). *Particle Deposition and Aggregation: Measurement, Modeling, and Simulation*, Butterworth-Heinemann, Oxford, England.

Evans, D. F. W., H. (1999). *The Colloidal Domain: Where Physics, Chemistry, Biology, and Technology Meet*, Wiley-VCH, New York.

Fleer, G. J., Cohen, Stuart, M. A., Scheutjens, J. M., Cosgrove, H. M., and Vincent, T. B. (1993). *Polymers at Interfaces*, Chapman and Hall, New York.

Gavaskar, A., Tatar, L., and Condit, W. (2005). "Cost and Performance Report: Nanoscale Zero-Valent Ion Technologies for Source Remediation." *CR-05-007-ENV*, Naval Facilities Engineering Command, Port Hueneme.

Gerlach, R., Cunningham, A. B., and Caccavo, F. (2000). "Dissimilatory Iron-Reducing Bacteria Can Influence the Reduction of Carbon Tetrachloride by Iron Metal." *Environmental Science & Technology*, 34(12), 2461–2464.

Goodman, C. M., McCusker, C. D., Yilmaz, T., and Rotello, V. M. (2004). "Toxicity of Gold Nanoparticles Functionalized with Cationic and Anionic Side Chains." *Bioconjugate Chemistry*, 15, 897–900.

Hamilton, W. A. "Microbially Influenced Corrosion in the Context of Metal Microbe Interactions." *Microbial Corrosion: Proceedings of the 4th International EFC Workshop*, Sequeira, Portugal, 3–17.

He, F., and Zhao, D. (2005). "Preparation and Characterization of a New Class of Starch-Stabilized Bimetallic Nanoparticles for Degradation of Chlorinated Hydrocarbons in Water." *Environmental Science & Technology*, 39, 3314–3320.

Heimenz, P. C. R. (1997). *Principles of Colloid and Surface Chemistry*, Marcel Dekker, New York.

Holmberg, K., Jönsson, B., Kronberg, B., and Lindman, B. (2003). *Surfactants and Polymers in Aqueous Solution*, John Wiley & Sons, West Sussex.

Illangasekare, T. H., Yates, D., N., and Armbruster, E. J. (1995). "Effect of Heterogeneity on Transport and Entrapment on Nonaqueous Phase Waste Products in Aquifers: An Experimental Study." *Journal of Environmental Engineering*, 121, 572–579.

Kanel, S. R., Manning, B., Charlet, L., and Choi, H. (2005). "Removal of Arsenic(III) from Groundwater by Nanoscale Zero-Valent Iron." *Environmental Science & Technology*, 39, 1291–1298.

Kim, J., and Grate, J. W. (2003). "Single-Enzyme Nanoparticles Armored by a Nanometer-Scale Organic/Inorganic Network." *Nano Letters*, 3(9), 1219–1222.

Koziarz, J., and Yamazaki, H. (1999). "Stabilization of Polyvinyl Alcohol Coating of Polyester Cloth for Reduction of Bacterial Adhesion." *Biotechnology Techniques*, 13(4), 221–225.

Lecoanet, H. F., Bottero, J.-Y., and Wiesner, M. R. (2004). "Laboratory Assessment of the Mobility of Nanomaterials in Porous Media." *Environmental Science & Technology*, 38(19), 5164–5169.

Lecoanet, H. F., and Wiesner, M. R. (2004). "Velocity Effects on Fullerene and Oxide Nanoparticle Deposition in Porous Media." *Environmental Science & Technology*, 38(16), 4377–4382.

Lien, H. L., and Zhang, W. X. (1999). "Transformation of Chlorinated Methanes by Nanoscale Iron Particles." *Journal of Environmental Engineering*, 125, 1042–1047.

Lien, H.-L., and Zhang, W.-X. (2005). "Hydrodechlorination of Chlorinated Ethanes by Nanoscale Pd/Fe Bimetallic Particles." *Journal of Environmental Engineering*, 131, 4–10.

Liu, Y., Choi, H., Dionysiou, D., and Lowry, G. V. (2005a). "Trichloroethene Hydrodechlorination in Water by Highly Disordered Monometallic Nanoiron." *Chemistry of Materials*, 17(21), 5315–5322.

Liu, Y., Majetich, S. A., Tilton, R. D., Sholl, D. S., and Lowry, G. V. (2005b). "TCE Dechlorination Rates, Pathways, and Efficiency of Nanoscale Iron Particles with Different Properties." *Environmental Science & Technology*, 39(5), 1338–1345.

Lowry, G. V., and Johnson, K. M. (2004). "Congener Specific Dechlorination of Dissolved PCBs by Microscale and Nanoscale Zero-Valent Iron in a Water/Methanol Solution." *Environmental Science & Technology*, 38(19) 5208-5216

Lowry, G. V., and Reinhard, M. (1999). "Hydrodehalogenation of 1- to 3-Carbon Halogenated Organic Compounds in Water Using Palladium Catalyst and Hydrogen Gas." *Environmental Science & Technology*, 33(11), 1905–1911.

Lowry, G. V., and Reinhard, M. (2000). "Pd-catalyzed TCE Dechlorination in Groundwater: Solute Effects, Biological Control, and Oxidative Catalyst Regeneration." *Environmental Science & Technology*, 34(15), 3217–3223.

Matsson, M. K., Kronberg, B., and Claesson, P. M. (2004). "Adsorption of Alkyl Polyglucosides on the Solid/Water Interface: Equilibrium Effects of Alkyl Chain Length and Head Group Polymerization." *Langmuir*, 20(10), 4051–4058.

Mattigod, S. V., Fryxell, G. E., Alford, K., Gilmore, T., Parker, K., Serne, J., and Engelhard, M. (2005). "Functionalized TiO2 Nanoparticles for Use for in Situ Anion Immobilization." *Environmental Science & Technology*, 39(18), 7306–7310.

McCurrie, R., A. (1994). *Ferromagnetic Materials: Structure and Properties*, Academic Press, London.

Mondal, K., Jegadeesan, G., and Lalvani, S. B. (2004). "Removal of Selenate by Fe and NiFe Nanosized Particles." *Industrial & Engineering Chemistry Research*, 43, 4922–4934.

Moore, A. M., De Leon, C. H., and Young, T. M. (2003). "Rate and Extent of Aqueous Perchlorate Removal by Iron Surfaces." *Environmental Science & Technology*, 37(14), 3189–3198.

Nakamae, K., Tanigawa, S., Nakano, S., and Sumiya, K. (1989). "The Effect of Molecular Weight and Hydrophilic Groups on the Adsorption Behavior of Polymers onto Magnetic Particles." *Colloids and Surfaces A*, 37, 379.

Neto, C. B., Bonini, M., and Baglioni, P. (2005). "Self-Assembly of Magnetic Nanoparticles into Complex Superstructures: Spokes and Spirals." *Colloids and Surfaces A.*, 269, 96–100.

Nurmi, J. T., Tratnyek, P. G., Sarathy, V., Baer, D. R., Amonette, J. E., Pecher, K., Wang, C., Linehan, J. C., Matson, D. W., Penn, R. L., and Driessen, M. D. (2005). "Characterization and Properties of Metallic Iron Nanoparticles: Spectroscopy, Electrochemistry, and Kinetics." *Environmental Science & Technology*, 39(5), 1221–1230.

Nutt, M. O., Hughes, J. B., and Wong, M. S. (2005). "Designing Pd-on-Au Bimetallic Nanoparticle Catalysts for Trichloroethene Hydrodechlorination." *Environmental Science & Technology*, 39(5), 1346–1353.

Oh, B. T., and Alvarez, P. J. J. (2002). "Hexahydro-1,3,5-Trinitro-1,3,5-Triazine (RDX) Degradation in Biologically Active Iron Columns." *Water Air and Soil Pollution*, 141(1–4), 325–335.

Phillips, D. H., Gu, B., Watson, D. B., Roh, Y., Liang, L., and Lee, S. Y. (2000). "Performance Evaluation of a Zerovalent Iron Reactive Barrier: Mineralogical Characteristics." *Environmental Science & Technology*, 34(19), 4169–4176.

Plaza, R. C., de Vicente, J. Gómez-Lopera, S., and Delgado, A. V. (2001). "Stability of Dispersions of Colloidal Nickel Ferrite Spheres." *Journal of Colloid Interface Science*, 242, 306–313.

Ponder, S., Darab, J., and Mallouk, T. (2000). "Remediation of Cr(IV) and Pb(II) Aqueous Solutions Using Supported Nanoscale Zero-valent Iron." *Environmental Science & Technology*, 34(12), 2564–2569.

Pourbaix, M. J. N. (1966). *Atlas of Electrochemical Equilibria in Aqueous Solutions*, Pergamon Press, New York.

Quinn, J., Geiger, C., Clausen, C., Brooks, K., Coon, C., O'Hara, S., Krug, T., Major, D., Yoon, W.-S., Gavaskar, A., and Holdsworth, T. (2005). "Field Demonstration of DNAPL Dehalogenation Using Emulsified Zero-Valent Iron." *Environmental Science & Technology*, 39(5), 1309–1318.

Roddick-Lanzilotta, A., and McQuillan, J. (1999). "An in Situ Infrared Spectroscopic Investigation of Lysine Peptide and Polylysine Adsorption to TiO2 from Aqueous Solutions." *Journal of Colloid Interface Science*, 217, 194–202.

Rose, J., Auffan, M., Thill, A., Decome, L., Masion, A., Orsiere, T., Botta, A., and Bottero, J.-Y. "Nanoparticles as Adsorbents (for Liquid Waste and Water Treatment)." *Franco-American Conference on Nanotechnologies for a Sustainable Environment*, Rice University.

Rosensweig, R. E. (1985). *Ferrohydrodynamics*, Cambridge University Press, New York.

Royal Academy of Engineering, T. R. S. (2004). "Nanoscience and Nanotechnologies: Opportunities and Uncertainties." Royal Academy of Engineering, London.

Ryan, J., and Elimelech, M. (1996). "Colloid Mobilization and Transport in Groundwater." *Colloids and Surfaces*, 107, 1–56.

Saleh, N., Phenrat, T., Sirk, K., Dufour, B., Ok, J., Sarbu, T., Matyjaszewski, K., Tilton, R. D., and Lowry, G. V. (2005a). "Adsorbed Triblock Copolymers Deliver Reactive Iron Nanoparticles to the Oil/Water Interface." *Nano Letters*, 5(12), 2489–2494.

Saleh, N., Sarbu, T., Sirk, K., Lowry, G. V., Matyjaszewski, K., and Tilton, R. D. (2005b). "Oil-in-Water Emulsions Stabilized by Highly Charged Polyelectrolyte-Grafted Silica Nanoparticles." *Langmuir*, 21(22), 9873–9878.

Saleh, N., Sirk, K., Phenrat, T., Dufour, B., Matyjaszewski, K., Tilton, R. D., and Lowry, G. V. (2007). "Surface Modifications Enhance Nanoiron Transport and NAPL Targeting in Saturated Porous Media." *Environmental Engineering Science*, 24 (1) 45–57.

Schrick, B., Hydutsky, B., Blough, J., and Mallouk, T. (2004). "Delivery Vehicles for Zerovalent Metal Nanoparticles in Soil and Groundwater." *Chemistry of Materials*, 16(11), 2187–2193.

Sitharaman, B., Bolskar, R. D., Rusakova, I., and Wilson, L. J. (2004). "Gd@C60[C (COOH)2]10 and Gd@C60(OH)x: Nanoscale Aggregation Studies of Two Metallofullerene MRI Contrast Agents in Aqueous Solution." *Nano Letters*, 4(12), 2373–2378.

Song, H., and Carraway, E. R. (2005). "Reductive of Chlorinated Ethanes by Nanosized Zero-Valent Iron: Kinetics, Pathways, and Effects of Reaction Conditions." *Environmental Science & Technology*, 39, 6237–6245.

Sorel, D., Warner, S. D., Longino, B. L., Honniball, J. H., and Hamilton, L. A. "Dissolved Hydrogen Measurements at a Permeable Zero-Valent Iron Reactive Barrier." *American Chemical Society Annual Conference*, San Diego, CA.

Strenge, K. (1993). "Structure Formation in Disperse Systems." *Coagulation and Flocculation: Theory and Applications*, B. Dobiáš, ed., Marcel Dekker, New York.

Stroo, H., Unger, M., Ward, C. H., Kavanaugh, M., Vogel, C., Leeson, A., Marquesee, J., and Smith, B. (2003). "Remediating Chlorinated Solvent Source Zones." *Environmental Science & Technology*, 37(11), 193a–232a.

Tungittiplakorn, W., Cohen, C., and Lion, L. W. (2005). "Engineered Polymeric Nanoparticles for Bioremediation of Hydrophobic Contaminants." *Environmental Science & Technology*, 39(5), 1354–1358.

Tungittiplakorn, W., Lion, L., Cohen, C., and Kim, J. (2004). "Engineered Polymeric Nanoparticles for Soil Remediation." *Environmental Science & Technology*, 38(5), 1605–1610.

U.S.DOE. (2005). "Guidance for Optimizing Ground Water Response Actions at Department of Energy Sites."

Veiseh, O., Sun, C., Gunn, J., Kohler, N., Gabikian, P., Lee, D., Bhattarai, N., Ellenbogen, R., Sze, R., Hallahan, A., Olson, J., and Zhang, M. (2005). "Optical and MRI Multifunctional Nanoprobe for Targeting Gliomas." *Nano Letters*, 5(6), 1003–1008.

Velegol, S. B., and Tilton, R. D. (2001). "A Connection between Interfacial Self-Assembly and the Inhibition of Hexadecyltrimethylammonium Bromide Adsorption on Silica by Poly-L-lysine." *Langmuir*, 17(1), 219–227.

Viessman, W., and Hammer, M. (1998). *Water Supply and Pollution Control*, Addison Wesley Longman, Inc., Menlo Park, CA.

Vikesland, P. J., Klausen, J., Zimmermann, H., Roberts, A. L., and Ball, W. P. (2003). "Longevity of Granular Iron in Groundwater Treatment Processes: Changes in Solute Transport Properties Over Time." *Journal of Contaminant Hydrology*, 64(1–2), 3–33.

Vogan, J. L., Focht, R. M., Clark, D. K., and Graham, S. L. (1999). "Performance Evaluation of a Permeable Reactive Barrier for Remediation of Dissolved Chlorinated Solvents in Groundwater." *Journal of Hazardous Materials*, 68(1–2), 97–108.

Wang, C. B., and Zhang, W. X. (1997). "Synthesizing Nanoscale Iron Particles for Rapid and Complete Dechlorination of TCE and PCBs." *Environmental Science & Technology*, 31(7), 2154–2156.

Water Science Technology Board, N. (2004). *Contaminants in the Subsurface: Source Zone Assessment and Remediation*, National Academies Press, Washington, D.C.

Wilkin, R. T., Su, C., Ford, R. G., and Paul, C. J. (2005). "Chromium-Removal Processes during Groundwater Remediation by a Zerovalent Iron Permeable Reactive Barrier." *Environmental Science & Technology*, 39(12), 4599–4605.

Williams, A. G. B., Gregory, K. B., Parkin, G. F., and Scherer, M. M. (2005). "Hexahydro-1,3,5-Trinitro-1,3,5-Triazine Transformation by Biologically Reduced Ferrihydrite: Evolution of Fe Mineralogy, Surface Area, and Reaction Rates." *Environmental Science & Technology*, 39(14), 5183–5189.

Yang, G. C., and Lee, H. L. (2005). "Chemical Reduction of Nitrate by Nanosized Iron: Kinetics and Pathways." *Water Research*, 39(5), 884–894.

Yao, K. M., Habibian, M. T., and O'Melia, C. R. (1971). "Water and Wastewater Filtration: Concepts and Applications." *Environmental Science & Technology*, 5, 1105–1102.

Membrane Processes

Mark R. Wiesner   *Duke University, Durham, North Carolina, USA*

Andrew R. Barron   *Rice University, Houston, Texas, USA*

Jérôme Rose   *CNRS-University of Aix-Marseille, Aix-en-Provence, France*

Over half a century since the invention of synthetic polymers and the asymmetric membrane, refinements and new developments in membrane technologies are active areas of research that are rapidly expanding our capabilities to restructure production processes, protect the environment and public health, and provide technologies for sustainable growth. Today, membrane technologies are playing an increasingly important role as unit operations for environmental quality control, resource recovery, pollution prevention, energy production, and environmental monitoring. Membranes are also key technologies at the heart of fuel cells and bio-separations.

The nanotechnology-inspired approach of building objects from the molecular scale, and the new materials that this approach inspires, have great potential for stimulating innovation in membrane science. Nanoscale control of membrane architecture may yield membranes of greater selectivity and lower cost in both water treatment and water fabrication. Nanomaterials may serve as the basis for new membrane processes based on passive diffusion or active transport. New materials may also be used in conjunction with membranes to create new nanomaterial/membrane reactors (NMRs). Multifunctional membranes that simultaneously separate and react or detect are possible. Nanotechnology is likely to improve the reliability and efficiency of membrane processes while broadening the range of applications.

In this chapter, we review the fundamentals of membrane processes as a basis for understanding where nanomaterials are likely to kindle innovations in membrane processes used to protect our environment and public health. We then examine several examples of nanomaterial-based approaches to improved membrane technologies.

Overview of Membrane Processes

A membrane, or more properly, a semipermeable membrane, is a thin layer of material that is capable of separating materials based on their physical or chemical properties when a driving force is applied across the membrane (Figure 9.1). Materials to be separated are introduced to the membrane on the feed or "concentrate" side where the portion of the materials rejected by the membrane accumulate. The concentration of rejected materials is typically highest near the membrane, setting up a concentration gradient for diffusion away from the membrane and back into the bulk concentrate. The "permeate" side of the membrane is enriched in materials that are able to move through the membrane more easily. The efficiency of membrane *rejection*, R, for a given component (also referred to as the efficiency of separation) is generally defined as 1.0 minus the ratio of the concentrations of that component in the permeate and feed:

$$R = 1 - \frac{c_{permeate}}{c_{feed}} \tag{1}$$

Water filtration presents a very simplified case where particles are rejected by the membrane, potentially accumulating as a cake on the concentrate side, while water passes through the membrane as permeate.

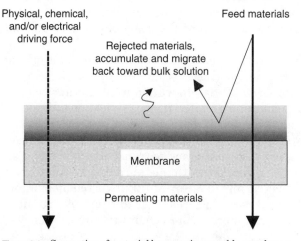

Figure 9.1   Separation of material by a semipermeable membrane under a driving force.

The accumulation of materials in, on, or near the membrane may have undesirable consequences for transport across the membrane resulting in a condition referred to as membrane fouling. For example, in water filtration, the formation of a cake on the membrane may impede the permeation of water under the applied driving force of pressure. The issue of membrane fouling will be taken up later in this chapter.

Membranes can be fabricated in many shapes and sizes. *Symmetric membranes* have an approximately uniform composition throughout their entire thickness so that a slice through any layer of the membrane parallel to its surface would look essentially identical. By contrast, *asymmetric membranes* consist of a thin membrane skin that is responsible for the separation of permeate from rejected species. The skin of an asymmetric membrane is supported by a layer (typically much thicker) that offers little resistance to transport and does not play a role in membrane selectivity. The surface chemistry and composition of different membranes can be quite variable depending on the application. Hydrophobic, uncharged membrane material may be desirable for some membrane distillation applications, while high charge density membranes are characteristic of polymer electrolyte membranes (PEMs) used in fuel cells. Dense SiO_2 membranes have been used for gas separations, while porous alumina membranes can be used to treat high temperature brines emanating from oil and gas wells.

Membranes are packaged as several *elements* grouped into units that are referred to as *modules*, *vessels*, or *stacks*, depending on the type of membrane and its application. The most frequently encountered element geometries are as flat sheets, capillary fibers, hollow fibers, tubes, or spiral wounds. The geometry of the membrane element is critical in determining the economics of the membrane process since it determines how much membrane area can fit into a given volume. For example, both hollow fiber and tubular membranes share a cylindrical geometry, but tubular membranes have a much larger diameter. Thus, the area of membrane available for mass transport per volume of module (referred to as the *packing density*) will be less for tubular membrane (Table 9.1). The cost of a membrane system tends to decrease as the packing density of the membrane module increases, as the costs of module housing, instrumentation and hook-up surrounding the module, and space for the installation are spread out over more membrane area. However, as the packing density increases, less space is available within the module to allow for other functions. For example, in water filtration, space must be provided in the module to allow for rejected materials to circulate freely without obstructing flow, or the feed stream to the membrane must be pretreated to remove these materials.

TABLE 9.1   Approximate Packing Densities for a Number of Membrane Geometries

Membrane geometry		Approximate packing density (m^2/m^3)
Capillary	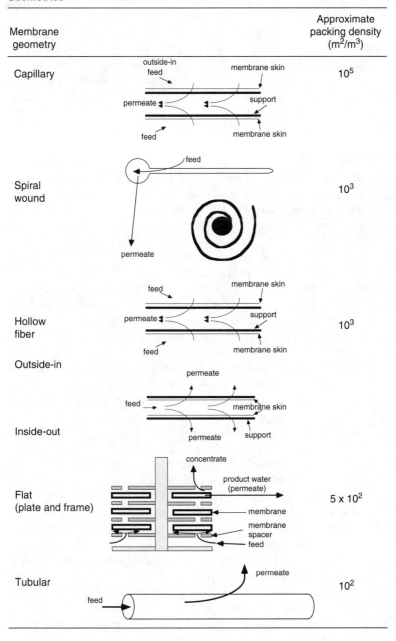	10^5
Spiral wound		10^3
Hollow fiber		10^3
Outside-in		
Inside-out		
Flat (plate and frame)		5×10^2
Tubular		10^2

Transport Principles
for Membrane Processes

The driving force for transport across a membrane may be physical (as in the case of pressure), chemical, or electrical, acting individually or in concert. In theory, a gradient across the membrane in any parameter that affects the chemical potential of a compound can be used as a driving force for transport across the membrane. Examples of driving forces and some corresponding membrane processes are listed in Table 9.2.

Pressure-driven membrane processes have been developed at large scale for the treatment of water (both potable and wastewater) and other fluids. In water filtration, pressure-driven membrane processes are typically differentiated based on the range of materials rejected. Microfiltration (MF) and ultrafiltration (UF) are pressure-driven processes that use *porous membranes* to separate micron-sized and nanometer-sized materials, respectively. Nanofiltration (NF) and reverse osmosis (RO) use *dense membranes* to separate solutes ranging from macromolecules, nanoparticles, and larger ions in the case of NF to simple salts and very small molecular weight organic compounds in the case of RO. Unlike pressure-driven processes in which solvent passes through the membrane, electrodialysis involves the passage of the solute through the membrane. In electrodialysis, ions pass through a semipermeable membrane under the influence of an electrical potential and leave the water behind, whereas in RO, water passes through the membrane leaving the ions behind. Polymer electrolyte membranes used in fuel cells allow for the transport protons under a concentration gradient, while rejecting the fuel (for example, hydrogen) that is introduced to a catalyst at the "concentrate" side of the membrane. In membrane distillation, solvent (e.g., water) evaporates and is transported across a porous membrane under the driving force of a temperature gradient.

The driving force for transport reflects the differences in the available energy on the two sides of the membrane. Analogous to water flowing downhill, material is transported from one side of the membrane to the other if and only if that transport decreases the total available (or free) energy of the system. Any spatial gradient in energy, E, can be interpreted

TABLE 9.2 Categorization of Membrane Processes by Driving Force

Driving force	Examples of membrane processes
Temperature gradient	Membrane distillation
Concentration gradient	Dialysis, pervaporation
Pressure gradient	Reverse osmosis, ultrafiltration
Electrical potential	Electrodialysis, electro-osmosis

as a force, F. The force is in the direction of decreasing energy so, for example, for movement in the x direction:

$$F_x = -\frac{dE}{dx} \qquad (2)$$

Applying this idea to a membrane system, the decrease in total available energy of the system (concentrate, membrane, and permeate) as a consequence of a substance moving across the membrane divided by the distance moved can be interpreted as the driving force for that movement.

In most membrane systems, transport is driven by an externally imposed gradient in a single type of energy. Although each type of energy is typically linked, in many systems, the imposed gradient in one type of energy is the only one that need be considered and the linkages or coupling with the other gradients can be ignored. For instance, pressure-driven MF and UF membranes do not reject solutes to an appreciable extent. In this case, the transmembrane gradient in pressure is the only significant factor affecting the decline in available energy of water and solutes as they cross from the feed to the permeate side of the membrane.

If two or more types of energy affect transport of a single component in the feed, then at any single moment in time the total force for transport can be approximated by adding the corresponding energy gradients for that component. The expressions for the available energy per mole of a substance associated with pressure, solution composition, and electrical energy are shown in Table 9.3.

In the table, V_i is the molar volume of i; $\overline{G}_i$ and $\overline{G}_i^o$ are the molar Gibbs free energy of i in the given system and at standard state, respectively; a_i is the chemical activity of i; R and T are the universal gas constant and the absolute temperature; z_i is the charge on species i (including sign); F is the Faraday constant; and Ψ is the electrical potential. $\overline{G}_i$ and $\overline{G}_i^o$ are also commonly written as μ_i and μ_i^0, in which case they are

TABLE 9.3   Expressions of Available Energy per Mole of Chemical Species

Type of energy	Expression for energy/mole of i	
Mechanical (pressure-based)	$\overline{E}_{p,i} = V_i P$	(9.3)
Chemical (concentration-based)	$\overline{E}_{chem,i} = \overline{G}_i = \overline{G}_i^o + RT\ln a_i$	(9.4)
Electrical	$\overline{E}_{elec,i} = z_i F\Psi$	(9.5)
Total available energy[a]	$\overline{E}_i = \overline{E}_{p,i} + \overline{E}_{chem,i} + \overline{E}_{elec,i}$	(9.6)

[a] In a typical membrane system; in other systems, other types of energy would have to be included

referred to as the chemical potential and the standard chemical potential of i, respectively. The sum of chemical and electrical potentials ($\overline{E}_{chemi} + \overline{E}_{eleci}$) is called the electrochemical potential.

Substituting Eqs. 3 through 5 into Eq. 6, and noting that the standard molar Gibbs energy of any species i is independent of the system conditions, the overall energy change accompanying transport of species i across a membrane per unit mass of i transported can be computed as:

$$\Delta \overline{E}_i = V_i \Delta P + \Delta(\overline{G}_i^o + RT \ln a_i) + z_i F \Delta \Psi$$

$$= V_i \Delta P + RT \Delta \ln a_i + z_i F \Delta \Psi \qquad (7)$$

where the Δ's are taken as the feed value minus the permeate value and are assumed to be small. As noted previously, a spatial gradient in energy can be interpreted as a force. Therefore, approximating ΔP, Δa_i, and $\Delta \Psi$ as the changes in those parameters across the thickness δ_m of the membrane, the driving force for transport of i across the membrane from feed to permeate, per mole of i, is:

$$F_i = \frac{\Delta \overline{E}_i}{\delta_m} = V_i \frac{\Delta P}{\delta_m} + RT \frac{\Delta \ln a_i}{\delta_m} + z_i F \frac{\Delta \Psi}{\delta_m} \qquad (8)$$

Simplifications to Eq. 8 in which one driving force dominates in the transport of a single component of a system give rise to several frequently encountered expressions such as Fick's law (concentration gradient), Ohm's law (only electrical potential), and D'Arcy's law (pressure gradient). Consider, for example, a single component ($i = 1$) where transport is driven by differences in concentration alone ($\Delta \Psi = \Delta P = 0$). The driving force per molecule for transport is obtained by dividing Eq. 8 by Avogadro's number, N_A. If we further assume that the molecules rapidly achieve a steady, average diffusion velocity, and that the molecules experience a force of resistance to movement, F_{resist}, which is proportional to this average velocity, v_{diff}, the driving force for diffusion must be balanced by this resistive force:

$$F_{resist} = v_{diff} f = F_{tot,1} = \frac{kT}{\delta_m} \frac{\Delta c_1}{c_1} \qquad (9)$$

where the Boltzmann constant, k, is equal to R/N_A and f is the proportionality factor relating resistive force to velocity. The diffusive flux is the product of the average diffusion velocity and the concentration of the diffusing component. Thus, rearranging Eq. 9, the diffusive flux can be expressed as:

$$J_1 = v_{diff} c_1 = \frac{kT}{f} \frac{\Delta c_1}{\delta_m} \qquad (10)$$

Eq. 10 is a restatement of Fick's law, where the concentration gradient is approximated by the difference in concentration across the membrane divided by the thickness of the membrane, and the diffusion coefficient, D, is equal to kT/f. Similarly, when a difference in pressure is the sole driving force for moving a fluid through a network of pores and this force is balanced by a resistance force that is proportional to fluid velocity ($F_{\Delta P} = F_{\text{resistance}}$), then we obtain an expression similar to D'Arcy's law:

$$v_{\text{fluid}} = J_{\text{fluid}} = \frac{V_i}{f_p} \frac{\Delta P}{\delta_m} \qquad (11)$$

where f_p is the friction coefficient for the pores. In each of these cases, the flux of one component in the system is shown to be proportional to a driving force.

However, in many systems of practical interest, energy gradients and the transport of multiple components do not act in isolation. The transport of one component in the system may affect the transport of another. One example is the flow of fluid through a membrane pore carrying a charge that gives rise to a flux of current known as the streaming current. Desalination by reverse osmosis is a second example. In RO desalination, a pressure gradient is applied to transport water across a semipermeable membrane that rejects much of the salts in solution. Calculation of the flux of either salt or water requires that gradients in both pressure and concentration be considered. Phenomena such as these are examples of coupled transport.

Irreversible thermodynamics addresses coupled transport in a generalization of expressions such as Fick's law that describes the flux of a component of a system as being proportional to a driving force. We generalize the relationship for flux as a function of driving force by considering the flux of any one component, i, as a function of a vector of driving forces, X. A Taylor series expansion from the point at which the driving forces disappear ($X = 0$ at thermodynamic equilibrium where the fluxes will also be equal to zero) to X yields the following result:

$$J_i(X) = J_i(0) + \sum_j \left.\frac{\partial J_i}{\partial X_j}\right|_{X=0} \cdot X_j + \sum_{j,k} \frac{1}{2!} \frac{\partial^2 J_j}{\partial X_j \partial X_k} : X_j X_k + O(X^3) \quad (12)$$

If the system is near equilibrium, then we can ignore the higher order terms and truncate the Taylor's series after the linear terms. This implies that the partial derivatives evaluated at $X = 0$ can be treated as constants. These constants are referred to as phenomenological coefficients, which relate the flux of component i to driving force j.

The result is the second postulate of irreversible thermodynamics, referred to as the Onsager principle, which states that for small deviations from equilibrium, the fluxes of materials across a membrane, J_i, can be expressed as linear combinations of all the pertinent driving forces, X_j [1, 2]. Thus, for the case of several driving forces, the flux of component i, J_i is calculated as:

$$J_i = \sum_i L_{ij} X_j \qquad (13)$$

where the L_{ij}, are the phenomenological coefficients.

Consider the case of transport across a membrane involving two molecular species (i and j) such as might occur in a diffusion dialysis separation. In this case, the two conjugate driving forces (X_i and X_j) are simply the gradient of the chemical potential for each of these species across the membrane. In the absence of gradients in pressure or electrical potential, and for dilute solutions, these driving forces are proportional to the concentration gradients for each species. The phenomenological coefficients for $i \neq j$ account for the possibility that the flux of i (J_i) may affect the flux of j (J_j), that is to say that these fluxes are *coupled*.

In the case of our two molecules transported by diffusion:

$$J_i = L_{ii} X_i + L_{ij} X_j$$

$$J_j = L_{ji} X_i + L_{jj} X_j$$

The matrix of phenomenological coefficients will be a square matrix since the flux of each species "i" is described as a linear sum of the gradients of the chemical potentials of all other species "j." We can begin to better appreciate the meaning of the phenomenological coefficients by rewriting Eq. 13 so that terms with $j = i$ are removed from the summation:

$$J_i = L_{ii} X_i + \sum_{k \neq i} L_{ik} X_k \qquad (14)$$

The first term in Eq. 14 describes the flux of component i due to a driving force such as a concentration gradient in i. In the case of Fick's law, L_{ii} is proportional to the diffusion coefficient. For the case of D'Arcy's law, L_{11} can be interpreted as the membrane hydraulic permeability. The second term in Eq. 14 describes the impact that the transport of other components may have on the flux of component i. The phenomenological coefficients in the summation are referred to as *coupling coefficients* since they relate the impact of a driving force for one component (and hence the transport of that component) on the transport of a second component. The Onsager reciprocal relationship further states that coupling effects are symmetric and thus:

$$L_{ik} = L_{ki} \qquad (15)$$

Application to dense membranes:
Reverse osmosis

Reverse osmosis membranes are permeable to water while rejecting, to a large extent, solutes present in the feed. If, in the absence of any applied pressure, pure water is placed on one side of such a membrane (1) and an aqueous salt solution on the other (2), water will is pass from side 1 to side 2 by osmosis resulting in a decrease in the concentration of salt in compartment 2. If the compartments separated by the membrane are open to the atmosphere and allow for the height of water to adjust in each compartment (as in Figure 9.2), water will diffuse until the driving force of an increasing pressure from a greater height of water in compartment 2 comes to equilibrium with the decreasing driving force for diffusion from 1 to 2.

At equilibrium, the total available energy per mole of water, $\overline{E}_w$, is equal on the two sides of the membrane, i.e., $\Delta \overline{E}_w = 0$. In the absence of an electrical potential, and further noting that water is uncharged ($z_w = 0$), according to Eq. 6, this condition is met when:

$$0 = V_w \Delta P + RT\Delta \ln a_w$$

$$V_w(P_2 - P_1) = -RT\ln \frac{a_{w,2}}{a_{w,1}} \tag{16}$$

where the subscripts 1 and 2 refer to the dilute and concentrated solutions, respectively. Eq. 15 indicates that if the activity of water is greater on one side of the membrane, the system can be in equilibrium only if the hydrostatic pressure is greater on the other side. If we take the activity of the pure water to be unity, and note that the pressure differential ($P_2 - P_{\text{pure water}}$) is known as the *osmotic pressure* of the solution, commonly designated as π, we obtain an expression for the osmotic pressure:

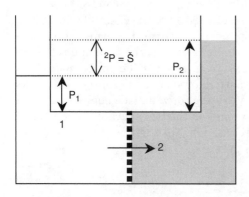

Figure 9.2 Pure water (1) diffuses across a salt-rejecting membrane resulting in a dilution of (2) and difference in pressure between the two compartments.

$$\pi = \frac{RT}{V_w}\ln a_{w,2} \tag{17a}$$

$$\pi \approx RT\sum c_s \tag{17b}$$

where π and a_w are the osmotic pressure and activity of water in the solution of interest and c_s is the molar concentration of salt. Eq. 17b applies if the solution is dilute where we can approximate the activity of water as being 1 minus the mole fraction of the salt in compartment 2 and note that $\ln(1-x) \approx -x$. Higher molecular weight solutes produce less osmotic pressure than do small molecular weight compounds. This can be illustrated by considering the case of a dilute solution of molecules with molecular weight $\overline{M}$ at a mass concentration C. The osmotic pressure can be approximated as:

$$\pi \cong \frac{C}{\overline{M}}RT \tag{18}$$

Eqs. 17a, 17b, and 18 are all ways of writing what is known as the van't Hoff equation. Note that Eq. 18 predicts that the osmotic pressure will decrease as the molecular weight of the species increases. For this reason, the osmotic pressure exerted by larger particles and macromolecules can typically be ignored. In contrast, small nanoparticles, while exerting an osmotic pressure much less than that of an ionic solute at equal mass concentration, may nonetheless result in a significant osmotic pressure.

Substituting Eq. 17a into Eq. 6, we obtain the following result for the difference in available energy of the water on the two sides of the membrane in the absence of an electrical potential:

$$\Delta \overline{E_w} = V_w\Delta P + RT\Delta \ln a_w = V_w(\Delta P - \Delta\pi) \tag{19}$$

In other words, a pressure ΔP greater than the osmotic pressure must be applied across the membrane to create a driving force sufficient to move water across the membrane. We now write the corresponding expression for the solute:

$$\Delta \overline{E_s} = V_s\Delta P + RT\Delta \ln a_s \tag{20}$$

Based on Eq. 17, for this system, $\Delta\pi \approx RT\Delta c_s$, so $\frac{\Delta\pi}{RT\Delta c_s} = 1$. Substituting that relationship into Eq. 20, and assuming that activity is approximated by concentration, we obtain:

$$\Delta \overline{E_s} = V_s\Delta P + \frac{\Delta\pi}{RT\Delta c_s}RT\Delta \ln c_s \tag{21}$$

Eqs. 19 and 21 indicate the amount of available energy that is released (dissipated) per mole of water and solute, respectively, that cross from the feed to the permeate side of the membrane. If J_i is the molar flux of species i, the product $J_i \Delta \overline{E}_i$ indicates the rate of energy dissipation associated with the transport of species i per unit area of membrane. Therefore, the overall rate of energy dissipation per unit area of membrane, Φ, considering the flux of both water and solute is:

$$\Phi = J_w \Delta \overline{E}_w + J_s \Delta \overline{E}_s \tag{22}$$

$$= J_w V_w (\Delta P - \Delta \pi) + J_s \left(V_s \Delta P + \frac{\Delta \ln c_s}{\Delta c_s} \Delta \pi \right)$$

Grouping terms in $\Delta \pi$ and ΔP, we obtain:

$$\Phi = (J_w V_w + J_s V_s) \Delta P + \left(\frac{J_s \Delta \ln c_s}{\Delta c_s} - J_w V_w \right) \Delta \pi$$

$$= J_V \Delta P + J_D \Delta \pi \tag{23}$$

Eq. 23 describes the rate of energy dissipation in our simple system as the sum of two terms. The first term on the right-hand side of Eq. 23 is the product of the total volume flux, J_V, of the components (water and solute) and a driving force, the difference in mechanical pressure across the membrane (ΔP). The second term is the product of the diffusive flux of solute relative to the flux of water, J_D, and a second driving force for transport, the osmotic pressure. Thus, in this two-component system, the rate of energy dissipation is described by two fluxes and their two corresponding conjugate driving forces. This is an important result since it reveals the appropriate fluxes and conjugate driving forces to be substituted into Eq. 13 yielding the result for reverse osmosis performance [3]:

$$J_V = L_V \Delta p + L_{VD} \Delta \pi \tag{24a}$$

$$J_D = L_{DV} \Delta p + L_D \Delta \pi \tag{24b}$$

where L_V corresponds to L_{11}, etc. Since we have two components, we end up with two equations and four phenomenological coefficients, however, by the Onsager reciprocal relationship (Eq. 25), $L_{VD} = L_{DV}$. In a three-component system, applying the Onsager reciprocal relationship we would have three equations and six coefficients, and so on. Numerous variations, expansions, and simplifications on Eqs. 24a and 24b have been developed in the literature. However, they virtually all share the feature that volume flux is proportional to a net pressure drop ($\Delta P - \Delta \pi$) while solute flux is proportional to the concentration difference across the membrane, which in turn is proportional to $\Delta \pi$.

For example, assuming dilute solutions, Eqs. 24a and 24b, known as the Kedem and Katchalsky model [3], can be expressed as:

$$J_V = L_V(\Delta p + \sigma \Delta \pi)$$

$$J_s = \omega \Delta \pi + (1 - \sigma)c_{lm}J_V$$
(25)

where $\sigma = \frac{\Delta p}{\Delta \pi}\Big|_{J_v=0} = \frac{L_{VD}}{L_V}$ is the reflection coefficient, c_{lm} is the log mean average concentration across the membrane, and $\omega = (L_V L_D - L_{VD}^2)/L_V$.

The reflection coefficient can be shown to be [4] a product of an equilibrium term expressing relative affinity (or exclusion) of salt with respect to the membrane, and a kinetic term that expresses the relative mobilities of water and salt and their potentially coupled migration.

The parameter permeability, P, of a membrane to a given molecule also reflects a multiplicative relationship between affinity and mobility. In the solution-diffusion model [5], permeability is defined as the product of the molecule's solubility in the membrane, K_s, and its diffusivity, D:

$$P = K_s D$$
(26)

This leads to an equation for water flux that is again similar to that obtained from irreversible thermodynamics:

$$J_w = A_w(\Delta p - \Delta \pi)$$
(27)

where A_w is the hydraulic permeability, $A_w = \frac{D_w C_{m,1}^w \bar{V}_w}{RT\delta_m}$, and D_w, $c_{m,1}^w$ are the diffusivity and concentration of water in the membrane, respectively, and δ_m is the thickness of the membrane. For solutes, the solution-diffusion model yields following expression for the flux of solute, J_s:

$$J_s = D_s K_s \frac{(C_{\text{feed}} - C_{\text{permeate}})}{\delta_m}$$
(28)

where K_s is the solubility coefficient for the solute.

Application to porous membranes: Fluid filtration

Materials with high diffusivities are not typically removed from fluids by porous membranes such as MF and UF membranes operated under a pressure differential. As a result, coupled transport of mass can typically be ignored. (However, the development of a streaming current when flow moves through a charged membrane may entail coupling, particularly for smaller pore sizes.) As a result, we need only consider the first term in Eq. 13, which we have shown reduces to a form resembling D'Arcy's law where fluid flux $J_{f,vol}$ (volume of fluid per unit area of membrane per unit time) is proportional to the difference in pressure across the membrane. This is typically expressed as:

$$J_{f,\text{vol}} = \frac{\Delta p}{\mu R_m}. \tag{29}$$

where Δp is the pressure drop across the membrane (the TMP), μ is the absolute viscosity of the fluid, and R_m is the hydraulic resistance of the membrane, with dimensions of reciprocal length. Membrane performance is often expressed as the ratio of permeate flux, J, to the pressure drop across the membrane, Δp. This quantity is called the specific permeate flux, with an initial value equal to $\frac{1}{\mu R_m}$.

If each pore is modeled as a capillary, permeate flux can be represented as Poiseuille flow through a large number of these capillaries in parallel. In each pore, the velocity of the fluid is assumed to be zero at the wall of the pore (termed the "no-slip" condition), and at a maximum value in the center of the pore. The no-slip condition at the pore wall is ultimately a consequence of an affinity between fluid molecules and with those of the membrane pore and leads to a parabolic velocity profile.

Using the Hagen-Poiseuille equation to describe flow through cylindrical membrane pores (idealized as such, or perhaps truly cylindrical) the following expression is obtained for the permeate flux through a membrane characterized by an effective pore radius of r_{pore}:

$$J = \frac{A_{\text{pore}} r_{\text{pore}}^2 \Delta P}{A_{\text{membrane}} 8 \mu \theta \delta_m} = \frac{\Delta P}{\mu R_m} \tag{30}$$

where $R_m = \frac{A_{\text{membrane}} 8 \theta \delta_m}{A_{\text{pore}} r_{\text{pore}}^2}$, A_{pore}/A_m is the ratio of the open pore area (A_{pore}) to the entire area of the membrane surface (A_m), θ is the pore tortuosity factor, and δ_m is the effective thickness of the membrane. Note that Eq. 30 predicts that flux should decrease with the square of decreasing pore size. If the assumptions of the Hagen-Poiseuille model hold, very high pressures would be required to induce flow through membranes with nanometer-sized pores.

Polarization phenomena and membrane fouling

The rejection of materials by a membrane leads to the accumulation of these materials near, on, or sometimes within the membrane. This can lead to a decrease in membrane performance. For example, ion exchange membranes used in electrodialysis have a polymeric support structure with fixed charged sites and water-filled passages. Charged functional groups on these membranes attract ions of opposite charge (counterions). This is accompanied by a deficit of like-charged ions (co-ions) in the membrane and results in a so-called Donnan potential and the exclusion of ions from ion exchange membranes with like-charged functional groups. When an electrical potential is applied across these membranes, ions migrate to the electrode of opposite charge. However, the ion exchange membrane rejects co-ions, resulting in boundary layers on either side of the membrane (referred to as *concentration polarization* layers) that are either enriched in co-ions (the feed side) of the membrane or have a deficit of these ions (the permeate side or "dialysate"). Because there are fewer ions on the dialysate side, there is an increase in electrical resistance that leads to an increase in power consumption to achieve separation.

A similar phenomenon occurs in RO where salts are rejected by the membrane, leading to a concentration polarization layer near the membrane. The concentration polarization layer increases the local osmotic pressure, resulting in the need for a higher pressure to overcome this osmotic pressure, as well as a lower rejection of salt by the membrane. The concentration profile of rejected species can be calculated from a mass balance on solute in a differential volume in the concentration polarization layer. For a simplified mass balance around the concentration polarization layer, the advective flux of solutes toward the membrane is balanced by diffusive back transport of solute:

$$v_w c = -D_B \frac{\partial c}{\partial y} \tag{31}$$

where v_w is the fluid velocity in the y direction (perpendicular to the membrane), and D_B is the Brownian diffusion coefficient of the solute. This expression can be integrated to yield an expression for the limiting permeate flux as a function of the bulk concentration c_{bulk}, the limiting wall concentration, c_{wall}, the diffusion coefficient for the solute, and the concentration-polarization layer thickness, δ_{cp},

$$J_{lim} = v_{w,lim} = \frac{D}{\delta_{cp}} \ln\left(\frac{c_{wall,lim}}{c_{bulk}}\right) \tag{32}$$

Eq. 32 predicts that the limiting permeate flux should decrease with decreasing D. Because the diffusion coefficient increases as particle size decreases, we can expect that when membranes are used to separate nanoparticles, the limiting permeate flux for membrane operation should be higher for these species in comparison with larger colloidal species, if there is no resistant cake deposited on the membrane. However, if a layer of nanoparticles deposits on the membrane, this may lead to a decrease in flux due to membrane fouling.

In pressure-driven processes, fouling may be manifest as either an increase in the pressure drop, ΔP, across the membrane, (called the transmembrane pressure or TMP) required to maintain a constant flux or, for a constant TMP, by a decrease in the permeate flux. This is illustrated in Figure 9.3 where the specific permeate flux (J/TMP) of a laboratory membrane system is seen to decline quickly over time due to fouling under conditions when the feed contains many foulants. When the membrane is taken out of service and washed with water, a portion of the fouling is reversed (reversible fouling). However, after multiple cycles of operation and washing, the amount of permeate flux that is recovered by washing decreases (irreversible fouling). At some point, membranes must be cleaned more aggressively with chemicals to at least partially reverse the "irreversible" fouling. Fouling is not reserved to pressure-driven membrane processes. The deposition of material on a membrane may change its functionality and therefore its ability to effectively separate compounds. Additional layers of material may impede mass transport originating from any driving force.

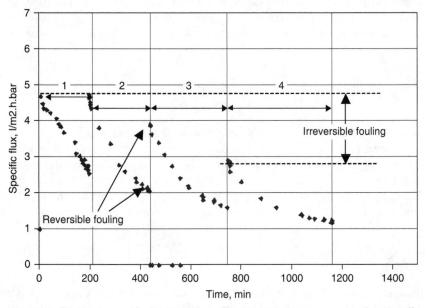

Figure 9.3 Decline in specific flux over time. Cleaning events reverse some, but not all, of the fouling.

The choice of chemicals used to clean a membrane is dependant on the chemical nature of the foulant. Dissolved organic materials are the primary foulants in many water-based membrane applications. Even when mass concentrations of particle exceed those of organic matter, organic matter may dominate fouling. Organic matter may form a gel layer on the surface of the membrane or adsorb to the surface or within the membrane matrix. In gas separations, gas-phase impurities may foul membranes by adsorption. The possibility of precipitative fouling is a serious concern in the operation of membranes designed specifically to remove scale-forming species (e.g., RO, membrane distillation, electrodialysis). Common foulants of concern for precipitation include calcium, magnesium, and iron (primarily ferrous), any of which might precipitate as a hydroxide, carbonate, or sulfate solid. Colloids of all kinds may accumulate on or near membranes, forming a cake. Biofouling of membranes is a key concern in applications in which membranes are in contact with an aqueous environment. Bacteria may colonize membranes forming biofilms that consist of the bacteria and compounds secreted by the bacteria.

In pressure-driven membranes, these potential causes of fouling can be represented mathematically by modifying Eq. 29 to include resistance terms for each of the causes of fouling, including those due to the changes in membrane permeability over time, $R_m(t)$, and the formation of a cake or biofilm, $R_c(t)$. One approach to describing the resistance of the cake, R_c, is to assume that the cake has a uniform structure with a specific resistance, $\hat{R}_c$, and thickness, δ_c (that may change over time). An expression for hydraulic permeability, such as the Kozeny equation, can then be used to predict the specific resistance, assuming that the cake is incompressible and is composed of uniform particles:

$$\hat{R}_c = \frac{180(1 - \varepsilon_c)^2}{d_p^2 \varepsilon_c^3} \tag{33}$$

where ε_c is the porosity of the cake, and d_p is the diameter of the particles that it comprises. This expression predicts that resistance to permeation by a deposited cake should increase as one over the square of the diameter of particles comprising the cake. Thus, in comparison with larger colloidal species, an accumulation of nanoparticles on the membrane should form a cake of relatively high specific resistance.

The effect of concentration polarization on permeate flux is primarily due to the accumulation of solutes near the membrane and the resultant osmotic pressure effect. It will typically be negligible for MF and UF that reject only particles or macromolecules. The concentration polarization term for RO and NF applications can be incorporated into an osmotic pressure term in the numerator, and the following expression is obtained:

$$J = \frac{(\Delta p - \sigma \Delta \Pi)}{\mu \{R_m(t) + R_c[\delta_c(t), \dots]\}} \tag{34}$$

Concentration polarization may be indirectly related to irreversible fouling through its effects on adsorption, cake formation, and precipitation. However, reductions in permeate flux (or increases in TMP) due directly to CP are completely reversible, making this resistance term quite different from the others.

Membranes in fuel cell applications

A polymer electrolyte membrane fuel cell (PEMFC) consists of a proton exchange membrane sandwiched between two layers of catalyst material. Hydrogen or an alternative fuel such as methanol reacts catalytically at the anode to form electrons and protons (Figure 9.4). The proton exchange membrane (PEM) selectively transports protons to the cathode where they react catalytically with oxygen to form water. The membrane must have a high rejection for the fuel on the anode or "feed" side of the membrane, and for the oxygen on the cathode side of the membrane. The most widely used membranes in fuel cells are perfluorosulfonate polymeric membranes, most widely marketed by Dow as the Nafion™ membrane. The perfluorosulfonic polymers are strongly hydrophobic with hydrophilic groups that can adsorb large quantities of water. They are highly conductive while fully hydrated at low

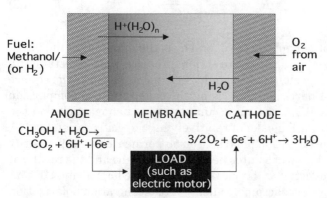

Figure 9.4 Diagram of a fuel cell, in this case using methanol.

temperatures. Perfluorosulfonic polymers are both electron insulators and excellent proton conductors. The weak attraction to the SO^{3-} group allows the H^+ ions in the sulfonic groups to move upon hydration [6]. At high relative humidities (greater than 80 percent), conductivity in the range of 3×10^{-2} S/cm [7] up to nearly 9×10^{-2} S/cm is typical for materials such as of Nafion 117 at room temperature [8]. Increasing operating temperature produces limited improvements in the conductivity of Nafion due to progressive dehydration [9] and a resultant decrease in the more mobile bulk water. Indeed, these membranes become nonconductive at operating temperatures above approximately 80°C. In addition, these perfluorosulfonate membranes are expensive, representing approximately 46 percent of the cost of current fuel cell stacks.

Other polymeric membranes have been investigated for their suitability in fuel cell applications. Fluorine-free, hydrocarbon-based membrane materials are less expensive and are commercially available. They contain polar groups with high water uptake. A hydrocarbon-based membrane, however, does not provide long-term thermal stability for fuel cell applications [10]. Polymeric materials such as polyetherketone [11], sulfonated polyaromatic polysulfones [12], grafted fluorinated polymers by radiation [13], and polyvinylidene fluoride and sulfonated polystyrene-co-divinylbenzene have all shown promise. However, these polymers also exhibit thermal, chemical, or mechanical instability, and/or they do not provide a protonic conductivity comparable to that of Nafion.

Although fuel cells are typically regarded as technologies for producing electricity, the prospects for improving fuel cell operation as water supply technology are particularly intriguing. In contrast with the conventional approach of *treating* water, fuel cells produce water in a manner that exemplifies the nanotechnology paradigm; *by assembling hydrogen and oxygen to fabricate water*. Cogeneration of electricity and water may in some instances be a cost-effective option, particularly when sources of high quality water are scarce or when the final quality of water required is demanding. A typical US household consumes approximately 4,200 kWh of electricity per year. The theoretical yield of water from a fuel cell is based on the stoichiometry of the reaction $(2H_2 + O_2 = 2H_2O)$ and the heat of formation [14]. The current generation of fuel cells is capable of producing approximately 1 liter/kWh of electricity, equivalent to some 10 liters of high-purity water per day for a typical US household without limitations on water quality degradation in the distribution system. However, one factor that limits the use of fuel cells for water production is the need to use much of the water produced by the fuel cell to keep the membranes hydrated and sustain proton conductivity.

Membrane Fabrication Using Nanomaterials

The performance of membranes is intimately linked to the materials they are made from. The composition of the membrane will determine important properties such as rejection (selectivity), propensity to foul, mechanical strength, and reactivity. Membrane composition may even determine the element geometries that are (or are not) possible and, of course, the cost of the membrane. It is therefore not surprising that the creation of new nanomaterials opens the door for many new approaches to fabricating and improving membranes. We will consider three cases. In the first case, we will consider examples where nanomaterials are used to make membranes. In the second case, we will examine enhancements in existing membrane materials achieved through the introduction of nanomaterials to create new composites. Finally, we will take a look at the use of nanomaterials to "mold" membranes in a process known as nanomaterial templating.

Membranes made from nanomaterials

In some instances, membranes may be made (nearly) entirely of products of nanochemistry. The advantages of these materials may include improved processing, due for example to the lower sintering temperature required when nanoparticles are used as precursors to ceramic membranes. Or they may include include improved performance linked directly with the properties of the materials.

Ceramic membranes derived from mineral nanoparticles. Mineral membranes have been made from a variety of mineral nanoparticle precursors. Commercially available ceramic membranes are typically made from metal oxides such as Al_2O_3, ZrO_2, and TiO_2 [15]. However, membranes can be made from many other nanomaterials ranging from gold [16, 17] to SiO_2 [18]. In most cases nanoparticles are deposited on a support surface and then calcined to create the membrane. Processes differ in the manner in which nanoparticle precursors are prepared. One common procedure for producing nanoparticle precursors to these membranes is to precipitate particles under controlled conditions creating a suspension or *sol* of nanoparticles that are deposited on a surface and dried to form a *gel*. This procedure is known as *sol-gel*.

Sol-gel involves a four-stage process: dispersion, gelation, drying, firing. A stable liquid dispersion or *sol* of the colloidal ceramic precursor is initially formed in a solvent with appropriate additives. In the case of alumina membranes, this first step may be carried out with 2-butanol or *iso*-propanol. By removing the alcohol, the polymerization of aluminum monomers occurs leading to a precipitate. This material is acidified, typically using nitric acid, to produce a colloidal suspension. By controlling the extent of aggregation in the colloidal sol, a gel of desired properties can be produced. The aggregation of colloidal particles in the sol may be controlled by adjusting the solution chemistry to influence the diffuse layer interactions between particles, by adding stabilizing agents such as surfactants, or through ultrasonification. Knowing that the properties of the gel will influence the permeability of the future membrane, it is clear that the gelation step is extremely important. This gel is then deposited, typically by a slip-cast procedure, on an underlying porous support. In variations on this procedure, functionalized surfaces may also be used to achieve a more ordered deposit or micelles can be used to direct film formation in specific geometries through self-assembly. In conventional sol-gel, the excess liquid is removed by drying and the final ceramic is formed by firing the resulting gel at higher temperatures. Drying and firing conditions have been shown to be very important in the structural development of the membranes, with higher drying rates resulting in more dense membrane films.

The sol-gel approach of reacting small inorganic molecules to form oligomeric and polymeric nanoparticles has several limitations such as difficulties in controlling the reaction conditions, and the stoichiometries, solubility, and processability of the resulting gel. It would thus be desirable to prepare nanoparticles in a one-pot bench-top synthesis from readily available, and commercially viable, starting materials, which would provide control over the products. One strategy for producing nanostructured membranes involves an environmentally benign alternative to the sol-gel process for ceramic membrane formation. Metal nanoparticles such as alumoxanes [19] and ferroxanes [20] can be produced based upon the reaction of boehmite, $[Al(O)(OH)]_n$, (or lepidicrocite in the case of the ferroxanes) with carboxylic acids [21]. The physical properties of such metal-oxanes are highly dependent on the identity of the alkyl substituents, R, and range from insoluble crystalline powders to powders that readily form solutions or gels in hydrocarbon solvents and/or water. Thus, a high degree of control over the nanoparticle precursors is possible. Metal-oxanes have been found to be stable over periods of at least years. Whereas the choice of solvents in sol-gel synthesis is limited, the solubility of the carboxylate metaloxanes is dependent on the identity of the carboxylic acid residue, which is almost unrestricted. The solubility of the metal-oxanes may therefore be readily controlled so as to make them compatible with a coreactant. Furthermore, the incorporation of metals into the metal-oxane core structure allows for atomic scale mixing of metals and formation of metastable phases. In the case of the aluminum-based alumoxanes, the low price of boehmite ($ 0.5 per kilogram) and the availability of an almost infinite range of carboxylic acids make these species ideal as precursors for ternary and doped aluminum oxides.

Application of a metal-oxane–based approach to creating ceramic membranes reduces the use of toxic solvents and energy consumption. By-products formed from the combustion of plasticizers and binders are minimized, and the use of strong acids eliminated. Moreover, the use of tailored nanoparticles and their deposition on a suitable substrate presents an extremely high degree of control over the nanostructure of the resulting sintered film. The versatility of the process can be used to tightly control pore-size distributions. The MWCO of the first generation of alumoxane-derived membranes is approximately 40,000 daltons [22], which is in the ultrafiltration range. Table 9.4 shows a comparison of the ceramic and sol-gel methods with that of the carboxylate alumoxanes for the synthesis of alumina and ternary aluminum oxides. The ease of modification of the alumoxanes suggests that a single basic coating system can be modified and optimized for use with a range of substrates.

There has been interest in using ceramics as electrolyte materials for proton exchange membrane fuel cells because of their thermal, chemical, and mechanical stability and their lower material costs [23]. However, traditionally ceramic membranes have exhibited comparatively small proton conductivities. The conductivities of silica glasses fired at 400 to 800°C is in the order of 10^{-6} to 10^{-3} S/cm [24]. The conductivities of silica, alumina, and titania sintered at 300 and 400°C are in the range of 10^{-7} to 10^{-3} S/cm [25].

However, recent work suggests that membranes derived from ferroxane nanoparticles may be attractive alternatives for such proton exchange membranes. With a conductivity of approximately 10^{-2} S/cm the ferroxane-derived membrane represents a large improvement over other ceramic materials prepared by the traditional sol-gel method, with conductivities close to that of Nafion (Table 9.5).

The protonic conductivity of these membranes varies as a function of the temperature at which they are sintered. For example, when ferroxane films are sintered at 300°C the resulting membranes display a

TABLE 9.4   Comparison of the Alumoxane and Sol-Gel Synthesis Methods

	Alumoxane	Sol-gel
Methodology	simple	complex
Atomic mixing	yes	yes
Metastable phases	yes	yes
Stability	excellent	fair
Solubility	readily controlled	difficult to control
Processability	good	good
Time	<8 h	>20 h
Cost	low	med.-high

TABLE 9.5   Representative Conductivities of Oxide Membranes and Nafion® Compared with Preliminary Results for Ferroxane-Derived Membrane. Conductivity is Reported at 100% Humidity and at 20°C

Material/Study	Conductivity (S/cm)
Nafion 117	
(Sumner, 1998)	0.06
(Sone, 1997)	0.09
(Kopitzke, 2000)	0.09
SiO_2 glasses	
(Nogami, 1998)	$10^{-6} - 10^{-3}$
$(SiO_2\text{-}P_2)_5$ glass	
(Tung and Hwang, 2004)	9.43E-3
Sol-gel Al_2O_3	
(Vichi, 1999)	6.0E-4
Alumoxane-derived Al_2O_3	
(Tsui et al.)	6.7E-4
Ferroxane-derived membranes	
(Tsui et al.)	0.03

higher conductivity (~0.03 S/cm) at all values of relative humidity compared with the green body and with the ferroxane sintered at 400°C (Figure 9.5). Sintering at higher temperatures likely results in the sacrifice of the small pores in the ceramic as it forms and a resulting decrease in conductivity. Under the conditions examined to date, proton conductivity of these membranes appears to be relatively insensitive to the relative humidity within the membrane. A membrane with high

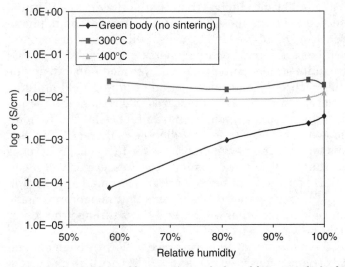

Figure 9.5 Conductivity of ferroxane green body and ferroxane-derived ceramics as a function of humidity.

proton conductivity at a low humidity would be highly desirable, allowing the fuel cell to operate at higher temperatures and with less difficulty in keeping the membrane hydrated. Similar to the Nafion membrane, the green body ferroxane material shows a strong dependence of proton conductivity on humidity.

This dependence on humidity implies a structure diffusion mechanism, while the transport mechanism in the case of the ferroxane-derived ceramics is less clear. However, it is likely to involve proton "hopping" between hydrogen-bonded water or hydroxyl groups along the oxide structure. By comparison, Nafion conductivity may decrease by 200 percent or more over the range of 100 percent RH to 80 percent RH [7, 9]. The structure diffusion mechanism that typically dominates proton transport in Nafion membranes is typically orders of magnitude larger than the hopping mechanism. Transport dominated by a hopping mechanism in the ferroxane-derived membranes would imply a potential for low methanol permeability, which has been confirmed, but may not explain the very high protonic conductivities that have been observed.

Fullerene-based membranes.   Fullerenes have unique properties of strength, ability to tailor size, flexibility in modifying functionality, and electron affinity, which have created much excitement around their potential for new membrane-based technologies. For example, the ability of fullerenes to act as electron shuttles has been considered as a possible basis for creating light-harvesting membranes using C_{60} or C_{70} contained in lipid bilayers [26] or incorporated into porous polymers [27]. The photocurrent density obtained from the C_{70}-bilayer system was observed to be about 40 times higher than that of the artificial system previously observed to be the most efficient [28]. In addition to photovoltaics, fullerenes may find uses in fuel cells. Fullerenes share some of the properties of the perfluorosulfonic polymers typically used in PEFCs in that they are quite stable, anhydrous, and yet modifiable in a wide variety of manners through the introduction of proton binding functions on the fullerene surface. These features make them interesting candidates for proton exchange membranes in fuel cells [29]. In addition to the unique properties that make fullerenes such as CNTs interesting materials for creating new membranes for fuel cells, fullerenes have also drawn interest as the basis for new pressure-driven membranes, particularly for the treatment of water. The small and controllable diameter of fullerene nanotubes suggests that membranes made from these materials in a fashion where fluid flows through the center of the CNT might be highly selective.

However, Eq. 30 predicts that the resistance to flow through a membrane composed of nanometer-sized pores should be very high and potentially prohibitive for practical applications. Surprisingly, molecular modeling first suggested that flow through pores composed of carbon nanotubes might not have the same limitations as those observed for other nanometer-sized pores [30]. Simulations indicate that water should be able to flow much faster through hydrophobic CNTs due to the formation of ordered hydrogen bonds. In the confined space of a nanotube, water is present in ordered crystalline domains. When the nanotube wall interacts significantly with the water, such as in the case of a small silica channel where silinol groups may anchor water molecules to the wall of the channel, the ordered water in the pore is thought to be less mobile. In contrast, the hydrophobic surface in the interior of a defect-free carbon nanotube appears to allow for a nearly frictionless flow that has been compared with the flow through the protein channel aquaporin-1 [30]. Visualization of water within CNTs confirms the lack of interaction between water molecules and the interior surface of CNTs [31].

Although these theoretical results suggest that CNTs are very promising materials for water filtration membranes, there are many challenges to be overcome in aligning CNTs and fabricating such a membrane. The problem of aligning membranes was first approached by filtering suspensions of SWNTs in a strong magnetic field [32]. A more promising approach is to grow arrays of CNTs on a substrate where nanoparticle catalysts for CNT growth have been arranged in a distinct pattern that defines the number and spacing of the resulting CNTs. The diameter of the nanotube is controlled by the size of the nanoparticle catalyst initially arranged on the substrate [33]. A working membrane of aligned CNTs requires that the spaces between CNTs be filled with a material that seals the membrane to flow between CNTs, allowing flow only through the interior of the CNTs. Among the approaches taken to accomplish this has been spin-coating the CNT arrays with a polymer solution [34] or filling the interstices of the aligned CNT with silicon nitride [35]. The permeate flux of water measured across a membrane of aligned multiwall CNTs with 7nm-diameter pores imbedded in a polystyrene matrix has been reported to be four to five orders of magnitude greater than that predicted by Eq. 30 [36]. However, smaller diameter CNTs may exhibit a bamboo-like structure [37] that impedes fluid flow [35]. Similar to these CNT-based membranes, fullerene-based membranes have also been made by grafting C_{60} onto the surface of track-etched membranes [38].

These fullerene-based membranes can be thought of as composite membranes in that they are composed of the fullerenes and at least one other material. The properties of these composite membranes reflect the sum of the properties of the components of the membrane. As such, they resemble conventional thin film composite membranes such as those used in RO. RO composite membranes rely on a thin layer of material (typically a polyamide) on the surface or skin of the membrane to provide the essential rejection characteristics of the membrane. The thicker, underlying layer (often polysulfone) serves as a support. The overall property of the membrane is approximated by the rejection characteristics of the skin plus the mechanical characteristics of the support. Similarly, in the current generation of aligned CNT membranes, the CNTs determine the transport properties of the membrane, while the support material envelopes rather than underlies the CNTs.

Nanocomposites: Modifications to existing materials with nanoparticles

The inherent limitations of temperature and water retention by fuel cell membranes made from perfluorosulfonic polymers (typically Nafion) have stimulated much research to develop nanocomposites that display high proton conductivity at high temperatures and low humidity. One approach has been to add nanoparticles designed to promote proton conductivity to polymer matrices with greater resistance to temperature than Nafion. The modification of polysulfonated membranes with solid acids in the form of silica [39] or zirconium phosphate [40] nanoparticles has resulted in membranes that can operate at higher temperatures, but still, with a lower conductivity than that of Nafion [41]. The electrical conductivity of several polymer-CNT blends has been evaluated [42]. While these materials may have some promise as electrode materials in fuel cells, their potential as fuel cell membranes has yet to be demonstrated. For example, poly(methyl methacrylate) (PMMA) nanocomposites containing MWNTs were found to have electric conductivities on the order of 10^{-4} to 10^{-2} S/cm [43].

Much consideration has also been given to improvements in the catalyst/membrane support materials used in fuel cells through the incorporation of fullerenes into these electrode/supports. SWNTs have been used to replace carbon black in fuel cell electrodes yielding an order of magnitude lower resistance to charge-transfer [44]. These electrodes can then be used as supports for the PEM. More efficient use of catalyst through the formation of nanoparticles with high ratio of surface area to volume has been an important element in reducing the costs of fuel cells. Nanoparticles of Pd [45] or Pt [46] catalyst assembled on a Nafion membrane have also been reported to increase methanol rejection by the membrane (reduced crossover) in direct methanol fuel cells.

Although often motivated by the need to improve fuel cells, there are also promising applications for these fullerene-polymer composites in pressure-driven membranes. The strength of CNTs, coupled with reported antibacterial properties, suggest that fullerene-polymer composites may find use in creating membranes that resist breakage or inhibit biofouling. The incorporation of C_{60} into polymeric membranes has been observed to affect membrane structure and rejection [47].

Polymer-TiO_2 nanoparticle composites have been created with the objective of creating antifouling membranes. These membranes would exploit the photocatalytic properties of TiO_2 to produce hydroxyl radicals that would then, in turn, oxidize organic foulants depositing on the membrane surface. Membranes decorated with TiO_2 nanoparticles have been fabricated in both UF [48] and RO [49] formats. These membranes are formed by self-assembly of the TiO_2 particles at functional sites (such as sulfone or carboxylate groups) on the membrane surface. Alternatively, the TiO_2 nanoparticles can be immobilized within the membrane matrix by introducing these materials as a mix during the process of membrane casting [50]. While both formats (decorated and immobilized) appear to mitigate fouling by bacterial growth, the decorated format appears to be more effective due to the higher amount of TiO_2 on the membrane surface [50].

An inherent conflict in membrane design must be overcome in developing a photocatalytic system for reducing membrane fouling. The economics of membrane module design dictate that a maximum of membrane area be contained within a given volume (high packing density). However, this design objective is in conflict with the objective of providing adequate illumination of the membrane surface to promote photocatalysis. There may be niche applications for photocatalytic membranes where a high packing density is not required to make the process economically feasible.

Membranes may also be modified by depositing particles on the surface for the purpose of simultaneous catalytic degradation, sensing, or simply improved selectivity. Nanoparticles with molecular imprints of a specific compound can be created and then attached to conventional membranes to impart a high specificity of separation for the imprinted molecule [51]. Catalytic nanoparticles (such as TiO_2 or iron) can be attached to membrane surfaces with the objective of degrading some compounds while physically separating others. However, the objective of creating a so-called reactive membrane is fraught with conflicts in the efficiency of the overall system. Reaction rate considerations tend to favor a membrane with a slow permeation rate (high residence time) to allow for sufficient conversion of the desired compound(s) as the fluid they are contained within flows across the membrane. On the other hand, the efficiency in fluid separation pushes the system toward a higher permeation rate and shorter residence time. The conflict between reaction rate and residence time is resolved for the case of reactive membranes that work on materials sorbed to the surface of the membrane. This might be the case for either the destruction of sorbed foulants or the accumulation and subsequent destruction of a contaminant for which the membrane is designed to have an enhanced adsorptive affinity.

Nanomaterial templating

Nanomaterials can also be used as "molds" or templates for membranes. In this case, the initial nanomaterials are no longer present in the membrane structure, but rather have imparted their structure to another material. Nanoparticles can be deposited on a substrate to form the initial template for a porous solid that will take on the mirror image of the initial deposit [52].

Particle size, stability, and/or depositional trajectories can be controlled to engineer template morphology and yield membranes with a desired structure. When particles deposit on a surface following very predictable trajectories, such as occurs when their transport is dominated by gravity, an electrical field, or laminar flow, their depositional trajectory is said to be ballistic. The templates formed by depositing particles on the surface via ballistic trajectories tend to be compact. In contrast, when particles follow more random, diffusive trajectories to deposition, they tend to form more open deposits that resemble dendrites or objects resembling small trees. Particle surface chemistry can also be altered to control template morphology. When particle-particle interactions are favorable ("sticky" particles), the resulting template tends to be dendritic. When particles are not sticky, compact deposits form. Electrostatic forces, dispersion forces, and capillary forces are likely to dominate particle stickiness during template formation. For example, if templates are formed by dip-coating a substrate in a suspension of particles and evaporating the suspending fluid, the deformation of the liquid-fluid interface due to trapped colloidal particles gives rise to capillary forces exerted on the particles. These forces are usually attractive [53]. Thus, by controlling solution chemistry (for example, ionic strength) and the conditions of template drying, the morphology of the template can be controlled. The self-assembly technique by capillary forces provides precise control of the thickness of the film through sphere size and concentration in solution. Control of template morphology is illustrated in SEM images of deposits of silica nanoparticles of 244 nm average diameter (Figure 9.6).

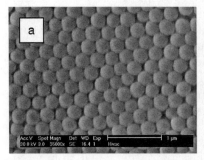 

Figure 9.6 Templates of silica particles with an average diameter of 244 nm formed from suspensions (a) ethanol (nonsticky) and (b) an aqueous solution 1.5 M NaCl (sticky).

Case a is a template formed from particles suspended in ethanol. In the case of ethanol, the particles experience a net repulsive electrostatic force in the bulk as they approach the glass surface on which they were deposited. Thus, these "nonsticky" particles form a compact deposit. In contrast, particles suspended in an aqueous solution of 1.5 M ionic strength (b) form a dendritic template.

The voids in these resulting templates are filled with a polymeric or inorganic material and upon etching (or buring) of the particles, a porous material with a three-dimensional structure is formed (Figure 9.7). The use of nanoparticles in the templating process allows for a high degree of control over chamber and pore size. The interior of the templated object can be subsequently functionalized or functionality can be introduced through the choice of material used to make the membrane. A high degree of control over both the structure and internal functionality of these membranes might be exploited to perform highly controlled reactions, with each chamber of the membrane serving as a reactor. Membrane selectivity can be modified with respect to both size exclusion and chemical affinity.

Alternatively, asymmetric templates can be formed via the sequential deposition of particle layers of nanoparticles from Langmuir-Blodgett films. A layer of small particles is first deposited onto the support, followed by a layer of larger particles. After casting the membrane around this template, a membrane with a more complex structure composed of distinct layers is formed (Figure 9.8).

CNTs have also been used as membrane templates. Beginning with an array of aligned carbon nanotubes, spaces are filled between the tubes as described earlier for the case of aligned CNT membranes. Using silicon

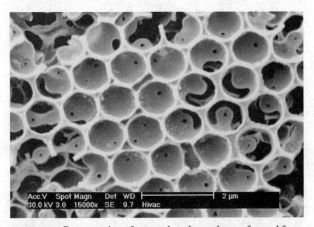

Figure 9.7 Cross section of a templated membrane formed from "nonsticky" particles forming honeycomb like structure.

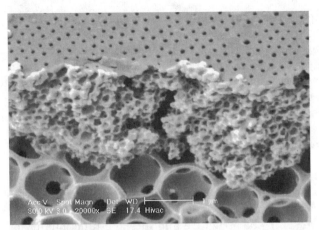

Figure 9.8 Asymmetric templated membrane produced from the Langmuir-Blodgett process.

nitride as the fill material, the CNT can then be subsequently oxidized, leaving behind a silicon nitride membrane with CNT-size pores [35].

Nanoparticle Membrane Reactors

Nanomaterials may also be used in conjunction with membranes as a nanomaterial/membrane reactor. In this case, the membrane serves only as a separation process to recover a nanomaterial that is introduced upstream for the purposes of adsorption, photocatalysis, disinfection, or some other function for which the nanomaterial is particularly well suited. The residence time of nanomaterials in the system is controlled to optimize their effectiveness. As nanomaterials lose their effectiveness (e.g., adsorption capacity is exhausted) they are removed from the system and regenerated.

The separation of nanoparticles used in such membrane reactors may present special challenges for separation due to nanoparticle size. Nano-sized materials are likely to have relatively small diffusivities compared with conventional solutes, but large diffusivities in comparison with larger colloids. While nanoparticles will therefore be more susceptible to concentration polarization (Eq. 32) compared with solutes, their osmotic pressure will be less than, for example, an equivalent mass concentration of ionic solutes due their relatively large molecular weight (Eq. 18). Unlike larger micron-sized particles that may form cakes on MF or UF membranes, the specific resistance of a cake formed by nanometer-sized particles will present a high specific resistance (Eq. 33).

Also, nanoparticles with a high degree of functionality per surface area may create cakes with a relatively high charge density. Coupling effects associated with the deformation of diffuse layers within these cakes and the flow of permeate across these cakes may be manifest as a significant electroviscous effect in which the viscosity of the fluid appears to be greater than the bulk viscosity due to the back-migration of ions.

Active Membrane Systems

Future convergence between nanochemistry and membrane science will likely yield a generation of active membrane systems. Nanomaterials might be used to develop membranes in the future with the capability to simultaneously sense and separate contaminants in a fashion that allows membranes to vary their selectivity as a function of the conditions in the feed stream. For example, self-regulating membranes might allow membranes to operate in a high permeability/low energy mode during periods where high rejections of small molecular weight materials are not required. Seasonal peaks in concentration of a given contaminant (for example, a target pesticide) would be detected by the membranes and trigger an increase in the membrane molecular weight cutoff.

Nanomaterials might also be incorporated into membranes to impart properties that are activated by an electrical or chemical signal. For example, a membrane composite that is capable of producing reactive oxygen in the presence of an electron donor might be activated by the introduction of such a compound with the purpose of periodically cleaning the membrane. Membranes might also be engineered to allow for local heating of the membrane skin with the purpose of promoting membrane distillation.

Living organisms are the ultimate nanotechnology. The ability of cell membranes to selectively transport materials, often against concentration gradient, and to avoid fouling is impressive. As the field of nanochemistry advances, engineered biomimetic systems based on selective transport or rejuvenating layers of self-organizing materials may be developed for performing critical separations in energy and environmental applications.

References

1. Onsager, L., *Reciprocal relations in irreversible processes I*. Physics Review, 1931a. 37: p. 405–426.
2. Onsager, L., *Reciprocal relations in irreversible processes II*. Physics Review, 1931b. 38: p. 2265–2279.
3. Kedem, O., and A. Katchalsky, *Thermodynamic analysis of the permeability of biological membranes in non-electrolytes*. Biochem. Biophys. Acta, 1958. 27: p. 229.
4. Spiegler, K.S., and O. Kedem, *Thermodynamics of hyperfiltration (reverse osmosis): criteria for efficient membranes*. Desalination, 1966. 1: p. 311.
5. Lonsdale, H.K., U. Merten, and Riley, R.L., *Transport Properties of Cellulose Acetate Osmotic Membranes*. Journal of Applied Polymer Science, 1965. 9: p. 1341–1362.

6. Eikerling, M., A.A. Kornyshev, and U. Stimming, *Electrophysical properties of polymer electrolyte membranes: a random network model.* Journal of Physical Chemistry B, 1997. 101(50): p. 10807–10820.
7. Sumner, J.J., et al., *Proton conductivity in Nafion117 and in a novel bis[(perfluoroalkyl)sulfonyl]imide ionomer membrane.* Journal of the Electrochemical Society, 1998. 148(1): p. 107–110.
8. Zawodzinski Jr., T.A., et al., *Determination of water diffusion coefficients in perfluorosulfonate ionomeric membranes.* J. Phys. Chem. B, 1991. 95: p. 6040–6044.
9. Sone, Y., P. Ekdunge, and D. Simonsson, *Proton conductivity of Nafion 117 as measured by a four-electrode AC impedance method.* J. Electrochem. Soc., 1996. 143(4): p. 1254–1259.
10. Brandon, N.P., S. Skinner, and B.C.H. Steele, *Recent advances in materials for fuel cells.* Annual Review of Materials Research, 2003. 33(1): p. 183–214.
11. Bailly, C., et al., *The sodium salts of sulfonated poly(aryl-ether-ether-ketone) (PEEK): preparation and characterization.* Polymer, 1987. 28(6): p. 1009–1016.
12. Nolte, R., et al., *Partially sulfonated poly(arylene ether sulfone)—A versatile proton conducting membrane material for modern energy conversion technologies.* Journal of Membrane Science, 1993. 83(2): p. 211–220.
13. Gupta, B., F.N. Buchi, and G.G. Scherer, *Cation exchange membranes by pre-irradiation grafting of styrene into FEP films. I. Influence of synthesis conditions.* Journal of Polymer Science Part A-Polymer Chemistry, 1994. 32(10): p. 1931–1938.
14. Wang, C.Y., *Fundamental Models for Fuel Cell Engineering.* Chem. Rev., 2004. 104(10): p. 4727.
15. Bhave, R.R., *Inorganic Membranes: Synthesis, Characteristics and Applications.* 1991, New York: Van Nostrand Reinhold.
16. Cai, X.M., et al., *Porous metallic films fabricated by self-assembly of gold nanoparticles.* Thin Solid Films, 2005. 491(1–2): p. 66–70.
17. Hu, X.G., et al., *Fabrication, characterization, and application in SERS of self-assembled polyelectrolyte-gold nanorod multilayered films.* Journal of Physical Chemistry B, 2005. 109(41): p. 19385–19389.
18. Fujii, T., et al., *The sol-gel preparation and characterization of nanoporous silica membrane with controlled pore size.* Journal of Membrane Science, 2001. 187(1–2): p. 171–180.
19. Callender, R.L., et al., *Aqueous synthesis of water soluble alumoxanes: environmentally benign precursors to alumina and aluminum-based ceramics.* Chemistry of Materials, 1997. 9: p. 2418–2433.
20. Rose, J., et al., *Synthesis and characterization of carboxylate-FeOOH nanoparticles (ferroxanes) and ferroxane-derived ceramics.* Chemistry of Materials, 2002. 14(2): p. 621–628.
21. Landry, C.C., and A.R. Barron, *From minerals to materials: synthesis of alumoxanes from the reactions of boehmite with carboxylic acids.* Journal of Materials Chemistry, 1995. 5(2): p. 331–341.
22. Cortalezzi, M.M., et al., *Characteristics of ceramic membranes derived from alumoxane nanoparticles.* Journal of Membrane Science, 2002. 205: p. 33–43.
23. Grahl, C.L., *Ceramic opportunities in fuel cells.* Ceramic Industry, 2002. 152(6): p. 35–39.
24. Nogami, M., R. Nagao, and C. Wong, *Proton conduction in porous silica glasses with high water content.* J. Phys. Chem. B, 1998. 102: p. 5772–5775.
25. Vichi, F.M., M.T. Colomer, and M.A. Anderson, *Nanopore ceramic membranes as novel electrolytes for proton exchange membranes.* Electrochem. Solid State Lett., 1999. 2(7): p. 313–316.
26. Bensasson, R.V., et al., *Transmembrane electron-transport mediated by photoexcited fullerenes.* Chemical Physics Letters, 1993. 210(1–3): p. 141–148.
27. Garaud, J.L., et al., *Photoinduced electron-transfer properties of porous polymer membranes doped with the fullerene C(60) associated with phospholipids.* Journal of Membrane Science, 1994. 91(3): p. 259–264.
28. Hwang, K.C., and D. Mauzerall, *Photoinduced electron-transport across a lipid bilayer mediated by C70.* Nature, 1993. 361(6408): p. 138–140.
29. Miyatake, K., and M. Watanabe, *Recent progress in proton conducting membranes for PEFCs.* Electrochemistry, 2005. 73(1): p. 12–19.

30. Hummer, G., J.C. Rasaih, and J.P. Noworyta, Nature, 2001. 414: p. 188–190.
31. Naguib, N., et al., *Observation of water confined in nanometer channels of closed carbon nanotubes.* Nano Letters, 2004. 4(11): p. 2237–2243.
32. Walters, D.A., et al., *In-plane-aligned membranes of carbon nanotubes.* Chemical Physics Letters, 2001. 338(1): p. 14–20.
33. Kanzow, H., C. Lenski, and A. Ding, Physical Review B, 2001. 6312: p. 5402.
34. Hinds, B.J., et al., *Aligned multiwalled carbon nanotube membranes.* Science, 2004. 303(5654): p. 62–65.
35. Holt, J.K., et al., *Fabrication of a carbon nanotube-embedded silicon nitride membrane for studies of nanometer-scale mass transport.* Nano Letters, 2004. 4(11): p. 2245–2250.
36. Majumder, M., et al., *Nanoscale hydrodynamics: enhanced flow in carbon nanotubes.* Nature, 2005. 438(7070): p. 44.
37. Wang, Y.Y., et al., *Hollow to bamboolike internal structure transition observed in carbon nanotube films.* Journal of Applied Physics, 2005. 98: p. 014312.
38. Biryulin, Y.F., et al., *Fullerene-modified dacron track membranes and adsorption of nitroxyl radicals on these membranes.* Technical Physics Letters, 2005. 31(6): p. 506–508.
39. Antonucci, P.L., et al., *Investigation of a direct methanol fuel cell based on a composite Nafion ¬Æ-silica electrolyte for high temperature operation.* Solid State Ionics, 1998. 125(1–4): p. 431–437.
40. Costamagna, P., et al., *Nafion ¬Æ115/zirconium phosphate composite membranes for operation of PEMFCs above 100¬†¬∞C.* Electrochimica Acta, 2002. 47(7): p. 1023–1033.
41. Boysen, D.A., et al., *Polymer solid acid composite membranes for fuel-cell applications.* Journal of the Electrochemical Society, 2000. 147(10): p. 3610–3613.
42. Wu, M., and L. Shaw, *Electrical and mechanical behaviors of carbon nanotube-filled polymer blends.* Journal of Applied Polymer Science, 2006. 99(2): p. 477–488.
43. Sung, J.H., et al., *Nanofibrous membranes prepared by multiwalled carbon nanotube/ poly(methyl methacrylate) composites.* Macromolecules, 2004. 37(26): p. 9899–9902.
44. Girishkumar, G., et al., *Single-wall carbon nanotube-based proton exchange membrane assembly for hydrogen fuel cells.* Langmuir, 2005. 21(18): p. 8487–8494.
45. Tang, H.L., et al., *Self-assembling multi-layer Pd nanoparticles onto Nafion(TM) membrane to reduce methanol crossover.* Colloids and Surfaces a-Physicochemical and Engineering Aspects, 2005. 262(1–3): p. 65–70.
46. Jiang, S.P., et al., *Self-assembly of PDDA-Pt nanoparticle/Nafion membranes for direct methanol fuel cells.* Electrochemical and Solid State Letters, 2005. 8(11): p. A574–A576.
47. Polotskaya, G., Y. Biryulin, and V. Rozanov, *Asymmetric membranes based on fullerene-containing polyphenylene oxide.* Fullerenes Nanotubes and Carbon Nanostructures, 2004. 12(1–2): p. 371–376.
48. Luo, M.L., et al., *Hydrophilic modification of poly(ether sulfone) ultrafiltration membrane surface by self-assembly of TiO_2 nanoparticles.* Applied Surface Science, 2005. 249(1–4): p. 76–84.
49. Kwak, S.Y., S.H. Kim, and S.S. Kim, *Hybrid organic/inorganic reverse osmosis (RO) membrane for bactericidal anti-fouling. 1. Preparation and characterization of TiO_2 nanoparticle self-assembled aromatic polyamide thin-film-composite (TFC) membrane.* Environmental Science & Technology, 2001. 35(11): p. 2388–2394.
50. Bae, T.H., and T.M. Tak, *Effect of TiO_2 nanoparticles on fouling mitigation of ultrafiltration membranes for activated sludge filtration.* Journal of Membrane Science, 2005. 249(1–2): p. 1–8.
51. Lehmam, A., H. Brunner, and G.E.M. Tovar, *Selective separations and hydrodynamic studies: a new approach using molecularly imprinted nanosphere composite membranes.* Desalination, 2002. 149(1–3): p. 315–321.
52. Cortalezzi, M.M., C. V., and M.R. Wiesner, *Controlling nanoparticle template morphology: effect of solvent chemistry.* Colloid and Interface Science, 2005. 283: p. 366–372.
53. Kralchevsky, P.A., et al., *Capillary meniscus interaction between colloidal particles attached to a liquid-fluid interface.* Journal of Colloid and Interface Science, 1992. 151(1): p. 79–94.

Nanomaterials as Adsorbents

Mélanie Auffan   *University of Aix-Marseille, Aix-en-Provence, France*
Heather J. Shipley   *Rice University, Houston, Texas, USA*
Sujin Yean   *Rice University, Houston, Texas, USA*
Amy T. Kan   *Rice University, Houston, Texas, USA*
Mason Tomson   *Rice University, Houston, Texas, USA*
Jerome Rose   *CNRS-University of Aix-Marseille, Aix-en-Provence, France*
Jean-Yves Bottero   *CNRS-University of Aix-Marseille, Aix-en-Provence, France*

Introduction

In recent years, drinking water regulations have continued to lower the maximum contaminant level (MCL) for pollutants. For instance, in 2002, the World Health Organization (WHO) decided to reduce the arsenic standard for drinking water from 50 µg/L to 10 µg/L. The stiffening of regulations generates strong demands to improve methods for removing pollutants from the water and controlling water-treatment residuals. Currently, a wide range of physico-chemical and biological methods are used and studied for the removal of organic and/or inorganic contaminants from polluted waters (Sheoran and Sheoran, 2006). Coagulation-flocculation, membrane processes, and adsorption are the most common methods of contaminant removal. The most efficient and low cost process for the removal of colloids and organics in water treatment is the use of inorganic salts as coagulation-flocculation agents such as Al_{13} (Bottero et al., 1980; Bottero et al., 1982) and Fe_{24} (Bottero et al., 1993; Bottero et al., 1994) polycation species. However, this approach has two disadvantages: a higher volume of sludge generated and difficulty in recovering the metals for reuse. For water treatment, it is an advantage to use a decontamination process that does not generate residuals. This is the case of the Magnetically Assisted Chemical Separation (MACS) process (Ngomsik et al., 2005). MACS is a useful decontamination technique widely used for water and liquid waste treatment involving superparamagnetic particles (iron oxide microspheres of 0.1 to 25 μm of diameter). Superparamagnetism is very important for recovering and regenerating particles after adsorption of the pollutant. With MACS, no residuals are produced and the microparticles can be reused. However, even if micron-sized adsorbents have an internal porosity that increases their specific surface area (SSA), the diffusion limitation within the particles leads to a decrease in adsorption efficiency. An efficient system to remove containments from solution would consist of particles with large surface area, small diffusion resistance, superparamagnetic properties, and high reactivity and affinity for adsorbates.

Research has shown that nanoparticles represent a new generation of environmental remediation technologies that could provide cost-effective solutions to some of the most challenging environmental cleanup problems: pollution monitoring (Riu et al., 2006), groundwater (Liu et al., 2005b), and soil remediation (Zhang, 2003). For instance, metal iron nanoparticles are used in contaminated aquifers and soils for the transformation and decontamination of a wide variety of environmental pollutants, such as chlorinated organic solvents, pesticides, and metals or metalloids. The use of magnetic nanoparticles is also becoming promising for the adsorption of polluted ions during water and industrial liquid waste treatments. Indeed, magnetic nanoparticles exhibit properties (large surface area, small diffusion resistance, and superparamagnetic properties) that make them excellent candidates for containment removal from polluted water. This chapter presents results that demonstrate the effectiveness of using magnetic nanoparticles, as nano-adsorbents, for the removal of organic and inorganic ions during water-treatment processes. Also, this chapter gives insight into the nano-size effect on adsorption efficiency and experimental techniques used to investigate the magnetic nanoparticle/pollutant interactions.

Adsorption at the Oxide Nanoparticles/Solution Interface

As the dimensions of metal oxide particles decrease to the nanometer range, there are significant changes in optical and electronic properties due to both quantum and size effects. This is mainly due to the increasing role of the surface in controlling the overall energy of the particles. For instance, dissolution of nanoparticles is a fascinating issue. For 1 mm macrocrystalline quartz, dissolution kinetics would be about 34×10^6 years, while nano-cube quartz with an edge of 10 nm should dissolve in 1.07 seconds under identical conditions; however this is not observed, in general (Bertone et al., 2003; Lasaga, 1998). The remarkable stability of nanoparticles in solution might be explained by the lack of defects on the surface, the strong surface passivation, or the altered surface composition due to adsorption of surfactants or oxidation, as seen with magnetite (Shipley et al., 2006). While many of the unusual physical properties of the nanocrystalline materials are now well known (Banfield and Navrotsky, 2003), the influence of size on adsorption, chemical reactivity, and on the nanoparticles/solution interface needs to be further understood.

Size effect on adsorption capacity

As a particle shrinks to the nanometer range, an increasing fraction of atoms are exposed to the surface, giving rise to excess energy. Consequently, nanoparticles are thermodynamically metastable compared to macrocrystalline materials. They tend to approach the minimum free energy state (equilibrium state) through several ways: phase transformation, crystal growth, surface structural changes, aggregation, and surface adsorption (Banfield and Navrotsky, 2003; Rusanov, 2005). Therefore, nanoparticles with a higher total energy should be more prone to adsorb molecules onto their surfaces in order to decrease the total free energy. Hence, adsorption should be favored on nanoparticles (Banfield and Navrotsky, 2003). This is the case for the adsorption of arsenic at the surface of iron oxide nanoparticles. For instance, magnetide with a diameter of 11 nm adsorbs 3 times more arsenic per nm^2 than does magnetide of 20 nm or 300 nm.

However, some experimental results have shown that the sorption capacity of inorganic nanoparticles is not always size-dependent for a given surface area. This is the case for the adsorption of organic acids (valeric, acetic, adipic, and oxalic acids) at the surface of titanium nanoparticles (Zhang et al., 1999). This study explicitly demonstrated that when corrected for surface area there is no size-dependence, while for a given mass of TiO_2 nanoparticles that are 6 nm in diameter are more prone to adsorb organic molecules than nanoparticles with 16 nm in diameter (Table 10.1).

In this case a greater interest of using nano-sized particles as adsorbents is their higher specific surface area (SSA). For instance, the SSA of an oxide nanoparticle of 10 nm in diameter is $\approx$100 times larger than the SSA of an oxide particle of 1 μm. It is well known that the surface hydroxyl groups are the chemically reactive entities at the surface of the solid in an aqueous environment. A higher SSA increases the number of available functional groups on the nanoparticle surface. Consequently, for a given mass, the maximum adsorption capacity of ions in solution is higher for nanoparticles than for micron-sized particles.

TABLE 10.1 Adsorption Capacity of Oxide Particles

Adsorbents mineral	Size (nm)	SSA* (m^2/g)	Adsorbates	Adsorption ($\mu mol/m^2$)	Capacity (mmol/g)	Experimental conditions	References
Fe_3O_4	300	3.7	As^{III}	5.62	0.021	$pH = 8.0$	(Yean et al., 2005)
Fe_3O_4	20	60	As^{III}	6.48	0.388	$pH = 8.0$	
Fe_3O_4	11.72	98.8	As^{III}	15.49	1.532		
				18.22	1.800	$pH = 8.0$	
Fe_3O_4	300	3.7	As^{V}	3.89	0.014	$pH = 4.8$	
				2.70	0.010	$pH = 6.1$	
Fe_3O_4	20	60	As^{V}	2.54	0.152	$pH = 4.8$	
				1.69	0.101	$pH = 6.1$	
				1.32	0.079	$pH = 8.0$	
Fe_3O_4	11.72	98.8	As^{V}	6.30	0.623	$pH = 8.0$	(Tamura et al., 1997)
				23.30	2.300	$pH = 7.0$ $IS = 0.1\ M$	
Fe_3O_4	1000	1.73	Co^{II}	1.6	0.003	$pH = 6.9$ $IS = 0.01\ M$	(Uheida et al., 2006)
Fe_3O_4	11.3	97.5	Co^{II}	58	5.655		
Fe_3O_4 + Chitosan	13.5	(86.43)	Co^{II}	–	0.467	$pH = 5.5$	(Chang et al., 2006)
Fe_3O_4 + Chitosan	13.5	(86.43)	Cu^{II}	–	0.374	$pH = 5.5$	(Chang and Chen, 2005)
α-Fe_2O_3	10000	15.9	Co^{II}	0.2	0.030	$pH = 6.5$ $IS = 0.1\ M$	(Tamura et al., 1997)
γ-Fe_2O_3	10	(114.50)	Co^{II}	37	3.601	$pH = 6.9$ $IS = 0.01\ M$	(Uheida et al., 2006)
γ-Fe_2O_3	6	174	$As^{III,V}$	10	1.810	$pH = 7.4$ $IS = 0.2\ M$	(Auffan et al., 2007)
β-$FeOOH$	2.6	330	As^{V}	5.4	1.79	$pH = 7.5$ $IS = 0.1\ M$	(Deliyanni et al., 2003)

HFO	—	600	As^V	2.2	1.34		(Wilkie and Hering, 1996)
TiO$_2$	6	260	Valeric acid	1.23	0.319	$pH = 4.0$ $IS = 0.01\,M$ pH = 3–4.5	
TiO$_2$	16	80	Valeric acid	1.51	0.121	pH = 3–4.5	(Zhang et al., 1999)
TiO$_2$	6	260	Acetic acid	1.59	0.415	pH = 3–4.5	
TiO$_2$	16	80	Acetic acid	1.66	0.133	pH = 3–4.5	
TiO$_2$	6	260	Adipic acid	1.08	0.281	pH = 2.5–4	
TiO$_2$	16	80	Adipic acid	1.08	0.086	pH = 2.5–4	
TiO$_2$	6	260	Oxalic acid	2.13	0.553	pH = 1–3	
TiO$_2$	16	80	Oxalic acid	2.13	0.169	pH = 1–3	

* SSA in parentheses is the calculated SSA from the particle size.

Adsorption/desorption experiments

The maximum adsorption capacity of an adsorbent is usually evaluated using adsorption isotherms. The results of adsorption experiments can be presented as a curve showing the amount of adsorbate taken up per unit weight or area of the adsorbent (in g or mol of adsorbates per g or m^2 of adsorbent) versus the equilibrium concentration remaining in solution. The pH, temperature, ionic strength, and concentration of the solid phase are maintained constant during the experiment. Numerous analytical forms of the adsorption isotherms have been derived (e.g., Langmuir, Freundlich, Linear) that invoke differing underlying assumptions, such as the existence of multiple adsorption layers, uniformity of the adsorption bond, and interactions between adsorption sites, to name a few (Sawyer et al., 2003; Stumm and Morgan, 1996). In macroscopic systems, adsorption data can be fitted with these equations in order to deduce thermodynamic and energetic information. However, they cannot lead to a thorough description of the different mechanisms in the case of a heterogeneous and finite surface.

Many theories and models have been used to express desorption hysteresis of hydrophobic organic compounds. These models express sorption and desorption as different processes, such as the two-compartment desorption model. The basic concept of the model is that both sorption and desorption are biphasic, consisting of two compartments, each with unique equilibrium and kinetic characteristics. A model that accomplishes this is the following equation (Chen et al., 1999; Cheng et al., 2005; Kan et al., 1998):

$$q = K_d^{1st}C_w + \frac{K_d^{2nd}fq_{max}^{2nd}C_w}{fq_{max}^{2nd} + K_d^{2nd}C_w}$$

where K_d^{1st} and K_d^{2nd} are solid-water distribution coefficients for the first and second compartment; and q_{max}^{2nd} (μg/g) is the maximum sorption capacity for the second compartment. The factor f ($0 \leq f \leq 1$) denotes the fraction of the second compartment that is filled at the time of exposure. At higher aqueous concentrations, q is related to aqueous concentration by a linear isotherm, $K_d^{1st}C_w$, and at lower concentrations by a Langmuir-type isotherm that becomes linear, $K_d^{2nd}C_w$ (Chen et al., 1999; Cheng et al., 2005; Kan et al., 1998).

Adsorption mechanisms at the atomic scale

A large body of work has been done on the adsorption mechanisms of ions onto metal oxide surfaces from solution (Al-Abadleh and Grassian, 2003; Arai et al., 2001; Cornell and Schwertmann, 1996; Manceau, 1995; Stipp et al., 2002; Waychunas et al., 1993; Waychunas et al., 1995). The surface hydroxyl groups ($\equiv$MOH) are the functional groups of metal oxides. They have a double pair of electrons with a dissociable hydrogen atom that enables the metal oxide to react with ions in solution and to form surface complexes. The following are acid-base equilibria reactions of metal oxides in water (K_1 and K_2: equilibrium constants that represent the concentration of surface species [moles/kg] of metal oxide) (Cornell and Schwertmann, 1996; Sawyer et al., 2003).

$$\equiv MOH \rightarrow \equiv MO^- + H^+ \qquad \text{where } K_1 = [\equiv MO^-][H^+]/[\equiv MOH]$$

$$\equiv MOH + H^+ \rightarrow \equiv MOH_2^+ \quad \text{where } K_2 = [\equiv MOH_2^+]/[\equiv MOH][H^+]$$

Surface sensitive techniques can be used to provide direct information about the mode of attachment of adsorbates to the surface of nanoparticles at the atomic scale. In general, electron-based spectroscopy, vibrational spectroscopy, and synchrotron-based X-ray techniques are among the common structural methods in use (see Chapter 4). Spectroscopic studies reveal that sorption on a crystalline surface may result from the molecular mechanisms shown in Figure 10.1.

It is difficult to precisely study sorption reactions without spectroscopic evidence. Spectroscopic data provide a better understanding of the adsorption mechanisms at the atomic scale and enable more meaningful interpretation of the adsorption isotherm experiments. In addition to the increasing importance and the remarkable improvement of the experimental surface methods, theoretical calculations are also becoming an indispensable tool in chemical research, particularly in the case of nano-systems. The computational and calorimetric techniques yield rich insight into the structure and the adsorption processes at the surface of the nanoparticles.

Nanomaterial-Based Adsorbents for Water and Wastewater Treatment

As an example of the use of nanomaterials as adsorbents in water treatment, we will consider the problem of arsenic removal from water. Arsenic in drinking water is a priority pollutant of considerable interest worldwide. In the Bangladesh crisis, approximately 30 to 36 million people are exposed to carcinogenic levels of arsenic (Bates et al., 1992; Hossain, 2006). Arsenic mobility in aqueous media and its removal by various water treatment technologies have been the subject of considerable study. Much of this work has addressed the adsorption of arsenic on iron-containing solids. The sorption of inorganic or organic contaminants at the surface of iron (hydr)oxide minerals may affect the mobility, reactivity, bioavailability, and toxicity of these particles in natural waters (Al-Abadleh and Grassian, 2003; Johnson and Hallberg, 2005; Stipp et al., 2002).

Outer-sphere complex: if the ion is attracted to the surface via long-range coulombic forces but retains its water of hydration.

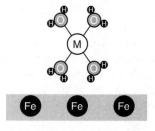

Inner-sphere complex: if the ion loses some of its hydration water molecules and becomes chemically bound to the surface via short-range forces.

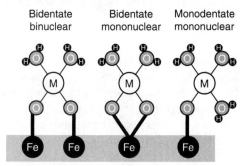

Bidentate binuclear   Bidentate mononuclear   Monodentate mononuclear

Surface precipitation: 3D species nucleate and grow on the surface.

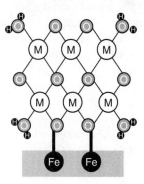

Figure 10.1   Modes of ligand coordination to the oxide surface.

Recently, several studies have focused on the removal of (in)organic contaminants from polluted water, using nano-sized iron oxide particles. Iron oxide magnetic nanoparticles have been chosen as promising adsorbents because: (1) they can be produced in large quantities using physicochemical methods, (2) it is expected that their adsorption capacity and affinity for pollutants is higher considering the larger surface area and possibly highly active surface sites, (3) the separation of metal-loaded magnetic nano-adsorbent from treated water can be achieved via an external magnetic field (Prodan et al., 1999), and (4) nanoparticles might be regenerated and reused (Ngomsik et al., 2005). Also, iron oxide nanoparticles can be surface-coated with organic compounds to increase their sorption efficiency and specificity for pollutants. In this case, it is worth noting that the covalent attachment of molecules to the surface of nanoparticles and the stability of the coating are important. In the following sections, we consider the role and effectiveness of nano-sized magnetic particles, including maghemite (γ-Fe_2O_3, a Fe^{III} oxide) and magnetite (Fe_3O_4, a mixed Fe^{II} and Fe^{III} oxide) in the adsorption of inorganic and organic ions from aqueous solution, focusing on the interactions between these nanoparticles, arsenite (As^{III}) and arsenate (As^V).

Iron oxide nanoparticles for metals and metalloids removal

Numerous papers have shown that iron oxides have a high affinity for arsenite and arsenate (Dixit and Hering, 2003; Pierce and Moore, 1982; Raven et al., 1998). Direct evidence of inner sphere adsorption of arsenic to iron oxides has been provided by extended X-ray absorption fine structure and infrared spectroscopy (XAS, FTIR) spectroscopy. This adsorption is mainly controlled by the surface properties of the absorbate, the pH on the solution, and the oxidation state of arsenic.

Maghemite nanoparticles for arsenic removal.   To study the efficiency of iron oxide nanoparticles to treat arsenic contaminated water, the adsorption mechanisms of arsenic ions onto maghemite nanoparticles (6 nm) (Jolivet et al., 2002) were quantitatively and qualitatively examined. The experimental data of a typical adsorption isotherm are presented in Figure 10.2. The maximum sorption capacity of arsenite was found to

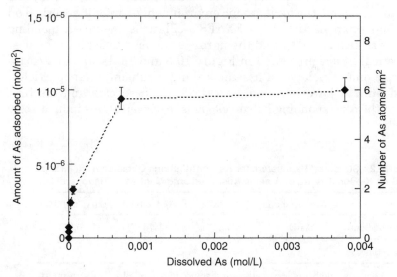

Figure 10.2 Experimental adsorption isotherm of As onto maghemite nanoparticles. IS = 0.2 M NaCl, pH = 7.4, adsorbent concentration is 27 mg·L^{-1}, contact time is 24 h at 298 K.

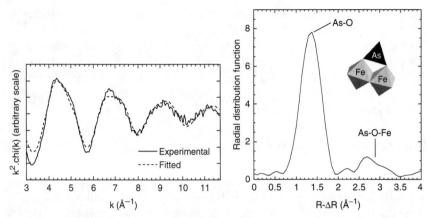

Figure 10.3 EXAFS As K-edge spectra (left) and its corresponding Fourier transform (right) of arsenic adsorbed onto maghemite nanoparticles. Experiments performed on the FAME beamline at the ESRF synchrotron (Grenoble, France).

be of the order of 1.0×10^{-5} mol·m^{-2} (or 1.8 mmol·g^{-1} of maghemite or ~6 As atoms/nm^2). This value can be compared (Table 10.4) with those obtained with β-FeOOH or HFO taking into account that the specific surface area of HFO is certainly lower than 600 m^2/g. The value of ~6 As atoms/nm^2 corresponds to a complete fulfillment of the surface of the maghemite nanoparticles.

EXAFS spectroscopy at the As K-edge (see Chapter 4, Rose et al.) was used to determine the local atomic environment of arsenic adsorbed on maghemite nanoparticles. The EXAFS oscillations, their corresponding Fourier transform (FT), and the parameters used to achieve the best theoretical fit are presented in Figure 10.3 and Table 10.2 and correspond to one monolayer of arsenic around maghemite nanoparticles. The dominant lower frequency on the EXAFS spectra and the first peak in the corresponding FT corresponds to oxygen atoms immediately

TABLE 10.2   Structural Parameters for As Contributions Obtained from Fitting EXAFS As K-edge Spectra of Arsenic Adsorbed onto Maghemite Nanoparticles

Samples	Atomic shells	N ± 20%	R(Å) ± 0.02 Å	σ (Å) ± 0.01 Å
As adsorbed onto maghemite nanoparticles	As-O	3.2	1.76	0.061
	As-Fe*	1.7	3.33	0.102

N represents the number of backscatters at distance R; σ, the Debye-Waller term, is the disorder parameter for the absorber-backscatterer pair.

* Corresponds to iron atoms in the second coordination sphere of As.

surrounding the arsenic. In agreement with the literature and the study of reference compounds, the presence of 3.2 oxygen atoms at 1.76 Å, highlights that arsenic remains under the trivalent form during adsorption at the maghemite nanoparticles surface and XAS experiments. The second peak on the FT corresponds to the presence of iron atoms in the second coordination sphere of As. Previous EXAFS studies have described the adsorption of As^V and As^{III} onto iron oxide using multiple As-Fe shells: As-Fe = 3.55–3.60 Å for monodentate mononuclear complex, As-Fe = 3.25–3.40 Å for bidentate binuclear complex, and As-Fe = 2.8–2.9 Å for bidentate mononuclear complex (Arai et al., 2001; Ladeira et al., 2001; Manceau, 1995; Manning et al., 1998; Manning and Goldberg, 1996; Pierce and Moore, 1982; Sherman and Randall, 2003; Thoral et al., 2005; Waychunas et al., 1993; Waychunas et al., 1995). In the case of arsenate, the distance As-Fe at 2.8–2.9 Å is controversial and several authors suggest that it corresponds to multiple scattering of the photoelectron involving O-O pairs with the $As^V O_4$ tetrahedron (Arai et al., 2001; Manceau, 1995; Manning et al., 1998; Waychunas et al., 1995; Waychunas et al., 1993). In our particular case, a contribution of 1.7 iron atoms at 3.33 Å from As is necessary to satisfactorily reproduce the experimental spectra. This As-Fe distance is very close to the average As^{III}-Fe^{III} inter-atomic distances of 3.25–3.40 Å attributed to As^{III}-O-Fe linkages through double corner sharing.

Magnetite nanoparticles for arsenic removal. The effectiveness of using nanoscale magnetite to treat arsenic-contaminated drinking water was investigated. The adsorption and desorption behavior of arsenite and arsenate to magnetite nanoparticles brings some very interesting responses on the size effect on the surface properties. Three types of magnetite nanoparticles were studied: 300 nm (3.7 m^2/g) and 20 nm (60 m^2/g) commercially made nano-magnetites, and 11.72 nm (98.8 m^2/g) monodispersed magnetite synthesized in the laboratory.

The adsorption of As^{III} to 20 nm and 300 nm magnetite nanoparticles is dependent on pH and SSA (Table 10.1). The adsorption of As^V to 20 nm and 300 nm nanoscale magnetite decreases with increasing pH, because arsenate ionicity is pH dependent and the surface of the magnetite is positively charged at pH values below 6.8 (the point of zero charge for magnetite). The adsorption isotherm (Figure 10.4) of As^{III} and As^V onto 11.72 nm magnetite reveals dramatic changes (Al-Abadleh and Grassian, 2003; Axe and Trivedi, 2002; Dixit and Hering, 2003; Zhang et al., 1999). Compared with other literature results (Dixit and Hering, 2003; Morel and Hering, 1993), the maximum adsorption densities for 11.72 nm magnetite are very high (~18 and ~23 µmol/m^2 or ~10 and ~15 As atoms/nm^2).

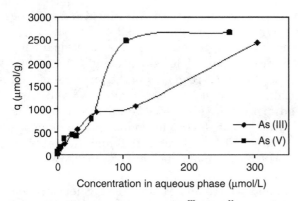

Figure 10.4   Plot of the adsorption of As^{III} and As^{V} to 11.72 nm magnetite at pH 8 (Yean et al., 2005).

Arsenite and arsenate adsorption experiments were also performed with 20 nm magnetite with Lake Houston water (pH 7.8) and 0.01 M $NaNO_3$ electrolyte solution (pH 8.0). The sorption of arsenite and arsenate to 20 nm magnetite was lower in the presence of Lake Houston water compared to the electrolyte-only solution. Literature has shown that arsenic sorption to mineral surfaces decreases in presence of competitive ions such as phosphate, sulfate, nitrate, natural organic matter (NOM), and molybdate (Appelo et al., 2002; Jackson and Miller, 2000; Manning and Goldberg, 1996; Munoz et al., 2002; Su and Puls, 2001; Swedlund and Webster, 1999; Violante and Pigna, 2002). Therefore, the decreased arsenite and arsenate adsorption to 20 nm magnetite was probably due to the NOM, since the other competing ion concentrations in Lake Houston water were negligible (Yean et al., 2005).

Desorption experiments were conducted on 20 nm and 300 nm nanoscale magnetite. Literature shows that desorption hysteresis occurs for many gas-solid and liquid-solid interactions (Ainsworth et al., 1994; Gao et al., 1993; Stumm and Morgan, 1996; Yin et al., 1997). Genc-Fuhrman et al. (2004) observed that arsenate desorption rate from activated neutralized red mud greatly increased when the pH increased (pH>9), but was irreversible at common environmental pH ranges. For 300 nm magnetite at pH 6.1, desorption of As^{III} and As^{V} appears to be hysteretic, only 20–25 percent of adsorbed As^{III} and As^{V} desorbs. From 20 nm magnetite, desorption of As^{III} and As^{V} was almost completely hysteretic, only 1 percent of the adsorbed As^{III} and As^{V} desorbs. Similar phenomena were seen at pH 4.8 and 8.0.

This study revealed that a critical size in the 12–20 nm range corresponds to a dramatic change of the surface structure. The very large adsorption density increase could be due to the formation of a porous surface lower than this critical size due to the large increase of the surface stress. This hypothesis could explain the desorption hysteresis, which increases as the size decreases.

Iron oxide–doped TiO$_2$ nanoparticles for the degradation of organic pollutants

Recently, photocatalytic oxidation with TiO$_2$ nanoparticles (6–20 nm) has been investigated as a promising water-treatment process. When irradiated with UV light, TiO$_2$ nanoparticles can adsorb and degrade a wide variety of environmental organic pollutants (Bianco Prevot and Pramauro, 1999; Bianco Prevot et al., 1999a; Bianco Prevot et al., 1999b; Pramauro et al., 1998). For instance, the strong affinity between the surface of TiO$_2$ nanoparticles for arsenic organic species (monomethylarsonic [MMA] and dimethylarsinic [DMA] acids) were shown by EXAFS (Jing et al., 2005). Results show that both MMA and DMA are covalently bounded to the surface of nanoparticles through bidentate (As$_{MMA}$-Ti = 3.32 Å) and monodentate (As$_{DMA}$-Ti = 3.37 Å) inner sphere complexes, respectively.

In treating organic pollutants in water, there are some problems that arise from using TiO$_2$ nanoparticles. Two limiting factors are the low efficiency of the utilization of visible light and the recombination between the photogenerated electrons and holes, even if the kinetics of degradation of organic pollutants are increased in presence of TiO$_2$ nanoparticles. It has been reported that doping TiO$_2$ nanoparticles with ions (Zn^{2+}, Mn^{2+}, Al^{3+}, K$^+$, Fe^{3+} . . .) at 0.1–0.5 percent may significantly increase the photocatalytic activity (Bessekhouad et al., 2004; Choi et al., 1994; Liu et al., 2005a). The doped ions act as charge separators of the photoinduced electron–hole pair and enhanced interfacial charge transfers. For instance, Liu et al., (2004) have studied the photocatalytic activity of TiO$_2$ nanoparticles doped with ZnFe$_2$O$_4$ (20 nm). The ZnFe$_2$O$_4$ doping strongly enhanced the photocatalytic activity of TiO$_2$ nanoparticles and improved the degradation of chlorinated pesticide (Liu et al., 2004).

The main problem is that the separation of the nano-photocatalyst from the polluted aqueous media is difficult and TiO$_2$ suspensions are not easily regenerated. To get around these difficulties, iron oxide nanoparticles (Fe$_3$O$_4$, CoFe$_2$O$_3$, or (Ba,S,Pb)Fe$_{12}$O$_{19}$) have recently been coated with TiO$_2$ to synthesize magnetic photocatalytic nanoparticles (Fu et al., 2006). In such core-shell nano-structures (20–100 nm), the magnetic core is useful for the recovery of nano-photocatalyst from the treated water stream by applying an external magnetic field while a TiO$_2$ outer shell is used to destroy organic contaminants in wastewaters (Fu et al., 2006). The topic of photocatalytic degradation of organic contaminants is discussed in detail in Chapter 5.

Interaction of organic pollutants
with carbon nanomaterials

Recently, interest has been focused on carbon nanomaterials, such as single wall nanotubes and fullerenes, because of their unique chemical, physical, and mechanical properties. Possible applications for these carbon nanomaterials are in energy, space, medicine, industry, and environmental applications, as well as concern about their environmental impact (Borm, 2002; Dagani, 2003). Fullerene (C_{60}) interactions with common environmental contaminants, such as naphthalene and 1,2-dichlorobenzene, have been studied on C_{60} large aggregates (20–50 µm), C_{60} small aggregates (1–3 µm), and nC_{60} (about 100 nm) (Figure 10.5). A linear isotherm ($K_d = 10^{2.39\pm0.02}C$), a Freundlich isotherm ($q =$

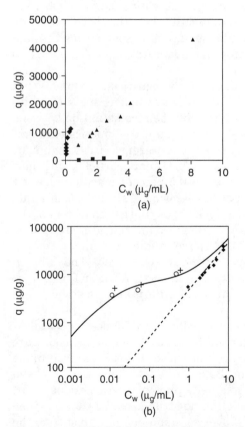

Figure 10.5 (a) Naphthalene adsorption to C_{60} large aggregates ■, C_{60} small aggregates ◆, nC_{60} ▲ (b) Adsorption and desorption of naphthalene with nC_{60} ◆; o and +, desorption of naphthalene from nC_{60}. Straight line, a linear isotherm in the form of q(µg/g) = 103.75 Cw (µg/mL); upper curve, model fitting curve assuming two-compartment desorption model. (Cheng et al., 2005).

**TABLE 10.3   DED Model Parameters for Naphthalene and 1,2-
Dichlorobenzene Adsorption and Desorption with nC_{60} (Cheng
et al., 2005)**

Sorbate	[a]$\text{Log } K_d^{1st}$(ml/g)	[a]$\text{Log } K_d^{2nd}$(ml/g)	[a]$\text{Log } q_{max}^{2nd}$($\mu$g/g)	R^2
Naphthalene	3.75±0.01	5.9±0.04	3.91±0.01	0.996
1,2-DCB	3.48±0.01	5.68±0.08	3.98±0.03	0.977

a adsorption data used to determine K_d^{1st} from first term in Eq. 6. Desorption
data and K_d^{1st} determined from adsorption data used to find K_d^{2nd} and q_{max}^{2nd} by
fitting the data to Eq. 6.

$10^{4.28\pm0.04}$ $C_w^{0.45\pm0.05}$), and a linear isotherm ($K_d = 10^{3.7\pm0.01}C$) were used
to fit the data of the C_{60} large aggregates, small aggregates, and the nC_{60}
forms, respectively. The data suggest that adsorption of naphthalene to
nC_{60} is similar to other forms of carbon. Desorption of the naphthalene
from the nC_{60} was highly hysterertic. The K_d value for desorption
increased by about two orders of magnitude higher than the corresponding adsorption value. These data were fitted with a two-compartment
desorption model (Kan et al., 1998) (Figure 10.5 and Table 10.3).
Adsorption and desorption experiments were also performed with 1,2-
dichlorobenzene to nC_{60}. The adsorption and desorption data were fitted
with a linear isotherm ($q = 10^{3.48\pm0.01}$ C_w) and with the two-compartment
desorption model, respectively (Table 10.2).

The experimental data show that 1,2-dichlorobenzene and naphthalene adsorption and desorption with nC_{60} are similar. The different
sizes of the C_{60} aggregates affect the adsorption of naphthalene by
orders of magnitude. The desorption of the environmental contaminants from nC_{60} exhibits hysteresis. Kinetic data also showed that desorption of naphthalene from C_{60} aggregates is composed of two
compartments: a labile compartment, where naphthalene is probably
adsorbed on the outside surface and can readily be desorbed, and a
resistant desorption compartment, where naphthalene may be
entrapped either in the C_{60} aggregates or in the surface crevices and
desorption is hindered.

There is great potential for the use of inorganic nanoparticles, such
as TiO_2, carbon or metallic nanoparticles (see Chapters 5 and 8), for
the treatment of organic pollutants in water treatment and the
environment. These nanoparticles have several advantages over the
current microparticles used, such as high surface areas, higher affinity and surface reactivity, faster degradation rates, photocatalytic
abilities, lower cost, and higher efficiency. With more fundamental
research and development of technological applications, these inorganic nanoparticles have the ability to be used for environmental
applications.

Inorganic nanoparticles with an organic shell

The combination of inorganic nanoparticles with an organic shell has attracted considerable attention because of the potential applications in many fields, such as separation processes, optoelectronics, catalysts and sensors, biotechnology, medical diagnostics, and therapy. Several nano-systems using iron oxide nanoparticles as the core and an organic layer on the surface have been studied to improve the water and liquid waste treatment. In such systems, the binding strength between the coating layer and the surface of nanoparticles and the affinity of the organic surface for specific pollutants play preponderant roles (Yamaura et al., 2002).

Organic-coated magnetite nanoparticles.   Attention has been given to the functionalization of the surface of magnetite nanoparticles to develop new nano-adsorbents for the removal of ions from solution. For instance, Liao et al. (2003) coated magnetite nanoparticles (12 nm) with polyacrylic acid (PAA). The PAA coating was chosen for its strong affinity for cationic solutes with large molecular weights. The PAA coating layer attaches to the surface of magnetite nanoparticles through covalent linkage (Liao et al., 2003). This nano-adsorbent exhibits a high adsorption capacity and fast adsorption and desorption rates (Huang et al., in press, Liao et al., 2003; Liu et al., 1999). However, the PAA surface layer is not effective for the adsorption of metallic ions. Therefore, magnetite nanoparticles (13.5 nm) were surface-functionalized with chitosan (Chang et al., 2006; Chang and Chen, 2005). Chitosan is a natural polysaccharide with many useful characteristics, such as hydrophilicity, antibacterial properties, and affinity for heavy metal ions. Vibrational spectroscopic experiments have demonstrated the covalent attachment of the carboxymethyl chitosan at the surface of magnetite nanoparticles through the carboxylate functions. The resulting chitosan-coated nanoparticles were efficient for the fast removal of Co^{II} (0.47 mmol/g) and Cu^{II} ions (0.34 mmol/g) from aqueous solution (Chang et al., 2006; Chang and Chen, 2005). As for the PAA-coated nanoparticles, the kinetics of adsorption and desorption reactions were fast—one minute to reach equilibrium for Cu^{II} ions. The fast removal of ion from solution can be explained by the small internal diffusion within the structure of the nanoparticles. The chitosan coating may prevent a direct contact between the contaminants and the surface of the nanoparticle. Moreover, magnetic analysis indicates that once chitosan-coated, magnetite nanoparticles remain superparamagnetic, which could allow them to be separated from water under a magnetic field and reused.

DMSA-coated maghemite nanoparticles.   To improve the efficiency of maghemite nanoparticles in the removal of arsenic from polluted waters, these nanoparticles were surface-coated with DMSA [meso-2,3-dimercaptosuccinic acid, $(SH)_2(CH)_2(COOH)_2$]. DMSA is an effective

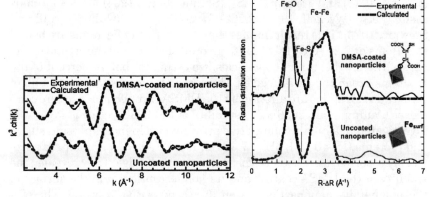

Figure 10.6 EXAFS Fe K-edge spectra (left) and its corresponding Fourier transform (right) of uncoated and DMSA-coated maghemite nanoparticles. Experiments performed on the beamlines 11.1 on the ELETTRA synchrotron (Trieste, Italy) (adapted from Auffan et al., 2006).

chelating agent, which has been used as a therapeutic antidote to heavy metal toxicity since the 1950s (Domingo, 1995). DMSA has been chosen as a coating agent for several reasons: (1) its lack of toxicity (Kramer et al., 2003); (2) its high efficiency to chelate heavy metals (Pb, Hg, Cd, As) (Domingo, 1995); and (3) because once DMSA is adsorbed on the surface of maghemite nanoparticles the colloidal suspension is stable over a broad range of pH values (3–11 pH) and ionic strengths (IS <0.35mol/L) (Fauconnier et al., 1997). The stability and the nature of the interactions between DMSA and the surface of the nanoparticles have been studied by EXAFS at the Fe K-edge (7.112 KeV) (Auffan et al., 2006). The EXAFS oscillations, their corresponding FT, and the parameters used to achieve the best theoretical fit are presented in Figure 10.6 and Table 10.4. The main difference between EXAFS results of the uncoated and DMSA-coated maghemite nanoparticles is the presence of 0.5 sulfur atoms in the first coordination sphere of iron (Fe-S = 2.21 ± 0.02 Å) after the coating. Such interatomic distance exists in iron sulfur minerals: 2.15 Å for pyrite (FeS_2)

TABLE 10.4   Structural Parameters for Fe Contributions Obtained from Fitting EXAFS Spectra of Uncoated and DMSA-Coated Maghemite Nanoparticles (adapted from Auffan et al., 2006)

Samples	Atomic shells	N ±20%	R(Å) ±0.02Å	σ (Å) ±0.01 Å
Uncoated maghemite	Fe-O	5.0	1.96	0.107
nanoparticles	Fe- -Fe	3.9	2.97	0.101
	Fe- -Fe	8.5	3.43	0.105
DMSA-coated maghemite	Fe-O	4.9	1.95	0.090
nanoparticles	Fe-S	0.5	2.21	0.010
	Fe- -Fe	4.8	2.96	0.108
	Fe- -Fe	7.3	3.44	0.094

(Bayliss, 1977) and greigite (Fe_3S_4) (Skinner et al., 1964), 2.24 Å for marcasite (FeS_2) (Wyckoff, 1963), 2.26 Å for chalcopyrite ($CuFeS_2$) (Hall and Stewart, 1973). This result implies that 40% $\pm 10\%$ of Fe atoms are bound to S atoms at 2.21 Å. By taking into account the size and the crystal structure of the maghemite nanoparticles, we estimate that 40 percent of the Fe atoms are in the surface layer. Therefore, it appears that almost all of the surface Fe atoms are affected by the DMSA through inner sphere complexes (Auffan et al., 2006). The stability of the DMSA coating layer was investigated in a highly competitive medium rich in inorganic salt, sugar, and proteins ($IS \approx 0.2$ M; pH $= 7.4$). Similar EXAFS experiments, as previously reported, have been performed at the Fe K-edge. They revealed the stability of the DMSA coating and its nondesorption from the surface of the maghemite nanoparticles even in a competitive solution. The efficiency of DMSA-coated maghemite nanoparticles to treat arsenic contaminated water was evaluated using a typical adsorption isotherm experiment. The maximum sorption capacity of arsenic was found to be of the order of $2.5\ 10^{-5}$ mol·m^{-2} (or 4.6 mmol·g^{-1} of maghemite or ~15 As atoms/nm^2). This value is comparatively higher than previously reported for maghemite nanoparticles alone without any coating.

All these results show that the functionalization of the surface of inorganic nanoparticles offer excellent opportunities for selective removal of a wide array of target compounds from polluted water. As illustrated, the combination of iron oxide nanoparticles with an organic compound surface provides advantages for water treatment processes that cannot be attained separately by the inorganic nanoparticles or organic compounds alone: pollutant specificity, fastest adsorption/desorption rates, and magnetic removal (Cumbal et al., 2003).

Concluding Remarks

In recent years, serious problems with water contamination have produced high demands to improve methods for contaminant treatment in water and liquid waste, along with controlling water-treatment residuals. This chapter has shown that oxide nanoparticles (<20 nm) have the potential to be an efficient system to remove contaminants from solution due to their high surface area, small diffusion resistance, high reactivity, and high affinity to adsorbates. This chapter also shows that a lot of opportunities can exist for a variety of hybrid organic/inorganic oxide nanomaterials in the field of water and liquid waste treatment. The hybrid nanomaterials can be created with specific coatings that are selective for contaminants, such as DMSA with its affinity for As, Cd, or Pb. The superparamagnetic properties of iron oxide nanoparticles might allow for the removal of nanoparticles from water or liquid waste using a magnetic field. Also, the well-defined crystalline structure makes it possible to regenerate the surface for reuse along with not creating residuals. These unique properties of nanoparticles could be used to create a small water-treatment system that provides a low cost process and localized treatment. This could be useful in poor regions such as Bangladesh, Vietnam, and Latin America, which are exposed to high levels of arsenic in groundwater.

Acknowledgements

We acknowledge the French national "ACI-ECCOdyn" program supported by the CNRS. We also acknowledge the financial support from National Science Foundation through the Center for Biological and Environmental Nanotechnology [EEC-0118007], United States Environmental Protection Agency (US EPA) Office of Research and Development (ORD)/National Center for Environmental Research (NCER)/Science to Achieve Result (STAR) program, US EPA Hazardous Substance Research Center/South and Southwest Region.

References

Ainsworth, C.C., Pilou, J.L., Gassman, P.L., and Van der sluys, W.G., 1994. Cobalt, cadmium and lead sorption on hydrous iron oxide: Residence time effect. *Soil Sci. Soc. Am.,* 58:1615–1623.

Al-Abadleh, H.A., and Grassian, V.H., 2003. Oxide surfaces as environmental interfaces. *Surface Science Reports,* 52(3–4):63.

Appelo, C.A.J., Van der Weiden, M.J.J., Tournassat, C., and Charlet, L., 2002. Surface complexation of ferrous iron and carbonate on ferrihydrite and the mobilization of arsenic. *Environ. Sci. Technol.,* 36(14):3096–3103.

Arai, Y., Elzinga, E.J., and Sparks, D.L., 2001. X-ray absorption spectroscopic investigation of arsenite and arsenate adsorption at the aluminum oxide-water interface. *Journal of Colloid and Interface Science,* 235(1):80.

Auffan, M.; Decome, L.; Rose, J.; Orsiere, T.; DeMeo, M.; Briois, V.; Chaneac, C.; Olivi, L.; Berge-Lefranc, J.L.; Botta, A.; Wiesner, M.R.; Bottero, J.Y., In vitro interactions between DMSA-coated maghemite nanoparticles and human fibroblasts: A physicochemical and cyto-genotoxical study. *Environ. Sci. Technol.* 2006, 40, 4367–4373.

Axe, L., and Trivedi, P., 2002. Intraparticle surface diffusion of metal contaminants and their attenuation in microporous amorphous Al, Fe, and Mn oxides. *Journal of Colloid and Interface Science,* 247(2):259.

Banfield, J.F., and Navrotsky, A., 2003. Nanoparticles and the environment. Reviews in mineralogy and geochemistry. *Geochimica et Cosmochimica Acta,* 67, Washington DC, 1753 pp.

Bates, M.N., Smith, A.H., and Hopenhayn-Rich, C., 1992. Arsenic ingestion and internal cancers: A review. *Am. J. Epidemiol.,* 135(5):462–476.

Bayliss, P., 1977. Crystal structure refinement of a weakly anisotropic pyrite. *American Mineralogist,* 62:1168–1172.

Belin, T., Guigue-Millot, N., Caillot, T., Aymes, D., and Niepce, J.C., 2002. Influence of grain size, oxygen stoichiometry, and synthesis conditions on the [gamma]-Fe_2O_3 vacancies ordering and lattice parameters. *Journal of Solid State Chemistry,* 163(2):459.

Bertone, J.F., Cizeron, J., Wahi, R.K., Bosworth, J.K., and Colvin, V.L., 2003. Hydrothermal synthesis of quartz nanocrystals. *Nano Letters,* 3:655–659.

Bessekhouad, Y., Robert, D., Weber, J.V., and Chaoui, N., 2004. Effect of alkaline-doped TiO_2 on photocatalytic efficiency. *Journal of Photochemistry and Photobiology A: Chemistry,* 167(1):49.

Bianco Prevot, A., and Pramauro, E., 1999. Analytical monitoring of photocatalytic treatments. Degradation of 2,3,6-trichlorobenzoic acid in aqueous TiO_2 dispersions. *Talanta,* 48(4):847.

Bianco Prevot, A., Pramauro, E., and de la Guardian, M., 1999a. Photocatalytic degradation of carbaryl in aqueous TiO_2 suspensions containing surfactants. *Chemosphere*, 39(3):493.

Bianco Prevot, A., Vincenti, M., Bianciotto, A., and Pramauro, E., 1999b. Photocatalytic and photolytic transformation of chloramben in aqueous solutions. *Applied Catalysis B: Environmental*, 22(2):149.

Borm, P.J.A., 2002. Particle toxicology: From coal mining to nanotechnology. *Inhalation Toxicol.*, 14:311–324.

Bottero, J.Y. et al., 1993. Surface and textural heterogeneity of fresh hydrous ferric oxides in water and in dry state. *Journal of Colloids and Interface Science*, 159:45–52.

Bottero, J.Y., Cases, J.M., Fiessinger, F., and Poirier, J.E., 1980. Studies of hydrolyzed aluminum chloride solutions. 1. Nature of aluminum species and composition of aqueous solutions. *J. Phys. Chem.*, 84(22):2933–2937.

Bottero, J.Y., Manceau, A., Villieras, F., and Tchoubar, D., 1994. Structure and mechanisms of formation of FeOOH(Cl) polycations. Langmuir, 10(1):316–320.

Bottero, J.Y., Tchoubar, D., Cases, J.M., and Fiessinger, F., 1982. Investigation of the hydrolysis of aqueous solutions of aluminum chloride. 2. Nature and structure by small-angle x-ray scattering. *J. Phys. Chem.*, 86(18):3667–3673.

Chang, Y.-C., Chang, S.-W., and Chen, D.-H., 2006. Magnetic chitosan nanoparticles: Studies on chitosan binding and adsorption of Co(II) ions. *Reactive and Functional Polymers*, in press, corrected proof.

Chang, Y.-C., and Chen, D.-H., 2005. Preparation and adsorption properties of monodisperse chitosan-bound Fe_3O_4 magnetic nanoparticles for removal of Cu(II) ions. *Journal of Colloid and Interface Science*, 283(2):446.

Chen, W., Kan, A.T., Fu, G., Vignona, L.C., and Tomson, M.B., 1999. Adsorption-desorption behaviors of hydrophobicorganic compounds in sediments of Lake Charles, Louisiana, USA. *Environ. Toxicol. Chem.*, 18(8):1610–1616.

Cheng, X., Kan, A.T., and Tomson, M.B., 2005. Uptake and sequestration of naphthalen and 1,2-dichlorobenzene by C_{60}. *Journal of Nanoparticle Research.*, 7:555–567.

Choi, W.Y., Termin, A., and Hoffmann, M.R., 1994. The role of metal ion dopants in quantum-sized TiO_2: Correlation between photoreactivity and charge carrier recombination dynamics. *J. Phys. Chem.*, 98:13669–13679.

Cornell, R.M., and Schwertmann, U., 1996. The iron oxide: Structure, properties, reactions, occurrence and uses, Germany.

Cumbal, L., Greenleaf, J., Leun, D., and SenGupta, A.K., 2003. Polymer supported inorganic nanoparticles: Characterization and environmental applications. *Reactive and Functional Polymers*, 54(1–3):167.

Dagani, R., 2003. Nanomaterials: safe or unsafe? *Chem. Eng. News*, 81(30).

Deliyanni, E.A., Bakoyannakis, D.N., Zouboulis, A.I., and Matis, K.A., 2003. Sorption of As(V) ions by akaganeite-type nanocrystals. *Chemosphere*, 50(1):155.

Dixit, S., and Hering, J.G., 2003. Comparison of arsenic(V) and arsenic (III) sorption onto iron oxide minerals: Implications for arsenic mobility. *Environmental Science & Technology*, 37(18):4182–4189.

Domingo, J.L., 1995. Prevention by chelating agents of metal-induced developmental toxicity. *Reproductive Toxicology*, 9(2):105.

Fauconnier, N., Pons, J.N., Roger, J., and Bee, A., 1997. Thiolation of Maghemite Nanoparticles by Dimercaptosuccinic Acid. *Journal of Colloid and Interface Science*, 194(2):427.

Fu, W., Yang, H., Chang, L., Yu, Q., and Zou, G., 2006. Preparation and photocatalytic characteristics of core-shell structure $TiO_2/BaFe_{12}O_{19}$ nanoparticles. *Materials Letters*, in press, uncorrected proof.

Gao, Y., Kan, A.T., and Tomson, M.B., 1993. Critical evaluation of desorption phenomena of heavy metals from natural sediments. *Environmental Science & Technology*, 37(24):5566–5573.

Genc-Fuhrman, H., Tjell, J.C., and McConchie, D., 2004. Adsorption of arsenic from water using activated neutralized red mud. *Environmental Science & Technology*, 38(8):2428.

Hall, S.R., and Stewart, Y.M., 1973. The crystal structure refinement of chalcopyrite, $CuFeS_2$. *Acta Cryst.*, B29:579–585.

Hossain, M.F., 2006. Arsenic contamination in Bangladesh—An overview. *Agriculture, Ecosystems & Environment*, 113(1–4):1.

Huang, S.-H., Liao, M.-H., and Chen, D.-H., Fast and efficient recovery of lipase by poly-acrylic acid-coated magnetic nano-adsorbent with high activity retention. *Separation and Purification Technology*, in press, corrected proof.

Jackson, B.P., and Miller, W.P., 2000. Effectiveness of phosphate and hydroxide for desorption of arsenic and selenium species from iron oxides. *Soil Sci. Soc. Am. J.*, 64:1616–1622.

Jing, C., et al., 2005. Surface complexation of organic arsenic on nanocrystalline titanium oxide. *Journal of Colloid and Interface Science*, 290(1):14.

Johnson, D.B., and Hallberg, K.B., 2005. Acid mine drainage remediation options: A review. *Science of the Total Environment*, 338(1–2):3.

Jolivet, J.-P., Tronc, E., and Chaneac, C., 2002. Synthesis of iron oxide-based magnetic nanomaterials and composites. *Comptes Rendus Chimie*, 5(10):659–664.

Kan, A.T., et al., 1998. Irreversible sorption of neutral hydrocarbons to sediments: Experimental observations and model predictions. *Environ. Sci. Technol.*, 32:892–902.

Kramer, H.J., Mensikova, V., Backer, A., Meyer-Lehnert, H., and Gonick, H.C., 2003. Interaction of dimercaptosuccinic acid (DMSA) with angiotensin II on calcium mobilization in vascular smooth muscle cells. *Biochemical Pharmacology*, 65(10):1741.

Ladeira, A.C.Q., Ciminelli, V.S.T., Duarte, H.A., Alves, M.C.M., and Ramos, A.Y., 2001. Mechanism of anion retention from EXAFS and density functional calculations: Arsenic (V) adsorbed on gibbsite. *Geochimica et Cosmochimica Acta*, 65(8):1211.

Lasaga, A.C., 1998. *Kinetic Theory and Applications in Earth Sciences*. Princeton Press, Princeton.

Liao, M.-H., Wu, K.-Y., and Chen, D.-H., 2003. Fast removal of basic dyes by a novel magnetic nano-adsorbent. *Chemistry Letters*, 32(6):488–489.

Liu, G., et al., 2004. Effect of $ZnFe_2O_4$ doping on the photocatalytic activity of TiO_2. *Chemosphere*, 55(9):1287.

Liu, G., et al., 2005a. The preparation of Zn^{2+}-doped TiO_2 nanoparticles by sol-gel and solid phase reaction methods respectively and their photocatalytic activities. *Chemosphere*, 59(9):1367.

Liu, T., et al., 1999. Bondlength alternation of nanoparticles Fe_2O_3 coated with organic surfactants probed by EXAFS. *Nanostructured Materials*, 11(8):1329.

Liu, Y., Choi, H., Dionysiou, D., and Lowry, G.V., 2005b. Trichloroethene hydrodechlorination in water by highly disordered monometallic nanoiron. *Chemistry of Materials*, 17(21):5315.

Manceau, A., 1995. The mechanism of anion adsorption on iron oxides: Evidence for the bonding of arsenate tetrahedra on free $Fe(O, OH)_6$ edges. *Geochimica et Cosmochimica Acta*, 59(17):3647.

Manning, B.A., Fendorf, S.E., and Goldberg, S., 1998. Surface structures and stability of arsenic(III) on goethite: Spectroscopic evidence for inner-sphere complexes. *Environmental Science & Technology*, 32(16):2383.

Manning, B.A., and Goldberg, S., 1996. Modeling arsenate competitive adsorption on kaolinite, montmorillonite and illite. *Clays Clay Miner.*, 44(5):609–623.

Morel, F.M., and Hering, J.G., 1993. Principles and applications of aquatic chemistry. *Journal of Hydrology*, 155, New York, 588 pp.

Munoz, J.A., Gonzalo, A., and Valiente, M., 2002. Arsenic adsorption by Fe(III)-loaded open-celled cellulose sponge. Thermodynamic and selectivity aspects. *Environ. Sci. Technol.*, 36(15):3405–3411.

Ngomsik, A.-F., Bee, A., Draye, M., Cote, G., and Cabuil, V., 2005. Magnetic nano- and microparticles for metal removal and environmental applications: A review. *Comptes Rendus Chimie*, 8(6–7):963.

Pierce, M.L., and Moore, C.B., 1982. Adsorption of arsenite and arsenate on amorphous iron hydroxide. *Water Research*, 16(7):1247.

Pramauro, E., Prevot, A.B., Vincenti, M., and Gamberini, R., 1998. Photocatalytic degradation of naphthalene in aqueous TiO_2 dispersions: Effect of nonionic surfactants. *Chemosphere*, 36(7):1523.

Prodan, D., et al., 1999. Adsorption phenomena and magnetic properties of [gamma]-Fe_2O_3 nanoparticles. *Journal of Magnetism and Magnetic Materials*, 203(1–3):63–65.

Raven, K.P., Jain, A., and Loeppert, R.H., 1998. Arsenite and arsenate adsorption on ferrihydrite: Kinetics, equilibrium, and adsorption envelopes. *Environmental Science & Technology*, 32(3):344.

Riu, J., Maroto, A., and Rius, F.X., 2006. Nanosensors in environmental analysis. *Talanta*, 69(2):288.

Rusanov, A.I., 2005. Surface thermodynamics revisited. *Surface Science Reports*, 58(5–8):111.

Sawyer, C.N., McCarty, P.L., and Parkin, G.B., 2003. *Chemistry for Environmental Engineering and Science.*, 5th ed.

Sheoran, A.S., and Sheoran, V., 2006. Heavy metal removal mechanism of acid mine drainage in wetlands: A critical review. *Minerals Engineering*, 19(2):105.

Sherman, D.M., and Randall, S.R., 2003. Surface complexation of arsenic(V) to iron(III) (hydr)oxides: Structural mechanism from ab initio molecular geometries and EXAFS spectroscopy. *Geochimica et Cosmochimica Acta*, 67(22):4223.

Shipley, H.J., Yean, S., Kan, A.T., and Tomson, M.B., 2006. Rice University.

Skinner, B.S., Erd, R.C., and Grimaldi, F.S., 1964. Griegite, the thiospinel of iron: A new mineral. *American Mineralogist*, 49:543–555.

Stipp, S.L.S., et al., 2002. Behaviour of Fe-oxides relevant to contaminant uptake in the environment. *Chemical Geology*, 190(1–4):321.

Stumm, W., and Morgan, J.J., 1996. Aquatic chemistry: 3rd ed. *Geochimica et Cosmochimica Acta*, 60, 1200 pp.

Su, C.M., and Puls, R.W., 2001. Arsenate and arsenite removal by zerovalent iron: Effects of phosphate, silicate, carbonate, borate, sulfate, chromate, molybdate, and nitrate relative to chloride. *Environ. Sci. Technol.*, 35 (22):4562–4568.

Swedlund, P.J., and Webster, J.G., 1999. Adsorption and polymerisation of silicic acid on ferrihydrite, and its effect on arsenic adsorption. *Water Res.*, 33(16):3413–3422.

Tamura, H., Katayama, N., and Furuichi, R., 1997. The Co_{2+} adsorption properties of Al_2O_3, Fe_2O_3, Fe_3O_4, TiO_2, and MnO_2 evaluated by modeling with the Frumkin isotherm. *Journal of Colloid and Interface Science*, 195(1):192.

Thoral, S., et al., 2005. XAS study of iron and arsenic speciation during Fe(II) oxidation in the presence of As(III). *Environmental Science & Technology*, 39(24):9478.

Uheida, A., Salazar-Alvarez, G., Bjorkman, E., Yu, Z., and Muhammed, M., 2006. Fe_3O_4 and [gamma]-Fe_2O_3 nanoparticles for the adsorption of Co_{2+} from aqueous solution. *Journal of Colloid and Interface Science*, in press, corrected proof.

Violante, A., and Pigna, M., 2002. Competitive sorption of arsenate and phosphate on different clay minerals and soils. *Soil Sci. Soc. Am. J.*, 66:1788–1796.

Waychunas, G.A., Davis, J.A., and Fuller, C.C., 1995. Geometry of sorbed arsenate on ferrihydrite and crystalline FeOOH: Re-evaluation of EXAFS results and topological factors in predicting sorbate geometry, and evidence for monodentate complexes. *Geochimica et Cosmochimica Acta*, 59(17):3655.

Waychunas, G.A., Rea, B.A., Fuller, C.C., and Davis, J.A., 1993. Surface chemistry of ferrihydrite: Part 1. EXAFS studies of the geometry of coprecipitated and adsorbed arsenate. *Geochimica et Cosmochimica Acta*, 57(10):2251.

Wilkie, J.A., and Hering, J.G., 1996. Adsorption of arsenic onto hydrous ferric oxide: Effects of adsorbate/adsorbent ratios and co-occurring solutes. *Colloids and Surfaces A: Physicochemical and Engineering Aspects*, 107:97.

Wyckoff, R.W.G., 1963. Crystal structure. 1:355.

Yamaura, M., Camilo, R.L., and Felinto, M.C.F.C., 2002. Synthesis and performance of organic-coated magnetite particles. *Journal of Alloys and Compounds*, 344(1–2):152.

Yean, S., et al., 2005. Effect of magnetic particle size on adsorption and desorption of arsenite and arsenate. *J. Matter. Res*, 20:3255–3264.

Yin, Y., Allen, H.E., Huang, C.P., and Sanders, P.F., 1997. Adsorption/desorption isotherms of Hg(II) by soil. *Soil Sci.*, 162:35–45.

Zhang, H., Penn, R.L., Hamers, R.J., and Banfield, J.F., 1999. Enhanced Adsorption of Molecules on Surfaces of Nanocrystalline Particles. *Journal of Physical Chemistry B*, 103(22):4656.

Zhang, W.X., 2003. Nanoscale iron particles for environmental remediation: An overview. *Journal of Nanoparticle Research*, 5:323–323.

IV

Potential Impacts of Nanomaterials

Toxicological Impacts of Nanomaterials

Nancy A. Monteiro-Riviere   *North Carolina State University, Center for Chemical Toxicology Research and Pharmacokinetics, Raleigh, NC*

Thierry Orsière   *University of Aix-Marseille, Marseille, France*

Keywords: toxicity; carbon nanotubes, fullerenes; iron-, cerium- and titanium-oxide nanoparticles; copper nanoparticles, gold nanoparticles; quantum dots, exposure toxicity

Introduction

Nanotechnology research has stimulated new interest in the role of particle size in determining toxicity. Nanoparticles may be more toxic than larger particles of the same substance (Lam et al., 2004) because of their larger surface area, high ratio of particle number to mass, enhanced chemical reactivity, and potential for easier penetration of cells (Gurr et al., 2005). To give some perspective to the role of size, consider that 2 g of 100 nm-diameter nanoparticles apportioned equally to the world's population corresponds roughly to a potential exposure of some 300,000 particles per individual (Hardman, 2006).

Due to the widespread use of nanomaterials, it is essential that we understand and investigate the biological effects of exposure for their medical, occupational health, and environmental effects (Dagani, 2003). Since their discovery, nanomaterials have been proposed for use in many biological applications, although little is known of their toxicity, potential mutagenic (carcinogenic and teratogenic) effects, or overall risk to human health. In order to avoid past mistakes made when new technological innovations, chemicals, or drugs were released prior to a broad-based risk assessment, information is needed now regarding the potential toxicological impact of nanomaterials on human health and the environment.

New nanomaterials should be thoroughly investigated for occupational safety during manufacture, exposure scenarios likely to be encountered by the consumer (e.g., commercial products, medicines, cosmetics), and postuse release and migration in the environment. Also, the biocompatibility of nanomaterials should be evaluated before incorporating these new structures into biomedical devices or implants produced by tissue engineering. This chapter describes the known toxicity of several classes of nanomaterials.

Fullerenes

Fullerenes or buckyballs (short for buckminsterfullerene) are molecular structures containing 60 carbon atoms (C_{60}). These spheres are made of carbon and each carbon is bonded to three neighboring carbons of about one nanometer in diameter. Three scientists first discovered fullerenes in 1985 while studying "clusters-aggregates of atoms" where they vaporized graphite with a laser in an atmosphere of helium gas to form stable carbon clusters. It was discovered that only a geometric shape could combine 60 carbon atoms into a spherical structure of hexagons and pentagons. This combination of structure was the basis of a geodesic dome designed by Buckminster Fuller for the 1967 Montreal Exhibition. Since the newly discovered molecule resembled this architectural structure, they were termed buckminsterfullerene (Kroto et al., 1985). In 1996, the Nobel Prize was awarded to Curl, Kroto, and Smalley for their discovery.

Although carbon has many beneficial applications, including prosthetic materials, dental implants, bone plates, and heart valves, the impacts of carbon in the form of fullerenes on human health are largely unknown. These materials possess unique chemical, mechanical, electrical, optical, magnetic, and biological properties that make them candidates for a variety of novel commercial and medical applications. One aspect of fullerenes that makes them particularly attractive for applications ranging from drug delivery to cosmetics is that they may be derivatized in an infinite number of variations to tailor the fullerene's properties to a given application.

Information regarding the biodistribution and metabolism of C_{60} is lacking and it has been difficult to study due to the low solubility of C_{60} in water coupled with a lack of sensitive analytical techniques for detecting fullerenes. In one study, water-soluble ^{14}C-labeled fullerenes were orally administered to rats. The fullerenes were not absorbed but were excreted in the feces. Intravenously administered fullerenes were distributed to various tissues in the body after one week. ^{14}C-labeled fullerene even penetrated the blood-brain barrier although acute toxicity was low (Yamago et al., 1995). A pharmacokinetic study used intravenous administration of another water-soluble fullerene, the bis (monosuccinimide) derivative of p,p'-bis(2-amino-ethyl)-diphenyl-C_{60} (MSAD-C_{60}), in rats. The concentration of fullerenes in blood plasma was observed to decrease over time in a fashion that was described by a polyexponential model with a terminal half-life of 6.8 hours. Fullerenes were extensively bound by proteins (99%) and fullerenes were not detected in the urine at 24 hours. These data confirm extensive tissue distribution and minimal clearance from the body. A 15-mg/kg dose of this fullerene derivative was tolerated by the rats, but a 25-mg/kg dose resulted in death after five minutes (Rajagopalan et al., 1996). Biodistribution was also studied with radio-labeled ^{99m}Tc-labeling of C_{60} $(OH)_x$ fullerenes in mice and rabbits, which showed a wide distribution in all tissues, with a significant percentage in the kidneys, bone, spleen, and liver by 48 hours (Qingnuan et al., 2002). These studies have demonstrated that "water-soluble" fullerenes can be distributed after nonoral routes of administration. Very little is known regarding fullerene metabolism or systemic elimination. One study has shown that fullerenol-1 can suppress levels of cytochrome P450–dependent monooxygenase *in vivo* in mice and mitochondrial oxidative phosphorylation *in vitro* liver microsomes (Ueng et al., 1997).

Carboxy fullerenes are potent free radical scavengers and have been both effective in reducing neuronal cell death and suggested to act as neuroprotective agents in mice and in mouse neocortical cultures. At 100 μM of a water-soluble carboxylic acid derivative fullerene fully blocked an N-methyl-D-aspartate receptor–mediated toxicity and reduced apoptotic neuronal death induced by serum deprivation or exposure to the Alzheimer's disease amyloid peptide $A\beta_{1-42}$ (Dugan et al., 1997). Other studies have shown that the immune system can process a water-soluble fullerene derivative conjugated to bovine and rabbit serum albumin and present the processed peptides for recognition by T cells to yield IgG antibodies (Chen et al., 1998).

A water-soluble carboxylic acid derivative fullerene (carboxyfullerene) has been shown to possess antioxidative properties that can suppress iron-induced lipid peroxidation in rat brains (Lin et al., 1999). Uncoated fullerenes have also been shown to induce oxidative stress in juvenile largemouth bass at 0.5 ppm and cause a significant increase in lipid peroxidation of the brain and glutathione depletion in the gills after 48 hours (Oberdorster, 2004). In addition, they also assessed the toxicity of stable fullerene suspension (nC_{60}) on fresh water crustaceans, a marine copepod, fathead minnow, and Japanese medaka fishes. It was noted that after 21 days of exposure, the daphnia had a delay in molting and a significant decrease in offspring production at 2.5 and 5.0 ppm of nC_{60}. They showed that the peroxisomal lipid transport protein PMP70 was reduced in the fathead minnow but not in the medaka, indicating alterations in the acyl-CoA pathways (Oberdörster et al., 2006).

Some C_{60} fullerene derivatives can interact with the active site of HIV-1 protease, suggesting antiviral activity (Friedman et al., 1993). C_{60} can also protect quiescent human peripheral blood mononuclear cells from apoptosis induced by 2-deoxy-D-ribose or tumor necrosis factor alpha plus cycloheximide (Monti et al., 2000). Studies with C_{60} solubilized by polyvinyl-pyrrolidone (PVP) coating in water and applied to the mouse midbrain cell differentiation system found inhibition of cell differentiation and proliferation. Harmful effects with C_{60} on mammalian embryos have also been noted (Tsuchiya et al., 1995, 1996; Friedman et al., 1993). Under some conditions, C_{60} can convert oxygen from the triplet to the singlet state (see Chapter 5), creating some concern for potential health risks through this mechanism.

Since skin is the largest organ of the body and most probably will come into contact with fullerenes, especially during the manufacturing process, the dermal toxicity of fullerenes has been the focus of some studies. Topical administration of 200 μg of fullerenes to mouse skin over a 72-hour period found no effect on either DNA synthesis or ornithine decarboxylase activity. The ability of fullerenes to act as tumor promoters has also been investigated. Repeated application to mouse skin after initiation with dimethlybenzanthracene (DMBA) for 24 weeks did not result in benign or malignant skin tumor formation. But promotion was observed with 12-0-tetradecanoylphorbol-13-acetate (TPA) resulting in benign skin tumors (Nelson et al., 1993). *In vitro* studies using ^{14}C-labeled underivatized C_{60} exposed to immortalized human keratinocytes depicted cellular incorporation of the label uptake at various times. Approximately 50 percent of the radio-labeled C_{60} was taken up within six hours, but it was unclear whether particles actually entered the cell or were associated with the cell surface. They also found no effect on the proliferation of immortalized keratinocytes and fibroblast (Scrivens and Tour, 1994).

There are conflicting reports as to the potential toxicity of fullerenes such as C_{60}. While C_{60} itself has essentially no solubility in water, it does have the ability to form aggregates, referred to as nC_{60}, through either organic solvent inclusion or after partial hydrolysis, that are stable colloidal suspensions in water (see Chapter 7). One hypothesis that has been put forward is that functionalizing a fullerene with carboxyl or hydroxyl groups should decrease the cytotoxic response in cells by altering membrane solubility and interactions with cellular membrane binding sites. In one study, four types of water-soluble fullerenes were assessed with respect to their toxicity to human carcinoma cells and dermal fibroblasts (a connective tissue cell). Fullerenes that had been derivatized through carboxylation, C_3, Na^+_{2-3} $[C_{60}O_{7-9}(OH)_{12-15}]$, or hydroxylation, $C_{60}(OH)_{24}$, (likely present as colloidal aggregates) showed much less toxicity compared to a variety of colloidal aggregates of C_{60}, nC_{60} thought to be derivatized to a much lesser extent (Sayes et al., 2004). However, interpretation of these results is complicated by the fact that aggregate size was not controlled in these experiments and the nC_{60} aggregate composition varied with respect to not only functional groups on the fullerene, but also residual solvent.

Fullerene derivatives have been shown to affect protein configurations, a factor that could play a role in dermatotoxic effects. A series of fullerene-substituted phenylalanine derivatives was prepared to compare their behavior to related functionalized fullerenes. The presence of the C_{60} substituent alters the conformation of the native peptide (e.g., from a random coil to a β-sheet), making the conditions under which conversion to an alpha helix occurs important. These studies showed no apparent toxicity to cells; however, early studies did not confirm that the peptide was incorporated into the cells (Yang and Barron, 2004). When human epidermal keratinocytes were exposed to agglomerates of a fullerene-based amino acid at concentrations of 0.4 mg/ml, clusters of C_{60} were observed within large cytoplasmic vacuoles (Figure 11.1). Also, agglomerates of the fullerene-based amino acid at 0.4 mg/ml has also been shown to accumulate along the

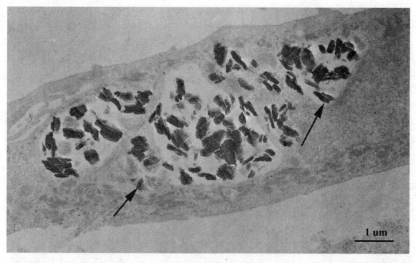

Figure 11.1 Transmission electron micrograph of a human epidermal keratinocyte that was exposed to 0.4 mg/ml of a functionalized fullerene for 48 hours. Arrows depict fullerene agglomerates within cytoplasmic vacuoles.

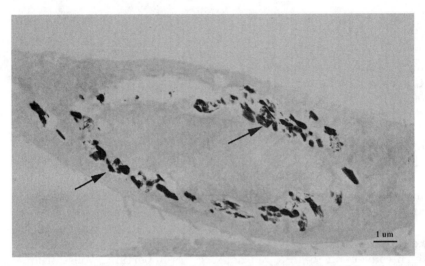

Figure 11.2 Transmission electron micrograph of a human epidermal keratinocyte that was exposed to 0.4 mg/ml of the fullerene-based amino acid for 24 hours. Arrows depict agglomerates in vacuoles around the periphery of the nucleus.

periphery of the nucleus in large vacuoles (Figure 11.2). Human epidermal keratinocytes exposed to fullerene-based amino acid solutions ranging from 0.04–0.004 mg/ml resulted in a significant decrease in viability and a statistically significant increase in IL-8 at a dose of 0.04 mg/ml over periods of 8, 12, and 24 hours. IL-6 and IL-1β were greater at 24 and 48 hours, but there was no significant TNF-α or IL-10 expression. (Rouse et al., 2006).

Studies were also conducted with a fullerene-substituted phenylalanine derivative of a nuclear localization peptide sequence to evaluate the effects on mechanical stressors such as repetitive flexing motion. Skin flexed for 60 or 90 minutes and dosed with the fullerene or left unflexed (control) was studied by confocal laser microscopy which depicted penetration after 60 and 90 minutes of flexion and 8 hours of flow-through diffusion and further penetration after 24 hours of flow-through diffusion. This study showed that mechanical flexion caused an increase in nanoparticle penetration through skin (Rouse et al., 2007).

In contrast with these studies suggesting toxicity of various fullerenes and their derivatives, other studies have concluded that fullerenes may have beneficial effects for cells. Numerous studies have also been published that show functionalized fullerenes to be therapeutically useful in the treatment of a number of diseases. A balanced benefit-risk assessment must be made before these substances are released for human use.

Single-Walled Carbon Nanotubes (SWCNT)

Carbon nanotubes (CNT), also known as buckytubes, are made up of a seamless single sheet of graphite rolled into cylindrical shells that range from one to tens of nanometers in diameter and up to several micrometers in length. They have numerous novel electrical and mechanical properties. They can be synthesized by electric arc discharge, laser ablation, or chemical vapor deposition (Ebbesen and Ajayan, 1992; Thess et al., 1966; Willems et al., 2000; see Chapter 3).

Due to the fibrous structure of SWCNT, they have been proposed for use in many tissue engineering applications. For biomedical applications, the SWCNT must be compatible with many different cell types. If SWCNT are used in composite materials and these materials degrade over time, then the SWCNT may be in direct contact with the tissue and may not be eliminated through the body. Therefore, it is crucial that SWCNT be biocompatible with the surrounding tissue. This is particularly important in regenerative medicine and tissue engineering where SWCNT are designed into scaffolds for cell support and growth.

Studies have shown acute lung toxicity by intratracheal instillation of SWCNT at high doses of 5 mg/kg in rats for 24 hours. Multifocal granulomas were observed, with a mortality rate of 15 percent that resulted from the mechanical blockage of the upper airways by the instillate due to a foreign body reaction and not to the SWCNT particulate (Warheit et al., 2004). Other studies conducted in mice exposed to unpurified SWCNT manufactured using three different methods and catalysts depicted dose-dependent persistent epithelioid granulomas and interstitial inflammation (Lam et al., 2004). These primary findings of multifocal granulomas and inflammation were dependent upon the type of particle used. Intratracheal installation has been considered to be an unrealistic route for normal human exposure and has several limitations because of the nonphysiologic rapid delivery of the nanotubes, the delivery of agglomerates, and bypassing the typical nose filtering mechanism (Driscoll et al., 2000; Muller et al., 2006). One important observation was that once SWCNT reach the lung, they were more toxic than carbon black or quartz dust, two known pulmonary toxicants (Lam et al., 2004). However, other studies have shown that carbon nanotubes are not toxic after four weeks following a single intratracheal installation in guinea pigs (Huczko et al., 2001). Pharyngeal aspiration with purified SWCNT in mice elicited progressive fibrosis and granulomas with two distinct morphologies. In addition, SWCNT penetrated the interstitial, tissue raising the possibility of translocation into the systemic circulation (Shvedova et al., 2005). This may be secondary to the particles' tendencies to self-aggregate when removed from controlled conditions. It must be stressed that inhalational exposure of particulate matter such as nanoparticles is fundamentally different than dermal or oral exposure, as the lung is designed to trap particulate matter.

Immortalized nontumorigenic human epidermal (HaCaT) cells exposed to SWCNT suggest that carbon nanotubes may be toxic to epidermal keratinocyte cultures (Shvedova et al., 2003). Significant cellular toxicity was found when unrefined SWCNT were exposed to cells for 18 hours. The inflammatory markers of irritation/inflammation were not conducted in this cell line. Previously, our lab has demonstrated significant differences in the toxicological response between immortalized versus primary keratinocytes (Allen et al., 2001). Gene expression profiling was conducted on human epidermal keratinocytes exposed to 1.0 mg/ml of SWCNT that showed a similar profile to alpha-quartz or silica. Alpha-quartz is considered to be the main cause of silicosis in humans. Genes not previously associated with these particulates from the structural protein and cytokine families were significantly expressed (Cunningham et al., 2005).

In addition to toxicity studies, data have been collected when SWCNT have been evaluated as therapeutic agents for drug and vaccine delivery (Smart et al., 2006). Drug delivery can be enhanced by many types of chemical vehicles, including lipids, peptides, and polyethylene glycol (PEG) derivatives. Strategies using SWCNT and SWCNT-streptavidin conjugates as biocompatible transporters have shown localization within human promyelocytic leukemia (HL60) cells, and human T (Jurkat) cells via endocytosis. Functionalized SWCNT exhibited little toxicity to the HL60 cells, but the SWCNT–biotin-streptavidin complex caused extensive cell death (Kam et al., 2004). Other studies have demonstrated that functionalized, water-soluble SWCNT derivatives modified with a fluorescent probe can translocate across the cell membrane of fibroblasts without causing toxicity (Pantarotto et al., 2004). The translocation pathway remains to be elucidated. It has been shown that the solubility of functionalized SWCNT is controlled by the length of the hydrocarbon side chain (Zeng et al., 2005). Other studies comparing SWCNT as purchased, purified SWCNT or functionalized with glucosamine showed significant effects on *in vitro* fibroblast cell function and demonstrated that chemical modifications can alter their cellular interactions (Nimmagadda et al., 2006). Three different water-dispersible SWCNT were exposed to fibroblasts to determine if the degree of functionalizations affects its response in cells. These investigators found that as the side-wall functionalization increases, SWCNT suspensions appear to be less cytotoxic and that functionalized tubes were less cytotoxic than the surfactant stabilized tubes (Sayes et al., 2006). This is consistent with the fullerene data reviewed above, but with some of the same complications in interpreting these results originating from aggregation effects.

Purified SWCNT did not stimulate the release of nitric oxide by murine macrophages in culture, their uptake was low and cell surface morphology was unchanged (Fiorito et al., 2006). The cytotoxicity of primary human umbilical vein endothelial cells in culture exposed to carbon nanotubes showed no toxicity based on cell viability and cell metabolic activity (Flahaut et al., 2006).

Carbon nanotubes carrying DNA or peptide molecules can serve as a potential delivery system in gene or peptide delivery (Gao et al., 2003). Other investigators studied the effect of SWCNT with the idea that they are biocompatible. These investigators have shown that SWCNT can inhibit cell proliferation and decrease cell adhesive ability in a dose- and time-dependent fashion using human embryo kidney HEK293 cells (Cui et al., 2005). Exogenous DNA can be introduced into mammalian cells by manipulating signal transduction by nanospearing, which is based on penetrating nickel embedded nanotubes into the cell membranes by magnetic fields (Cai et al., 2005). Mouse peritoneal macrophage-like cells can ingest SWCNT in a surfactant without changes in viability or population growth (Cherukuri et al., 2004). Iron-rich SWCNT have been shown to cause a significant loss of intracellular thiols and accumulation of lipid peroxides in macrophages (Kagan et al., 2006). Studies have also shown activation of oxidative stress and nuclear transcription factor-kB in immortalized keratinocytes (Manna et al., 2005). Nucleic acid encapsulated SWCNT have been located within cytoplasmic vacuoles of myoblast stem cells and have been shown to be persistent with Raman scattering and fluorescence spectra in mammalian cells for up to three months in culture, which makes these DNA-SWCNT function as long-term cellular biomarkers or sensors (Heller et al., 2005).

Multi-Walled Carbon Nanotubes (MWCNT)

MWCNT toxicity has also been addressed in a primary human keratinocyte cell culture model. Human neonatal epidermal keratinocytes exposed to 0.1, 0.2, and 0.4 mg/ml of MWCNT for 1, 2, 4, 8, 12, 24, and 48 hours depicted MWCNT within the cytoplasmic vacuoles of human epidermal keratinocytes. These MWCNT exhibited typical base mode growth; very little disordered carbon, and were well ordered and aligned. Using transmission electron microscopy, MWCNT were predominantly located in vacuoles in the cytoplasm of the keratinocyte and were found up to 3.6 µ in length (Figure 11.3). At 24 hours, 59 percent of the human keratinocytes contained MWCNT, compared to 84 percent by 48 hours at the 0.4 mg/ml dose. The viability of these cells decreased with an increase in MWCNT concentration. IL-8, an early biomarker for irritation, increased with time and concentration (Monteiro-Riviere et al., 2005a). Proteomic analysis conducted in human epidermal keratinocytes exposed to MWCNT showed altered expression of 36 proteins after 24 hours and 106 altered proteins after 48 hours relative to controls. These protein alterations suggested dysregulation of

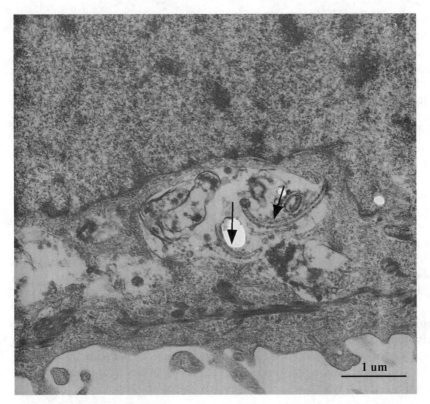

Figure 11.3 Transmission electron micrograph of human epidermal keratinocyte exposed to 0.4 mg/ml of MWCNT for 24 hours. Arrows depict localization of MWCNT within cytoplasmic vacuoles.

intermediate filament expression, cell cycle inhibition, altered vesicular trafficking/exocytosis, and membrane scaffold protein down-regulation (Monteiro-Riviere, 2005; Monteiro-Riviere et al., 2005b; Witzmann and Monteiro-Riviere, 2006).

These data showed that MWCNT, not derivatized nor optimized for biological applications, were capable of both localizing and initiating an irritation response in skin cells. These data suggest a significant dermal hazard after topical exposure to select nanoparticles should they penetrate the stratum corneum barrier. One may question that the exposure level ranging from 0.1 mg/ml to 0.4 mg/ml is quite high. However, studies involving the handling of SWCNT in four different field sites showed that estimated exposure to SWCNT on left- and right-hand glove samples ranged from 0.2 mg to 6.0 mg of CNT per glove (Maynard et al., 2004). Also, the amount of carbon nanotubes deposited on the glove was so vast that they remained visible on the glove at the end of the sampling period. Therefore, the exposure levels studied above are realistic and within the range seen after occupational exposure.

Hat-stacked carbon nanofibers that resembled multi-walled carbon nanotubes were implanted in rat subcutaneous tissue. They produced a granulation and an inflammatory response that were similar to foreign body granulomas. These carbon nanofibers were found within the macrophages without severe inflammation, necrosis, or degeneration of tissue (Yokoyama et al., 2005).

The effects of pristine and oxidized MWCNT at 400 μg/ml were compared using human T cells. In contrast with results using SWCNTs that showed less toxicity with greater degrees of functionality on the CNT, these results showed greater toxicity and decreased cell viability with the oxidized MWCNT (Bottini et al., 2006). Some investigators tested the influence of grinding on the dispersion of nanotubes and on the inflammatory potential of ground or intact MWCNT intratracheally administered in rats at three different doses. Ground nanotubes were dispersed in the lung and induced an inflammatory and fibrotic response, while the intact nanotubes caused an agglomeration in the airways along with granulomas in the bronchial lumen and inflammation of the alveoli. These investigators also conducted *in vitro* studies with peritoneal macrophages and found only the ground nanotubes induced TNF-α (Muller et al., 2005; 2006). Other investigators evaluated 10-nm to 20-nm MWCNT in alveolar macrophages after 6 hours *in vitro* and showed necrosis, degeneration, and rarefaction of the nuclear matrix at 3.06 μg/cm^2 dose of the 10-nm MWCNT (Jia et al., 2005).

Complications in Screening Assays Using Carbon-Based Materials

When conducting any carbon nanomaterial toxicology studies, the nature of the controls (reference material of known toxicity) employed is of major concern. Carbon black (CB), a nanoscale particulate, has often been recommended as a negative (nontoxic) control when conducting toxicity studies. Use of CB has grown out of its previous role in serving as a particulate control in inhalational toxicology studies. However, caution must be taken when utilizing CB because of its ability to adsorb colorimetric chemical markers used in biochemical assays. Carbon blacks are not porous materials engineered to be sorbents, but by virtue of their very small primary particle sizes, can possess significant surface areas for adsorption. CB and its characterization and composition are extremely important.

For example, ultrafine carbon black commonly used for *in vivo* inhalation studies with gross and microscopic endpoints may not be

suitable for use in cell culture because of this interference with viability and cytokine assays (Monteiro-Riviere and Inman, 2006). These
studies with the neutral red (NR) assay, which is a widely utilized viability test in cell culture experiments, showed that carbon can adsorb
viability marker dyes such as NR (3-amino-7-dimethylamino-2
methylphenazine hydrochloride) from cell culture media, interfering
with the absorption spectra causing false readings. Interference
studies with the NR assay protocol using different CB sources (Fisher,
Cabot, and Degussa varieties) in the absence of human epidermal
keratinocytes demonstrated a false negative signal indicating the
presence of viable cells. CB was found to adsorb NR dye and generate a signal in the assay suggesting high cell viability when in fact
cells were not even present. Similar studies were also shown for the
MTT (3-[4,5-dimethyl-2-thiazol]-2,5-diphenyl-2H-tetrazolium bromide)
viability assay. MTT is a colorimetric metabolic assay based on mitochondrial dye conversion to assess viability. Tetrazolium salt is used to
assess the activity of various dehydrogenase enzymes where the tetrazolium ring is cleaved in active mitochondria, demonstrating the presence of living cells. In the presence of CB, the tetrazolium dye desolved
from the cells is adsorbed by the CB, thereby reducing the absorbance
reading and giving a false positive signal for reduced viability and thus
cytotoxicity.

These studies have shown how carbon adsorbents may interfere with
in vitro cytotoxicity assays. CB may also adsorb the constituents of the
grow media as well as proteins and growth factors, thereby preventing
the cells from receiving their proper nutrients and growth factors. CB
may adsorb other soluble components that could alter pH and cell viability. Ultrafine CB may act as an adsorbent that could potentially bind
compounds during the manufacturing process (Monteiro-Riviere and
Inman, 2006). The type of CB used in a study and its characterization
and composition is extremely important. Their rational use as particulate controls in inhalational studies may not carry over into other organ
or exposure systems.

Table 11.1 lists a number of reports on carbon-based nanomaterials
toxicity.

Titanium Dioxides

Titanium dioxide (TiO_2) particles larger than 100 nm are generally considered to be biologically inert in both humans and animals (Bernard
et al., 1990; Chen and Fayerweather, 1988; Hart and Hesterberg, 1998;
Lindenschmidt et al., 1990; Ophus et al., 1979). Thus, they have been
widely used as a food colorant (Lomer et al., 2002) and as a white pigment (Nordman et al., 1986) in sunscreens and in cosmetic creams

TABLE 11.1 Negative Impacts of Carbon-Based Nanomaterials on Living Mammalian Cells and Organisms

Type of nanomaterials	Organisms, or cell types, or organelles	Effects observed	References
Fullerene			
C_{60} water suspension	Rat (IV administration)	Blood-brain-barrier penetration	Yamago et al., 1995
	Mouse (skin applications)	Benign tumors formation following initiation with TPA	Nelson et al., 1993
MSAD-C_{60}	Rat (intravenous administration)	Extensive distribution and minimal clearance, death after 5 min. at 25 mg/kg	Rajagopalan et al., 1996
$C_{60}(OH)_x$	Mice and rabbits	Distribution in the kidneys, bone, spleen, and liver by 48 hrs	Qingnuan et al., 2002
Fullerenol-1	Mice	Decrease in the cytochrome P450 monooxygenase levels	Ueng et al., 1997
	Liver microsomes	Decrease in mitochondrial oxidative phosphorylation	Ueng et al., 1997
C_{60}-PVP	Mouse midbrain cell differentiation system	Inhibition of cell differentiation and proliferation	Tsuchiya et al., 1995, 1996; Friedman et al., 1993
Colloidal and derivatized C_{60}	Human carcinoma cells and dermal fibroblasts	Toxicity	Sayes et al., 2004
Functionalized fullerenes	Human epidermal keratinocytes	Localized in cytoplasmic vacuoles; expression of cytokines	Rouse et al., 2006
SWCNT			
Underivatized SWCNT	Rat lung (intratracheal instillation)	Multifocal granulomas, mortality rate of 15% (5 mg/kg in rats for 24 hrs): mechanical blockage of the upper airways by the instillate due to a foreign body reaction and not to the SWCNT particulate	Warheit et al., 2004
	Mice lung (intrathecal instillation)	Epithelioid granulomas and interstitial inflammation	Lam et al., 2004

(Continued)

TABLE 11.1 Negative Impacts of Carbon-Based Nanomaterials on Living Mammalian Cells and Organisms *(Continued)*

Type of nanomaterials	Organisms, or cell types, or organelles	Effects observed	References
	Human epidermal (HaCaT) cells	Cellular toxicity	Shvedova et al., 2003
	Human epidermal keratinocytes	Genes induction (structural proteins and cytokines)	Cunningham et al., 2005; Zhang et al., 2007
	Human embryo kidney HEK293 cells	Decrease in cell proliferation and cell adhesion	Cui et al., 2005
SWCNT– biotin-streptavidin complex	Human promyelocytic leukemia (HL60) cells and human T (Jurkat) cells	Endocytosis, intracytoplasmic localization, extensive cell death	Kam et al., 2004
Functionalized carbon nanotubes	Fibroblasts	Increase in the side-wall functionalization decrease cytotoxicity	Sayes et al., 2006
Iron-rich SWCNT	Macrophages	Loss of intracellular thiols and accumulation of lipid peroxides	Kagan et al., 2006
	Immortalized keratinocytes	Activation of oxidative stress and nuclear transcription factor-kB	Manna et al., 2005
MWCNT	Human epidermal keratinocytes	Internalization in vacuoles; concentration dependent decrease in viability; concentration- and time-dependent increase in IL-8 production	Monteiro-Riviere et al., 2005a
		Proteomic analysis: dysregulation of intermediate filament expression, cell cycle inhibition, altered vesicular trafficking/exocytosis, and membrane scaffold protein down-regulation	Monteiro-Riviere 2005; Witzmann and Monteiro-Riviere, 2006

	Intratracheal administration in rats	Agglomeration in the airways along with granulomas in the bronchial lumen and inflammation of the alveoli	Muller et al., 2005, 2006
	Alveolar macrophages	Necrosis, degeneration, and rarefaction of the nuclear matrix	Jia et al., 2005
Hat-stacked carbon nanofibers	Implantation in rat subcutaneous	Granulation and an inflammatory response; translocation to macrophages without severe inflammation, necrosis, or degeneration of tissue	Yokoyama et al., 2005
Oxidized MWCNT	Human T cells	Toxicity and decrease in cell viability	Bottini et al., 2006
Ground MWCNT	Intratracheal administration in rats	Dispersion in the lung and induction of an inflammatory and fibrotic response	Muller et al., 2005, 2006

Negative impacts of nanomaterials on living microorganisms are illustrated in Chapter 12.
IV: intravenous; MSAD-C_{60}: bis (monosuccinimide) derivative of p,p'-bis(2-amino-ethyl)-diphenyl-C_{60}; C_{60}-PVP: C_{60} solubilized by polyvinyl-pyrrolidone (PVP) coating in water; TPA: 12-0-tetradecanoylphorbol-13-acetate.

(Gelis et al., 2003). On the other hand, TiO_2 is also a well-known photocatalyst (see Chapter 5).

TiO_2 absorbs UVA light, catalyzing the generation of reactive oxygen species, such as superoxide anion radicals, hydrogen peroxide, free hydroxyl radicals, and singlet oxygen in aqueous media. These hydroxyl radicals are known to initiate oxidation. Using chemical methods, it has been shown that TiO_2 sunscreen samples catalyze the photooxidation of some organic substrate (e.g., phenol) (Dunfort et al., 1997). Because of its photocatalytic properties, TiO_2 has been applied in the environment and wastewater as a disinfectant. Recently, TiO_2 was used as a photosensitizer for photodynamic therapy for endobronchial and esophageal cancers.

Several studies have shown that the cytotoxicity of nano-sized TiO_2 was very low or negligible as compared with other nanoparticles (Peters et al., 2004; Yamamoto et al., 2004; Zhang et al., 1998), and size alone was not the effective predictor of cytotoxicity (Yamamoto et al., 2004). Other studies have evaluated the effects of five different nanoscaled particles (PVC, TiO_2, SiO_2, Co, Ni) on endothelial cell function and viability. TiO_2 nanoparticles did not show significant cytotoxic or inflammatory effects, although some proinflammatory effects were observed in human endothelial cells (Peters et al., 2004).

As TiO_2 reflects and scatters UVB and UVA in sunlight, nano-sized TiO_2 is used in numerous sunscreens. However, it has been noted that sunlight-illuminated TiO_2 catalyses DNA damage both *in vitro* and in human cells. These results may be relevant to the overall effects of sunscreens because without UV irradiation, DNA damage was not observed (Dunfort et al., 1997).

TiO_2 nanoparticles have been shown to be photogenotoxic, but little information has been published on the genotoxic properties of unirradiated TiO_2 nanoparticles. Studies have reported that TiO_2 nanoparticles can induce apoptosis and micronuclei formation, these changes are known to reveal numerical and/or structural chromosomal damage in Syrian hamster embryo fibroblast (Rahman et al., 2002).

In the absence of photoactivation, nanoscale TiO_2 (10 and 20 nm) in the anatase form can induce lipid peroxidation and oxidative DNA damage, and increase cellular nitric oxide and hydrogen peroxide levels in BEAS-2B, a human bronchial epithelial cell line. Fpg (formamidopyrimidine [fapy]-DNA glycosylase)-digestible DNA adducts were detected in treatment with anatase TiO_2 (10 nm) particles, but oxidative DNA damage was not detected with >200 nm particles. The size of the particles and the crystalline form are extremely important factors as 200 nm rutile size particles can induce oxidative DNA damage. Concomitantly, levels of cellular melanodialdehyde (MDA), a major end-product of lipid peroxidation, increased following 10 and 20 nm anatase TiO_2 treatments but not with >200 nm anatase or 200 nm rutile treatments. These results indicated that nanoscale anatase could induce oxidative damage to lipids and DNA. Finally, cell treatments with anatase TiO_2 nanoparticles showed an increase in nitric oxide levels and hydrogen peroxide that probably leads to chromosomal damage. Iron and titanium oxide nanoparticles have also been shown to distort mitochondria, leading to a mitochondrial activity disruption that increases oxidative burst (Long et al., 2006). These results are strongly correlated with the size of the particles, the smaller particles being the more destructive. Thus, the anatase form of TiO_2 could induce reactive oxygen species and chromosome damage *in vitro* in the absence of UV photoactivation. Futhermore, rutile and anatase forms differ in their oxidative properties following UV irradiation (Gurr et al., 2005).

The pulmonary effects of TiO_2 particles have been well documented *in vivo*. Several investigators have shown that 20–30 nm TiO_2 particles cause pulmonary inflammation in laboratory animals (Ferin et al., 1990; Ferin et al., 1992; Oberdoerster et al., 1994; Oberdoerster et al., 1995). Rat cells did not respond to 1-µm TiO_2 particles in suspension (Stringer et al., 1996), which strongly suggested that an enhancement of biological reactivity and/or cytotoxicity occurs as the particle size decreased from the micrometer to the nanometer range (Cheng et al., 2004). When the inflammatory responses following intratracheal instillation of ultrafine particles of Co, Ni, and TiO_2 in male rats were compared, TiO_2 particles were found to be the least toxic of these three materials (Zhang et al., 1998).

Untreated TiO_2 particles (hydrophilic surface) and silanized TiO_2 particles (hydrophobic surface) were administered to rats and the inflammatory and genotoxic lung effects were recorded. Animals exposed to untreated and silanized TiO_2 particles showed no signs of inflammation with TNF-α, fibronectin, and surfactant phospholipids (Rehn et al., 2003). Immunohistochemical detection of 8-oxoguanine (8-oxoGua) by a polyclonal antibody in the DNA of individual lung cells was conducted on frozen sections of the left lobe of the lung. The amount of 8-oxoGua, a marker of DNA damage, was at the same level as that of the controls (Rehn et al., 2003). In contrast, nano-sized TiO_2 without UV irradiation has been shown to cause chronic pulmonary inflammation in rats in the presence of alveolar macrophages (Oberdörster et al., 1992). Inhaled ultrafine titanium dioxide particles induced the production of reactive oxygen species in human in rat and human alveolar macrophages (Rahman et al., 1997).

Other investigators have evaluated the acute lung toxicity in rats of intratracheal instilled pigment grade TiO_2 particles (rutile type approximately 300 nm), nanoscale TiO_2 rods (anatase 200 nm $\times$ 35 nm) or nanoscale TiO_2 dots (anatase-10 nm) compared with a positive control particle-type, quartz (1 or 5 mg/kg of the various particle-types). They showed that exposures to nanoscale TiO_2 rods or nanoscale TiO_2 dots produced transient inflammation and cell injury by 24 hours, and was not different from the pulmonary effects of larger-sized TiO_2 particles (Warheit et al., 2006). These results suggest that nanoscale particles were not more cytotoxic or inflammogenic to the lung compared to larger-sized particles of similar composition.

Rat lung cells exposed to ultrafine Ni, Co, and TiO_2 particles of similar diameter indicated that ultrafine ($\leq$ 100 nm) Ni was much more potent than Co or TiO_2 that caused inflammation, suggesting that free radicals produced by transition metals can induce lung inflammation (Zhang et al., 1998). Other studies indicate that nanometer scale size and the surface area of nanoparticles are more important than mass (Cheng et al., 2004). Considerable evidence has been produced indicating that surface area rather than mass should be used as a metric for ultrafine particle toxicity (Brown et al., 2001).

These studies illustrate the complex relationship between toxicity and particle characteristics, including surface coatings and size, which makes generalized statements (e.g., smaller particles are more toxic) incorrect for some substances (Tsuji et al., 2006). Furthermore, if DNA damage occurs, this could suggest a carcinogenic process that is not considered when conducting acute or subacute toxicology studies. Special attention must be focused on titanium dioxide nanoparticles as TiO_2 was classified in 2006 as a possible carcinogen by International Agency for Research on Cancer (IARC) for humans.

Iron Oxides

Iron in its cationic states (Fe^{2+} and Fe^{3+}) is essential for normal cell function and growth. However, chelation of intracellular iron can result in increased apoptosis and DNA fragmentation (Fukuchi et al., 1994; Porreca et al., 1994). Increases in intracellular unbound iron results in oxidative stress and injury to the cells by causing the formation of reactive oxygen species (ROS), which may lead to cell death (Arbab et al., 2003; Emerit et al., 2001; Gutteridge et al., 1982). Though iron plays an important role in virtually all living organisms—primarily through electron transport due to its ability to change valence—it has a rather limited bioavailability and, in some situations, it can be toxic to cells. For this reason, it is necessary for organisms to sequester iron in a nontoxic form. In the human body (including the brain), as well as in most organisms, iron is stored primarily in the core of the iron storage protein ferritin. The ferritin protein is a hollow spheroid shell 12 nm in diameter made up of 24 subunits, and is normally occupied by the iron biomineral ferrihydrite—a hydrated iron oxide ($5Fe_2O_3$ $9H_2O$) that generally contains only Fe (III). It is in this form that most of the iron in the body is stored (Dobson, 2001).

The reaction of the magnetic particles in a magnetic force has been used in applications including drug targeting, bioseparation, and cell sorting. Cell labeling with magnetic nanoparticles is an increasingly common method for *in vitro* cell separation as well as for *in vivo* imaging due to their signal amplification properties in magnetic resonance imaging (MRI). Magnetic cell labeling is very promising for therapy, by allowing for targeted magnetic intracellular hyperthermia (Ito et al., 2001, 2005). All these applications require that cells efficiently capture the magnetic nanoparticles either *in vitro* or *in vivo*. For *in vitro* studies, magnetic labeling only needs cellular uptake by the endocytosis pathway, whereas *in vivo*, high affinity ligands needs to be grafted onto nanoparticles surface for specific cellular interactions (Wilhelm et al., 2003; Zhang et al., 2002). The primary problem encountered by all particles used *in vivo* is the adsorption of biological elements, especially proteins (Portet et al., 2001; Ramge et al., 2000). Once the particles are injected into the bloodstream, they are rapidly coated by plasma proteins, a process known as opsonization, which is critical in dictating the disposition of the injected particles (Davis et al., 1997). Normally, opsonization renders the particles recognizable by the body's major defense system, the reticuloendothelial system (Araujo et al., 1999; Berry et al., 2003; Kreuter et al., 1994).

The role of coating iron nanoparticles on the internalization efficiency has been investigated in a series of studies by Wilhelm et al. (2003). These authors compared cell uptake of anionic maghemite nanoparticles (AMNP), which were coated with DMSA (meso-2,3-dimercaptosuccinic acid), bovine serum albumin (BSA), or dextran. They quantified particle uptake using new complementary magnetic assays, magnetophoresis, and electron spin resonance. After one hour of incubation in mouse macrophages or human ovarian tumor HeLa cells with bare AMNP, adhesion of the anionic nanoparticles on the plasma membrane was seen mainly in the form of clusters. A few minutes later, densely confined AMNP were located in various morphological forms within endosomes and lysosomes. As shown in Figure 11.4, similar clusters on the cell membrane and endosomes containing nanoparticles have been observed when human fibroblasts were exposed for two hours to 0.1 g/l DMSA-coated nanomaghemite. The anionic properties of the particles are important in the binding and uptake efficiency. Following preincubation of AMNP with bovine serum albumin, the linkage of bovine serum albumin onto the AMNP strongly reduced the binding and the internalization of the particles. Uptake of dextran-coated iron oxide was three times lower than that of anionic nanoparticles in HeLa cells. DMSA-coated nanomaghemite interactions with the plasma membrane

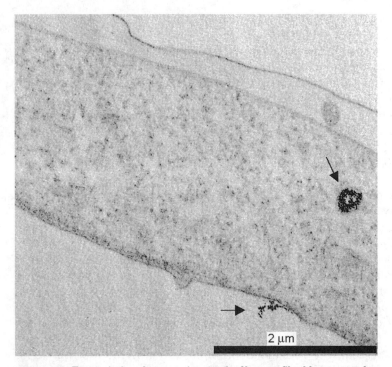

Figure 11.4 Transmission electron micrograph of human fibroblast exposed to 0.1 g/l DMSA-coated nanomaghemite for 2 hours. Arrows depict DMSA-coated nanomaghemite clustered on the cell surface and within a cytoplasmic vacuole.

are triggered by saturable reactive sites that have not been identified. These reactive sites seem ubiquitous. Thus, it is likely that the adsorption of anionic nanoparticles on the cell membrane does not depend on cell specificity. Anionic nanoparticles are characterized by surface negative charges, mainly due to carboxylate groups and by the absence of any steric coating, suggesting that electrostatic interactions govern the adsorption of the anionic nanoparticles onto the cell membrane. As discussed in the next section on quantum dots, cellular uptake of anionic quantum dots was also greater compared to neutral or cationic species.

It is known that plasma membranes possess large negatively charged domains, which should repel anionic nanoparticles. AMNP binds to the cell surface in clusters probably because of their repulsive interactions with the large negatively charged domains of the cell surface. In addition, nanoparticles bound to the cell surface present a reduced charge density that may cause aggregation with other free nanoparticles (Wilhelm et al., 2003). Finally, one can hypothesize that the high efficiency of anionic nanoparticles cell uptake is related first to a nonspecific process of nanoparticles adsorption on the cell membrane and, second, to the formation of nanoparticles clusters. Albumin coating on AMNP hampers their interactions with the membrane, probably due to a steric effect that reduces the accessibility of nanoparticles for the positively charged binding sites on the cell membrane and on the other part diminishes the aggregation of the nanoparticles. Thus, the nonspecific adsorption of nanoparticles on the cell membrane is considerably reduced by albumin coating and, as a consequence, cell internalization is limited to the fluid phase endocytotic pathway.

Further research is needed to determine if the electrostatic forces bound between cationic sites on the cell membrane and anionic nanoparticles are due to their internalization or through a physical mechanism. How are cationic sites on the cell, present at a low density, able to attract anionic nanoparticles? A full explanation for the surprisingly high level of cell uptake that is achieved using anionic nanoparticles is lacking.

The magnetic resonance contrast properties of this DMSA-coated nanomaghemite, free of any dextran coating with a negative surface charge, has been studied (Billotey et al., 2003). The uptake of nanoparticles in macrophages was quantified using electron spin resonance. The precise determination of particle load in cells allows for quantitative comparison in contrast to the cell's internalized particles vs. dispersed isolated particles, and to demonstrate a drastic decrease of longitudinal relaxivity due to cell internalization. The effect on longitudinal relaxivity is explained by a saturation of the relaxing effect of the particles confined within the micrometric endosomes.

Berry and coworkers (2003) have investigated the *in vitro* influence of dextran- or albumin-coated iron oxide nanoparticles in fibroblasts, as compared to those underivatized, using bromodeoxyuridine (BrdU) uptake, light microscopy, scanning electron microscopy, fluorescence of the cytoskeletal filaments (F-actin and vinculin), and clathrin localization. Their results strengthened the role of the shell core of iron oxide nanoparticles in cell responses. Indeed, dextran-derivatized, albumin-derivatized, and underivatized plain magnetite nanoparticles did not induce the same effect on actin filaments and cell proliferation. Whereas dextran-derivatized and underivatized magnetite nanoparticles induced a strong inhibition of the BrdU incorporation with disruption of the F-actin and vinculin filaments, albumin-derivatized nanomagnetite did not have this effect. Nevertheless, the albumin-derivatized nanomagnetite did induce vacuole formation in the cell. The dextran shell on the particles can be broken down, yielding particle chains and aggregates that may influence cell processes (Berry et al., 2003; Jordan et al., 1996). This stresses the importance of the nanoparticle's shell and, consequently, the importance of its stability over time, as the molecular interaction between nanoparticles and the environment (Berry et al., 2004a and 2004b). However, these observations are not in accordance with other studies in which the cell response to dextran-starch–derivatized iron oxide particles showed no measurable lethality (Babincova et al., 2001; Lubbe et al., 1999).

Recently, we have shown that DMSA-coated maghemite nanoparticles *in vitro* induced a moderate but significant decrease in the metabolic mitochondrial activity (MTT assay) on human fibroblast from concentrations ranging from 10^{-6} to 10^{-3} g/L, whereas a slight increase was observed at 10^{-1} g/L. Almost all the surface Fe atoms were affected by the DMSA through chemical linkage, and the DMSA coating was stable in biological media (Auffan et al., 2006). Underivatized Fe_2O_3 nanoparticles induce a dramatic decrease in the metabolic activity and proliferation of human cells (MSTO-211H) (Brunner et al., 2006). This strong difference seems to be related to the DMSA coating. Interestingly, we evidenced a lack of genotoxicity (no DNA strand break using the Comet assay) of DMSA-coated maghemite nanoparticles that could be associated with the lack of evidence of nanoparticles in the cytoplasm or inside mitochondria and nucleus. Nevertheless, we could point out that NmDMSA damage occurs in very few cells at large DMSA-coated maghemite nanoparticles concentration (0,1 g/L). Although the absence of DNA break is not proof of the absence of carcinogenic effect, complete DMSA coating should increase the biocompatibility of iron oxide nanoparticles. Experiments with partial surface coating of nanomaghemites should be led to confirm this protective effect.

A combination of morphological, immunological, and genetic methods demonstrated that the synthesized supraparamagnetic iron oxide nanoparticles (magnetite) derivatized with transferrin encouraged an enhanced cell response as compared to underivatized particles. These influences included an increase in proliferation, altered cytoskeletal organization, cell signaling, and production of an extracellular matrix (Berry et al., 2004a and 2004b). Of interest was the lack of internalization of the transferrin-derivatized particles, despite attachment to the cell membrane. Whereas underivatized magnetite particles inhibit DNA replication, transferring-derivatized nanomagnetite induces cell proliferation. Accordingly, transferrin nanomagnetites induce an up-regulation in a wide variety of genes, including those involved in cytoskeleton, extracellular matrix, replication, and cell signaling, whereas underivatized nanoparticles down-regulated them (Berry et al., 2004a).

The effects of magnetic particles using the ferumoxides (Feridex IV; Berlex Laboratories, Wayne, NJ)-PLL complex for magnetic cell labeling on the long-term viability, function, metabolism, and iron utilization were studied in mammalian cells. The major finding of this study is that the intracellular endosomal magnetic labeling of human mesenchymal stem cells and HeLa cells with ferumoxides combined with the appropriate dilution of PLL, caused no short- or long-term toxic effects on cell viability or proliferation and no increase in the rate of apoptosis or ROS formation (Arbab et al., 2004).

Studies with different contrast agents have shown that the biodistribution depends on the size, charge, and thickness of the coating of the particles (Chouly et al., 1996). Particles that have a hydrophobic surface are efficiently coated with plasma components are thus rapidly removed from the circulation, whereas particles that are more hydrophilic can resist this coating process and are cleared more slowly (Berry et al., 2003, Gaur et al., 2002). In the literature, the most common coatings are derivatives of dextran, polyethylene glycol (PEG), polyethylene oxide (PEO), poloxamers, and polyoxamines (Lacava et al., 2001). In addition, manipulation of particles to ensure the capability of reaching and interacting with target cells would also improve with synthesis of particles with a small hydrodynamic radius (ideally <0 nm) (Berry et al., 2003; Gref et al., 1994; Moghimi et al., 2001).

Mammalian stem cells can be efficiently labeled with a Food and Drug Administration–approved contrast agent, ferumoxides (Feridex IV; Berlex Laboratories, Wayne, NJ), combined with different nonviral transfection agents, such as poly-L-lysine (PLL). These cells can then be used for cellular magnetic resonance (MR) imaging (Arbab et al., 2003; Frank et al., 2002; Frank et al., 2003). Ferumoxides, a dextran-coated superparamagnetic iron oxide (SPIO) nanoparticle, has been used as an MR contrast agent for hepatic imaging (Ferrucci and Stark, 1990; Pouliquen et al., 1991). After intravenous injection, most SPIO nanoparticles accumulate in the Kupffer cells in the liver and in the reticuloendothelial system in the spleen (Ferrucci and Stark, 1990; Pouliquen et al., 1991). Ultrasmall SPIO nanoparticles are used as blood-pool imaging agents and for imaging the lymphatic system. They have a long intravascular half-life and are taken up by the reticuloendothelial system and tissue macrophages. Dextran-coated iron oxide nanoparticles are biodegradable and are metabolized by cells. Subsequently, the iron is introduced into the normal plasma iron pool and can be incorporated into the hemoglobin of red blood cells or used for other metabolic processes (Arbab et al., 2003; Weissleder et al., 1989, 1990). Dendrimer-coated SPIOs (magnetodendrimers) have been used for cellular labeling and remained viable for 10 days (Bulte et al., 2001).

Paramagnetic or modified dextran-coated SPIOs have been used to label cells *ex vivo*, providing researchers with the ability to monitor the migration of these cells with MR imaging (Arbab et al., 2003; Bulte et al., 1999; Josephson et al., 1999; Lewin et al., 2000). The migration of magnetically labeled progenitor oligodendrocytes following implantation into the lateral ventricles of demyelinated rats was identified *in vivo* by using MR imaging up to six weeks following transplantation (Bulte et al., 2001).

MR imaging of the central nervous system can be performed with superparamagnetic iron oxide nanoparticle MRI contrast agents instead of gadolinium-based contrast agents. These iron oxide–based agents include the laboratory preparation MION-46 (Weissleder et al., 1990), the clinically approved agent ferumoxides (Jung CW and Jacobs, 1995; Ros et al., 1995), the investigational agents ferumoxtran-10 (Combidex; Advanced Magnetics Inc., Cambridge, MA) (Anzai et al., 2003; Hoffman et al., 2000; Nguyen et al., 1999), and the new agent ferumoxytol (Ersoy et al., 2004; Prince et al., 2003). These iron oxide contrast agents consist of an iron oxide core with a variable coating that determines cellular uptake and biological half-life. Ferumoxide, a superparamagnetic iron oxide, is 60 to 185 nm in size, incompletely coated with dextran, and rapidly opsonized and endocytosed. The other agents are ultrasmall superparamagnetic iron oxides that are completely coated either with native dextran (ferumoxtran-10) or with a semi-synthetic carbohydrate (ferumoxytol) ([Muldoon et al., 2005). These agents have particle diameters of 20 to 50 nm, show little opsonization and endocytosis, and have long plasma half-lives of 14 to 30 hours (Jung et al., 1995). Despite the growing use of the iron oxide–based MRI contrast agents in the CNS, little is known about the long-term effects of these agents. The distribution and toxicity of superparamagnetic iron oxide magnetic nanoparticles were studied in rat brains and cerebral tumors (Muldoon et al., 2005). Iron has been implicated in formation of free radicals, which would be harmful to neural tissues, particularly in cases of stroke or neurological disease (Dobson, 2004; Moss and Morgan, 2004). The cellular loading experiments specifically tested for formation of reactive oxygen species. No evidence of *in vitro* toxic effects was found from high efficiency iron loading and no increase in reactive oxygen species (Arbab et al., 2004).

Chronic exposure to iron nanoparticles over time is of concern. It has been thought that the particles would dissolve and superparamagnetic iron oxide signal would decrease through natural metabolic pathways (Muldoon et al., 2005). It was shown that different iron oxide particles may be detected in the CNS over extended periods, including strong staining in CNS parenchymal cells. This agrees with previous ultrastructural studies showing intracellular localization of iron particles (Muldoon et al., 1999; Neuwelt et al., 1994). In human biopsy specimens, iron uptake was demonstrated in cells morphologically identical to cells in adjacent sections that stained for glial fibrillary acidic protein (Neuwelt et al., 2004a). Additional studies are needed to demonstrate that the iron labeling is still in the implanted cells at the end of a study, rather than taken up secondarily by macrophages or reactive astrocytes (Muldoon et al., 2005).

Use of the transient osmotic disruption of the blood brain barrier to achieve global delivery throughout an arterial distribution in the CNS has been used showing minimal pathological sequelae to ferumoxtran-10, ferumoxytol, or ferumoxides in rats (Muldoon et al., 2005). The only toxicities reported were mild vacuolization in the corpus callosum. Nevertheless, this is likely the result of the blood brain barrier disruption procedure itself because it was found when each iron oxide agent was used. Indeed, osmotic blood brain barrier disruption is known to have some procedural related toxicity in rats due to the high level of disruption (Remsen et al., 1999). Permeability of the blood-brain barrier has been also investigated using silica-overcoated magnetic nanoparticles containing rhodamine B isothiocyanate (RITC) within a silica shell of 50-nm thickness (Kim et al., 2006). After intraperitoneal administration of these iron oxide nanoparticles in mice for four weeks, they were detected in the brain, indicating that such nanosized materials can penetrate the blood-brain barrier without disturbing its function or producing apparent toxicity.

In humans, disruption of normal iron metabolism in the brain is a characteristic of several neurodegenerative disorders, including Alzheimer's disease, Parkinson's disease, and progressive supranuclear palsy (Dobson, 2001). Thus, distribution, clearance, and toxicity of iron oxide nanoparticles in the brain have particular relevance. For example, excess iron accumulation is known to occur in Alzheimer's disease patients, particularly in plaques, with total iron levels elevated in the hippocampus, amygdala, nucleus basalis of Meynert, and the cerebral cortex (Beard et al., 1993; Fisher et al., 1997; Lovell et al., 1998). These elevated iron levels do not correlate with elevated levels of ferritin (Fisher et al., 1997). Though ferritin is the primary mechanism for iron storage in the brain, over the past decade experimental work has demonstrated the presence of another form of iron in human brain tissue, biogenic magnetite (Fe_3O_4), along with maghemite (γFe_2O_3, an oxidation product of magnetite with very similar magnetic properties) (Kirschvink et al., 1992).

It has been suggested that biogenic magnetite may be present in Alzheimer's disease plaques, senile plaques, and aberrant tau filaments extracted from progressive supranuclear palsy tissue (Quintana et al., 2000). High-resolution transmission electron microscopy and electron energy loss spectroscopy have been used to examine ferritin in paired helical filaments from Alzheimer's disease tissue and ferritin bound to aberrant tau filaments in neurodegenerative progressive supranuclear palsy. The results gave a preliminary indication of the presence of a cubic iron oxide within the ferritin protein cage with spectra similar to synthetic magnetite/maghemite standards (Quintana et al., 2000). Magnetite and maghemite nanoparticles, however, also have been shown to have a substantial effect on free radical generation (Chignell and Sik, 1998; Scaiano, 1997). Experimental results have demonstrated that iron-oxygen complexes may be a more effective catalyst for free radical damage in brain tissue than the Fenton reaction (Schafer et al., 2000). These effects are achieved through strong, local magnetic fields generated by biogenic magnetite particles that stabilize triplet states during biochemical reactions taking place nearby (Dobson, 2001). This leads to the production of membrane-damaging free radicals and changes in reaction yields. Even relatively weak magnetic fields can have a strong influence on reaction yields (Brocklehurst et al., 1996). In addition, Fe (II) in magnetite can be readily oxidized (forming maghemite) and this process, together with local magnetic field effects, may influence L-amyloid production and aggregation (Dobson, 2001). This is particularly relevant considering studies have shown that iron promotes aggregation of L-amyloid peptides *in vitro* and that L-amyloid potentiates free radical formation by stabilizing ferrous iron (Dobson, 2001; Mantyh et al., 1993; Yang et al., 1999).

Based on the involvement of iron in the pathogenesis of important neurodegenerative diseases in humans, the neurological toxicity of iron nanoparticles should be carefully addressed, using long-term animal exposure, for each type of iron oxide nanoparticles, including those differing in terms of coating.

Cerium Dioxides

Cerium oxide has been widely investigated because of its multiple applications—as a catalyst, an electrolyte material of solid oxide fuel cells, a material of high refractive index, and an insulating layer on silicon substrates. The transport and uptake of industrially important cerium oxide nanoparticles into human lung fibroblasts has been measured *in vitro* after exposing thoroughly characterized particle suspensions to fibroblasts with four separate size fractions and concentrations ranging from 100 ng/g to 100 μg/g of fluid (100 ppb to 100 ppm). At physiologically relevant concentrations, a strong dependence of the amount of incorporated ceria on particle size was reported, while nanoparticle number density or total particle surface area was of minor importance (Ludwig et al.). In fact, the rapid formation of agglomerates in the liquid was strongly favored for small particles due to a high number density, while larger ones are mainly unagglomerated (Ludwig et al., 2005). In fact, the biological uptake process on the surface of the cell was faster than the physical transport to the cell. By comparing the colloid stability of a series of oxide nanoparticles, one can hypothesize that untreated oxide suspensions rapidly agglomerate in biological fluids. Thus, the transport and uptake kinetics at low concentrations may be extended to other industrially relevant materials.

Cytotoxicity of insoluble ceria nanoparticles on rodent and human cells (mesothelioma MSTO-211H and 3T3 fibroblast cells) has been assessed using the MTT assays as biomarker of cell viability and DNA content as biomarker of cell proliferation. After three days of exposure, these parameters showed a concentration-dependent decrease, and the cells did not die at 30 ppm. After six days, a small increase in DNA content was reported for ceria concentrations above 7.5 ppm (Brunner et al., 2006). Thus, the moderate but significant cytotoxic effect was transient. This could be explained by the fact that an initial stress arose from nanoparticle incorporation, but sealing or detoxification of the nanoparticles in compartments may have helped to recover full cell viability. Such a sealing mechanism would agree with the results from Limbach et al. (2005), who found that cells incorporate the ceria nanoparticles into vesicles. The formation of small vesicles presumably containing part of the absorbed materials would suggest a mechanism where cells exclude the toxic intruders (Brunner et al., 2006). Further research is needed to fully elucidate the mechanism of such recovery.

Most of the *in vitro* studies describe the toxicity of nanoparticles. Nevertheless, some attention has been given to some types of particles that can protect cells from various forms of lethal stress, as discussed in Chapter 12. It has been shown that nanoparticles composed of cerium oxide or yttrium oxide can protect nerve cells from oxidative stress and that the neuroprotection is independent of particle size (Schubert et al., 2006). Thus, ceria and yttria nanoparticles act as direct antioxidants to limit the amount of reactive oxygen species required to kill the cells.

Copper Nanoparticles

Manufactured copper nanoparticles are industrially produced and are mainly used as inks, metallic coating, polymers/plastics, and additives in lubricants. They have been added to lubricant oil to reduce friction and wear or deposited on the surface of graphite to improve the charge-discharge property. Fluoropolymer conjugated copper nanoparticles have also been employed as bioactive coatings that can cause inhibition of microbial growth (*Saccharomyces cerevisiae*, *Escherichia coli*, *Staphylococcus aureus*, and *Listeria*).

Excessive intake of copper would lead to hemolysis, jaundice, hepatocirrhosis, change in lipid profile, oxidative stress, renal dysfunction, and even death (Björn et al., 2003; Galhardi et al., 2004). Consequently, in the human body, copper is maintained in homeostasis. Nevertheless, there is very little published data related to the toxicity of the copper nanoparticles on mammalian systems.

Using an electrospray technique to generate synthetic nanoparticles in the range of 8–13 nm with monodisperse aerosol particles of approximately the same concentration (5×10^5 cm^{-3}), it has been shown that nano-Cu is a potent inducer in the production of IL-8 in an epithelial cell line (Cheng 2004). Mice exposed to ion-copper, and to nanoscale (25 nm) and micro-sized particles of copper via the gastrointestinal tract had LD50 values of 413 mg/kg, 110 mg/kg, and 5000 mg/kg for nano-copper, ionic-copper, and micro-copper particles, respectively. Ionic-copper and copper nanoparticles were moderately toxic whereas copper microparticles were practically nontoxic, accordingly to the Hodge and Sterner Scale. The kidney of micro-copper exposed mice was similar to control, but nano-copper exposed mice appeared bronze-colored. Copper microparticles caused a slight change in spleens compared to controls, but nanocopper particles (of the similar mass) caused severe atrophy and color changes of spleens. This suggests that spleen is one of the target organs for nanoscale copper particles. Viscera (e.g., kidney, spleen, and liver) of mice exposed to copper nanoparticles were gravely affected. In addition, nanoparticles induced a dose-dependent change. Pathological observations were noted in the kidney, especially to the renal proximal tubules glomeruli lumen of Bowman's capsules. These changes are typical of glomerulonephritis. Biochemical parameters (blood urea nitrogen [BUN], creatinine–total bile acid [TBA], and alkaline phosphatase [ALP]) that depict renal and hepatic functions were significantly greater in nanoparticles-exposed mice than in the controls (Chen et al., 2005).

The great difference in the LD50 of copper nano- and microparticles is thought to be linked to the difference in the specific areas of these particles. Indeed, the specific surface areas for the 23.5 nm and 17 µm copper particles are 2.95×10^5 and 3.99×10^2 cm^2/g, respectively. The particle number (µ/g) for micro-copper is 44 µ/g, while nano-copper is 1.7×10^{10} µ/g. Compared with the copper microparticles, copper nanoparticles (of the same mass) possess much higher collision probability with bio-substances *in vivo*. For the copper nanoparticles, the huge specific surface area leads to the ultrahigh reactivity.

When copper nanoparticles reach the stomach, they react drastically with hydrogen ions (H$^+$) of gastric juices and can be quickly transformed into ionic states. This chemical process undoubtedly results in an overload of ionic copper *in vivo*. Finally, it has been shown that kidney, liver, and spleen are target organs for copper nanoparticles.

Gold Nanoparticles

Many biological applications using gold nanoparticles have emerged during the last decades. Indeed, these nanoparticles could be used as transfection vectors (Muangmoonchai et al., 2002; O'Brien et al., 2002;

Sandhu et al., 2002), DNA-binding agents (Gearheart et al., 2001; McIntosh et al., 2001; Wang et al., 2002), protein inhibitors (Fischer et al., 2002), and spectroscopic markers (Gole et al., 2001; Otsuka et al., 2001; Park et al., 2002; Weizmann et al., 2001). These nanoparticle systems, however, have not been well evaluated to determine their interactions with cells beyond the designated functions.

Cationically functionalized mixed monolayer–protected gold clusters can be used to mediate DNA translocation across the cell membrane in mammalian cells at levels much higher than polyethyleneimine, a widely used transfection vector (Sandhu et al., 2002). Nevertheless, the toxicity of the nanoparticles was observed at concentrations only twofold in excess of that found to have maximal transfection activity (Sandhu et al., 2002). The cytotoxicity of cationic (quaternary ammonium functionalized) and anionic (carboxylate-substituted) gold nanoparticles (2 nm core particles) has been evaluated in red blood cells, Cos-1 cells, and bacterial cultures (*Escherichia coli*). LC50 levels were determined only with the cationic gold nanoparticles. Virtually indistinguishable toxicity profiles have been reported for the two mammalian cells, while a two- to threefold increase is required as a lethal concentration for the bacterial cultures (Goodman et al., 2004). The small variation in LC50 values reported suggests that the mechanism of toxicity is similar for all three cell types, and this implies that the nanoparticles should interact with the cells in a passive process (Goodman et al., 2004). Thus, toxicity should not be caused by cellular uptake via endocytosis as the endocytotic pathway is actively utilized by Cos-1 cells, but not erythrocytes. Additional experiments using a vesicle-disruption assay have demonstrated that the specific charge pairing of the nanoparticles and lipid bilayers mediates membrane interaction and lysis. Thus, the toxicity of the gold nanoparticles seems to be related to their interactions with the cell membrane, a feature initially mediated by their strong electrostatic attraction to the negatively charged bilayer and that the core shell is of great importance in the toxicity properties of the nanoparticles (Goodman et al., 2004).

Water-soluble gold nanoparticles have been functionalized with a Tat protein-derived peptide sequence, with the ability to reach the nucleus. The particles were tested *in vitro* in human fibroblasts by light and transmission electron microscopy, and the biocompatibility and efficacy of the functionalization with the translocated peptide allowed particles to transfer across the cell membrane and locate within the nucleus. Only a slight decrease in MTT metabolic activity and/or proliferation was observed following 24-hour exposure at 10 µM (de la Fuente, 2005).

Gold nanoparticles can be used as X-ray contrast agents. There have been few fundamental improvements in clinical X-ray contrast agents in more than 25 years, and the chemical platform of tri-iodobenzene has not changed. In mice intravenously injected with gold nanoparticles

1.9 nm in diameter, gold cleared the blood more slowly than iodine agents did. Retention in the liver and spleen was low with elimination by the kidneys. With 10 mg/ml of gold initially in the blood, mouse behavior was unremarkable and neither blood plasma analytes nor organ histology revealed any evidence of toxicity 11 days and 30 days after injection (de la Fuente et al., 2005).

Globally, any high level of toxicity associated with the gold nanoparticles is expected because gold as well as silver colloids have a history in medicines and natural therapeutics. Nevertheless, the smaller size merits investigation, as do the ligand shells that can be combined with the metal core. Table 11.2 lists a number of reports on metal nanomaterials toxicity.

Quantum Dots

Quantum dots (QD) have widespread potential applications in biology and medicine, including semiconductor nanocrystals used as fluorescent probes in ultrasensitive bioassays, biological staining, and diagnostics. QD are ideal for fluorescent biolabeling because they have long-term photostability, high brightness, and they can be labeled with several colors for multi-target studies. QD can be synthesized with different core semiconductor materials, including cadmium selenide (CdSe), cadmium telluride, indium phosphide, or indium arsenide, however, the latter three have not been shown to be useful conjugates for biological purposes. CdSe core QD has been used in biological labeling due to their bright fluorescence, narrow emission, broad UV excitation, and high photostability (Bruchez et al., 1998; Jovin, 2003; Watson et al., 2003). Compared to organic dyes such as rhodamine, they can be 20 times as bright and 100 times more stable against photobleaching. When QD are embedded in biological fluids and tissues, their excitation wavelengths can be compromised, so their excitation and wavelength should be selected based on their specific application (Lim et al., 2003). Customized QD may have multiple layers with one or more surface coatings. The core can consist of CdSe with a shell or "cap," such as ZnS, that will reduce leaching of the toxic core metals (Derfus et al., 2004). The unique properties of QD have shown promise for numerous biological applications for detection and imaging, however, their toxicology in humans is not known.

There are numerous reports of QD cytotoxicity in the literature, some being conducted by the laboratories that synthesize QD for customized applications. Interpretation of these studies make it difficult because of the heterogeneity of these QD preparations from different laboratories (core composition, coatings, size, etc.), the use of different cell lines, and a variety of endpoints of cytotoxicity. QD can be purchased commercially, but their toxicity in cells is unknown.

TABLE 11.2 Negative Impacts of Metal Nanomaterials on Living Mammalian Cells and Organisms

Type of nanomaterials	Organisms, or cell types, or organelles	Effects observed	References
Titanium dioxide			
	Human endothelial cells	Proinflammatory effects	Peters et al., 2004
	Human cells	DNA damage induction by sunlight-illuminated TiO_2	Dunfort et al., 1997
	Syrian hamster embryo fibroblast	Apoptosis induction and micronuclei formation	Rahman et al., 2002
	Laboratory animals	Pulmonary inflammation	Ferin et al., 1990; Ferin et al., 1992; Oberdoerster et al., 1994; Oberdoerster et al., 1995 Oberdoerster et al., 1992
	Rat and human alveolar macrophages	Production of reactive oxygen species	Rahman et al., 1997
Anastase TiO_2	Human bronchial epithelial cell line	Lipid peroxidation and oxidative DNA damage; increase in cellular nitric oxide and hydrogen peroxide levels; increase in cellular MDA levels	Gurr et al., 2005
Iron oxide			
	Mouse macrophages or human ovarian tumor HeLa cells	Adhesion to plasma membrane and internalization efficiency of dextran-, albumin- and DMSA-coated iron oxide nanoparticles.	Wilhelm et al., 2003
	Human CNS	Detection in CNS and staining of CNS parenchymal cells	Muldoon et al., 2005
	Human and rodent cells	Decreased MTT activity and DNA content	Brunner et al., 2006

(Continued)

TABLE 11.2 Negative Impacts of Metal Nanomaterials on Living Mammalian Cells and Organisms (*Continued*)

Type of nanomaterials	Organisms, or cell types, or organelles	Effects observed	References
Magnetite	Human fibroblasts	Comparison between dextran-, albumin-derivatized and underivatized magnetite cytotoxicity (inhibition of the BrdU incorporation, disruption of the F-actin and vinculin filaments)	Berry et al., 2003
Transferrin-derivatized magnetite	Human fibroblasts	Increase in cell proliferation, alteration of cytoskeletal organization, and upregulation of proteins involved in cell signalization and extracellular matrix	Berry et al., 2004a and 2004b
Ferumoxides	Intravenous injection	Accumulation in the Kupffer cells in the liver and in the reticuloendothelial system in the spleen	Ferrucci and Stark, 1990; Pouliquen et al., 1991
Silica-overcoated magnetic nanoparticles containing RITC within the silica shell	Intraperitoneal injection in mice	Penetration through the blood-brain barrier without disturbing its function or producing apparent toxicity	Kim et al., 2006
Cerium oxide	Human lung fibroblasts	Size-dependent internalization of the particles	Limbach et al., 2005
	Human and rodent cells	Decreased MTT activity and DNA content	Brunner et al., 2006
Copper	Epithelial cell line	Inducer in production of IL-8	Cheng, 2004
	Mice (Gastrointestinal exposure)	Target organs of copper nanoparticles: kidney, spleen, and liver	Chen et al., 2005

| Gold | Red blood cells, Cos-1 cells and *Escherichia Coli* | *In vitro* comparison of cationic (quaternary ammonium functionalized) and anionic (carboxylate-substituted) gold nanoparticles | Goodman et al., 2004 |
| | Human fibroblast cell line | Slight decrease in cell metabolic activity and/or proliferation induced by a water-soluble gold nanoparticles functionalized with a Tat protein-derived peptide sequence | de la Fuente, 2005 |

Negative impacts of nanomaterials on living microorganisms are illustrated in Chapter 12.
MDA: melanodialdehyde; DMSA: meso-2,3-dimercaptosuccinic acid; CNS: central nervous system; MTT: 3-[4,5-dimethylthiazol-2-yl]
-2,5-diphenyltetrazolium bromide; PEG: polyethylene glycol.

Depending upon the applications, hydrophilic coatings may be added to increase the solubility for biological media or conjugated to macromolecules for drug delivery and diagnostics (Michalet et al., 2005). These nanometer sized conjugates are water soluble and biocompatible. When QD are prepared in organic solvents, they are not compatible for biological applications. By using mercaptoacetic acid for solubilization and covalent protein attachment, they become water-soluble. This allows for free carboxyl groups to be available for covalent coupling to biomolecules such as proteins, peptides, and nucleic acids. When labeled with the protein transferrin, they undergo a receptor-mediated endocytosis in cultured HeLa cells, and when conjugated to immunomolecules, they can recognize specific antibodies or antigens (Chan and Nie, 1998). Other studies have shown that CdSe-core/ZnS-shell QD conjugated to a targeting peptide could be internalized by HeLa cells and tracked in live cells for more than ten days with no signs of toxicity (Jaiswal et al., 2003). This study suggested that QD were nontoxic at doses suitable for long-term imaging studies and as vectors for drug delivery.

The effect of two different dosages of PEG-silane-QD (8 and 80 nM) on phenotypic and whole genome expression of exposed human fibroblasts were examined. These QD did not induce any statistically significant changes in cell cycle and were associated with minimal apoptosis/necrosis in human skin fibroblast at both doses. There was a down-regulation of genes involved in controlling the mitotic M-phase spindle formation and cytokinesis. Importantly, they did not activate genes associated with immune and inflammatory responses (Zhang et al., 2006).

Recently, we reported that exposure of skin to commercially available QD differing in core/shell shape, hydrodynamic size, and surface coatings resulted in penetration of the intact stratum corneum barrier with localization of QD in the underlying epidermal and dermal layers as early as eight hours after topical application (Ryman-Rasmussen et al., 2006). This study has indicated that skin is a potential route of exposure for QD. However, the toxicity of QD to skin cells is unknown and the potential for QD to cause inflammation had not been addressed. Therefore, we conducted a study to determine the cytotoxic and inflammatory potential of QD in skin cells (human epidermal keratinocytes [HEK]). Soluble QD of two sizes, QD 565 and QD 655, were obtained with three different surface coatings: polyethylene glycol (PEG), PEG-amines, or carboxylic acids. The QD were studied with laser scanning confocal microscopy on live human epidermal keratinocytes and transmission electron microscopy to verify QD uptake in cells by 24-hour MTT viability and inflammatory cytokines IL-1β, IL-6, IL-8, IL-10, and TNF-α was assessed at 24 and 48 hours. The results indicated that all QD were intracellularly located in HEK within 24 hours, independent of QD

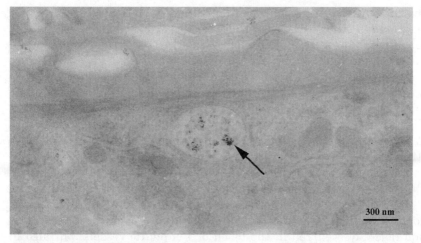

Figure 11.5 Transmission electron micrograph of human epidermal keratinocyte exposed to QD 655 with a carboxylic acid surface coating at 24 hours. Arrow depicts QD within a cytoplasmic vacuole.

size and surface coating (Figure 11.5). However, the cytotoxic and inflammatory effects of QD were dependent upon their surface coating, with a modulatory role for size on cytokine release. Quantum dots and other nanostructures are expected to have high inflammatory potential due to a large surface area to volume ratio (Oberdorster et al., 2005). Similar to the preferential uptake of anionic ferric nanoparticles in cells discussed above, our studies showed carboxylated anionic QD accumulated the most in cells and were most cytotoxic. This study is the first account of QD uptake, cytotoxicity, and inflammation in human epidermal keratinocytes (Ryman-Rasmussen et al., 2007).

Other studies have shown how QD can act as fluorescence labels for biological and biomedical cellular imaging. Wu et al. (2003) synthesized specific immunofluorescent probes by linking QD to streptavidin and IgGs to label breast cancer marker Her2 on the surface of fixed and live cancer cells. They conducted labeling efficiency studies of these probes in targets such as cell surface receptors, cytoskeleton components, and nuclear image antigens at different cellular locations, including cell surface, intracellular, and intranuclear. Multiphoton fluorescence imaging of water-soluble QD in mice were conducted to compare conventional fluorescein isothiocyanate conjugated to dextran beads. They were found to be superior and achieved greater depths using less power, especially in high scattering skin and adipose tissue. Mice showed no noticeable ill effects (Larson et al., 2003). Noninvasive imaging of four different QD coatings were tested in mice. The QD circulating half-lives were less than 12 minutes for amphiphilic poly (acrylic acid), short chain methoxy-PEG, or long chain carboxy-PEG QD, but 70 minutes for long chain methoxy-PEG QD. These studies also demonstrated that QD remained fluorescent for four months in mice (Ballou et al., 2004), indicating potential long-term exposure.

Other explorations have been conducted on QD of less than 10 nm on the feasibility of *in vivo* targeting to specific tissues and cell types that could escape the reticuloendothelial system filter. These investigators showed the selective targeting of peptide-coated QD in the vasculature of lungs and tumors in mice after intravenous injection that demonstrated it is possible to target QD in living mammals. These QD showed homing specificity for the relevant vascular sites but did not see accumulation of fluorescence within their targeted tissue (Akerman et al., 2002). This was in contrast to what was seen earlier with two tumor-homing peptides F3 and LyP-1 coupled to fluorescein where accumulation not only occurred in the blood or lymphatic vessels, but also in tumor cells. Nanocrystals can be encapsulated in phospholipid block polymer-copolymer micelles to increase their biocompatibility in Xenopus embryos and were found to be stable and nontoxic. QD encapsulated in phospholipid micelles were injected in frog oocyte cells for real-time tracking in embryonic development (Dubertret et al., 2002). Bioconjugated QD probes with an ABC triblock copolymer linked to a tumor-targeting ligands and drug functionalities were studied in human prostate cancer growing in nude mice. This study indicated that these QD accumulated at tumors both by the enhanced permeability and retention of tumor sites and by the antibody binding to cancer specific cell surface biomarkers (Gao et al., 2004).

Derfus et al. (2004) specifically designed a cytotoxicity study with QD at relevant concentrations for imaging applications and found that MTT viability decreased in human hepatocytes after treatment with uncoated UV irradiated or chemically oxidized CdSe QD. A dose-dependent increase in MTT cytotoxicity was noted in mercaptoundecanoic acid-coated CdSe QD of three different sizes in African green monkey kidney cells (Vero cells), HeLa cells, and human hepatocytes (Shiosahara et al., 2004). Other studies have shown that positively charged cadmium tellerium (CdTe) QD coated with cysteineamine were shown to be cytotoxic at concentrations used for confocal imaging in N9 murine microglial and rat pheochromocytoma (PC12) cells (Lovric et al., 2005). Thus, some QD preparations have demonstrated cytotoxicity in some cell lines at doses relevant for biomedical applications. Additionally, cells undergoing rapid migration or those with high membrane turnover are potentially vulnerable to nonselective uptake of QD by association with the cell membrane (Parak et al., 2002).

It has been noted that QD size, surface charge, and chemical composition, along with cell-dependent properties, play a role in cytotoxicity.

Studies have shown that in cell culture, QD with a cadmium-selenium core could be rendered nontoxic with additional coatings. In contrast, when exposed to air or ultraviolet radiation, they were toxic. This toxicity correlated with Cd^{+2} release in the culture medium, since hepatocytes are highly cadmium-sensitive. Cadmium release and cytotoxicity could be attenuated by coating QD with zinc sulfide or bovine serum albumin prior to exposure to an oxidative environment. These observations illustrated that oxidative microenvironments, QD chemical composition, and cellular sensitivity can all contribute to QD cytotoxicity. This would also suggest that QD exposure to unintended physical or chemical stressors during occupational or environmental settings could modulate toxicity.

There are also indications that the size and charge of QD, together with their intracellular localization, may play a role in their cytotoxicity. Positively charged CdTe-core, cysteineamine-capped QD were cytotoxic to PC12 and N9 cells (Lovrić et al., 2005). In both cases, the 2.3 nm QD were significantly more cytotoxic than the 5.2 nm QD. This cytotoxicity could be partially attenuated by N-acetylcysteine but not by α-tocopherol, suggesting a mechanism other than free-radical generation resulting from Cd^{+2} release. The 5.2-nm QD localized to the cytoplasm of N9 cells, whereas the 2.3 nm localized to the nucleus. Nuclear localization of cationic QD has been reported previously in mouse fibroblast (3T3) cells (Bruchez et al., 1998). Whether or not this potentiated cytotoxicity of 2.3-nm QD is directly related to nuclear localization is unknown. It has been noted that CdSe/ZnS QD can cause nicking of supercoiled plasmid DNA (Green and Howman, 2005).

In summary, QD offer great potential as therapeutic and diagnostic agents. Contrary to initial reports, QD can be cytotoxic at concentrations used in biomedical imaging applications in some cell types. QD size, charge, shape, and chemical composition, together with cell-specific factors, can play a role in cytotoxicity. All of these physiochemical properties play a role in the toxicity of QD and are important determinants for risk assessment. Table 11.3 lists a number of reports on QD nanomaterials toxicity.

Exposure and Risk Assessment

This chapter has reviewed the literature related to specific nanoparticle types, which often dictates the kind of studies conducted based on their intended biomedical applications (e.g., drug delivery, imaging, tissue engineering). However, more general principles governing the safety of all particles scenarios need to be evaluated to establish specific nanotoxicology safety and testing guidelines after occupational exposure during manufacture, exposure in academic research laboratories, or environmental exposure, either from manufacturing waste or postconsumer

TABLE 11.3   Negative Impacts of Quantum Dots on Living Mammalian Cells and Organisms

Type of nanomaterials	Organisms, or cell types, or organelles	Effects observed	References
Quantum dots (QD)	Porcine skin	Penetration of the stratum corneum barrier with localization of QD in the underlying epidermal and dermal layers	Ryman-Rasmussen et al., 2006
	Mouse fibroblast (3T3) cells	Nuclear localization of cationic QD	Bruchez et al., 1998
QD 565 and QD 655 coated with polyethylene glycol (PEG), PEG-amines, or carboxylic acids	Human epidermal keratinocytes	Nondependent size and surface coating internalization; surface coating–dependent cytotoxicity; size and surface coating–dependent cytokine release. Carboxylic acid QDs accumulated the most in cells and were most cytotoxic.	Ryman-Rasmussen et al., 2007
PEG-silane-QD	Human fibroblasts	Down-regulation of genes involved in controlling the mitotic M-phase spindle formation and cytokinesis; no activation of genes associated with immune and inflammatory responses	Zhang et al., 2006
CdSe QD	Human hepatocytes	Decrease in MTT viability after treatment with uncoated UV irradiated or chemically oxidized CdSe QD	Derfus et al., 2004
Mercaptoundecanoic acid-coated CdSe QD	Vero cells, HeLa cells, human hepatocytes	Increase in MTT Cytotoxicity	Shiosahara et al., 2004
Cadmium tellerium (CdTe) QD coated with cysteineamine	N9 murine microglial and rat pheochromocytoma (PC12) cells	Cytotoxicity induced by positively charged QDs. Nuclear localization of the 2.3 nm but not the 5.2 nm QD	Lovrić et al., 2005

Negative impacts of nanomaterials on living microorganisms are illustrated in Chapter 12.
MTT: 3-[4,5-dimethylthiazol-2-yl]-2,5-diphenyltetrazolium bromide.

use (Monteiro-Riviere and Ryman-Rasmussen, 2006). Risk is defined as "exposure" times "hazard." The studies reviewed above demonstrate that many nanomaterials can produce biological effects; hence hazard exists. Exposure is a function of both source and portal of entry into the body. Studies have shown injection, dermal, inhalation, and oral absorption; hence exposure occurs. What is our risk of toxicity to nanomaterials?

A great deal of attention has been focused on inhalational toxicology of carbon nanomaterials due to their particulate nature and their anthropogenic occurrence in diesel exhaust. However, it is obvious that many nanomaterials produced for biomedical or cosmetic applications do not result in inhalational exposure. Engineered nanoparticles are produced in liquid phase or close gas phase reactors; therefore inhalation would have to occur in the solid state. Since many engineered nanomaterials are prepared and processed in liquids, there is a high probability that dermal absorption and oral ingestion may be the more relevant exposure route during the manufacturing process or during accidental spills, shipping, and handling (Colvin, 2003). Prevention of dermal exposure by the use of gloves is underscored by a study indicating that some workers had as much as 7 mg of nanotube material deposited on gloves in areas directly contacting nanotubes (Baron et al., 2002). This study had shown that gloves could offer dermal protection from nanotube material, with the caveat being that the permeability of different types of gloves to specific nanomaterials is unknown.

Environmental Impact

Nanotechnology is producing significant quantities of nanomaterials that could potentially be released into the environment and migrate through porous media found in groundwater aquifers and water-treatment plant filters. Deliberate release of nanomaterials has been suggested for environmental remediation to detoxify pollutants in the soil and groundwater. Nanomaterials such as titanium dioxide and zinc oxide can be activated by light, and could help to remove organic contaminants from various media. Additional studies are needed in this arena before remediation can be recommended. There must be sufficient evidence to clearly demonstrate that there are no adverse health effects to humans and wildlife. There is little data available on evaluating fullerene toxicity in aqueous systems. We have previously discussed how surface characteristics of many nanomaterials, including fullerenes, modify their interactions with biological systems. Monomeric C_{60} has extremely low solubility in water and is unlikely to have a significant effect in aqueous systems, as discussed in Chapter 7. However, environmentally modified or functionalized nanomaterials could exhibit different transport behavior. Functionalized fullerenes that are used to facilitate dispersal in water have shown to have the highest mobility and can migrate 10 meters in unfractured sand aquifers (Lecoanet et al., 2004). Studies have also shown that fullerol toxicity can increase in the presence of light (Pickering and Wiesner, 2005). These studies underline the importance of basic nanomaterials chemistry to predict potential adverse effects.

Conclusion

Many of the studies discussed in this chapter clearly indicate that nanomaterials interact with biological systems. Composition, size, and surface property are important attributes that are needed to predict biological effects. It is intriguing to realize that although materials of vastly different chemical compositions have similar interactions with biological systems—the clearest example being preferential cellular uptake of anionic iron oxide nanoparticles and QD. Size also seems to be an important factor. The precise nature of toxicity seen is a function of the chemistry of the particle. The physical chemical properties of many nanoparticle surface modifications may be the factor that determines their ultimate safety. A great deal of literature exists for those nanomaterials with commercial applications, including titanium dioxide particles for sunscreens and iron oxide particles or QD for imaging. Similarly, pivotal disposition and cell targeting data are often available for those nanomaterials intended for use in drug delivery, such as seen with QD, fullerenes, and carbon nanotubes. There is a serious lack of information about human health and environmental implications of manufactured nanomaterials. This emerging field of nanotoxicology will continue to grow as new products are produced. The need for toxicology studies will increase for use in risk assessment. Knowledge of exposure and hazard are needed for understanding risks associated with nanomaterials. This chapter has explored the beginning threads of nanomaterial toxicology for a variety of nanomaterials.

References

Akerman, M.E., Chan, W.C.W., Laakkonen, P., Bhatia, S.N., and Ruoslahti, E. (2002) "Nanocrystal targeting in vivo," Proceeding of the National Academy of Sciences *Natl. Acad. Sci.*, 99(20):12617–12621.

Allen, D.G., Riviere, J.E., and Monteiro-Riviere, N.A. (2001) "Cytokine induction as a measure of cutaneous toxicity in primary and immortalized porcine keratinocytes exposed to jet fuels, and their relationship to normal human epidermal keratinocytes," *Toxicology Letters.*, 119:209–217.

Anzai, Y., Piccoli, C.W., Outwater, E.K., Stanford, W., Bluemke, D.A., Nurenberg, P., Saini, S., Maravilla, K.R., Feldman, D.E., Schmiedl, U.P., Brunberg, J.A., Francis, I.R., Harms, S.E., Som, P.M., and Tempany, C.M. (2003) "Evaluation of neck and body metastases to nodes with ferumoxtran 10-enhanced MR imaging: phase III safety and efficacy study," *Radiology*, 228:777–788.

Araujo, L., Lobenberg, R., and Kreuter, J. (1999) "Influence of the surfactant concentration on the body distribution of nanoparticles," *J. Drug Targeting*, 6(5):373–385.
Arbab, A., Bashaw, L., Miller, B., Jordan, E., Lewis, B., Kalish, H., and Frank, J. (2003) "Characterization of biophysical and metabolic properties of cells labelled with superparamagnetic iron oxide nanoparticles and transfection agent for cellular MR imaging," *Radiology*, 229:838–846.
Arbab, A.S., Yocum, G.T., Kalish, H., Jordan, E.K., Anderson, S.A., Khakoo, A.Y., Read E.J., and Frank, J.A. (2004) "Efficient magnetic cell labelling with protamine sulphate complexed to ferumoxides for cellular MRI," *Blood*, 104:1217–1223.
Auffan, M., Decome, L., Rose, J., Orsière, T., De Méo, M., Briois, V., Chaneac, C., Olivi, L., Bergé-Lefranc, J-L., Botta A., Wiesner, M., and Bottero, J-Y. (2006) In vitro interactions between DMSA-coated maghemite nanoparticles and human fibroblasts: a physico-chemical and cyto-genotoxical study. *Environmental. Sciences. Technology.*, 40(14); 4367–4373.
Babincova, M., Leszczynska, D., Sourivong, P., and Babinec, P. (2001) "Selective treatment of neoplastic cells using ferritin-mediated electromagnetic hyperthermia," *Medical. Hypothermia.*, 54(2):177–179.
Ballou, B., Langerhholm, B.C., Ernst, L.A., Bruchez, M.P., and Waggoner, A.S. (2004). "Noninvasive imaging of quantum dots in mice," *Bioconjugate Chemistry.*, 15:79–86.
Baron, P.A., Maynard, A., and Foley, M. (2002) "Evaluation of aerosol release during the handling of unrefined single walled carbon nanotube material," *NIOSH Dart*-02-191.
Beard, J.L., Connor, J.R., and Jones, B.C. (1993) "Iron in the brain," *Nutr. Rev.,*? 51:157–170.
Bernard, B.K., Osheroff, M.R., Hofmann, A., and Mennear, J.H. (1990) "Toxicology and carcinogenesis studies of dietary titanium dioxide-coated mica in male and female Fischer 344 rats," *J. Toxicology. Environmental. Health*, 29(4):417–429.
Berry, C., Charles, S., Wells, S., Dalby, M., and Curtis, A. (2004a) "The influence of transferrin stabilised magnetic nanoparticles on human dermal fibroblasts in culture," *Internat. J. Pharmacology.*, 269:211–225.
Berry, C., Wells, S., Charles, S., Aitchison, G., and Curtis, A. (2004b) "Cell response to dextran-derivatized iron oxide nanoparticles post internalisation," *Biomaterials*, 25(23):5405–5413.
Berry, C., Wells, S., Charles, S., and Curtis, A. (2003) "Dextran and albumin derivatized iron oxide nanoparticles: influence on fibroblasts in vitro," *Biomaterials*, 24:4551–4557.
Billotey, C., Wilhelm, C., Devaud, M., Bacri, J.C., Bittoun, J., and Gazeau, F. (2003) "Cell internalization of anionic maghemite nanoparticles: quantitative effect on magnetic resonance imaging," *Mag. Res. Med.*, 49:646–654.
Björn, P.Z., Hermann, H.D., Max, L., Heide, S., Barabara, K.G., Hartmut, D. (2003) "Epidemiological investigation on chronic copper toxicity to children exposed via the public drinking water supply," *Sci. Total Environ.* 302:127–144.
Bottini, M., Bruckner, S, Nika, K., Bottina, N., Bellucci, S., Magrini, A., Bergamaschi, A., and Mustelin, T. (2006) "Multi-walled carbon nanotubes induce T lymphocyte apoptosis," *Toxicology Letters. Lett.* 160:121–126.
Brocklehurst, B., and McLauchlan, K.A. (1996) "Free radical mechanism for the effects of environmental electromagnetic fields on biological systems," *Int. J. Radiation. Biology.*, 69:3–24.
Brown, D.M., Wilson, M.R., MacNee, W., Stone, V., and Donaldson, K. (2001) "Size-dependent proinflammatory effects of ultrafine polystyrene particles: a role of surface area and oxidative stress in enhanced activity of ultrafines," *Toxicol. Applied. Pharmacology.*, 175:191–199.
Bruchez, M., Moronne, M., Gin, P., Weiss, S., and Alivisatos, A.P. (1998) "Semiconductor nanocrystals as fluorescent biological labels," *Science,* 281:2013–2016.
Brunner, T., Wick, P., Manser, P., Spohn, P., Grass, R., Limbach, L., Bruinink, A., and Stark, W. "In vitro cytotoxicity of oxide nanoparticles: comparison to asbestos, silica, and the effect of particle solubility," *Environmental. Science. Technology.*, 2006; DOI: 10.1021/es052069i.
Bulte, J.M.W., Zhang, S.C., van Gelderen, P., Herynek, V., Jordan, E.K., Duncan, I.D., and Frank, J.A. (1999) "Neurotransplantation of magnetically labelled oligodendrocyte

progenitors: magnetic resonance tracking of cell migration and myelination," *Proc. Natl. Acad. Sci. USA,* 96:15256–15261.

Bulte, J.W., Douglas, T., Witwer, B., Zhang, S.C., Strable, E,, Lewis, B.K., Zywicke, H., Miller, B., van Gelderen, P., Moskowitz, B.M., Duncan, I.D., and Frank, J.A. (2001) "Magnetodendrimers allow endosomal magnetic labelling and in vivo tracking of stem cells," *Nature. Biotechnol.,* 19(12):1141–1147.

Cai, D., Mataraza, J.M., Qin, Z.H., Huang, Z., Huang, J., Chiles, T.C., Carnahan, D., Kempa, K., and Ren, Z. (2005) "Highly efficient molecular delivery into mammalian cells using carbon nanotube spearing," *Nature Methods,* 2:449–454.

Chan, W.C.W., and Nie, S. (1998) "Quantum dot bioconjugates for ultrasensitive noniso-topic detection," *Science,* 281:2016–2018.

Chen, J.L., and Fayerweather,W.E. (1988) "Epidemiologic study of workers exposed to tita-nium dioxide," *J. Occupational. Medicine.,* 30(12):937–942.

Chen, B.-X., Wilson, S.R., Das, M., Coughlin, D.J., and Erlanger, B.F. (1998) "Antigenicity of fullerenes: antibodies specific for fullerenes and their characteristics," *Proc. Natl. Acad. Sci. USA,* 95:10809–10813.

Chen, Z., Meng, H., Xing, G., Chen, C., Zhao, Y., Jia, G., Wang, T., Yuan, H., Ye, C., Zhao, F., Chai, Z., Zhu, C., Fang, X., Ma, B., and Wan L. (2006) Acute toxicological effects of copper nanoparticles in vivo. *Toxicology Letters.* 163(2):109–120.

Cheng, M.D. (2004) "Effects of nanophase materials ($\leq$ 20 nm) on biological responses," *J. Environ. Sci. Health,* A39:2691–2705.

Cherukuri, P., Bachilo, S.M., Litovsky, S.H., and Weisman, R.B. (2004) "Near-infrared fluorescence microscopy of single-walled carbon nanotubes in phagocytic cells," *J. American. Chemical. Society.,* 126:15638–15639.

Chignell, C.F., and Sik, R.H. (1998) "Effect of magnetite particles on photoinduced and nonphotoinduced free radical processes in human erythrocytes," *Photochemistry. Photobiology.,* 68(4):598–601.

Chouly, C., Pouliquen, D., Lucet, I., Jeune, J.J., and Jallet, P. (1996) "Development of super-paramagnetic nanoparticles for MRI: effect of particle size, charge and surface nature on biodistribution," *J. Microencapsul.,* 3:245–255.

Colvin, V.L. (2003) "The potential environmental impact of engineered nanomaterials," *Nature Biotechology.,* 21:1166–1170.

Cui, D., Tian, F., Ozkan, C.S., Wang, M., and Gao, H. (2005) "Effect of single wall carbon nanotubes on human HEK293 cells," *Toxicology Letters,* 155:73–85.

Cunningham, M.J., Magnuson, S.R., and Falduto, M.T. (2005) "Gene expression profiling of nanoscale materials using a systems biology approach," *Toxicological Sciences* 84(S-1):9.

Dagani, R. (2003) "Nanomaterials: safe or unsafe?" *Chemical Engineering News,* 81: 30–33.

Derfus, A.M., Chan, W.C.W., and Bhatia, S. (2004) "Probing the cytotoxicity of semicon-ductor quantum dots," *Nano Letters.,* 4(1):11–18.

Dobson, J. (2001) "Nanoscale biogenic iron oxides and neurodegenerative disease," *FEBS Letters,* 496:1–5.

Dobson, J. (2004) "Magnetic iron compounds in neurological disorders," *Ann. NY Acad. Sci.,* 1012:183–192.

Driscoll, K.E., Costa, D.L., Hatch, G., Henderson, R., Obersorster, G., Salem, H., and Schlesinger, R.B. (2000) "Intratracheal instillation as an exposure technique for the eval-uation of respiratory tract toxicity: uses and limitations," *Toxicological Sciences.,* 55:24–35.

Dubertret, B., Skourides, P., Norris, D.J., Noireaux, V., Brivanlou, A.H., and Libchaber, A. (2002) "In vivo imaging of quantum dots encapsulated in phosopholipid micelles," *Science,* 298:1759–1762.

Dugan, L., Turetski, D.M., Du, C., Lobner, D., Wheeler, M., Almli, C.R., Shen, C.K.-F., Luh, T.-Y., Choi, D.W., and Lin, T.-S. (1997) "Carboxyfullerenes as neuroprotective agents," *Proc. Natl. Acad. Sci.,* 94:9434–9439.

Dunford, R., Salinaro, A., Cai, L., Serpone, N., Horikoshi, S., Hidaka, H., and Knowland, J. (1997) "Chemical oxidation and DNA damage catalysed by inorganic sunscreen ingre-dients," *Federation Experimental Biology Society Lett.,* 418(1/2):87–90.

Ebbesen, T.W., and Ajayan, P.M. (1992) "Large-scale synthesis of carbon nanotubes," *Nature,* 358:220.

Emerit, J., Beaumont, C., and Trivin, F. (2001) "Iron metabolism, free radicals, and oxidative injury," *Biomedical. Pharmacotherapy.,* 55:333–339.

Ersoy, H., Jacobs, P., Kent, C.K., and Prince, M.R. (2004) "Blood pool MR angiography of aortic stent-graft endoleak," *Am. J. Roentgenology.,* 182:1181–1186.

Ferin, J., Oberdoerster, G., Penney, D.P., Soderholm, S.C., Gelein, R., and Pipper, H.C. (1990) "Increased pulmonary toxicity of ultrafine particles ? 1. particles clearance, translocation, morphology," *J. Aerosol Sci.,* 21:381–384.

Ferin, J., Oberdoerster, G., and Penney, D.P. (1992) "Pulmonary retention of ultrafine and fine particles in rats," *Am. J. Respiratory. Cell Mol. Biol.,* 6:535–542.

Ferrucci, J.T., and Stark, D.D. (1990) "Iron oxide-enhanced MR imaging of the liver and spleen: review of the first 5 years," *Am. J. Roentgenol.,* 155:943–950.

Fiorito, S., Serafino, A., Andreola, F., and Bernier, P. (2006) "Effects of fullerenes and single-wall carbon nanotubes on murine and human macrophages," *Carbon,* 44:1100–1105.

Fischer, P., Gotz, M.E., Danielczyk, W., Gsell, W., and Riederer, P. (1997) "Blood transferrin and ferritin in Alzheimer's disease," *Life Sci.,* 60:2273–2278.

Fischer, N. O., McIntosh, C. M., Simard, J. M., and Rotello, V. M. (2002) Inhibition of chymotrypsin through surface binding using nanoparticle-based receptors. *Proc. Natl. Acad. Sci.* 99:5018–5023.

Flahaut, E., Durrieu, M.C., Remy-Zolghadri, M., Bareille, R., and Baquey, C. (2006) "Investigation of the cytotoxicity of CCVD carbon nanotubes towards human umbilical vein endothelial cells," *Carbon,* 44:1093–1099.

Frank, J.A., Miller, B.R., Arbab, A.S., Zywicke, H.A., Jordan, E.K., Lewis, B.K., Bryant, L.H., and Bulte, J.W. (2003) "Clinically applicable labelling of mammalian cells and stem cells by combining superparamagnetic iron oxides and commonly available transfection agents," *Radiology,* 228:480–487.

Frank, J.A., Zywicke, H., Jordan, E.K., Mitchell, J., Lewis, B.K., Miller, B., Bryant, L.H., and Bulte, J.W. (2002) "Magnetic intracellular labelling of mammalian cells by combining (FDA-approved) superparamagnetic iron oxide MR contrast agents and commonly used transfection agents," *Acad. Radiology.,* 9(suppl 2):S484–S487.

Friedman, S.H., DeCamp, D.L., Sijbesma, R.P., Srdanov, G., Wudl, F., and Kenyon, G.L. (1993) "Inhibition of the HIV-1 protease by fullerene derivatives: model building studies and experimental verification," *J. Am. Chemical. Society.,* 115:6506–6509.

Fukuchi, K., Tomoyasu, S., Tsuruoka, N., and Gomi, K. (1994) "Iron deprivation-induced apoptosis in HL-60 cells," *FEBS Lett.* 350:139–142.

Galhardi, C.M., Diniz, Y.S., Faine, L.A., Rodrigues, H.G., Burneiko, R.C., Ribas, B.O., and Novelli, E.L. (2004). "Toxicity of copper intake: lipid profile, oxidative stress and susceptibility to renal dysfunction," *Food Chem. Toxicol.* 42:2053–2060.

Gao, X., Cui, Y., Levenson, R.M., Chung, L.W.K., and Nie, S. (2004) "In vivo cancer targeting and imaging with semiconductor quantum dots," *Nature Biotechnology.,* 22(8):969–976.

Gao, H., Kong, Y., Cui, D., Ozkan, C.S. (2003) "Spontaneous insertion of DNA oligonucleotides into carbon nanotubes," *Nano. Lett.,* 3:471–473.

Gaur, U., Sahoo, S.K., De, T.K., Ghosh, P.C., Maitra, A., and Ghosh, P.K. (2002) "Biodistribution of fluoresceinated dextran using novel nanoparticles evading reticuloendothelial system," *Int. J. Pharmacology.,* 202:1–10.

Gearheart, L. A., Ploehn, H. J., and Murphy, C. J. (2001) "Oligonucleotide adsorption to gold nanoparticles: a surface enhanced Raman spectroscopy study of intrinsically bent DNA," *J. Physical. Chemistry.,* 105:12609–12615.

Gelis, C., Girard, S., Mavon, A., Delverdier, M., Paillous, N., and Vicendo, P. (2003) "Assessment of the skin photoprotective capacities of an organo-mineral broad-spectrum sunblock on two ex vivo skin models," *Photodermatology. Photoimmunology. Photomedicine.,* 19(5):242–253.

Gole, A., Dash, C., Soman, C., Sainkar, S. R., Rao, M., and Sastry, M. (2001) "On the preparation, characterization, and enzymatic activity of fungal protease-gold colloid bioconjugates," *Bioconjugate Chem.,* 12:684–690.

Goodman, C., McCusker, C., Yilmaz, T., and Rotello, V. (2004) "Toxicity of gold nanoparticles functionalized with cationic and anionic side chains," *Bioconjugate Chem.,* 15:897–900.

Gref, R., Minamitake, Y., Peracchia, M.T., Trubetskoy, V., Torchilin, V., and Langer, R. (1994) "Biodegradable long-circulating polymeric nanospheres," *Science*, 18(263):1600–1603.

Gurr, J.-R., Wang, A., Chen, C.-H., and Jan, K.-Y. (2005) "Ultrafine titanium dioxide particles in the absence of photoactivation can induce oxidative damage to human bronchial epithelial cells," *Toxicology*, 213:66–73.

Gutteridge, J.M., Rowley, D.A., and Halliwell, B. (1982) "Superoxide-dependent formation of hydroxyl radical and lipid peroxidation in the presence of iron salt: detection of catalytic iron and antioxidant activity in extracellular fluids," *Biochemistry. J.*, 206:605–609.

Hardman, R. (2006) "A toxicologic review of quantum dots: toxicity depends on physico-chemical and environmental factors," *Environmental Health Perspectives*, 114(2):165–172.

Hart, G.A., and Hesterberg, T.W. (1998) "In vitro toxicity of respirable-size particles of diatomaceous earth and crystalline silica compared with asbestos and titanium dioxide," *J. Occupational. Environmental. Medicine.*, 40(1):29–42.

Heller, D.A., Baik, S., Eurell, T.A., and Strano, M.S. (2005) "Single-walled carbon nanotube spectroscopy in live cells: towards long-term labels and optical sensors," *Adv. Materials*, 17:2793–2799.

Hoffman, H.T., Quets, J., Toshiaki, T., Funk, G.F., McCulloch, T.M., Graham, S.M., Robinson, R.A., Schuster, M.E., and Yuh, W.T. (2000) "Functional magnetic resonance imaging using iron oxide particles in characterizing head and neck adenopathy," *Laryngoscope*, 110:1425–1430.

Huczko, A., Lange, H., Calko, E., Grubek-Jaworska, H., and Droszcz, P. (2001). "Physiological testing of carbon nanotubes: are they asbestos-like?" *Fullerene Sci. Technol.*, 9:251–254.

Ito, A., Shinkai, M., Honda, H., and Kobayashi, T. (2001) "Heat-inducible TNF alpha gene therapy combined with hyperthermia using magnetic nanoparticles as a novel tumor-targeted therapy," *Cancer Gene Ther.*, 8(9):649–654.

Ito, A., Shinkai, M., Honda, H., and Kobayashi, T. (2005) "Medical application of functionalized magnetic nanoparticles," *J. Biosci. Bioeng.*, 100:1–11.

Jaiswal, J.K., Mattoussi, H., Mauro, J.M., and Simon, S.M. (2003) "Long-term multiple color imaging of live cells using quantum dot bioconjugates," *Nature Biotech.*, 21:47–51.

Jia, G., Wang, H., Yan, L., Wang, X., Pei, R., Yan, T., Zhoa, Y., and Guo, X. (2005) "Cytotoxicity of carbon nanomaterials: single-wall nanotube, multi-wall nanotube, and fullerene," *Environ. Sci. Technol.*, 39:1378–1383.

Jordan, A., Wust, P., Scholz, R, Tesche, B., Fahling, H., Mitrovics, T., Vogl, T., Cervos-Navarro, J., and Felix, R. (1996) "Cellular uptake of magnetic fluid particles and their effects on human carcinoma cells exposed to AC magnetic fields in vitro," *Int. J. Hyperthermia*, 12(6):705–722.

Josephson, L., Tung, C.H., Moore, A., and Weissleder, R. (1999) "High-efficiency intracellular magnetic labelling with novel superparamagnetic-tat peptide conjugates," *Bioconjug. Chem.*, 10:186–191.

Jovin, T.M. (2003) "Quantum dots finally come of age," *Nat. Biotechnol.*, 21:32–33.

Jung, C.W., and Jacobs, P. (1995) "Physical and chemical properties of superparamagnetic iron oxide MR contrast agents: ferumoxides, ferumoxtran, ferumoxsil," *Magn. Res. Imaging*, 13:661–674.

Kagan, V.E., Tyruina, Y.Y., Tyurin, V.A., Konduru, N.V., Potapovich, A.I., Osipov, A.N., Kisin, E.R., Schwegler-Berry, D., Mercer, R., Castranova, V., Shvedova, A.A. (2006). "Direct and indirect effects of single walled carbon nanotubes on RAW 264.7 macrophages: role of iron," *Toxicology Letters*, 165(1):88–100.

Kam, N.W.S., Jessop, T.C., Wender, P.A., and Dai, H. (2004) "Nanotube molecular transporters: internalization of carbon nanotube-protein conjugates into mammalian cells," *J. Am. Chem. Soc.*, 126:6850–6851.

Kim, J.S., Yoon, T.-J., Yu, K.N., Kim, B.G., Park, S.J., Kim, H.W., Lee, K.H., Park, S.B., Lee, J.-K., and Cho, M.H. (2006) "Toxicity and tissue distribution of magnetic nanoparticles in mice," *Toxicol. Sci.*, 89(1):338–347.

Kirschvink, J.L., Kobayashi-Kirschvink, A., and Woodford, B.J. (1992) "Magnetite bio-mineralization in the human brain," *Proc. Natl. Acad. Sci. USA*, 89:7683–7687.

Kreuter, J. (1994) "Drug targeting with nanoparticles," *Eur. J. Drug Metab. Pharmacok.*, 19(3):253–256.

Kroto, H.W., Hearth, J.R., O'Brien, S.C., Curl, R.F., and Smalley, R.E. (1985) "C_{60}: Buckminsterfullerene," *Nature*, 318:162–163.

Lacava, L.M., Lacava, Z.G.M., Da Silva, M.F., Silva, O., Chaves, S.B., Azevedo, R.B., Pelegrini, F., Gansau, C., Buske, N., Sabolovic, D., and Morais, P.C. (2001) "Magnetic resonance of a dextran-coated magnetic fluid intravenously administered in mice," Biophys. J., 80:2483–2486.

Lam, C.W., James, J.T., McCluskey, R., and Hunter, R.L. (2004) "Pulmonary toxicity of single wall carbon nanotubes in mice 7 and 90 days after intratracheal instillation," Toxicological Sciences, 77:126–134.

Larson, D.R., Zipfel, W.R., Williams, R.M., Clark, S.W., Bruchez, M.P., Wise, F.W., and Webb, W.W. (2003) "Water-soluble quantum dots for multiphoton fluorescence imaging in vivo," Science, 300:1434–1436.

Lecoanet, H.F., Bottero, J.Y., and Wiesner, M.R. (2004) "Laboratory assessment of the mobility of nanomaterials in porous media," Environ. Sci. Technol., 38:5164–5169.

Lewin, M., Carlesso, N., Tung, C.H., Tang, X.W., Cory, D., Scadden, D.T., and Weissleder, R. (2000) "Tat peptide-derivatized magnetic nanoparticles allow in vivo tracking and recovery of progenitor cells," Nat. Biotechnol., 18:410–414.

Lim, Y.T., Kim, S., Nakayama, A., Stott, N.E., Bawendi, M.G., and Frangioni, J.V. (2003) "Selection of quantum dot wavelength for biomedical assays and imaging," Mol. Imaging, 2:50–64.

Limbach, L.K., Li, Y., Grass, R.N., Brunner, T.J., Hintermann, M.A., Muller, M., Gunther, D., and Stark, W.J. (2005) "Oxide nanoparticle uptake in human lung fibroblasts: effects of particle size, agglomeration, and diffusion at low concentrations," Environ. Sci. Technol., 39(23):9370–9376.

Lin, A.M.Y., Chyi, B.Y., Wang, S.D., Yu, H.-H., Kanakamma, P.P., Luh, T.-Y., Chou, C.K., and Ho, L.T. (1999) "Carboxyfullerene prevents iron-induced oxidative stress in rat brain," J. Neurochem., 72:1634–1640.

Lindenschmidt, R.C., Driscoll, K.E., Perkins, M.A., Higgins, J.M., Maurer, J.K., and Belfiore, K.A. (1990) "The comparison of a fibrogenic and two nonfibrogenic dusts by bronchoalveolar lavage," Toxicol. Appl. Pharmacol., 102(2):268–281.

Lomer, M.C., Thompson, R.P., and Powell, J.J. (2002) "Fine and ultrafine particles of the diet: influence on the mucosal immune response and association with Crohn's disease," Proc. Nutr. Soc., 61(1):123–130.

Long, T., Saleh, N., Phenrat, T., Swartz, C., Parker, J., Lowry, G., Veronesi, B. "Metal oxide nanoparticles produce oxidative stress in CNS microglia and neurons." Society of Toxicology Annual Meeting, San Diego, CA; March 5–9, 2006.

Lovell, M.A., Robertson, J.D., Teesdale, W.J., Campbell, J.L., and Markesbery, W.R. (1998) "Copper, iron and zinc in Alzheimer's disease senile plaques," J. Neurol. Sci., 158:47–52.

Lovrić, J., Bazzi, H., Cuie, Y., Fortin, G.R.A., Winnik, F.M., and Maysinger, D. (2005) "Differences in subcellular distribution and toxicity of green and red emitting CdTe quantum dots," Journal of Molecular Medicine, 83(5):377–385.

Lubbe, A.S., Bergemann, C., Brock, J., and McClure, D.G. (1999) "Physiological aspects in magnetic drug targeting," J. Magn. Mater., 194:149.

Manna, S.K., Sarkar, S., Barr, J., Wise, K., Barrera, E.V., Jejelowo, O., Rice-Ficht, A.C., and Ramesh, G.T. (2005) "Single-walled carbon nanotube induces oxicative stress and activates nuclear transcription factor-κB in human keratinocytes," Nano Letters., 5(9):1676–1684.

Mantyh, P.W., Ghilardi, J.R., Rogers, S., Demaster, E., Allen, C.J., Stimson, E.F., and Maggio, J.E. (1993) "Aluminum, iron, and zinc ions promote aggregation of physiological concentrations of beta-amyloid peptide," J. Neurochemistry., 61:1171–1174.

Maynard, A.D., Baron, P.A., Foley, M., Shvedova, A.A., Kisin, E.R., and Castranova, V. (2004) "Exposure to carbon nanotube material: aerosol release during the handling of unrefined single-walled carbon nanotube material," J. Toxicol. Environ. Health, Part A., 67:87–107.

McIntosh, C. M., Esposito, E. A., Boal, A. K., Simard, J. M., Martin, C. T., and Rotello, V. R. (2001) "Inhibition of DNA transcription using cationic mixed monolayer protected gold clusters," J. Am. Chem. Soc., 123:7626–7629.

Michalet, X., Punaud, F.F., Bentolila, L.A., Tsay, J.M., Doose, S., Li, J.J., Sundaresan, G., We, A.M., Gambhir, S.S., and Weiss, S. (2005) "Quantum dots for live cells, in vivo imaging, and diagnostics," Science, 307:538–544.

Moghimi, S.M., Hunter, A.C., and Murray, J.C. (2001) "Long-circulating and target-specific nanoparticles: theory to practice," *Pharmacological. Review.*, 53:283–318.

Monteiro-Riviere, N.A. (2005) "Multi-walled carbon nanotube exposure in human epidermal keratinocytes: localization and proteomic analysis," *Proceedings of the 229th American Chemical Society*, Division of Industrial and Engineering Chemistry.

Monteiro-Riviere, N.A., Nemanich, R.A., Inman, A.O., Wang, Y.Y., and Riviere, J.E. (2005a) "Multi-walled carbon nanotube interactions with human epidermal keratinocytes," *Toxicology. Letters.*, 155:377–384.

Monteiro-Riviere, N.A., Wang, Y.Y., Hong, S.M., Inman, A.O., Nemanich, R.J., Tan, J., Witzmann, F.A., and Riviere, J.E. (2005b) "Proteomic analysis of nanoparticle exposure in human keratinocyte cell culture," *Toxicological. Sciences.*, 84(S-1):447.

Monteiro-Riviere, N.A., and Ryman-Rasmussen, J.P. (2006) "Toxicology of nanomaterials," in J.E. Riviere (ed.) *Biological Concepts and Techniques in Toxicology: An Integrated Approach*, New York: Taylor and Francis Publishers, Chapter 12, pp. 217–233, 2006.

Monteiro-Riviere, N.A., and Inman, A.O. (2006) "Challenges for assessing carbon nanomaterials toxicity to the skin," *Carbon*, 44:1070–1078.

Monti, D., Moretti, L., Salvioli, S., Straface, E., Malorni, W., Pellicciari, R., Schettini, G., Bisaglia, M., Pincelli, C., Fumelli, C., Bonafe, M., and Franceschi, C. (2000) "C60 carboxyfullerene exerts a protective activity against oxidative stress-induced apoptosis in human peripheral blood mononuclear cells," *Biochemical. Biophy. Res. Commication.*, 277:711–717.

Moos, T., and Morgan, E.H. (2004) "The metabolism of neuronal iron and its pathogenic role in neurological disease: review," *Annuals. NY Acad. Sci.*, 1012:14–26.

Muangmoonchai, R., Wong, S. C., Smirlis, D., Phillips, I. R., and Shephard, E. A. (2002) "Transfection of liver in vivo by biolistic particle delivery: its use in the investigation of cytochrome P450 gene regulation," *Molecular Biotechnol.*, 20:145–151.

Muldoon, L.L., Pagel, M.A., Kroll, R.A., Roman-Goldstein, S., Jones, R.S., and Neuwelt, E.A. (1999) "A physiologic barrier distal to the anatomic blood-brain barrier in a model of transvascular delivery," *Am. J. Neuroradiology.*, 20:217–222.

Muldoon, L.L., Sàndor, M., Pinkston, K.E., and Neuwelt, E.A. (2005) "Imaging, distribution, and toxicity of superparamagnetic iron oxide magnetic nanoparticles in the rat brain and cerebral tumor," *Neurosurgery*, 57:785–796.

Muller, J., Huaux, F., Moreau, N., Misson, P., Heilier, J.G., Delos, M., Arras, M., Fonseca, A., Nagy, J.B., and Lison, D. (2005) "Respiratory toxicity of multi-wall nanotubes," *Toxicol. Applied. Pharmacology.*, 207:221–231.

Muller, J., Huaux, F., and Lison, D. (2006) "Respiratory toxidity of nanotubes: how worried should we be?" *Carbon*, 44:1048–1056.

Nelson, M.A., Frederick, E.D., Bowden, G.T., Hooser, S.B., Fernando, Q., and Carter, D.E. (1993) "Effects of acute and subchronic exposure of topically applied fullerene extracts on the mouse skin," *Toxicology Industrial Health*, 9:623–630.

Neuwelt, E.A., Varallyay, P., Bago A.G., Muldoon, L.L., Nesbit, G., and Nixon, R. (2004a) "Imaging of iron oxide nanoparticles by MR and light microscopy in patients with malignant brain tumors," *Neuropathology. Applied. Neurobiology.*, 30:456–471.

Neuwelt, E.A., Weissleder, R., Nilaver, G., Kroll, R.A., Roman-Goldstein, S., Szumowski, J., Pagel, M.A., Jones, R.S., Remsen, L.G., McCormick, C.I., Shannon, E.M., and Muldoon, L.L. (1994) "Delivery of virus-sized iron oxide particles to rodent CNS neurons," *Neurosurgery*, 34:777–784.

Neuwelt, E.A. (2004b) "Mechanisms of disease: the blood-brain barrier," *Neurosurgery*, 54:131–142.

Nguyen, B.C., Stanford, W., Thompson, B.H., Rossi, N.P., Kernstine, K.H., Kern, J.A., Robinson, R.A., Amorosa, J.K., Mammone, J.F., and Outwater, E.K. (1999) "Multicenter clinical trial of ultrasmall superparamagnetic iron oxide in the evaluation of mediastinal lymph nodes in patients with primary lung carcinoma," *J. Magnetic Resonance. Imaging*, 10:468–473.

Nimmagadda, A., Thurston, K., Nollert, M.U., and McFetridge, P.S. (2006) "Chemical modification of SWNT alters *in vitro* cell-SWNT interactions," *J. Biomed Material Research*, 76A:614–625.

Nordman, H., and Berlin, M. (1986) "Titanium," in: Friberg, L., Nordberg, G.F., and Vouk, V.B. (Eds.), *Handbook on the Toxicology of Metals*, vol. II. Elsevier, Amsterdam, pp. 595–609.

Oberdörster, G., Ferin, J., Gelein, R., Soderholm, S.C., and Finkelstein, J. (1992) "Role of the alveolar macrophage in lung injury: studies with ultrafine particles," *Environ. Health Perspect.*, 97:193–199.

Oberdörster, G., Ferin, J., and Lehnert, B.E. (1994) "Correlation between particle size, *in vivo* particle persistence, and lung injury," *Environ. Health Perspect.*, 102 (suppl 5):173–179.

Oberdoerster, G., Gelein, R.M., Ferin, J., and Weiss, B. (1995) "Association of particulate air-pollution and acute mortality-involvement of ultrafine particles," *Inhahalational. Toxicol.*, 7(1):111–124.

Oberdörster, E. (2004) "Manufactured nanomaterials (fullerenes, C_{60}) induce oxidative stress in the brain of juvenile largemouth bass," *Environ. Health Perspect.*, 112:1058–1062.

Oberdörster, E., Zhu, S, Blickley, T.M., McClellan-Green, P., and Haasch, M.L. (2006) "Ecotoxicology of carbon-based engineered nanoparticles: effects of fullerene (C_{60}) on aquatic organisms," *Carbon*, 44:1112–1120.

O'Brien, J., amd Lummis, S. C. (2002) "An improved method of preparing microcarriers for biolistic transfection," *Brain Res. Protoc.*, 10:12–15.

Ophus, E.M., Rode, L.E., Gylseth, B., Nicholson, G., and Saeed, K. (1979) "Analysis of titanium pigments in human lung tissue," *Scand. J. Work Environ. Health*, 5:290–296.

Otsuka, H., Akiyama, Y., Nagasaki, Y., and Kataoka, K. (2001) "Quantitative and reversible lectin-induced association of gold nanoparticles modified with alpha-lactosylomegamercapto-poly(ethylene glycol)," *J. Am. Chem. Soc.*, 123:8226–8230.

Pantarotto, D., Briand, J.P., Prato, M., and Bianco, A. (2004) "Translocation of bioactive peptides across cell membranes by carbon nanotubes," *Chem. Commun.*, 7(1):16–17.

Parak, W.J., Boudreau, R., Le Gros, M., Gerion, D., Zanchet, D., Micheel, C.M., Williams, S.C., Alivisatos, A.P., and Larabell, C. (2002) "Cell motility and metastatic potential studies based on quantum dot imaging of phagokinetic tracks," *Adv. Mater.*, 14(12):882–885.

Park, S. J., Taton, T. A., and Mirkin, C. A. (2002) "Array-based electrical detection of DNA with nanoparticle probes," *Science*, 295:1503–1506.

Peters, K., Unger, R.E., Kirkpatrick, C.J., Gatti, A.M., and Monari, E. (2004) "Effects of nanoscaled particles on endothelial cell function in vitro: studies on viability, proliferation and inflammation," *J. Mater. Sci. Mater. Med.*, 15:321–325.

Pickering, K.D., and Wiesner, M.R. (2005) "Fullerol-sensitive production of reactive oxygen species in aqueous solution," *Environ. Sci. Technol.*, 39:1359–1365.

Porreca, E., Ucchino, S., Di Febbo, C., Di Bartolomeo, N., Angelucci, D., Napolitano, A.M., Mezzetti, A., and Cuccurullo, F. (1994) "Antiproliferative effect of desferrioxamine on vascular smooth muscle cells in vitro and in vivo," *Arterioscler Thromb.*, 14:299–304.

Portet, D., Denoit, B., Rump, E., Lejeunne, J.J., and Jallet, P. (2001) "Nonpolymeric coatings of iron oxide colloids for biological use as magnetic resonance imaging contrast agents," *J. Colloids Interface Sci.*, 238:37–42.

Pouliquen, D., Le Jeune, J.J., Perdrisot, R., Ermias, A., and Jallet, P. (1991) "Iron oxide nanoparticles for use as an MRI contrast agent: pharmacokinetics and metabolism," *Magn. Reson. Imaging*, 9:275–283.

Prince, M.R., Zhang, H.L., Chabra, S.G., Jacobs, P., and Wang, Y. (2003) "A pilot investigation of new superparamagnetic iron oxide (ferumoxytol) as a contrast agent for cardiovascular MRI," *J. X-Ray Sci. Technol.*, 11:231–240.

Qingnuan, L., Yan, X., Xiaodong, Z., Ruili, L., Qieqie, D., Xiaoguang, S., Shaoliang, C., and Wenxin, L. (2002) "Preparation of ^{99m}Tc-$C_{60}(OH)_x$ and its biodistribution studies," *Nuclear Med. Biol.*, 29:707–710.

Quintana, C., Lancin, M., Marhi,c C., Perez, M., Martin-Benito, J., Avila, J., Carrascosa, J.L. (2000) "Initial studies with high resolution TEM and electron energy loss spectroscopy studies of ferritin cores extracted from brains of patients with progressive supranuclear palsy and Alzheimer disease," *Cell. Mol. Biol.*, 46:807–820.

Rahman, Q., Lohani, M., Dopp, E., Pemsel, H., Jonas, L., Weiss, D.G., and Schiffmann, D. (2002) "Evidence that ultrafine titanium dioxide induces micronuclei and apoptosis in Syrian hamster embryo fibroblasts," *Environ. Health Perspect.*, 110(8): 797–800.

Rahman, Q., Norwood, J., and Hatch, G. (1997) "Evidence that exposure of particulate air pollutants to human and rat alveolar macrophages lead to different oxidative stress," *Biochem. Biophys. Res. Commun.*, 240:669–672.

Rajagopalan, P., Wudl, F., Schinazi, R.F., and Boudinot, F.D. (1996) "Pharmacokinetics of a water-soluble fullerenes in rats," *Antimicrob. Agents Chemother,* 40:2262–2265.

Ramge, P., Unger, R.E., Oltrogge, J.B., Zenker, D., Begley, D., Kreuer, J., and von Briesen, H. (2000) "Polysorbate-80 coating enhances uptake of polybutylcyanoacrylate (PBCA)-nanoparticles by human and bovine primary brain capillary endothelial cells," *Eur J Neurol.,* 12:1931–2934.

Rehn, B., Seiler, F., Rehn, S., Bruch, J., and Maier, M. (2003) "Investigations on the inflammatory and genotoxic lung effects of two types of titanium dioxide: untreated and surface treated," *Toxicol. Appl. Pharmacol.,* 189(2):84–95.

Remsen, L.G., Pagel, M.A., McCormick, C.I., Fiamengo, S., Sexton, G., and Neuwelt, E.A. (1999) "The influence of anesthetic choice, PaC02, and other factors on osmotic blood-brain barrier disruption in rats with brain xenografts," *Anesth. Analg.,* 88:559–567.

Ros, P.R., Freeny, P.C., Harms, S.E., Seltzer, S.E., Davis, P.L., Chan, T.W., Stillman, A.E., Muroff, L.R., Runge, V.M., and Nissenbaum, M.A. (1995) "Hepatic MR imaging with ferumoxides: a multicenter clinical trial of the safety and efficacy in the detection of focal hepatic lesions," *Radiology,* 196:481–488.

Rouse, J.G., Yang, J., Barron A.R., and Monteiro-Riviere N.A. (2006) "Fullerene based amino acid nanoparticle interactions with human epidermal keratinocytes," *Toxicology In Vitro,* 20:1313–1320.

Rouse, J.G., Yang, J., Ryman-Rasmussen, J.P., Barron, A.R., and Monteiro-Riviere, N.A. (2007): Effects of mechanical flexion on the penetration of fullerene amino acid-derivatized peptide nanoparticles through skin. *Nano Letters* 7:155–160.

Ryman-Rasmussen, J.P., Riviere, J.E., and Monteiro-Riviere, N.A. (2006) "Penetration of intact skin by quantum dots of diverse physiochemical properties," *Toxicological Sciences,* 91(1):159–165.

Ryman-Rasmussen, J.P., Riviere, J.E., and Monteiro-Riviere, N.A. (2007) " Surfacecoatings determine cytotoxicity and irritation potential of quantum dot nanopaticles in epidermal keratinocytes," *Journal of Investigative Dermatology* 127:143–153.

Sandhu, K. K., McIntosh, C. M., Simard, J. M., Smith, S. W., and Rotello, V. M. (2002) "Gold nanoparticle-mediated transfection of mammalian cells," *Bioconjugate Chem.,* 13:3–6.

Sayes, C.M., Fortner, J.D., Guo, W., Lyon, D., Boyd, A.M., Ausman, K.D., Tao, YJ., Sitharaman, B., Wilson, L.J., Hughes, J.B., West, J.L., and Colvin, V.L. (2004) "The differential cytotoxicity of water-soluble fullerenes," *Nano Lett.,* 4:1881–1887.

Sayes, C.M., Liang, F., Hudson, J.L., Mendez, J., Guo, W., Beach, J.M., Moore, V.C., Doyle, C.D., West, J.L., Billups, W.E., Ausman, K.D., and Colvin, V.L. (2006) "Functionalization density dependence of single-walled carbon nanotubes cytotoxicity in vitro," *Toxicol. Lett.,* 161:135–142.

Scaiano, J.C., Monahan, S., and Renaud, J. (1997) "Dramatic effect of magnetite particles on the dynamics of photogenerated free radicals," *Photochem. Photobiol.,* 65:759–762.

Schafer, F.Q., Qian, S.Y., and Buettner, G.R. (2000) "Iron and free radical oxidations in cell membranes," *Cell. Mol. Biol.,* 46:657–662.

Schubert, D., Dargusch, R., Raitano, J., and Chan S-W. (2006) "Cerium and yttrium oxide nanoparticles are neuroprotective," *Biochemical and Biophysical Research Communications,* 342:86–91.

Schulz, J. Hohenberg, H., Pflucker, F, Gartner, E, Will, T., Pfeiffer, S.,Wepf, R.,Wendel, V., Gers-Barlag., H., and Wittern, K.P. (2002) "Distribution of sunscreens on skin," *Adv. Drug. Delivery. Reviews.,* 54 (Suppl) S157–S163.

Scrivens, W.A., and Tour, J.M. (1994) "Synthesis of ^{14}C-labeled C_{60}, its suspension in water, and its uptake by human keratinocytes," *J. Am. Chem. Soc.,* 116:4517–4518.

Shiosahara, A., Hoshino, A., Hanaki, K., Suzuki, K., and Yamamoto, K. (2004) "On the cytotoxicity caused by quantum dots," *Microbiology. Immunology.,* 48(9):669–675.

Shvedova, A.A., Castranova, V., Kisin, E.R., Schwegler-Berry, D., Murray, A.R., Gandelsman, V.Z., Maynard, A., and Baron, P. (2003) "Exposure to carbon nanotube materials: assessment of nanotube cytotoxicity using human keratinocytes cells," *J Toxicol Environ Health A,* 66:1909–1926.

Shvedova, A.A., Kisin, E.R., Mercer, R., Murray, A.R., Johnson, V.J., Potapovich, A.I., Tyurina, Y.Y., Gorelik, O., Arepalli, S., Schwegler-Berry, D., Hubbs, A.F., Antonini, J., Evans, D.E., Ku, B.K., Ramsey, D., Maynard, A., Kagan, V.E., Castranova, V., and

Baron, P. (2005) "Unusual inflammatory and fibrogenic plmonary responses to single-walled carbon nanotubes in mice," *Am. J. Physiology. Lung Cell. Molecular. Physiology.*, 289:L698–L708.

Smart, S.K.. Cassady, A.I., Lu, G.Q., Martin, D.J. (2006) "The biocompatibility of carbon nanotubes," *Carbon*, 44:1034–1047.

Stringer, B.K., Imrich, A., and Kobzik, L. (1996) "Lung epithelial cells (A549) interaction with unopsonized environmental particulates: quantitation of particle-specific binding and IL-8 production," *Exp. Lung Res.*, 22:495–508.

Thess, A., Lee, R., Nikolaev, P., Dai, H., Petit, P., Robert, J., Xu, C., Lee, Y.H., Kim, S.G., Rinzler, A.G., Colbert, D.T., Scuseria, G.E., Tomanek, D., Fischer, J.E., and Smalley, R.E. (1996) "Crystalline ropes of metallic carbon nanotubes," *Science*, 273:483–487.

Tremblay, J.F. (2002) "Mitsubishi chemical aims at breakthrough," *Chemical and Engineering News*, 80:16–17.

Tremblay, J.F. (2003) "Fullerenes by the ton," *Chem. & Eng. News*, 81:13–14.

Tsuchiya, T., Oguri, I., Yamakoshi, Y.N., and Miyata, N. (1996) "Novel harmful effects of [60]fullerene on mouse embryos in vitro and in vivo," Federation of Experimental Biology *Letters*, 393:139–145.

Tsuchiya, T., Yamakoshi, Y.N., and Miyata, N. (1995) "A novel promoting action of fullerene C_{60} on the chondrogenesis in rat embryonic limb bud cell culture system," *Biochemical. Biophysical. Research. Communications.*, 206:885–894.

Tsuji, J.S., Maynard, A.D., Howard, P.C., James, J.T., Lam, C.-W., Warheit, D.B., and Santamaria, A.B. (2006) "Research strategies for safety evaluation of nanomaterials, Part IV: risk assessment of nanoparticles," *Toxicological Sciences* 89(1):42–50.

Ueng, T.H., Kang, J.J., Wang, H.W., Chen, Y.W., and Chiang, L.Y. (1997) "Suppression of microsomal cytochrome P450-dependent monooxygenases and mitochondrial oxidative phosphorylation by fullerenol, a polyhydroxylated fullerene C_{60}," *Toxicol. Lett.*, 93:29–37.

Wang, G., Zhang, J., and Murray, R. W. (2002) "DNA binding of an ethidium intercalator attached to a monolayer-protected gold cluster," *Analytical. Chemistry.*, 74:4320–4327.

Warheit, D.B., Laurence, B.R., Reed, K.L., Roach, D.H., Reynolds, G.A.M., and Webb, T.R. (2004) "Comparative pulmonary toxicity assessment of single wall carbon nanotubes in rats," *Toxicological. Sciences.*, 77:117–125.

Warheit, D.B., Webb, T.R., Sayes, C.M., Colvin, V.L., and Reed, K.L. (2006) "Pulmonary instillation studies with nanoscale TiO_2 rods and dots in rats: toxicity is not dependent upon particle size and surface area," *Toxicological. Sciences Tox. Sci.*, doi:10.1093/ toxsci/kfj140.

Watson, A., Wu, Xingyong and Bruchez, M. (2003) "Lighting up cells with quantum dots," *BioTechniques*, 34(2):296–303.

Weissleder, R., Elizondo, G., Wittenberg, J., Rabito, C.A., Bengele, H.H., and Josephson, L. (1990) "Ultrasmall superparamagnetic iron oxide. characterization of a new class of contrast agents for MR imaging," *Radiology*, 175:489–493.

Weissleder, R., Stark, D.D., Engelstad, B.L,, Bacon, B.R., Compton, C.C., White, D.L., Jacobs, P., and Lewis, J. (1989) "Superparamagnetic iron oxide: pharmacokinetics and toxicity," *Am. J. Roentgenology.*, 152:167–173.

Weizmann, Y., Patolsky, F., and Willner, I. (2001) "Amplified detection of DNA and analysis of single base mismatches by the catalyzed deposition of gold on Au nanoparticles," *Analyst*, 126:1502–1504.

Wilhelm, C., Billotey, C., Roger, J., Pons, J.N., Bacri, J.-C., and Gazeau, F. (2003) "Intracellular uptake of anionic superparamagnetic nanoparticles as a function of their surface coating," *Biomaterials*, 24:1001–1011.

Willems, I, Konya, Z., Colomer, J.F., Van Tendeloo, G., Nagaraju, N., Fonseca, A., and Nagy, J.B. (2000) "Control of the outer diameter of thin carbon nanotubes synthesized by catalytic decomposition of hydrocarbons," *Chem. Phys. Lett.*, 317:71.

Witzmann, F.A., and Monteiro-Riviere, N.A. (2006) "Multi-walled carbon nanotube exposure alters protein expression in human keratinocytes," *Nanomedicine: Nanotechnology, Biology and Medicine* 2:158–168.

Wu, X., Liu, H., Liu, J., Haley, K.N., Treadway, J.A., Larson, J.P., Ge, N., Peale, F., and Bruchez, M.P. (2003) "Immunofluorescent labeling of cancer marker Her2 and other cellular targets with semiconductor quantum dots," *Nature. Biotechnology.*, 21:41–46.

Yamago, S.,Tokuyama, H., Nakamura, E., Kikuchi, K., Kananishi, S., Sucki, K., Nakahara, H., Enomoto, S., and Ambe, F. (1995) "In-vivo biological behavior of a water-miscible fullerene-C-14 labeling, absorption, distribution, excretion and acute toxicity," *Chem. Biol.,* 2:385–389.

Yamamoto, A., Honma, R., Sumita, M., and Hanawa, T. (2004) "Cytotoxicity evaluation of ceramic particles of different sizes and shapes," *J. Biomedicine. Material. Research.,* 68A (2),244–256.

Yang, E.Y., Guo-Ross, S.X., and Bondy, S.C. (1999) "The stabilization of ferrous iron by a toxic beta-amyloid fragment and by an aluminum salt," *BrainResearch,* 839:221–226.

Yang, J., and Barron, A.R. (2004) "A new route to fullerene substituted phenylalanine derivatives," *Chemical. Communication.,* 21(24):2884–2885.

Yokoyama, A., Sato, Y., Nodasaka, Y, Yamamoto, S., Kawasaki, T., Shindoh, M., Kohgo, T., Akasaka, T., Uo, M., Watari, F., and Kazuyuki, T. (2005) "Biological behavior of hat-stacked carbon nanofibers in the subcutaneous tissue in rats," *Nano Letters.,* 5:157–161.

Zeng, L., Zhang, L., and Barron A.R. (2005) "Tailoring aqueous solubility of functional-ized single-wall carbon nanotubes over a wide pH range through substituent chain length," *Nano Letters.,* 5(10):2001–2004.

Zhang, Q., Kusaka, Y., Sato, K., Nakakuki, K., Kohyama, N., and Donaldson, K. (1998) "Differences in the extent of inflammation caused by intratracheal exposure to three ultrafine metals: role of free radicals," *J. Toxicol. Environ. Health,* 53(6):423–438.

Zhang, Y., Kohler, N., and Zhang, M. (2002) "Surface modification of superparamagnetic magnetite nanoparticles and their intracellular uptake," *Biomaterials,* 23:1553–1561.

Zhang, T., Stilwell, J.L., Gerion, D., Ding, L., Elboudwarej, O., Cooke, P.A., Gray, J.W., Alivisatos, A.P., and Chen, F.F. (2006). "Cellular effect of high doses of silica-coated quantum dot profiled with high throughput gene expression analysis and high content cellomics measurements," *Nano Letters,* 6(4):800–808.

Zhang, L., Zeng, L., Barron A.R., and Monteiro-Riviere, N.A. "Biological interaction of func-tionalized single-walled carbon nanotubes in human keratinocytes," *International Journal of Toxicology,* 26:110–113.

Ecotoxicological Impacts
of Nanomaterials

Delina Y. Lyon   *Rice University, Houston, Texas*

Antoine Thill   *Commissariat de l'Energy Atomique, Saclay (Paris), France*

Jerome Rose   *CNRS-University of Aix-Marseille, Aix-en-Provence, France*

Pedro J. J. Alvarez   *Rice University, Houston, Texas*

Keywords: ecotoxicology, toxicology, antibacterial, reactive oxygen species (ROS), bacteria, developmental toxicity, antimicrobial, uptake, biotransformation.

Introduction

The widespread production of engineered nanomaterials started in the 1980s, and their rapid incorporation into a variety of consumer products and applications is outpacing the development of appropriate regulations to mitigate potential risks associated with their release to the environment. Therefore, several research initiatives have been recently started to improve our understanding of the transport, fate, reactivity, and ecotoxicity of several nanomaterials that have a relatively high probability of environmental release.

Many of the inorganic nanomaterials, such as TiO_2, ZnO, and quantum dots, are likely to be found in the environment due to their manufacture or intended application. Nano-sized titanium dioxide (a.k.a. anatase, TiO_2), a good opacifier, is used as a pigment in paints, paper, inks, and plastics. In electronics, crystalline SiO_2 works as both a semiconductor and an electrical insulator. The ceramic nature of ZnO allows its use as a pigment and a semiconductor. Nano-scale TiO_2, SiO_2, and ZnO offer greater surface area than their bulk counterparts, allowing for improved performance in established applications. Quantum dots (QDs) are semiconductors that display narrow fluorescence or absorption bands due to the quantum constraints imposed on electrons by the finite size of the material. Applications of QDs include medical imaging and sensors. Some nanomaterials are intentionally dispensed into the environment, such as zerovalent iron nanoparticles, which have been applied at more than 20 sites for the *in situ* remediation of groundwater contaminated with chlorinated solvents. Commercial applications of such inorganic nanomaterials currently or will soon include nano-engineered titania particles for sunscreens and paints, silica nanoparticles as solid lubricants, and other reagents for groundwater remediation.

Organic nanomaterials, such as fullerenes and carbon nanotubes, are also being produced in increasing amounts. For example, buckminsterfullerene (C_{60}) is being used in applications ranging from cosmetics to drug delivery vectors to semiconductors while carbon nanotubes composites are used in tires. Frontier Carbon built a plant to mass produce C_{60} on the scale of tons per year [9]. The economy of fullerene production indicates that fullerene-containing products will soon become widely available. Although C_{60} is relatively insoluble in water, it does not precipitate completely when coming into contact with the aquatic environment. C_{60} can form stable nanoscale suspended aggregates (nC_{60}), whose concentration can reach up to 100 mg/L [4, 10]. Fullerols (hydroxylated fullerenes) are highly photosensitive and generate ROS that may be used for bio-oxidations [11]. Both fullerols and carboxyfullerenes can be used in medical applications as drugs or for diagnostic drug delivery [12]. These derivatized molecules are more soluble in water than their parent fullerene, implying greater potential interaction with organisms.

In the environmental technology industry alone, nanotechnologies hold great promise for reducing waste production, cleaning up industrial contamination, providing potable water, and improving the efficacy of energy production and use. On the other hand, the environment will be increasingly prone to suffer pollution from nanomaterials in consumer products such as sunscreens, detergents, and cosmetics, as well from accidental releases during production, transportation, and disposal operations. The manufacture, use, and disposal of engineered nanomaterials are not currently regulated by any government, although the US House Scientific Committee has prioritized legislation of nanotechnology research [13–15]. There has also been movement toward including environmental and health issues in the European Union and Japanese research budgets for nanotechnology. The current European budget for research in these areas is approximately $7.5 million, a much smaller share of their total nanotechnology research budget.

Many examples in modern history illustrate the unintended consequences of initially promising technologies, including the blind release of "beneficial" chemicals into the environment, such as asbestos or DDT. These examples forewarn us of potential environmental impacts of some nanomaterials, which deserve more attention and research [16–18]. Furthermore, the large intellectual and financial investments in nanotechnology demand that it be publicly accepted and sustainable [19]. The backlash against genetically modified crops resulted in a huge setback for agriculture. A similar backlash against nanotechnology would result in the delay of beneficial nanomaterials coming to market.

The matter of determining whether or not a substance is "dangerous" involves not only determining any hazards presented by the material such as toxicity, but also to what degree the material contacts living organisms. Currently, the degree to which cellular processes and ecosystem health may be impacted by nanomaterials, let alone specific toxicity mechanisms, remain largely unknown. This chapter discusses the known and postulated interactions between nanomaterials and non-mammalian biological indicators, specifically microbes, and how these relationships foreshadow the potential effects of nanomaterial releases into the environment.

Why Study the Effects of Nanomaterials on Microorganisms?

Microbes are present in almost every environment on earth, and their flexibility and adaptability allow them to survive under seemingly unlivable conditions, such as anaerobic, high heat, or extreme cold conditions. Microbes as a whole produce the majority of the biomass in aquatic systems. While plants are the primary biomass in terrestrial systems, their survival depends on the activity of microbes for the breakdown of dead matter and the recycling of needed nutrients. Microbes play key roles in the cycling of carbon, nitrogen, phosphorous, and other minerals. Microbial ecotoxicology is therefore a particularly important consideration because microorganisms serve as the basis of food webs and the primary agents for global biogeochemical cycles. Microorganisms are also important components of soil health and could serve as potential mediators of transformations that affect nanoparticle mobility and toxicity in the environment.

One benefit of evaluating microbial toxicity is the ability to extrapolate the observed effects of chemicals on microbes to other higher level organisms. Quantitative structure activity relationships (QSARs) are one way to calculate the impact on other organisms based on chemical structure [20]. QSARs incorporate mathematical relationships between the structure of a chemical and its likelihood to increase the toxicity of a compound. Each chemical component is given a numerical value, and the sum of its components determines its toxicity. One can also use similar calculations to extrapolate the toxicity of a chemical to an organism based on its toxicity to an unrelated organism [20].

Methods to Assess Ecotoxicity

As yet, a comprehensive ecotoxicity study of any nanomaterial has not been performed. Most studies have analyzed nanomaterial impacts on a type of organism in isolation. The methods for looking at ecosystems impact as a whole are established and are reviewed in several texts [20–22], so this review is by no means exhaustive.

There are many levels at which one can analyze the impact of a chemical, from a single biochemical reaction up to an entire ecosystem with all of its complexity. Most ecotoxicity tests relate to survival, mutation, and reproduction. The majority of the current research on nanomaterials has examined their impact on biochemical reactions in a cell up to the survivability of whole multicellular organisms. In order to make comparisons between chemicals and organisms for risk analysis, several benchmark measurements have been established. The most common is the LC_{50}, or lethal concentration of chemical that kills 50 percent of an exposed population as compared to a control. Another common measurement is the EC_{50}, or effective concentration of a chemical that elicits some response in 50 percent of the population. The response examined can be reproductive capacity, growth, respiration, or any number of endpoints. Tests of both the LC_{50} and EC_{50} must be carefully controlled and performed in a standardized manner to ensure comparability between experiments and laboratories.

There are numerous methods and organisms at various trophic levels to examine LC_{50} or EC_{50}. Part of the challenge in ecotoxicology is to find appropriate biomarkers, or physiological processes that respond in a sensitive manner to chemical exposure, and bioindicators, or organisms in an ecosystem that reflect the health of that environment. Most of the studies in nanomaterial toxicology have only looked at biomarkers and not bioindicators of any specific ecosystem. Many pollutants have a specific biomarker, since the chemicals elicit a specific response from organisms [22]. There are also some established organisms that can be relatively easily cultured and used to assess toxicity. Table 12.1 lists some of the commonly used organisms, their common name, and common endpoints.

Bacteria are among the easiest and least expensive organisms to culture, and they are relatively sensitive to many of the same compounds

TABLE 12.1   Common Biomarkers Used in Ecotoxicology

Organism	Common name	Common toxicological endpoint
Vibrio fischeri	Bacterium	Loss of fluorescence
Selenastrum capricornutum	Algae	Growth inhibition
	Terrestrial plants	Rate of seedling emergence and growth
Daphnia magna	Water flea	Reproductive capacity, loss of mobility, death
Eisenia fetida	Earthworm	Reproductive capacity, death
Danio rerio	Fish	Developmental malformations, death
Xenopus laevis	Frog	Developmental malformations

as higher level organisms, such as nC_{60}, which is antibacterial (Figure 12.1) as well as cytotoxic [23]. Their high surface to volume ratio makes them sensitive to small concentrations of chemicals. Microbes are present in almost all environments and form the basis of food webs and element cycles. For these reasons, microbes make excellent test organisms, and there are a plethora of microbial toxicity tests available. As with other organisms, bacteria can be analyzed for nonlethal endpoints relating to metabolic activity, reproduction, and mutation. Enzyme tests are used primarily with soil microbial communities to look at the activity of common enzymes in soil, such as dehydrogenases. Bioluminescent tests use the bioluminescence of certain bacteria, such as *Vibrio fischeri*, as indicators of microbial health. Diminished bioluminescence after exposure

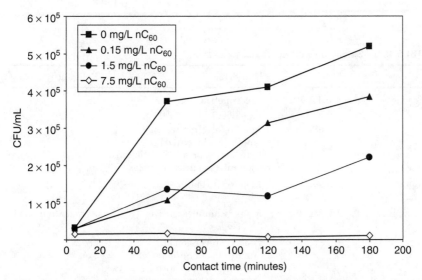

Figure 12.1   Inhibition of *Escherichia coli* growing on rich medium plates following exposure to different nC_{60} concentrations as a function of contact time.

to a chemical suggests the chemical has antimicrobial activity. Growth inhibition tests use plating methods or spectrophotometric methods to determine the amount of growth in a culture. Some environmental contaminants can be mutagenic and cause changes in the genetic code of receptor organisms. These include carcinogens that affect somatic cells, such as benzene, and teratogens that affect germ cells, such as thalidomide.

To examine a compound's potential mutagenicity, several assays determine the compound's ability to damage microbial DNA. Most of the tests, such as the Ames test and Mutatox test, look at the frequency of mutations caused by the chemical that restore some previously damaged function to the microbe. Table 12.2 summarizes some of the more popular mutagenicity tests available, along with other common microbial assays that are used to assess the general toxicity of a compound. However, only a few of these assays have been applied to nanomaterials [24].

Developmental toxicity tests for eukaryotic, multicellular organisms are also commonly used to identify xenobiotics that may alter ecosystems. These tests analyze impacts on the embryonic development of higher order organisms. Full life cycle chronic toxicity tests that expose all life stages of an organism to a toxicant can be quite long and costly. Short-term acute developmental toxicity tests, however, can be used to predict chronic effects with far less time and cost, allowing for broader evaluation of factors that influence toxicity. Furthermore, due to the higher sensitivity of organisms in their embryonic stage than during

TABLE 12.2 Selected Bacterial Toxicity Assays

Test type	Test name	Test principle	Reference
Enzyme	MetPAD & MetPLATE	Colorimetric assay for β-galactosidase activity	25
	MitoScan	Monitor electron transfer inhibition by NADH production or consumption in submitochondrial particles	26
Growth inhibition	*Pseudomonas putida*	Monitor turbidity of cell culture	27
Bioluminescent	Microtox	Monitor luminescence of *Vibrio fischeri*	www.azurenv.com
Mutagen	Ames	Look for reversion of mutations in *Salmonella typhimurium* restoring ability to produce histidine	28, 29
	Mutatox	Look for *V. fischeri* with increased luminescence	www.azurenv.com

adulthood, exposure levels that elicit no adverse developmental effects are likely to be safe, and determining such levels is important for developing criteria and regulations to protect environmental and public health. Additional advantages of these tests include easy operation, high sensitivity, and multiple sensible endpoints.

The early life stage (ELS) test using zebrafish (*Danio rerio*) embryo is a widely used tool [30–33], with many studies showing ELS to be the most sensitive stage to assess toxicological effects [31]. The generation time of zebrafish is short, and their embryos are transparent and develop externally, which facilitates the observation of all the changes and movements during development [34]. ELS tests using zebrafish embryos can also provide toxicological indexes to help evaluate the sublethal effect of pollutants and discern their toxicological effect categories (e.g., neural toxicity or genotoxicity) [31–33]. Finally, as zebrafish share many cellular and physiological characteristics with higher vertebrates, the toxicological results of zebrafish embryo development are easily compared to mammalian developmental toxicity studies [34].

Another well understood and validated developmental test is FETAX (frog embryo teratogenesis assay-*Xenopus*), which is a four-day whole embryo developmental toxicity test that utilizes the embryos of the South African clawed frog, *Xenopus laevis* [35]. This assay similarly covers the embryogenesis periods during which cell division, interaction (induction), migration, differentiation, and selective cell death occur. Abnormal development or even death can occur when any of these mechanisms are interrupted. The endpoints of a FETAX assay are the 96-h LC_{50} (i.e., the concentration that is lethal to 50 percent of the embryos), the 96-h EC_{50} (the concentration that triggers malformation in 50 percent of the embryos), and the minimum concentration to inhibit growth (MCIG). The FETAX assay is based on the whole embryo and not on embryo parts or cultured cells. Thus, the endpoints are holistic in the sense that they account for the important cell and molecular mechanisms. The FETAX assay can also be used to screen for teratogenicity. The teratogenic index (TI) is defined as the 96-h LC_{50} (embryo death) divided by the 96-h EC_{50} (embryo malformation) [36]. TI values of 1.5 to 2.0 indicate that a material is a potential teratogen and materials with TI values >2.0 should be considered for further teratogenicity testing.

The methods discussed above have all been developed to analyze the toxicity of chemicals. Nanomaterials behave neither like chemicals nor like particles, thus there are several obstacles to simply employing the established protocols. One problem unique to nanomaterial toxicity is the propensity of certain nanomaterials to induce cell aggregation, which results in misleading results in growth inhibition tests that use spectrophotometric methods or plating methods. In the case of spectrophotometric methods, the aggregated cells will settle, causing the culture to lose turbidity but not necessarily cell viability. For plating methods, aggregation can result in lower cell counts due to the uneven spreading of the cells. In some cases, data are collected to determine a dose response. It is difficult to determine what exactly a "dose" is for nanomaterials. A weight per volume measurement, like mg/L, does not accurately reflect the actual surface area available to the organism. Furthermore, if a cell takes up the nanomaterial, the amount assimilated is difficult to measure.

Bioavailability and Cellular Uptake of Nanoparticles

The bioavailability of a chemical, or its likelihood to interact with and affect an organism, is a function of the amount, state, and stability of the chemical and the sensitivity of the receptor. Bioavailability is typically determined by physicochemical factors such as contaminant hydrophobicity, volatility, phase, and solubility, as well as reaction conditions, such as mixing intensity. It is often governed by mass transfer phenomena such as desorption from soil or suspended solids, dissolution from non-aqueous-phase liquids (NAPLs), or diffusion from nanopores to name a few examples. Bioavailability affects both dose and exposure duration, and thus dictates toxicity. Organisms that do not interact with a material cannot be directly damaged by that material. The bioavailability of manufactured nanomaterials has not yet been quantitatively evaluated, although several mathematical and experimental approaches are available to evaluate bioavailability of environmental pollutants from a biodegradation and bioremediation perspective [37].

Fate and transport of nanomaterials

The manufacture, use, and disposal of nanomaterials ensure that, in some form or another, they all eventually end up in the environment. The fate and transport of nanomaterials are addressed in part in Chapter 7. Most nanomaterials will end up either in the water or in the soil, and while these materials may remain inert, the majority of them will interact with their new surroundings. Nanomaterials can bind to soil, be transported or diluted in water, or react with water or other chemicals. Weathering of nanomaterials can also occur. The phase in which the nanomaterial resides will determine how it interacts with organisms in the environment. In soil-water systems, the main property that determines bioavailability is desorption of the chemical from soil particles [38].

One phenomenon that may reduce nanomaterial bioavailability in porous media is its propensity to attach to surfaces or to form aggregates.

For example, particles that are readily transported and attach to mineral surfaces may be less mobile in porous media such as groundwater aquifers or sand filters used in potable water treatment. Intuitively, the assumption is often made that nanoparticles will be more mobile in porous media due to their small size, which is less amenable to straining than larger particles. However, all other factors being equal, smaller particles could be less mobile due to their relatively large diffusivity that produces more frequent contacts with the surfaces of aquifer porous media [39].

Bioavailability and uptake of nanomaterials into prokaryotic cells

It has been well established that nanomaterials may behave differently from their bulk counterparts, which is the subject of Chapters 3 through 7. If nanomaterials have altered chemical behavior, one can also assume that their interactions with biological entities will be different from their bulk counterparts. The increase in surface area of the nanomaterial versus the bulk material implies greater bioavailability, but the actual effects on biological systems remains to be seen.

One major protective layer for bacteria is the cell envelope, which includes the cell membrane, cell wall, and capsule. The capsule is an outer layer of polysaccharides that protects the cell and allows it to adhere to surfaces. The cell wall, made up of peptidoglycan, provides a rigid structure that protects against mechanical stress and osmotic pressure. The plasma membrane, composed of phospholipids and proteins, functions as a selective permeability barrier, allowing the passage only of small uncharged molecules. Other compounds have specialized transport mechanisms. The plasma membrane is also the site of cell respiration. An agent that is able to pass through the bacterial envelope has the potential to inflict a great amount of damage. The mechanisms of cell penetration by nanomaterials is one of the most intriguing and challenging topics under debate. The models proposed for cell penetrating peptides (CPPs) might be extrapolated to some nanomaterials (Figure 12.2). First is the inverse micelle model proposed by Derossi and coworkers [40], which assumes that positively charged proteins interact with the negatively charged cell membrane. This destabilizes the membrane and forms an inverted micelle (Figure 12.2C), which eventually collapses back to more stable planar bilayers and releases the protein (or possibly nanomaterial) into the cytoplasm. Another model is the barrel-stave model proposed by Gazit and coworkers [41] which assumes that amphipathic peptides adopt helical conformations when bound to the membrane, a common feature for many CPPs. These α-helices further associate into barrels and then insert themselves perpendicularly into the membrane,

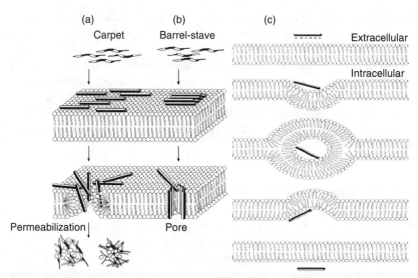

Figure 12.2 Illustration of the possible interactions of peptides with phospholipid membranes, and different mechanisms for translocation that could be extrapolated to the uptake of some nanomaterials. The inverted micelle model for CPP translocation: (a) The carpet model, (b) The pore formation model (barrel-stave), (c) The inverted micelle model [3]. Source: Yang, J. Fullerene-Based Aminoacids. *Chemistry* 2006.

allowing the hydrophobic sides of helices to face the alkyl chains of fatty acids and consequently let the hydrophilic sides form a pore (Figure 12.2B). Although the pore can let small molecules and ions diffuse freely, it cannot explain the cargo delivery properties of CPPs. The third model proposed by Matsuzaki et al. is the carpet model [42] (Figure 12.2A), which suggests that aggregates of peptides coat the cell surface and thereby disrupt the membrane structure to make cracks. The second step involves the rotation of the peptide leading to the interaction between the hydrophobic residues of the peptide with the hydrophobic core of the membrane. In the third step, a small disruption in the lipid packing occurs. It is the opened disruptions that form passages for peptides and possibly nanomaterials. While studies are required to define the detailed mechanism of cellular uptake of nanomaterials, it is postulated that the hydrophobic nature of some nanomaterials facilitates transport through the hydrophobic midlayer of cell membranes into the cytoplasm. The small size of nanomaterials, which are popularly defined as being smaller than 100 nm in one dimension, increases the likelihood of their entry into the cell although entry or contact is not necessary for observed toxicity, as discussed in the section entitled "Possible Mechanisms for Antimicrobial Activity." However, the average bacterium is only around 1 μm in length, and the incorporation of a 100 nm

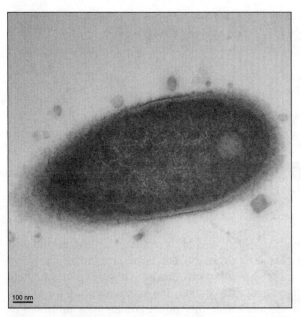

Figure 12.3   TEM image of nC_{60} particles associating with Bacillus subtilis Bar size is 100 nm.

particle seems unwieldy. Thus, larger nanomaterials are likely to attach to bacteria surfaces (Figure 12.3) and possibly compromise the integrity and functions of the cell membrane. There are several examples in the literature of nanomaterials entering bacteria. Carboxyfullerenes have been shown to puncture the membranes of some bacteria, but they were not shown to actually enter the bacteria [43]. In one study with *Pseudomonas aeruginosa*, silver nanoparticles up to 80 nm in diameter were able to traverse the inner and outer cell membrane (Figure 12.4) [8]. Conjugated CdSe quantum dots can be used to label bacteria according to their strain or metabolic activity, and these particles of less than 5 nm have been reported to enter bacterial cells [44]. Bacteria are themselves capable of making nanoparticles, and this concept is elaborated upon in the section entitled "Biotransformation of Nanomaterials by Microbes."

One concern with materials toxic to microbes and lower level organisms is the potential for bioaccumulation and biomagnification. If the compound's method of entry is unknown, then its accumulation is termed bioaccumulation [20]. One way to predict whether a compound is prone to bioaccumulate is to consider its hydrophobicity. The amount of compound in an organism depends on that organism's lipid content and the octanol/water coefficient, or K_{ow}, of the chemical. A large K_{ow} indicates a higher likelihood of a compound to bioaccumulate since a

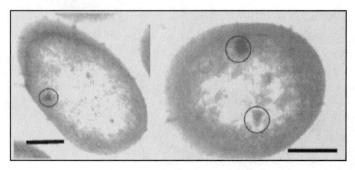

Figure 12.4   These TEM images of ultrathin cross sections of *P. aeruginosa* cells show Ag nanoparticles, averaging 48 nm in size. The scale bars represent 220 nm, and the circles indicate nanoparticles [8].

high K_{ow} indicates a lipophilic compound [20]. A lipophilic compound will reside in the fatty tissues of an organism and is less likely to be excreted. Once a chemical has bioaccumulated in an organism, it is likely to be biomagnified through the food chain. One method for examining bioaccumulation is to monitor the concentration of a chemical in an organism over time as compared to the background concentration in the growth medium. If a compound is simply entering the organism by passive methods, like diffusion, this is considered bioconcentration, and there are established quantitative structure-activity relationships (QSARs) based on K_{ow} to predict bioconcentration. To date, there have been no studies that show bioaccumulation or biomagnification of toxic nanomaterials. Based on their known hydrophobicity, fullerenes and carbon nanotubes make good candidates for materials that bioaccumulate.

Nanomaterial Interaction with Microbial Cell Components

The impact of nanomaterials on living cells can be broken down into the interactions between the nanomaterial and the individual cell components. The overall observed effect may be the same for a number of nanomaterials, but the underlying mechanism behind their behavior can be traced to their impact on a cell component. The interactions of different nanomaterials with proteins, nucleic acids, and membranes are discussed below. Some of these interactions are engineered, such as the labeling of proteins or DNA to monitor and visualize cells. Bacteria themselves can also be used as targeting mechanisms to deliver nanoparticles [45]. The interactions that are more relevant to environmental study are those that are unintentional and can impact the organism and microbial community at large.

Membranes

The first interaction between a material and a prokaryotic cell is at the membrane interface. The components of the bacterial cell envelope were discussed in the section entitled "Bioavailability and Uptake of Nanomaterials into Prokaryotic Cells." The bacterial cell membrane is a semipermeable barrier that serves as a locus for important biochemical activities and cellular functions. These include compartmentalization (i.e., serving as a physical barrier), regulation of the movement of materials (e.g., substrate uptake and waste excretion), energy transduction (i.e., electron transport phosphorylation), access to information (i.e., the membrane harbors response proteins that signal changes in environment), and intercellular interactions (e.g., the membrane participates in cellular attachment and gene transfer).

Some nanomaterials are made to target the cell membrane, where a cell can be most easily labeled. Some nanoparticles can embed themselves in the cell membrane [46]. Fullerene derivatives were able to incorporate into artificial lipid bilayer membranes [47], and carboxyfullerenes were able to puncture bacterial membranes [48]. Often, though, the membrane and cell wall are simply considered an obstacle for chemical penetration. Most of the studies showing transport of nanomaterials into a cell involve eukaryotic cells. Nanoparticles can be functionalized with peptides that allow translocation across eukaryotic membranes [49], and nanotubes can act as drug delivery devices that poke through a cell's membrane [50]. Some nanomaterial-membrane interactions are harmful. Oberdörster published an early study that pointed to possible negative impacts of nanomaterials on the health of aquatic organisms [6]. This pioneering study concluded that nC_{60} exerted oxidative stress and caused severe lipid peroxidation in fish brain tissue (Figure 12.5). Whether oxidative stress was the result of reactive oxygen species (ROS) produced by nC_{60} or by the cellular immune response system was not investigated. In other words, the ROS generated could result from the induced response of mitochondria and other eukaryotic cellular constituents (a.k.a. oxidative burst) [51, 52, 53], or that other free radicals besides ROS cause the damage (e.g., organic radicals generated by interactions between fullerene and cellular electron carriers).

Proteins

There is a lot of interest in fusing nanomaterials with biomolecules, taking advantage of the nanomaterial properties of proteins [54]. Proteins can be used as a template or a backbone onto which nanoparticles can be deposited. For example, viral protein capsules can be used as templates for the synthesis of gold nanoparticles, among others

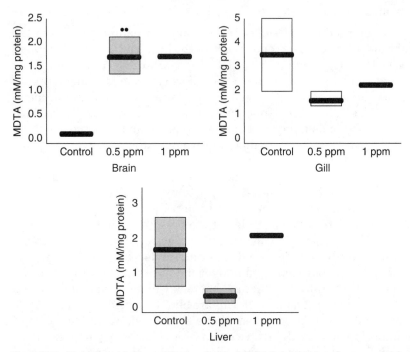

Figure 12.5 Lipid peroxidation, as represented by MDTA (1,1,3,3-tetraethoxy propane), of brain, gill, and liver of largemouth bass after 48-hour exposure to nC_{60}. Heavy black bands are the means, thinner lines are the medians, boxes represent 25th and 75th percentiles, error bars indicate minimum and maximum, and circles represent outliers [6].

[55–57]. Compounds that interact with proteins may alter protein structure, although this is not necessarily detrimental. A fullerene water suspension, named HyFn or nC_{60}, interacts with proteins and stabilizes them [58]. Fullerenes were also able to inhibit beta-amyloid peptide aggregation [59]. Nanomaterials can interact with enzymes to mediate redox reactions. An electrode functionalized with fullerenes can activate glucose oxidase to oxidize glucose [60]. Cytochrome P450, a redox enzyme, can react with C_{60} [61, 62]. Nanomaterials can also be used to target and label proteins, enhancing the fields of medical imaging. Quantum dots, used to fluorescently label cells, are expressly desirable for their interaction with cellular components [63].

DNA

In the field of nanotechnology, nanomaterial-nucleic acid interactions can be divided into intentional versus unintentional, which are often detrimental. Intentional nanomaterial-DNA interactions include using nanomaterials to label DNA or to destroy unwanted DNA. Nucleotides can be tagged with nanoparticles, such as cadmium selenide, that act as labeling agents for bioimaging applications [64, 65]. DNA itself can be treated as a nanoparticle and used as a template on which other nanoparticles are made [66]. As discussed in Chapter 11, photosensitive nanomaterials that generate reactive oxygen species upon exposure to light can be used for photodynamic therapy, targeting cells to be destroyed [67]. As with nanomaterials being made to traverse the cell membrane, iron oxide nanoparticles can be modified into nonviral nanoparticle gene transfection vectors to carry genetic information into cells [68].

Other literature focuses on the detrimental effects of nanomaterial-DNA interactions. Fullerenes are able to bind DNA [69], deforming the strand to potentially have a negative effect on the stability and function of the molecule [70]. Quantum dots, which will hopefully be applied in medical imaging, can nick supercoiled DNA [71]. Some nanomaterials indirectly damage DNA due to ROS production. For titanium dioxide nanoparticles, such as those used in sunscreen, nicking of supercoiled DNA results from the oxygen radicals generated by the light-sensitive nanoparticle, as also discussed in Chapters 6 and 11 [72]. Fullerenes can cleave double-stranded DNA, upon exposure to light [73]. This cleavage often occurs at guanine bases, which are most susceptible to oxidative damage [74, 75]. Not all fullerene derivatives have been shown to have this activity, and the type of fullerene derivative dictates its effect on cells [76].

Antibacterial Activity of Nanomaterials

Table 12.3 lists a number of nanomaterials and studies of their antimicrobial activity. Several different species of bacteria were used in these studies, although the studies were performed on isolated cultures and not on microbial communities that can exhibit more complex synergistic and antagonistic interactions. Some materials, like TiO_2, were already established as antimicrobial agents, and the nanocrystalline forms follow suit. Nano-TiO_2 in membranes inhibits fouling by *E. coli* when the system is placed under UV illumination [77]. Another promising application for nanomaterials is to behave as an antiviral agent [78–80].

The proposed mechanisms for the antimicrobial activity vary, especially between fullerene derivatives. $C_{60}(OH)$ generates singlet oxygen, and can behave as a potent oxidizing agent in biological systems [11]. Other fullerenes respond to illumination, such as C_{60} coated with polyvinylpyrrolidone (C_{60}-PVP), which generates singlet oxygen [83]. Singlet oxygen in turn causes lipid peroxidation, which oxidizes linoleate [84]. Other nanomaterials rely on illumination or, as in the case of

TABLE 12.3   Negative Impacts of Selected Nanomaterials on Organisms

Type of nanomaterial	Specific nanomaterial	Effects observed	References
Fullerene	C_{60} water suspension	Antibacterial Cytotoxic to human cell lines Taken up by human keratinocytes Stabilizes proteins	[24, 81] [82] [58]
	C_{60} encapsulated in polyvinylpyrrolidone, cyclodextrins, or polyethylene glycol	Damages eukaryotic cell lines Antibacterial	[83] [84, 85] [86] [87]
	Hydroxylated fullerene	Oxidative eukaryotic cell damage	[87]
	Carboxyfullerene (malonic acid derivatives)	Bactericidal for Gram-positive bacteria Cytotoxic for human cell lines	[48, 88, 89] [90] [91]
	Fullerene derivatives with pyrrolidine groups	Antibacterial Inhibits cancer cell proliferation Cleaves plasmid DNA	[92] [93], [94] [95] [73]
	Other alkane derivatives of C_{60}	Antimutagenic Cytotoxic Induces DNA damage in plasmids Inhibits protein folding Antibacterial Accumulates in liver of rats	[76] [75] [59] [96] [97]
	metallofullerene	Accumulates in liver of rats	[98]
Metal oxide	TiO_2—anatase, rutile	TiO_2 accelerates solar disinfection of *E. coli* through photocatalytic activity and reactive oxygen species	[99, 100]
		Surfaces coated with TiO_2 photocatalytically oxidize *E. coli*, *Micrococcus luteus*, *B. subtilis*, and *Aspergillus niger*	[101]
	MgO	Antibacterial activity against *B. sutbilis* and *Staphylococcus aureus*	[102]
	CeO	Antimicrobial effect on *E. coli*	[5]
	ZnO	Antibacterial activity against *E. coli* and *B. subtilis*	[103, 104]

magnetic nanoparticles, on the application of a magnetic field to generate heat [105]. Light-absorbing gold nanoparticles attached to bacteria allow the use of lasers to target the bacteria to be killed [106]. No irradiation was needed for certain fullerene derivatives and TiO_2 nanoparticles to produce hydrogen peroxide [93, 107].

Possible mechanisms for antimicrobial activity

Some general mechanisms behind the antimicrobial activity are membrane permeability/osmotic stress, interruption of energy transduction, genotoxicity, and formation of reactive oxygen species (ROS), which are discussed in the paragraphs that follow.

Membrane disruption.   A nanomaterial that physically damages a cell can be bactericidal if it comes into contact with the cell. If the membrane of a bacterium is compromised, the cell may repair itself or, if the damage is severe, the cell may leak and eventually die. There are few examples in the literature of nanomaterials directly damaging membranes and causing bacterial cell death. One case is the exposure of Gram-positive bacteria to carboxyfullerene, which resulted in the puncturing of the bacteria and cell death [48]. Another way in which the membrane can be compromised is if the lipid components of the membrane are altered. ROS can damage membranes by causing lipid peroxidation, therefore nanomaterials that generate ROS can indirectly damage membranes. This type of mechanism is reviewed in the section on ROS.

Interrupting energy transduction.   The bacterial cell envelope not only retains cell shape and protects it from osmotic stress, but it is also the locus of electron transport phosphorylation and energy transduction. These processes could be disrupted if membrane integrity is compromised or if a redox-sensitive nanomaterial contacts membrane-bound electron carriers and withdraws electrons from the transport chain (Figure 12.6). There is one such case cited in the literature in which fullerene derivatives were able to inhibit *E. coli* respiration of glucose [93, 96]. Oxidation-reduction reactions at bacterial surfaces can be observed by monitoring the redox state of nanoparticles bearing transition metals. For example,

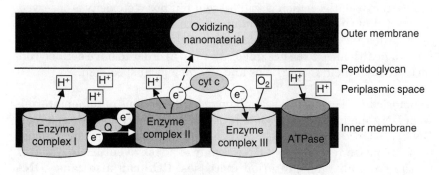

Figure 12. 6   Chemiosmosis in the Gram-negative bacterial membrane.

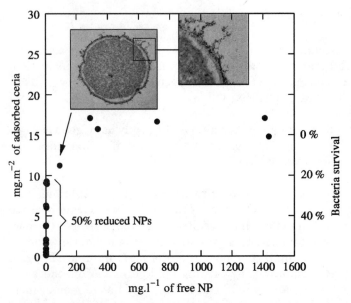

Figure 12.7   Adsorption isotherm of bare CeO_2 NPs on *E. coli* bacteria. Fifty percent of cerium atoms in the adsorbed nanoparticles (see TEM image) are reduced from CeIV to CeIII [5].

it is possible to assess the redox state of iron (FeIII → FeII) and cerium (CeIV → CeIII) atoms in the nanoparticles after contact with living organisms [5]. The localization of the nanoparticles is important in the case of oxidation induced damage on the cells. This has been observed for the contact between cerium oxide nanoparticles, which are strong oxidants, and *E. coli*. Upon direct contact between the bare CeO_2 nanoparticles and the *E. coli* membrane, a strong electrostatic interaction is observed that leads to the sorption of the nanoparticles onto the cell membrane (Figure 12.7). In this case, the ceria nanoparticles have an oxidative action directly at the cell membrane that is coupled to cytotoxicity for the cells. If the nanoparticles are covered with organic molecules, the oxidative action is still present but no cytotoxicity is observed. The fact that the cytotoxic mechanisms are still unclear illustrates that studies dealing with toxicity of nanoparticles for the environment and human health have to be designed carefully in order to have defined contact between the nanoparticles and the biological specimen.

Genotoxicity.   The earlier section on the interactions of nanomaterials with DNA mentioned that some fullerenes can cleave double-stranded DNA [73]. Once the DNA is cleaved, the cell can either repair the damage or it can cease reproducing. DNA repair is a risky venture in any cell, as the possibility for mutation increases. ROS can also cause DNA

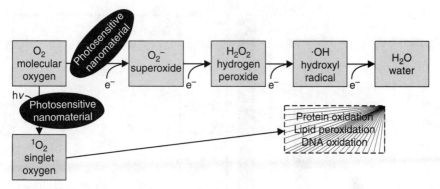

Figure 12.8   Evolution of reactive oxygen species.

damage, as mentioned in the following section. There have been few if any studies on the genotoxicity of nanomaterials using Ames tests or any other established protocol.

Reactive oxygen species (ROS) production.   By far the most researched and publicized mechanism for nanomaterial antimicrobial activity is the production of ROS. Once ROS are generated, they can trigger a chain of radical formation that affects every cell component (Figure 12.8). Understanding these effects aids in the engineering of ROS-mediated disinfection systems. The potential ecotoxicity of nano-sized TiO_2, SiO_2, and ZnO water suspensions has been investigated using two different bacteria (*Bacillus subtilis* and *Escherichia coli*) [2]. All three materials are reputed to produce ROS and so their potential toxicity was tested under both light and dark conditions [108–110]. Illumination enhanced the toxicity of the nanoparticle suspensions tested, with the exception of ZnO, yet antibacterial activity was noted under both light and dark conditions for all three materials (Figure 12.9). The greater inhibition noted in the presence of light supports the notion that the antibacterial activity of TiO_2 and SiO_2 is related to ROS production, which occurs only in the presence of light. Yet cell death, though less pronounced, also occurred in the dark, indicating that an additional mechanism is involved. Similar results were obtained in mammalian cytotoxicity assays with nano-TiO_2, where oxidative stress was observed in the dark under nonphotocatalytic conditions [107]. This underscores the need for further research on nanomaterial-cell interactions and cytotoxicity mechanisms that could prevail in the dark and factors that increase toxicity to enhance risk management.

Oxidation reactions can damage the cell membrane by oxidizing double bonds on fatty acid tails of membrane phospholipids in a process known as lipid peroxidation. Lipid peroxidation affects membrane permeability and fluidity, making cells more susceptible to osmotic stress or

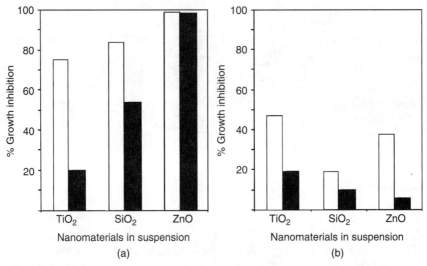

Figure 12.9 The effect of illumination on antibacterial activity of nano-sized particle suspensions toward (a) *B. subtilis* and (b) *E. coli* (Symbols: □ light, ■ dark) at predetermined antibacterial concentrations. Plotted are growth inhibitions averaged over three repeats of the experiment, all sets showed similar values (2).

hindering nutrient uptake [111]. Furthermore, the chemiosmotic potential of the cell membrane could be disrupted, leading to a loss in energy production. The peroxidized fatty acids then degrade to harmful aldehydes that can trigger longer duration damage of proteins and other molecules. Such redox interactions could in turn generate more free radicals that perpetuate the formation of highly reactive epoxides that could further compromise the integrity of the cell membrane and damage DNA. A study of the toxicity of water-soluble fullerenes on fresh water crustaceans, a marine copepod, fathead minnow, and Japanese medaka fishes showed that the peroxisomal lipid transport protein PMP70 was reduced in the fathead minnow, indicating alterations in the acyl-CoA pathways [112]. There is some debate as to whether lipid peroxidation occurs in bacteria, as bacterial lipids are mainly monounsaturated and thus unreactive to the lipid peroxidation chain reaction [113, 114].

Nucleic acids are very susceptible to oxidative stress, leading to DNA strand breaks, cross-linking within the molecule or with other molecules, and adducts of the bases or sugars [111]. While there are more than 20 known DNA adducts, 8-hydroxyguanine and 2-hydroxyadenine are two of the more common DNA lesions that have strong mutagenic potentials [115]. DNA damage can result from either the incorporation of a damaged nucleotide or direct oxidation. Free iron, bound strongly by certain DNA sequences, allows hydroxyl radical production via Fenton chemistry close to the nucleic acids, resulting in localized DNA damage [116].

Protein oxidation can occur in a variety of ways. Iron-sulfur clusters, which act as cofactors in many enzymes, can be damaged by superoxide. ROS cause the loss of an iron atom that can then go on to bind DNA and catalyze the Fenton chemistry mentioned previously [114]. Hydrogen peroxide oxidizes sulfur in the amino acids cysteine and methionine, which leads to disulfide bonds with other cysteines or sulfinic acids. Hydroxyl radicals can generate amino acid radicals, often in more sensitive amino acids like tyrosine and tryptophan. Oxidation of amino acids can also lead to protein carbonyl derivatives [117]. Damaging individual enzymes in a metabolic pathway can cause the entire pathway to fail, sometimes leading to auxotrophy.

The antibacterial activity of nC_{60} has already been established under light conditions [10]. ROS have been implicated in the antibacterial mechanism of PVP/C_{60} [83], but ROS have not been confirmed as the main factors responsible for the bactericidal activity of any of the Fullerene Water Suspensions (FWS). In a previous study of nC_{60} with bacteria, toxicity was not affected by the presence or absence of light that is needed to stimulate ROS production by nC_{60} [24]. This study was repeated under dark conditions where no leaks of light were allowed during setup or sampling. Again, these tests did not yield different results to those performed under ambient light conditions. In addition to light, oxygen is a critical precursor to ROS generation. In unpublished research, nC_{60} inhibited the growth of both *E. coli* and *B. subtilis* under anaerobic and fermentative conditions where O_2 was absent. All these lines of evidence indicate that photocatalyzed ROS production is probably not the sole antibacterial mechanism associated with nC_{60}. Further research is warranted to elucidate toxicity mechanisms and identify manufacturing and derivatization processes that decrease toxicity.

Indirect harmful effects

In some cases, the nanomaterial does not exert an antibacterial effect directly on the organism, but it participates in reactions and behavior that result in antibacterial activity. Cerium dioxide nanoparticles are charged in solution, and whenever they have an opposite charge compared to microorganisms, a strong electrostatic interaction is observed [5]. This can lead to the charge reversal of the cells, instability, and cell aggregation. The nanoparticles must reach a critical concentration before aggregation of the cells is observed. The aggregation of cells can be problematic for living organisms. For example, photosynthetic microalgae depend on sunlight, and their aggregation and resulting sedimentation to deep water is an indirect harmful effect of nanoparticles. Another potential indirect

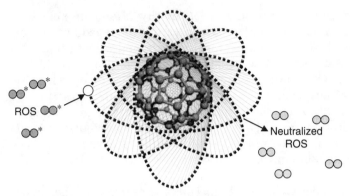

Figure 12.10 Fullerene behaving as an antioxidant.

harmful effect is sorption and sequestration of limiting nutrients by nano-materials, which would hinder growth.

Antioxidant properties of nanomaterials

This chapter has focused on the more publicized aspects of nanomate-rials, such as their interactions with cellular components and ROS gen-eration. The cytotoxicity of water soluble, photosensitive fullerenes has been associated with oxidative stress, based on the observation of greater toxicity and ROS production with higher levels of irradiation [75, 83]. However, in apparent opposition to this notion, there are numerous observations of fullerenes acting as antioxidants (Figure 12.10). Indeed, the antioxidant properties of C_{60} have been compared with those of vita-mins C and E in *preventing* lipid peroxidation induced by superoxide and hydroxyl radicals [118, 119]. This apparent dichotomy underscores the need for research on nanomaterial-cellular interactions and the result-ing effects on metabolic processes and cell physiology as a function of dose and exposure conditions. These theoretical interactions could serve as a guide for advanced microscopic and chemical analyses of cell con-stituents to elucidate toxicity mechanisms and discern physiological characteristics that confer bacterial resistance to fullerenes toxicity. It is plausible, for example, that cells possessing a high concentration of antioxidants (e.g., reduced glutathione) or enzymes that destroy toxic oxygen species (e.g., catalase, peroxidase, and superoxide dismutase) might be less susceptible to nanomaterial toxicity.

Biotransformation of Nanomaterials by Microbes

As mentioned previously, bacteria themselves are able to produce any number of nanomaterials, and thus it would be expected that they be able to transform some nanomaterials. The biocatalytic production of nanomaterials offers a cheap and efficient method to produce materials that would be difficult to manufacture chemically [120]. There are a number of publications detailing the production of inorganic nanomaterials by bacteria. The most published example is the ability of certain magnetotactic bacteria to produce magnetic nanoparticles, or magnetosomes [121]. *Magnetospirillum magnetotacticum* can synthesize nano- and micro-ZnO rods [122], and *E. coli* incubated with sodium sulfide and cadmium chloride can generate nanoparticles of cadmium sulfide [123]. *Pseudomonas stutzeri* is able to biosynthesize silver-based nanocrystals [124]. In biofilms containing sulfate-reducing bacteria, sphalerite (ZnS) nanoparticles are formed [125]. Using the principles found in cell biology, genetically engineered proteins for inorganics (GEPIs) can be produced to aid in the manufacture of functional nanostructures [126].

In contrast, there is a dearth of literature documenting the transformation of manufactured nanomaterials by microbial processes. It is expected that reduction and oxidation (redox) reactions, which dominate the degradation of organic compounds, might also be involved here. Redox processes are also the basis of various precipitation and dissolution reactions that influence the sequestration and mobility of inorganic metals. Thus redox reactions might be important for the transformation and fate of engineered nanoparticles. Redox reactions are often mediated by microorganisms, either directly through enzymatic activity or indirectly through the production of biogenic oxidants (e.g., reactive oxygen species produced by lignocellulytic fungi) or reductants, such as surface-associated $Fe(II)$, a common abiotic reductant in natural systems that can be produced by iron-reducing bacteria. Although it has not been reported in the literature, it is possible that biomolecules that are used to collect metals, like siderophores, could affect metal oxide nanomaterials and act as a method of compartmentalizing toxic metal oxides. Whether nanomaterials could be transformed by such abiotic redox processes in the environment to an appreciable extent is unknown, and would likely depend on the thermodynamic feasibility and kinetic facility of electron transfer.

To date, no systematic evaluations of fullerene transformation via biochemical mechanisms have been recorded. There is, however, extensive literature on organisms such as lignocellulytic fungi that possess nonspecific extracellular biocatalytic capabilities to degrade recalcitrant organics [127–131]. Furthermore, citations confirm fullerene oxidation via chemically-based, "model" enzyme systems [61]. Specifically, Fenton's chemistry is considered to be a chemical model for hydroxylation of carbon atoms mediated by fungal enzymes such as cytochrome P450, peroxidases, and laccases [132]. It is plausible that fullerenes could be oxidized by such extracellular fungal enzymes with relaxed specificity. Furthermore, the antioxidant properties of fullerenes and their high propensity to accept electrons suggest that microbially produced reductants, such as reduced glutathione and cobalamin, might also transform fullerenes.

Factors Mitigating Nanomaterial/ Organismal Interactions

For nanomaterials to present a risk, there must be both a potential for exposure to these materials and a hazard, such as toxicity, that results following exposure. Besides the size of the nanomaterial, there are environmental physicochemical parameters, such as ionic strength, pH, temperature, and redox conditions, that affect the way nanomaterials interact with microorganisms. Physicochemical properties, such as size, coating, and derivatization, can alter the behavior of the nanomaterial, and different microorganisms might react differently to the same nanomaterial.

Physicochemical factors

The physicochemical parameters of a system can completely alter the behavior of a compound. Increasing the ionic strength of a medium can cause nanoparticles, like fullerenes, to agglomerate and fall out of suspension, thus rendering them nontoxic [10, 24, 39]. Similar results have been reported for cerium oxide nanoparticle uptake by human fibroblasts [133]. The pH, while usually in a limited range in the environment, can affect the organisms' health and predispose it to susceptibility to the nanomaterial. It can also affect the chemical's valence state, thus affecting its toxicity. The temperature of a soil or water environment can affect the rate at which a compound is metabolized. The concept of Q_{10} refers to the tenfold increase in metabolic activity for every 10°C increase in temperature. As with pH, the temperature can affect organismal health and the reactivity of the nanomaterial. One important temperature-dependent abiotic reaction is sorption, which can influence the length of a contaminated plume in groundwater. As discussed above, redox conditions can affect the toxicity and mobility of a nanomaterial. Depending on the mechanism, the proper redox conditions can render a compound nontoxic. Thus redox reactions might be important for the transformation and fate of engineered nanoparticles.

Material properties

The properties of the nanomaterial, such as its size, charge, and derivatization, also dictate its interaction with microorganisms. As mentioned previously, the limit of particle size that can be taken up by a bacterium is not clear. Depending on the nanoparticle, one study places the limit at 5 nm while another places it at 80 nm [8, 44]. The surface area to volume ratio can also be important. Theoretical considerations suggest that smaller nanoparticles are likely to be more toxic due to their larger specific surface area, which is conducive to greater bioavailability. The size of TiO_2 particles affects their photocatalytic activity [134]. In a separate study, the antibacterial activity of different sizes of TiO_2, SiO_2, and ZnO water suspensions was investigated, but nominal particle size did not affect the observed antibacterial activity [2]. In the study of nC_{60}, larger nC_{60} particles (>100 nm in diameter) appeared to be 100 times less toxic than smaller (>100 nm) particles [135]. This increase in antibacterial activity was disproportionate to the increase in surface area. Thus, as mentioned before, factors that promote coagulation and precipitation of nanoparticles in the environment, such as increases in salt concentration, are likely to mitigate ecotoxicity.

The charge of the particle affects its ability to interact with cells. One study showed that cationic fullerene derivatives were bacteriostatic while anionic derivatives were not [96]. These differences in charge were the product of different functional groups being placed on the fullerene. The level of fullerene derivatization can affect the other factors of size and charge. Hydroxylated C_{60} is not toxic while underivatized nC_{60} has antibacterial activity [24]. In the case of fullerenes, it has been suggested that derivatization decreases toxicity (Figure 12.11) although the effects of derivatization provoke numerous changes in the physical

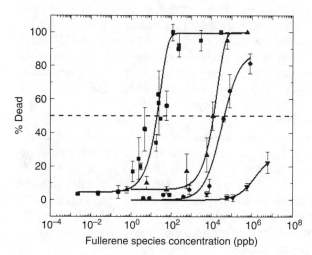

Figure 12.11 The differential cytotoxicity of nC_{60} (■), C3(▲), $Na^{+}{}_{2-3}[C_{60}O_{7-9}(OH)_{12-15}]^{(2-3)-}$(●), and $C_{60}(OH)_{34}$ (▼) in human dermal fibroblasts [4].

characteristics of these materials, including aggregation state, hydrophobicity, and reactivity, that have not been controlled in studies to date [4]. Derivatization increases solubility and the potential for transport and exposure, so the decreased toxicity may be offset by a higher potential for the nanomaterial to reach a receptor. There is as yet not enough information to draw any consensus on the effects of derivatization on overall risk to human and environmental health.

It is also important to consider other factors that change material properties, such as its physical morphology, the method of fabrication, and the age of the nanomaterial. The morphology or crystallinity of ZnO particles affects its photocatalysis [136], and this in turn may affect its antibacterial activity by increasing or decreasing the amount of ROS produced. Although manufacturing protocols may produce nanomaterials in powder form, they often form micro- or macroscale aggregates in aqueous suspension [2, 137]. This has been demonstrated for TiO_2, ZnO, and SiO_2. The mechanism used to suspend the nanoparticles can also be used to control the size of the resultant suspension. For example, the production of nC_{60} can be altered to tailor the size of the nC_{60} clusters [135].

Organismal properties

Properties of the organism itself could affect how it interacts with nanomaterials. Different physiological characteristics, including its developmental stage, could confer different susceptibilities to any toxic properties of the nanomaterial. There are few developmental toxicity studies on nanomaterials [1, 112]. One study investigated the developmental toxicity of buckminsterfullerene aggregates (nC_{60}) and fullerol [$C_{60}(OH)_{16-18}$] to aquatic organisms, using zebrafish (*Danio rerio*) as a model [1]. Early life stage parameters such as zebrafish embryo survival, hatching rate, heartbeat, and pericardial edema were noted and described within 96 hours of exposure. nC_{60}/THF (C_{60} suspended in water after tetrahydrofuran evaporation) at 1.5 mg L^{-1} delayed zebrafish embryo and larva development, decreased its survival and hatching rate, and caused pericardial edema (Figure 12.12). Toxicity was mitigated by adding the antioxidant GSH, which suggests that a free radical–induced mechanism or another form of oxidative stress plays a role in developmental toxicity and that an organism with increased antioxidant capacity would not suffer as readily from nC_{60} exposure. Another study examining the effect of nC_{60} on fresh water crustaceans noted that after 21 days of exposure, the daphnia had a delay in molting and a significant decrease in offspring production at 2.5 and 5.0 ppm [112].

Whether an organism is eukaryotic or prokaryotic determines the mechanisms the cell may use to resist a toxic agent. For example, certain

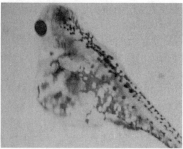

Figure 12.12 Zebrafish larva with pericardial edema, 84 (left) and 108 (right) hours postfertilization. [1]

antimicrobial agents target cell processes that occur only in bacteria and not in other organisms. Different types of bacteria have been shown to have different reactions to the same nanomaterial. Gram-positive bacteria are susceptible to carboxyfullerene while Gram-negative bacteria are not [48]. As mentioned previously, the biochemical capabilities of a cell, such as antioxidants or ROS-destroying enzymes, can better its chances of withstanding nanomaterial exposure. However, further dose-response studies are needed to understand the chemical and physical factors that exacerbate or mitigate the sublethal toxicity of various nanomaterials, and provide a basis for developing water quality standards to protect aquatic life.

Summary and Conclusions

While the concerns of this chapter are the potential deleterious effects of nanomaterials, there are obvious benefits to their use. Besides commercial and medical uses, there are potential applications of nanomaterials for environmental remediation, such as the use of tailored nanopolymers that enhance the solubility and thus bioavailability of recalcitrant organic compounds [138]. Zerovalent iron nanoparticles (nanoiron) and other metal catalysts are relatively advanced environmental nanotechnologies [139]. They have been applied at more than 20 sites for the *in situ* remediation of amenable groundwater contaminants. Metal and metal-oxide nanoparticles (e.g., nanoiron, magnetite, and titanium dioxide) have all been proposed for use in groundwater remediation [140–142], water treatment [143, 144], and removal of toxic contaminants from air streams [145]. Biodegradable nanocomposites offer a way to make plastics and polymers without using petroleum [146]. There are also numerous papers detailing the use of TiO_2 as a water disinfecting agent [99, 147–149].

Most of the studies on nanomaterial toxicity have been performed on eukaryotic cell lines or animal models, as discussed in Chapter 11. The environmental toxicity research has focused on specific bacterial strains under controlled laboratory conditions. The dosing has also been acute, and the experiments were conducted for a short duration. All of the published studies documenting the antimicrobial activity of fullerenes have used single bacterial isolates and not examined the impact on synergistic and antagonistic microbial interactions that are important for the functioning of microbial communities, including nutrient cycling and supporting the food web [24, 43, 83]. The ability to control the antimicrobial activity of nanomaterials allows both their use in controlled situations and an insight to ways to detoxify that material. For example, although photosensitive nanomaterials can exert oxidative stress in the dark [107], their antimicrobial activity can be accentuated under conditions in which light penetrates the system [83].

Responsible uses of manufactured nanomaterials in commercial products and environmental applications, as well as prudent management of the associated risks, require a better understanding of their mobility, bioavailability, and impacts on a wide variety of organisms. Microbial ecotoxicology is a particularly important consideration, not only to elucidate antibacterial mechanisms that could be extrapolated to eukaryotic cells, but also because microorganisms are the foundation of all known ecosystems, serving as the basis of food webs and the primary agents for global biogeochemical cycles. Responding to questions on environmental impacts of nanotechnology in the early stages of its development may result in better, safer products and less long-term liability for industry. Indeed, due diligence is the standard demanded by the law if not the public. Thus, improved understanding of nanomaterial-microbial interactions as well as their impact to higher-order organisms is important to ensure that nanotechnology evolves as a tool to improve material and social conditions without exceeding the ecological capabilities that support them.

References

1. Zhu, L.; Li, Y.; Duan, Z.; Chen, W.; Alvarez, P. J. Developmental toxicity in zebrafish embryos after exposure to manufactured nanomaterials: buckminsterfullerene aggregates (nC$_{60}$) and fullerol. *Environmental Toxicology and chemistry.* 2007 (in press).
2. Adams, L. K.; Lyon, D. Y.; Alvarez, P. J. J. Comparative eco-toxicity of nanoscale TiO$_2$, SiO$_2$, and ZnO water suspensions. *Water Research* 2006, *40*, 3527–3532.
3. Lundberg, P.; Langel, Ü. A brief introduction to cell-penetrating peptides. *Journal of Molecular Recognition* 2003, *16*, 227–233.
4. Sayes, C.; Fortner, J.; Guo, W.; Lyon, D.; Boyd, A.; Ausman, K.; Tao, Y.; Sitharaman, B.; Wilson, L.; Hughes, J.; West, J.; Colvin, V.L. The differential cytotoxicity of water-soluble fullerenes. *Nano Letters* 2004, *4*, 1881–1887.

5. Thill, A.; Spalla, O.; Chauvat, F.; Rose, J.; Auffan, M.; Flank, A. M. Cytotoxicity of CeO$_2$ nanoparticles for *Escherichia coli*. A physico-chemical insight of the cytotoxicity mechanism. *Environmental Science and Technology* 2006, *40*, 6151–6156.
6. Oberdörster, E. Manufactured nanomaterials (fullerenes, C$_{60}$) induce oxidative stress in the brain of juvenile largemouth bass. *Environmental Health Perspectives* 2004, *112*, 1058–1062.
7. Yang, J. Fullerene-based aminoacids. *Chemistry* 2006.
8. Xu, X. H.; Brownlow, W. J.; Kyriacou, S. V.; Wan, Q.; Viola, J. J. Real-time probing of membrane transport in living microbial cells using single nanoparticle optics and living cell imaging. *Biochemistry* 2004, *43*, 10400–10413.
9. Tremblay, J.-F. Mitsubishi chemical aims at breakthrough. *Chemical & Engineering News* 2002, *80*, 16–17.
10. Fortner, J. D.; Lyon, D. Y.; Sayes, C. M.; Boyd, A. M.; Falkner, J. C.; Hotze, E. M.; Alemany, L. B.; Tao, Y. J.; Guo, W.; Ausman, K. D.; Colvin, V. L., Hughes, J. B. C$_{60}$ in water: nanocrystal formation and microbial response. *Environmental Science and Technology* 2005, *39*, 4307–4316.
11. Vileno, B.; Lekka, M.; Sienkiewicz, A.; Marcoux, P.; Kulik, A. J.; Kasas, S.; Catsicas, S.; Graczyk, A.; Forro, L. Singlet oxygen (^{1}D$_g$)-mediated oxidation of cellular and subcellular components: ESR and AFM assays. *Journal of Physics: Condensed Matter* 2005, *17*, S1471–S1482.
12. Wilson, L. J. Medical applications of fullerenes and metallofullerenes. *The Electrochemical Society Interface* 1999, 24–28.
13. Hogue, C. Regulating chemicals: concerns regarding REACH, nanomaterials, biomonitoring voiced at GlobalChem meeting. *Chemical & Engineering News* 2005, *83*, 53–58.
14. Jones, R. M. Nanotechnology legislation on fast track. *FYI* 2003, *38*.
15. Eilperin, J. Nanotechnology's big question: safety—some say micromaterials are coming to market without adequate controls. *Washington Post* 2005, A11.
16. Colvin, V. L. The potential environmental impact of engineered nanomaterials. *Nature Biotechnology* 2003, *21*, 1166–1170.
17. Service, R. F. Is nanotechnology dangerous? *Science* 2000, *290*, 1526–1527.
18. Service, R. F. Nanomaterials show signs of toxicity. *Science* 2003, *300*, 243–243.
19. Malakoff, D. Nanotechnology research. Congress wants studies of nanotech's 'dark side.' *Science* 2003, *301*, 27.
20. Schuurmann, G; Markert, B. Ecotoxicology: ecological fundamentals, chemical exposure, and biological effects. 1998.
21. Markert, B. A.; Breure, A. M.; Zechmeister, H. G. bioindicators and biomonitors: principles, concepts, and applications. *Trace Metals and other Contaminants in the Environment* 2003, *6*.
22. Jamil, K. Bioindicators and biomarkers of environmental pollution and risk assessment. 2001.
23. Sayes, C. M.; Fortner, J. D.; Guo, W.; Lyon, D.; Boyd, A. M.; Ausman, K. C.; Tao, Y. J.; Sitharaman, B.; Wilson, L. J.; Hughes, J. B.; West, J. L.; Colvin, V. L. The differential cytotoxicity of water-soluble fullerenes. *Nano Letters* 2004, *4*, 1881–1887.
24. Lyon, D. Y.; Fortner, J. D.; Sayes, C. M.; Colvin, V. L.; Hughes, J. B. Bacterial cell association and antimicrobial activity of a C$_{60}$ water suspension. *Environmental Toxicology and Chemistry* 2005, *24*, 2757–2762.
25. Bitton, G.; Koopman, B.; Agami, O. MetPAD: a bioassay for rapid assessment of heavy metal toxicity in wastewater. *Water Environment Research* 1992, *64*, 834–836.
26. Knobeloch, L. M.; Blondin, G. A.; Lyford, S. B.; Harkin, J. M. A rapid bioassay for chemicals that induce pro-oxidant states. *Journal of Applied Toxicology* 1990, *10*, 1–5.
27. Juvonen, R.; Martikainen, E.; Schultz, E.; Joutti, A.; Ahtiainen, J.; Lehtokari, M. A battery of toxicity tests as indicators of decontamination in composting oily wastes. *Ecotoxicology and Environmental Safety* 2000, *47*, 156–166.
28. Ames, B.; Maron, D. Revised methods for the *Salmonella* mutagenicity test. *Mutation Research* 1983, *113*, 173–215.
29. Ames, B.; McCann, J.; Yamasaki, E. Methods for detecting carcinogens and mutagens with the *Salmonella*/Mammalian-Microsome Mutagenicity Test. *Mutation Research* 1975, *31*, 347–364.

30. Leeuwen, V.; Grootelaar, E.; Niebeek, G. Fish embryos as teratogenicity screens: a comparison of embryotoxicity between fish and birds. *Ecotoxicology & Environmental Safety* 1990, *20*, 42–52.
31. Hallare, A. V.; Köhler, H.-R.; Triebskorn, R. Developmental toxicity and stress protein responses in zebrafish embryos after exposure to diclofenac and its solvent, DMSO. *Chemosphere* 2004, *56*, 659–666.
32. Schulte, C.; Nagel, R. Test acute toxicity in the embryo of zebrafish, *Brachydanio rerio*, as an alternative to the acute fish test: preliminary results. *ATLA* 1994, *22*, 12–19.
33. Zhu, L.; Shi., S. Utilization of embryo development technique of *Brachydanio rerio* to evaluating toxicity on various chemicals. *Chinese Journal of Applied Ecology* 2002, *13*, 252–254.
34. Reimers, M. J.; Flockton, A. R.; Tanguay, R. L. Ethanol- and acetaldehyde-mediated developmental toxicity in zebrafish. *Neurotoxicology and Teratology* 2004, *26*, 769–781.
35. ASTM. Standard guide for conducting the frog embryo teratogenesis assay—Xenopus. Designation E 1439–91. *Annual Book of ASTM Standards* 1998, *11.5*, 825–836.
36. Bantle, J. A.; Dumont, J. N.; Finch, R. A.; Linder, G.; Fort, D. J. Atlas of abnormalities, a guide for the performance of FETAX. 1998.
37. Alvarez, P. J. J.; Illman, W. Bioremediation and natural attenuation of groundwater contaminants: process fundamentals and mathematical models. 2005.
38. Hurst, C. J.; Crawford, R. L.; McInerney, M. J.; Knudsen, G. R.; Stetzenbach, L. D. Manual of Environmental Microbiology. 2002.
39. Brant, J.; Lecoanet, H.; Wiesner, M. R. Aggregation and deposition characteristics of fullerene nanoparticles in aqueous systems. *Journal of Nanoparticle Research* 2005, *7*, 545–553.
40. Derossi, D.; Calvet, S.; Trembleau, A.; Brunissen, A.; Chassaing, G.; Prochiantz, A. Cell internalization of the third helix of the antennapedia homeodomain is receptor-independent. *Journal of Biological Chemistry* 1996, *271*, 18188–18193.
41. Gazit, E.; Lee, W. J.; Brey, P. T.; Shai, Y. Mode of action of the antibacterial cecropin B2: a spectrofluorometric study. *Biochemistry* 1994, *33*, 10681–10692.
42. Matsuzaki, K.; Sugishita, K.; Miyajima, K. Interactions of an antimicrobial peptide, magainin 2, with lipopolysaccharide-containing liposomes as a model for outer membranes of Gram-negative bacteria. *FEBS Letters* 1999, *449*, 221–224.
43. Tsao, N.; Luh, T. Y.; Chou, C. K.; Chang, T. Y.; Wu, J. J.; Liu, C. C.; Lei, H. Y. In vitro action of carboxyfullerene. *Journal of Antimicrobial Chemotherapy* 2002, *49*, 641–649.
44. Kloepfer, J. A.; Mielke, R. E.; Nadeau, J. L. Uptake of CdSe and CdSe/ZnS quantum dots into bacteria via purine–dependent mechanisms. *Applied and Environmental Microbiology* 2005, *71*, 2548–2557.
45. Diao, J. J.; Hua, D.; Lin, J.; Teng, H. H.; Chen, D. Nanoparticle delivery by controlled bacteria. *Journal of Nanoscience and Nanotechnology* 2005, *5*, 1749–1751.
46. Jang, H.; Pell, L. E.; Korgel, B. A.; English, D. S. Photoluminescence quenching of silicon nanoparticles in phospholipid vesicle bilayers. *Journal of Photochemistry and Photobiology A-Chemistry* 2003, *158*, 111–117.
47. Guldi, D. M.; Hungerbuhler, H. Electron transfer studies with fullerenes incorporated into artificial lipid bilayer membranes. *Research on Chemical Intermediates* 1999, *25*, 615–621.
48. Tsao, N.; Luh, T.; Chou, C.; Chang, T.; Wu, J.; Liu, C.; Lei, H. In vitro action of carboxyfullerene. *Journal of Antimicrobial Chemotherapy* 2002, *49*, 641–649.
49. Koch, A. M.; Reynolds, F.; Merkle, H. P.; Weissleder, R.; Josephson, L. Transport of surface-modified nanoparticles through cell monolayers. *Chembiochem* 2005, *6*, 337–345.
50. Martin, C. R.; Kohli, P. The emerging field of nanotube biotechnology. *Nature Reviews Drug Discovery* 2003, *2*, 29–37.
51. Block, M. L.; Hong, J. S. Microglia and inflammation-mediated neurodegeneration: multiple triggers with a common mechanism. *Progress in Neurobiology* 2005, *76*, 77–98.
52. Fernandes, M. A.; Santos, M. S.; Vicente, J. A.; Moreno, A.J.; Velena, A.; Dubrus, G.; Oliveira, G. R. Effects of 1,4-dihydropyridone derivatives on mitochondrial bioenergetics and oxidative stress: a comparative study. *Mitochondrion* 2003, *3*, 47–59.

53. Long, T.; Saleh, N.; Phenrat, T.; Swartz, C.; Parker, J.; Lowry, G.; Veronesi, B. Metal oxide nanoparticles produce oxidative stress in CNS microglia and neurons. *Society of Toxicology Annual Meeting* March 5–9, 2006.
54. Tatke, S. S.; Renugopalakrishnan, V.; Prabhakaran, M. Interfacing biological macro-molecules with carbon nanotubes and silicon surfaces: a computer modelling and dynamic simulation study. *Nanotechnology* 2004, *15*, S684–S690.
55. Slocik, J. M.; Naik, R. R.; Stone, M. O.; Wright, D. W. Viral templates for gold nanoparticle synthesis. *Journal of Materials Chemistry* 2005, *15*, 749–753.
56. Kramer, R. M.; Li, C.; Carter, D. C.; Stone, M. O.; Naik, R. R. Engineered protein cages for nanomaterial synthesis. *Journal of the American Chemical Society* 2004, *126*, 13282–13286.
57. Zhang, S. Fabrication of novel biomaterials through molecular self-assembly. *Nature Biotechnology* 2003, *21*, 1171–1178.
58. Rozhkov, S. P.; Goryunov, A. S.; Sukhanova, G. A.; Borisova, A. G.; Rozhkova, N. N.; Andrievsky, G. V. Protein interaction with hydrated C(60) fullerene in aqueous solutions. *Biochemical and Biophysical Research Communications* 2003, *303*, 562–566.
59. Kim, J. E.; Lee, M. Fullerene inhibits beta-amyloid peptide aggregation. *Biochemical and Biophysical Research Communications* 2003, *303*, 576–579.
60. Patolsky, F.; Tao, G.; Katz, E.; Willner, I. C_{60}-mediated bioelectrocatalyzed oxidation of glucose with glucose oxidase. *Journal of Electroanalytical Chemistry* 1998, *454*, 9–13.
61. Hamano, T.; Mashino, T.; Hirobe, M. Oxidation of [60]fullerene by cytochrome-P450 chemical-models. *Journal of the Chemical Society-Chemical Communications* 1995, 1537–1538.
62. Foley, S.; Curtis, A. D. M.; Hirsch, A.; Brettreich, M.; Pelegrin, A.; Seta, P.; Larroque, C. Interaction of a water soluble fullerene derivative with reactive oxygen species and model enzymatic systems. *Fullerenes Nanotubes and Carbon Nanostructures* 2002, *10*, 49–67.
63. Wang, F.; Tan, W. B.; Zhang, Y.; Fan, X. P.; Wang, M. Q. Luminescent nanomaterials for biological labelling. *Nanotechnology* 2006, *17*, R1–R13.
64. Wang, J. Nanomaterial-based amplified transduction of biomolecular interactions. *Small* 2005, *1*, 1036–1043.
65. Xiao, Y.; Barker, P. E. Semiconductor nanocrystal probes for human metaphase chromosomes. *Nucleic Acids Research* 2004, *32*, e28.
66. Ito, Y.; Fukusaki, E. DNA as a 'nanomaterial.' *Journal of Molecular Catalysis B-Enzymatic* 2004, *28*, 155–166.
67. Wang, S. Z.; Gao, R. M.; Zhou, F. M.; Selke, M. Nanomaterials and singlet oxygen photosensitizers: potential applications in photodynamic therapy. *Journal of Materials Chemistry* 2004, *14*, 487–493.
68. Xiang, J. J.; Tang, J. Q.; Zhu, S. G.; Nie, X. M.; Lu, H. B.; Shen, S. R.; Li, X. L.; Tang, K.; Zhou, M.; Li, G. Y. IONP-PLL: a novel non-viral vector for efficient gene delivery. *Journal of Gene Medicine* 2003, *5*, 803–817.
69. Takenaka, S.; Yamashita, K.; Takagi, M.; Hatta, T.; Tanaka, A.; Tsuge, O. Study of the DNA interaction with water-soluble cationic fullerene derivatives. *Chemistry Letters* 1999, *28*, 319–320.
70. Zhao, X.; Striolo, A.; Cummings, P. T. C_{60} binds to and deforms nucleotides. *Biophysical Journal* 2005, *89*, 3856–3862.
71. Green, M.; Howman, E. Semiconductor quantum dots and free radical induced DNA nicking. *Chemical Communications (Cambridge)* 2005, 121–123.
72. Wakefield, G.; Green, M.; Lipscomb, S.; Flutter, B. Modified titania nanomaterials for sunscreen applications—reducing free radical generation and DNA damage. *Materials Science and Technology* 2004, *20*, 985–988.
73. Takenaka, S.; Yamashita, K.; Takagi, M.; Hatta, T.; Tsuge, O. Photo-induced DNA cleavage by water-soluble cationic fullerene derivatives. *Chemistry Letters* 1999, 321–322.
74. Nakamura, E.; Tokuyama, H.; Yamago, S.; Shiraki, T.; Sugiura, Y. Biological activity of water-soluble fullerenes. Structural dependence of DNA cleavage, cytotoxicity, and enzyme inhibitory activities including HIV-protease inhibition. *Bulletin of the Chemical Society of Japan* 1996, *69*, 2143–2151.

75. Tokuyama, H.; Yamago, S.; Nakamura, E. Photoinduced biochemical activity of fullerene carboxylic acid. *Journal of the American Chemical Society* 1993, *115*, 7918–7919.

76. Babynin, E. V.; Nuretdinov, I. A.; Gubskaia, V. P.; Barabanshchikov, B. I. Study of mutagenic activity of fullerene and some of its derivatives using His+ reversions of *Salmonella typhimurium* as an example. *Genetics of Microorganisms* 2002, *38*, 453–457.

77. Kwak, S. Y.; Kim, S. H.; Kim, S. S. Hybrid organic/inorganic reverse osmosis (RO) membrane for bactericidal anti-fouling. 1. Preparation and characterization of TiO_2 nanoparticle self-assembled aromatic polyamide thin-film-composite (TFC) membrane. *Environmental Science and Technology* 2001, *35*, 2388–2394.

78. Marchesan, S.; Da Ros, T.; Spalluto, G.; Balzarini, J.; Prato, M. Anti-HIV properties of cationic fullerene derivatives. *Bioorganic & Medicinal Chemistry Letters* 2005, *15*, 3615–3618.

79. Piotrovsky, L.; Dumpis, M.; Poznyakova, L.; Kiselev, O.; Kozeletskaya, K.; Eropkin, M.; Monaenkov, A. Study of the biological activity of the adducts of fullerenes with poly(N-vinylpyrrolidone). *Molecular Materials* 2000, *13*, 41–50.

80. Schinazi, R.; Sijbesma, R.; Srdanov, G.; Hill, C.; Wudl, F. Synthesis and virucidal activity of a water-soluble, configurationally stable, derivatized C_{60} fullerene. *Antimicrobial Agents and Chemotherapy* 1993, *37*, 1707–1710.

81. Sayes, C. M.; Fortner, J. D.; Guo, W.; Lyon, D.; Boyd, A.; Ausman, K. D.; Tao, Y. J.; Sitharaman, B.; Wilson, L. J.; Hughes, J. B.; West, J. L.; Colvin, V. L. The differential cytotoxicity of water-soluble fullerenes. *Nano Letters* 2004, *4*, 1881–1887.

82. Scrivens, W. A.; Tour, J. M. Synthesis of ^{14}C-labeled C_{60}, its suspension in water, and its uptake by human keratinocytes. *Journal of the American Chemical Society* 1994, *116*, 4517–4518.

83. Kai, Y.; Komazawa, Y.; Miyajima, A.; Miyata, N.; Yamakoshi, Y. [60] Fullerene as a novel photoinduced antibiotic. *Fullerenes, Nanotubes, and Carbon Nanostructures* 2003, *11*, 79–87.

84. Sera, N.; Tokiwa, H.; Miyata, N. Mutagenicity of the fullerene C_{60}-generated singlet oxygen dependent formation of lipid peroxides. *Carcinogenesis* 1996, *17*, 2163–2169.

85. Tsuchiya, T.; Oguri, I.; Yamakoshi, Y.; Miyata, N. Novel harmful effects of [60]fullerene on mouse embryos in vitro and in vivo. *FEBS Letters* 1996, *393*, 139–145.

86. Tabata, Y.; Ikada, Y. Biological functions of fullerenes. *Pure Applied Chemistry* 1999, *71*, 2047–2053.

87. Kamat, J. P.; Devasagayam, T. P.; Priyadarsini, K. I.; Mohan, H. Reactive oxygen species mediated membrane damage induced by fullerene derivatives and its possible biological implications. *Toxicology* 2000, *155*, 55–61.

88. Tsao, N.; Kanakamma, P. P.; Luh, T. Y.; Chou, C. K.; Lei, H. Y. Inhibition of *Escherichia coli*-induced meningitis by carboxyfullerene. *Antimicrobial Agents and Chemotherapy* 1999, *43*, 2273–2277.

89. Tsao, N.; Luh, T.; Chou, C.; Wu, J.; Lin, Y.; Lei, K. Inhibition of group A streptococcus infection by carboxyfullerene. *Antimicrobial Agents and Chemotherapy* 2001, *45*, 1788–1793.

90. Yang, X.; Fan, C.; Zhu, H. Photo-induced cytotoxicity of malonic acid [C-60]fullerene derivatives and its mechanism. *Toxicology in vitro* 2002, *16*, 41–46.

91. Cusan, C.; Da Ros, T.; Spalluto, G.; Foley, S.; Janto, J.; Seta, P.; Larroque, C.; Tomasini, M.; Antonelli, T.; Ferraro, L.; Prato, M. A new multi-charged C-60 derivative: synthesis and biological properties. *European Journal of Organic Chemistry* 2002, *17*, 2928–2934.

92. Da Ros, T.; Prato, M. Easy access to water-soluble fullerene derivatives via 1,3-dipolar cycloadditions of azomethine ylides to C_{60}. *Journal of Organic Chemistry* 1996, *61*, 9070–9072.

93. Mashino, T.; Usui, N.; Okuda, K.; Hirota, T.; Mochizuki, M. Respiratory chain inhibition by fullerene derivatives: hydrogen peroxide production caused by fullerene derivatives and a respiratory chain system. *Bioorganic and Medicinal Chemistry* 2003, *11*, 1433–1438.

94. Mashino, T.; Nishikawa, D.; Takahashi, K.; Usui, N.; Yamori, T.; Seki, M.; Endo, T.; Mochizuki, M. Antibacterial and Antiproliferative activity of cationic fullerene derivatives. *Bioorganic and Medicinal Chemistry Letters* 2003, *13*, 4395–4397.
95. Bosi, S.; Da Ros, T.; Castellano, S.; Banfi, E.; Prato, M. Antimycobacterial activity of ionic fullerene derivatives. *Bioorganic and Medicinal Chemistry Letters* 2000, *10*, 1043–1045.
96. Mashino, T.; Okuda, K.; Hirota, T.; Hirobe, M.; Nagano, T.; Mochizuki, M. Inhibition of *E. coli* growth by fullerene derivatives and inhibition mechanism. *Bioorganic and Medicinal Chemistry Letters* 1999, *9*, 2959–2962.
97. Bullard-Dillard, R.; Creek, K. E.; Scrivens, W. A.; Tour, J. M. Tissue sites of uptake of ^{14}C-labeled C_{60}. *Bioorganic Chemistry* 1996, *24*, 376–385.
98. Cagle, D. W.; Kennel, S. J.; Mirzadeh, S.; Alford, J. M.; Wilson, L. J. In vivo studies of fullerene-based materials using endohedral metallofullerene radiotracers. *Proceedings of the National Academy of Science U S A* 1999, *96*, 5182–5187.
99. Rincon, A.; Pulgarin, C. Effect of pH, inorganic ions, organic matter and H_2O_2 on *E. coli* K12 photocatalytic inactivation by TiO_2. Implications in solar water disinfection. *Applied Catalysis B: Environmental* 2004, *51*, 283–302.
100. Rincon, A.; Pulgarin, C. Bactericidal action of illuminated TiO_2 on pure *Escherichia coli* and natural bacterial consortia: post-irradiation events in the dark and assessment of the effective disinfection time. *Applied Catalysis B: Environmental* 2004, *49*, 99–112.
101. Wolfrum, E. J.; Huang, J.; Blake, D. M.; Maness, P. C.; Huang, Z.; Fiest, J.; Jacoby, W. A. Photocatalytic oxidation of bacteria, bacterial and fungal spores, and model biofilm components to carbon dioxide on titanium dioxide-coated surfaces. *Environmental Science and Technology* 2002, *36*, 3412–3419.
102. Huang, L.; Li, D. Q.; Lin, Y. J.; Wei, M.; Evans, D. G.; Duan, X. Controllable preparation of Nano-MgO and investigation of its bactericidal properties. *Journal of Inorganic Biochemistry* 2005, *99*, 986–993.
103. Sawai, J.; Igarashi, H.; Hashimoto, A.; Kokugan, T.; Shimizu, M. Effect of particle size and heating temperature of ceramic powders on antibacterial activity of their slurries. *Journal of Chemical Engineering of Japan* 1996, *29*, 251–256.
104. Sawai, J.; Igarashi, H.; Hashimoto, A.; Kokugan, T.; Shimizu, M. Effect of ceramic powder slurry on spores of *Bacillus subtilis*. *Journal of Chemical Engineering of Japan* 1995, *28*, 556–561.
105. Hilger, I.; Hiergeist, R.; Hergt, R.; Winnefeld, K.; Schubert, H.; Kaiser, W. A. Thermal ablation of tumors using magnetic nanoparticles—An in vivo feasibility study. *Investigative Radiology* 2002, *37*, 580–586.
106. Zharov, V. P.; Mercer, K. E.; Galitovskaya, E. N.; Smeltzer, M. S. Photothermal nanotherapeutics and nanodiagnostics for selective killing of bacteria targeted with gold nanoparticles. *Biophysical Journal* 2006, *90*, 619–627.
107. Gurr, J. R.; Wang, A. S.; Chen, C. H.; Jan, K. Y. Ultrafine titanium dioxide particles in the absence of photoactivation can induce oxidative damage to human bronchial epithelial cells. *Toxicology* 2005, *213*, 66–73.
108. Fubini, B.; Hubbard, A. Reactive oxygen species (ROS) and reactive nitrogen species (RNS) generation by silica in inflammation and fibrosis. *Free Radical Biology and Medicine* 2003, *34*, 1507–1516.
109. Yeber, M. C.; Rodriguez, J.; Freer, J.; Duran, N.; Mansilla, H. D. Photocatalytic degradation of cellulose bleaching effluent by supported TiO_2 and ZnO. *Chemosphere* 2000, *41*, 1193–1197.
110. Kubo, M.; Onodera, R.; Shibasaki-Kitakawa, N.; Tsumoto, K.; Yonemoto, T. Kinetics of ultrasonic disinfection of Escherichia coli in the presence of titanium dioxide particles. *Biotechnology Progress* 2005, *21*, 897–901.
111. Cabiscol, E.; Tamarit, J.; Ros, J. Oxidative stress in bacteria and protein damage by reactive oxygen species. *International Microbiology* 2000, *3*, 3–8.
112. Oberdörster, E.; Zhu, S.; Blickley, T. M.; McClellan-Green, P.; Haasch, M. L. Ecotoxicology of carbon-based engineered nanoparticles: Effects of fullerene (C_{60}) on aquatic organisms. *Carbon* 2006, *44*, 1112–1120.
113. Bielski, B. H.; Arudi, R. L.; Sutherland, M. W. A study of the reactivity of HO_2/O_2^- with unsaturated fatty acids. *Journal of Biological Chemistry* 1983, *258*, 4759–4761.

114. Imlay, J. A. Pathways of oxidative damage. *Annual Review of Microbiology* 2003, *57*, 395–418.
115. Inoue, M.; Kamiya, H.; Fujikawa, K.; Ootsuyama, Y.; Murata-Kamiya, N.; Osaki, T.; Yasumoto, K.; Kasai, H. Induction of chromosomal gene mutations in Escherichia coli by direct incorporation of oxidatively damaged nucleotides: New evaluation method for mutagenesis by damaged DNA precursors in vivo. *Journal of Biological Chemistry* 1998, *273*, 11069–11074.
116. Rai, P.; Cole, T. D.; Wemmer, D. E.; Linn, S. Localization of Fe(2+) at an RTGR sequence within a DNA duplex explains preferential cleavage by Fe(2+) and H_2O_2. *Journal of Molecular Biology* 2001, *312*, 1089–1101.
117. Stadtman, E. Free Radical Mediated Oxidation of Proteins. *Free Radicals, Oxidative Stress, and Antioxidants: Pathological and Physiological Significance* 1998, 51–65.
118. Corona-Morales, A. A.; Castell, A.; Escobar, A.; Drucker-Colín, R.; Zhang, L. Fullerene C_{60} and ascorbic acid protect cultured chromaffin cells against levodopa toxicity. *Journal of Neuroscience Research* 2003, *71*.
119. Wang, C.; Tai, L. A.; Lee, D. D.; Kanakamma, P. P.; Shen, C. K.-F.; Luh, T.-Y.; Cheng, C. H.; Hwang, K. C. C_{60} and water-soluble fullerene derivatives as antioxidants against radical-initiated lipid peroxidation. *Journal of Medicinal Chemistry* 1999, *42*, 4614–4620.
120. Bhattacharya, D.; Gupta, R. K. Nanotechnology and potential of microorganisms. *Critical Reviews in Biotechnology* 2005, *25*, 199–204.
121. Schuler, D. Formation of magnetosomes in magnetotactic bacteria. *Journal of Molecular Microbiology and Biotechnology* 1999, *1*, 79–86.
122. Kumar, N.; Dorfman, A.; Hahm, J. I. Fabrication of optically enhanced ZnO nanorods and microrods using novel biocatalysts. *Journal of Nanoscience and Nanotechnology* 2005, *5*, 1915–1918.
123. Sweeney, R. Y.; Mao, C.; Gao, X.; Burt, J. L.; Belcher, A. M.; Georgiou, G.; Iverson, B. L. Bacterial biosynthesis of cadmium sulfide nanocrystals. *Chemistry and Biology* 2004, *11*, 1553–1559.
124. Klaus, T.; Joerger, R.; Olsson, E.; Granqvist, C.-G. Silver-based crystalline nanoparticles, microbially fabricated. *Applied Physical Microbiology* 1999.
125. Labrenz, M.; Druschel, G. K.; Thomsen-Ebert, T.; Gilbert, B.; Welch, S. A.; Kemner, K. M.; Logan, G. A.; Summons, R. E.; Stasio, G. D.; Bond, P. L.; Lai, B.; Kelly, S. D.; Banfield, J. F. Formation of sphalerite (ZnS) deposits in natural biofilms of sulfate-reducing bacteria. *Science* 2000, *290*, 1744–1747.
126. Sarikaya, M.; Tamerler, C.; Jen, A. K.; Schulten, K.; Baneyx, F. Molecular biomimetics: nanotechnology through biology. *Nature Materials* 2003, *2*, 577–585.
127. Tien, M.; Kirk, T. K. Lignin-degrading enzyme from hymenomycete *Phanerochaete chrysosporium* buds. *Science* 1983, *221*, 661–663.
128. Tien, M.; Kirk, T. K. Lignin Peroxidase of *Phanerocheate chrysosporium*. *Methods of Enzymology* 1988, *161*, 238–249.
129. Piontek, K.; Smith, A. T.; Blodig, W. Lignin peroxidase structure and function. *Biochemical Society Transactions* 2001, *29*, 111–116.
130. Hofrichter, M.; Scheibner, K.; Schneegab, I.; Fritsche, W. Enzymatic combustion of aromatic and aliphatic compounds by manganese peroxidase from *Nematoloma frowardii*. *Applied and Environmental Microbiology* 1998, *64*, 399–404.
131. Hofrichter, M.; Scheibner, K.; Schneegab, I.; Ziegenhagen, D.; Fritsche, W. Mineralization of synthetic humic substances by manganese peroxidase from the white-rot fungi *Nematoloma frowardii*. *Applied Microbiology and Biotechnology* 1998, *49*, 584–588.
132. Stienbuchel, A. Biopolymers. Lignin, humis substances and coal. 2001, *1*.
133. Limbach, L.; Li, Y.; Grass, R.; Brunner, T. J.; hintermann, M.; Muller, M.; Gunther, D.; Stark, W. Oxide nanoparticle uptake in human lung fibroblasts: effects of particle size, agglomeration, and diffusion at low concentrations. *Environmental Science and Technology* 2005, *39*, 9370–9376.
134. Jang, H. D.; Kim, S. K.; Kim, S. J. Effect of particle size and phase composition of titanium dioxide nanoparticles on the photocatalytic properties. *Journal of Nanoparticle Research* 2001, *3*, 141–147.

135. Lyon, D. Y.; Adams, L. K.; Falkner, J. C.; Alvarez, P. J. J. Antibacterial activity of fullerene water suspensions: effects of preparation method and particle size. *Environmental Science and Technology* 2006, *40*, 4360–4366.
136. Li, D.; Haneda, H. Morphologies of zinc oxide particles and their effects on photocatalysis. *Chemosphere* 2003, *51*, 129–137.
137. Hristovski, K.; Zhang, Y.; Koeneman, B. A.; Chen, Y.; Westerhoff, P.; Capco, D. G.; Crittenden, J. Nanomaterials in water environments: potential applications, treatments, fate and potential biological consequences. *230th ACS National Meeting* 2005.
138. Tungittiplakorn, W.; Cohen, C.; Lion, L. W. Engineered polymeric nanoparticles for bioremediation of hydrophobic contaminants. *Environmental Science and Technology* 2005, *39*, 1354–1358.
139. Masciangioli, T.; Zhang, W. X. Environmental technologies at the nanoscale. *Environmental Science and Technology* 2003, *37*, 102A–108A.
140. Liu, D.; Johnson, P. R.; Elimelech, M. Colloid deposition dynamics in flow through porous media: role of electrolyte concentration. *Environmental Science and Technology* 1995, *29*, 2963–2973.
141. McCormick, M. L.; Adriaens, P. Carbon tetrachloride transformation on the surface of nanoscale biogenic magnetite particles. *Environmental Science and Technology* 2004, *38*, 1045–1053.
142. Mattigod, S. V.; Fryxell, G. E.; Alford, K.; Gilmore, T.; Parker, K.; Serne, J.; Engelhard, M. Functionalized TiO_2 nanoparticles for use for in situ anion immobilization. *Environmental Science and Technology* 2005, *39*, 7306–7310.
143. Lee, J.; Choi, W.; Yoon, J. Photocatalytic degradation of N-nitrosodimethylamine: mechanism, product distribution, and TiO_2 surface modification. *Environmental Science and Technology* 2005, *39*, 6800–6807.
144. Ferguson, M. A.; Hoffmann, M. R.; Hering, J. G. TiO_2-photocatalyzed As(II) oxidation in aqueous suspensions: reaction kinetics and effects of adsorption. *Environmental Science and Technology* 2005, *39*, 1880–1886.
145. Esterkin, C. R.; Negro, A. C.; Alfano, O. M.; Cassano, A. E. Air pollution remediation in a fixed bed photocatalytic reactor coated with TiO_2. *AlChE* 2005, *51*, 2298–2310.
146. Pandey, J. K.; Kumar, A. P.; Misra, M.; Mohanty, A. K.; Drzal, L. T.; Singh, R. P. Recent advances in biodegradable nanocomposites. *Nanoscience and Nanotechnology* 2005, *5*, 497–526.
147. Otaki, M.; Hirata, T.; Ohgaki, S. Aqueous microorganisms inactivation by photocatalytic reaction. *Water Science and Technology* 2000, *42*, 103–108.
148. Watts, R. J.; Kong, S.; Orr, M. P.; Miller, G. C.; Henry, B. E. Photocatalytic inactivation of coliform bacteria and viruses in secondary wastewater effluent. *Water Research* 1995, *29*, 95–100.
149. Wei, C.; Lin, W. Y.; Zainal, Z.; Williams, N. E.; Zhu, K.; Kruzic, A. P.; Smith, R. L.; Rajeshwar, K. Bactericidal activity of TiO_2 photocatalyst in aqueous media: toward a solar-assisted water disinfection system. *Environmental Science and Technology* 1994, *28*, 934–938.

Assessing Life-Cycle Risks of Nanomaterials

Christine Ogilvie Robichaud *Duke University, Durham, North Carolina, USA*

Dicksen Tanzil *Bridges to Sustainability, Golder Associates Inc., Houston, Texas, USA*

Mark R. Wiesner *Duke University, Durham, North Carolina, USA*

Life-Cycle Impacts and Sustainability

The applications of nanomaterials discussed in previous chapters illustrate some of nanotechnology's many positive contributions to sustainability—an integrated approach to actions that takes into account the social, economic, and environmental consequences of these actions for current and future generations. The design of nanomaterials and the infrastructure for fabricating and using these materials require a systems approach that extends environmental responsibility upstream from the end-use of the material to the manufacturing of the nanomaterials.

In Chapter 1, we presented a road map for this book that generally follows the sources and environmental effects of nanomaterials and establishes the idea of weighing costs and benefits, which is shown again in Figure 13.1. The boxes in the conceptual schematic frame up the questions we are posing concerning the nature and uses of nanomaterials, as well as the research being done to answer these questions. They are not meant to represent all the components of a complete life cycle; instead, the arrows in this diagram can be read as saying that one box "feeds" or "influences" the next one. These pieces of information will be the necessary inputs to an assessment of the risk and the life-cycle impacts of nanomaterials.

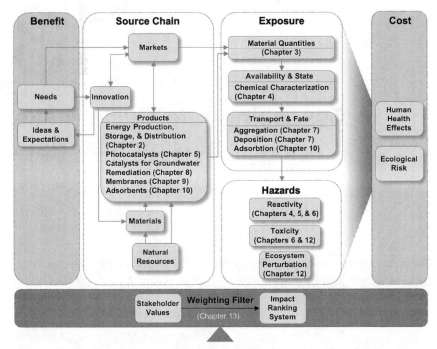

Figure 13.1 Conceptual diagram of the links between the costs and benefits due to nanomaterials in the environment, and how the subjects covered in the chapters of this book help us understand these links.

Applying life-cycle thinking and risk assessment to nanomaterials

In this chapter, we will consider ways to ascertain the life-cycle impacts and risks of nanomaterials. Recognizing that the production of these particles will have a wide array of impacts on the environment, some foreseeable and some not, some positive and some not, we must search for a way to move forward responsibly with this group of technologies that boasts so much positive potential. To that end, we will introduce the concepts and goals of life-cycle assessment and risk assessment, review varying approaches to framing a risk assessment, and present an example of how an existing risk assessment tool can be applied to look at impacts based on the pieces of information we already know.

This example application will illustrate how employing risk assessment methods used in industry can be directionally helpful; however, it is important to recognize that full impacts will not be adequately captured solely via application of existing assessment methods to these novel materials. The composition of nanomaterials may closely resemble known bulk materials, but their properties do not hold constant as size scales down. Instead, they exhibit distinct properties due to their small size, shape, and surface functionalities. As the surface area to volume ratio increases, electrical and magnetic properties change and quantum effects begin to dominate, changing the way substances interact. This should be no surprise, as it is precisely because of the advantages afforded by these differences that nanomaterials are engineered in the first place. Based on the novel characteristics and interactions of nanomaterials, we may learn that the particular properties and behaviors used until now to characterize the impacts of a chemical are inadequate in describing impacts of nanomaterials.

Indeed, the elemental atomic composition and structure of a substance appear to be insufficient identifiers for many nanomaterials. As a result, the risk assessment and life cycle impact communities must grapple with assessment criteria that will more adequately account for factors unique to nanomaterials, and any risk these materials may pose.

Life-cycle and risk assessments are instruments that tie together the insight of several disciplines in order to provide a framework for decision making. Now, just as the nano-industry is burgeoning, is the time to employ such decision making tools. We are beginning to have enough information to make meaningful assessments of the risks and benefits posed throughout the life cycle of these materials, but there are still many choices to be made. At this early stage, the industry still has the flexibility to intelligently and responsibly apply these technologies while minimizing negative environmental impact.

Concepts from life-cycle assessment and risk assessment

An examination of the repercussions of activities and products throughout a life cycle of production, use, disposal, and reuse is at the heart of approaches for sustainability planning and decision making. For any given industrial product, the life-cycle stages of resource extraction, raw material production, product manufacturing, transportation, use, and end-of-life can all be associated with significant costs and benefits to the manufacturers, customers, environment, and other stakeholders.

Life-cycle assessment and life-cycle thinking. Life-cycle assessment (LCA) provides a framework for identifying and evaluating the life-cycle impacts of a product, process, or activity. Typically, impacts on human health, ecosystem health, and effects of pollutant depositions in all environmental media are evaluated for each stage of the life cycle. For a new technology application in an area such as nanotechnology, LCA offers a context for looking at potential life-cycle implications, optimizing its economic, environmental, and societal benefits, and minimizing risks.

Formal LCA. While there are many variations in LCA methodologies, arguably the most broadly accepted is one formalized in the ISO-14040 series of standards [1, 2]. Often referred to as the "formal LCA" or "full LCA," the ISO methodology guides the quantitative assessment of environmental impacts throughout a product's life cycle. Typically used in comparing product or technology alternatives, the formal LCA involves the following phases:

(a) Goal and scope definition
Defining purpose, comparison basis, and boundaries, including the life-cycle stages and impacts that require detailed assessment

(b) Inventory
Developing "life-cycle inventory" (LCI) where information on the inflows and outflows of materials and energy (including resource use and emissions) for each relevant life-cycle stage are collected and tabulated

(c) Impact assessment
Analyzing LCI data and grouping the impacts into several impact categories, such as resource depletion, human and ecosystem toxicity, global warming potential, smog formation potential, eutrophication potential, and so on—this phase is often termed "life-cycle impact assessment" (LCIA)

(d) Interpretation
Evaluating, identifying, and reporting the needs and opportunities to reduce negative impacts and optimize the environmental benefits of the products

A major challenge in conducting the formal LCA is to obtain reliable and available data for LCI. Data collection can easily become an enormous and expensive task, especially if data specific for the exact materials and processes involved in the product life cycle are sought. Generic data for various raw material classes and life-cycle activities, however, are available in much of the commercially available LCA software. Once the inventory is developed, the life-cycle environmental impacts can be determined using various existing LCIA methodologies [3–5]. The life-cycle impacts can then be weighted based on criteria such as geographical relevance, relevance to the decision, and stakeholder value.

Although not explicitly part of the formal LCA, life-cycle economic implications are often considered. The "eco-efficiency assessment" developed by BASF [6], for example, balances weighted environmental impacts with the "total cost of ownership," which covers the costs of production, use, and disposal of the product. Results from LCA can also be expressed in eco-efficiency metrics that normalize impacts by revenue or monetary value-added [7, 8].

Efforts have also been made to integrate social impact considerations into LCA. The eco-efficiency assessment of BASF has recently been extended to include social impacts [9]. Individual indicators on a product's impacts on human health and safety, nutrition, living conditions, education, workplace conditions, and other social factors are assessed and scored relative to a reference (usually the product being replaced).

Streamlined LCA. The life-cycle thinking formalized in LCA also lies behind strategies, guidelines, and screening methodologies that may be used in the early stages of product and process design. Often called "streamlined LCA" [10], they typically rely on the use of a two-dimensional matrix, with life-cycle stages on one dimension and impacts on the other (see Figure 13.2). The streamlined LCA is typically qualitative, with a relative score or color-coded rating assigned for each impact at each stage of the life cycle. The qualitative assessment allows the consideration of impacts not typically considered in the formal LCA, such as nuisance, public perception, market position, and so on. Supported by a question-driven approach, such assessment can help identify problems and point out where improvements are most needed [11].

Application to emerging technologies

LCA in its various forms has been used to identify and communicate the potential costs and benefits of new technology applications. For example, formal LCA was applied to compare the environmental and human health impacts of the genetically modified herbicide-tolerant sugar-beet crop with those of the conventional crop [12]. The study demonstrated the benefits of the herbicide-tolerant crop in lowering emissions and their associated environmental and human health impacts in various

Life Cycle stages	Material choice	Energy use	Solid emissions	Liquid emissions	Gaseous emissions
Resource extraction and raw material production	3	2	1	4	2
Manufacturing	3	4	3	4	4
Product delivery	3	4	2	4	4
Product use	4	3	4	4	4
Refurbishment, recycling, disposal	2	1	1	4	4

Figure 13.2 Example of a streamlined life-cycle assessment matrix, scoring the environmental impacts of today's automobiles along the life-cycle stages, compared to those in the 1950s. Higher scores indicate larger improvements, while low scores indicate where attentions are needed. Adapted from Graedel, 1998 [10].

stages of the life cycle (primarily herbicide manufacturing, transport, and field operations), while recognizing that the life-cycle costs and benefits of genetically modified crops in general need to be assessed on a case-by-case basis.

LCA has also been used to examine the impacts of replacing traditional materials with engineered nanoparticles. Lloyd et al. [13, 14] applied LCA to examine the potential economic cost and environmental implications of nanotechnology applications in automobiles, specifically the use of nanocomposites [13] and the use of nanotechnology to stabilize platinum-group metal particles in automotive catalysts [14]. The authors utilized a variant of LCA called "economic input-output LCA" (EIO-LCA) developed by their group at Carnegie Mellon University, where environmental impacts are estimated using publicly available US data on various industrial sectors involved in the product's life cycle (see www.eiolca.net).

Both studies demonstrated the benefits of the nanotechnology applications in reducing resource use and emissions along the life cycle compared to conventional practices. Nevertheless, these applications of LCA toward nanotechnology applications focus primarily on their potential benefits. Risks posed by the nanomaterials themselves were not addressed.

Risk assessment

Risk assessment is the task of characterizing a level of risk, usually in terms of a relative score or ranking. As with LCA, the goal is to provide information that will help evaluate alternatives and arrive at decisions; indeed, categories of risk may be included as elements of an LCA. Generally speaking, hazards are identified and characterized, exposure likelihood is evaluated, and both are quantified. These risk-determining factors are then combined to characterize relative risk [15].

Defining risk.   Risk has both objective and subjective dimensions. From an expert's point of view, risk has a statistical meaning: the product of the probability of an undesirable event and some measure of the magnitude of its negative consequences. Hazard (the potential to cause harm) is combined with probability of exposure to arrive at risk, a quantification of the likelihood that such harm will occur [16]. A communal perception of risk, on the other hand, is more subjective. People combine their knowledge, experiences, and anecdotal evidence shaping their perception of the likelihood and unpleasantness of a particular outcome [17]. Both objective and subjective views affect the public's acceptance of a new technology and should be considered in decision making.

This chapter is primarily concerned with the mathematical assessment of risk. However, as the subjects of hazard identification and risk assessment are explored, it is important to maintain an appreciation for the role of risk perception and public opinion on the decision making process.

Systemic bias.   The kernel of the risk assessment is the definition and mathematical combination of exposure and hazard, weighted so that risk factors considered as most important are valued higher and a relevant ranking system can be created. It is easy to see how such a process of expressing some objective reality in terms of a score necessarily imparts a subjective filter, or value system, in order to weight and combine the factors that determine the risk. Such a filter has been introduced in the general life-cycle schematic in Figure 13.1, where the weighting filter is used to inform the balance between benefits and costs of the nanomaterials' life cycle.

Biases are a natural part of risk assessment and need to be understood for the results of the assessment to adequately inform a decision making process. The parameters chosen to represent hazard and exposure, as well as the way in which the collected values for these parameters are combined, may vary between systems and significantly affect the conclusions of the entire assessment. These inherent biases are depicted in Figure 13.3. Out of the universe of hazards potentially posed by nanomaterials and factors that may affect exposure, one inevitably

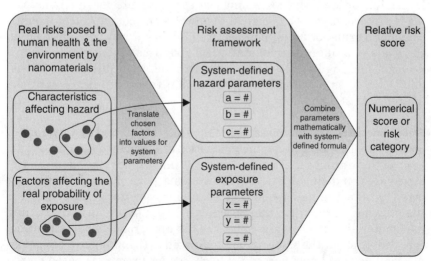

Figure 13.3   Pictorial representation of the steps and biases involved with mapping real-world risks to a relative risk score via a risk assessment framework.

must choose a subset to be included in the risk assessment model. While they may be selected because they are believed to be most critical to accurately capturing the hazard or exposure, they are sometimes selected on the basis of data availability or importance to specific group of stakeholders. Once selected, all the risk factors are defined in terms of measurable parameters, entered into the risk assessment methodology, and combined mathematically to produce a relative risk rating. The process of combining the parameters introduces another layer of bias because the relative weight of the different values decides which parameters have the greatest influence on final risk.

For example, an important factor that defines the hazard of a material is toxicity, a complex concept that can range in interpretation and can refer to multiple endpoints such as primary irritation, cytotoxicity, or genotoxicity. In many cases the toxicity is measured by recording the LC_{50}, which is the lethal concentration of a material administered to a population of rats or fish at which half of them die. However, this response represents a culmination of responses at the level of the cell and organisms, any one of which may have relevance in characterizing toxicity (see Chapter 6). In fact, selecting LC_{50} as *the* parameter to represent toxicity presents a bias. Whether the parameter is appropriate and adequate depends on the nature of the hazard and its utility in decision making.

We will further consider the manner in which we parameterize and score risk. However, as we explore the idea of risk assessment pertaining to nanomaterials it is, at the outset, easy to imagine a variety of risk assessment methods that deliver equally valid results. Depending on the decision making process being affected, a risk assessment protocol might be designed to address an economic, legal, political, engineering, or social perspective [15]. Whatever the perspective, the goals of a quantitative, statistical risk assessment are to inform decision making by allowing comparison between different options, and to serve as a critical input for a plan to communicate with the appropriate stakeholders [18].

Stakeholder bias.   Numerous stakeholders may be affected by and/or influence the application and commercialization of engineered nanomaterials. Each of these groups may characterize the importance of various parameters based on their respective value systems, none of which are without bias [19]. Startup nanotechnology companies see the great benefits of the growing markets for their products. Users of products containing nanomaterials will enjoy the benefit of the new products but will surely care deeply about the safety of being exposed to them. Insurance companies will need to know how much to charge for premiums based on the risk they are taking on when underwriting the nanotechnology industry.

The general public will both experience positive changes enabled by nano-materials and be impacted in some way by nanomaterials entering the air, water, soil, and organisms around them. Different stakeholder groups with their varying value systems and priorities see the costs and bene-fits through different lenses.

The public at large, including even a few influential groups, can be particularly effective in determining the trajectory of nanotechnologies. In the case of another emerging technology full of unknowns—genetically modified foods (GMOs)—market leaders failed to anticipate and react to societal perceptions of the risks (largely in Europe and Canada) versus benefits from these products. The resulting public back-lash led to regulatory limitations that continue to inhibit the develop-ment of this technology and have brought genetic modification of food to a virtual standstill in Europe. This example is not intended to be illus-trative of an extended analogy between the technology of nanomateri-als and GMOs, but to highlight the crucial role of public perception and communication, especially in the case of uncertainty. Financial indus-try forecasters, yet another stakeholder group, recognize this impor-tance; even as one strategic advising firm predicted the immense economic impact of nanotechnology in a 2005 report, they cautioned that "poor handling of risk by any player could result in perception problems that would affect entire markets" [20].

Regardless of which lens is chosen to view the risk, there is a need for information about the likelihood of hazard and exposure associated with nanomaterials. Once quantified, additional challenges lie in inter-preting and managing risk. Acceptance and risk avoidance behaviors often depend on the nature of the risk itself and may vary over time. For example, "publicity bias" refers to a more recently realized risk, such as a tornado, that may be overestimated when compared to a greater risk whose last manifestation has faded from memory, such as a pandemic plague [16]. In this case, the perceived probability of expo-sure outweighs the influence of hazard magnitude on the perceived risk. Alternatively, the opposite can happen, with people focusing heav-ily on hazard while downplaying the influence of exposure. They over-estimate their actual danger by highly emphasizing the severity of the potential effects without considering the likelihood that they will actu-ally experience the effects. This propensity is termed "compression bias," wherein rare or dreaded risks are overestimated when compared with the impact of common risks [21]. A familiar example of such mis-calculation would be the widespread fear of dying in a plane crash versus the relative acceptance of dangers inherent to the more mun-dane act of driving a car. Risk perception research has shown many other predictable tendencies, such as the preference for controllable risks over ambiguous, uncontrollable ones. An example of this would be the acceptance of comparatively large risk of food-borne illness from eating raw fish such as sushi or raw oysters, which many people purposely choose, as compared to the recent public outcry and aversion to eating beef due to the arguably less likely risk of contracting mad cow disease. Our very behavior reveals inherent calculations of the trade-offs between perceived risks and benefits that may or may not be consistent over time. We perform a rather intricate balancing act between factors such as temporality, ambiguity, controllability, and dread level to arrive at an instantaneous decision regarding preferences of risk. Better information on risk and reduced uncertainty are needed to improve the quality of decisions. However, generalizing individual preferences of risk to predict behavior may not be meaningful. Moreover, preferences for risk expressed in a political or social context may not coincide with decisions that would otherwise appear to be "optimal" based on available information. The public is becoming increasingly influential in shaping policy, and decreasingly reliant on a strictly science-based decision making regime regarding the risks of new technologies [22]. Nonetheless, information on risk is needed, whether the public makes informed choices between risks and benefits or chooses to put in place policies to avoid risk where information is thought to be lacking.

We will not attempt to condense the vast body of work addressing the complex behaviors and choices related to risk perception, but rather will simply summarize with the following two concepts:

- "Riskiness means more to people than 'expected number of fatalities.'" [21]
- Because no stakeholder group is without its bias, an environmental risk assessment process needs to include iterative feedback loops to allow for the influence of a variety of valid viewpoints [19].

While there are major gaps in our ability to characterize the risks of nanomaterials, a lack of information will not preclude stakeholders from taking actions that will determine the future landscape of risk for better or worse. One recent editorial summarized the importance of active participation by researchers in helping to inform the decision making process: "The time has come for the environmental research community to come together with the environmental regulators to address [impacts from the manufacture and use of nanomaterials]. Failure to undertake this will almost certainly ensure that the media steers public understanding of and confidence in nanotechnology, leading to unsubstantiated anecdotes and wild conjecture potentially forming the basis for an ill-informed debate with outcomes that may be wholly disproportionate to the risks" [23].

Assessing the risks of nanomaterials

A predicament is created by an urgent need for guidance in developing risk management strategies for a developing nanomaterials industry at a time when there are little data to support decision making. Our approach must therefore be incremental, allowing us to build toward the "ideal" risk assessment methodology as data become available.

Paths toward model development.   We will take as our starting point existing risk assessment tools, typically created to assess commonplace bulk chemicals. We will adapt available information on nanomaterials to correspond to required inputs for these tools as they are now used, and then modify the tools over time to incorporate criteria that are more specific to nanomaterials. Although this approach may initially miss much of the nuance that makes nanomaterials different from bulk materials, this will allow us to deliver an evolving set of guidelines in the very short term while efforts mature in producing information on nanomaterials exposure, hazards, and appropriate evaluation criteria. This approach can be visualized as a convergence of efforts in developing tools and obtaining information on nanomaterials (Figure 13.4).

Collecting new data.   One example of an "upper path" approach employs expert elicitation to assemble a list of the nanomaterial properties expected to govern the type and extent of the materials' impacts [24]. Kara Morgan of the US Food and Drug Administration carried out just such an investigation, delivering a thorough snapshot of what was

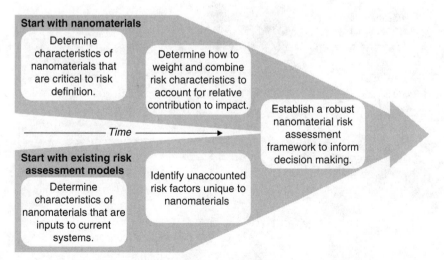

Figure 13.4   Pictorial representation of the path to constructing a nanomaterial risk assessment framework.

known about potential risk factors as of 2005. These variables are synthesized into an influence diagram linking the components of this complex problem, tying together the different variables that are expected to have causal relationships in determining the risk associated with the nanomaterial. With a recognition that published data on nanomaterials are limited, the application of such a decision analysis tool is not intended to yield numerical probabilities or weighted risk results. Instead its utility is in identifying the variables necessary to solidify a risk assessment model, with the intent to inform research agendas [24]. Figure 13.5 shows an overview of the resulting influence diagram, wherein each box represents a choice that influences the subsequent boxes, depicted as related by directional arrows.

Such a preliminary risk framework is also useful in assessing the current state of nanotechnology research funding to determine whether it aligns with the knowledge gaps that are most likely to affect risk. A 2006 assessment found that the focus areas of nanotechnology funding to date could be grouped into four major categories—exposure pathways, toxicity, fate, and transport—but that nanoparticle behavior in specific media and the global life-cycle impact of nanoparticles were still inadequately understood [25]. The full benefit of the influence diagram approach is realized as research results are delivered and the skeleton of the risk framework is filled out. Progressive generations of the diagram incorporate the adjusted relationship pathways that are

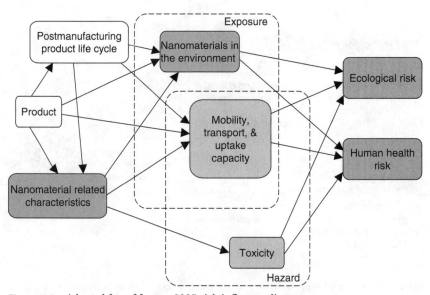

Figure 13.5   Adapted from Morgan 2005 risk influence diagram.

eventually articulated by coefficients describing the strength of the links as inputs to the model.

Working with existing risk assessment models.   Current tools for risk assessment and life-cycle analysis may provide useful guidance for short-term risk assessments for nanomaterials as we await the results of studies of exposure and hazards. Physical and chemical properties, transport and mobility behaviors, projected physical quantities, and manufacturing practices must be understood for chemicals and the materials used to produce them in order to evaluate their impacts. In the spirit of beginning to fit nanomaterials into established impact evaluation systems, we will present an example of a risk assessment as seen through the lens of an insurance company. This relative risk assessment of five nanomaterial manufacturing processes will demonstrate how an existing insurance risk analysis tool can be applied to nanotechnology manufacturing [26]. Because of the established dearth of conclusive data regarding nanomaterials, the fabrication processes are initially assessed independent of the nanomaterials themselves. After the methodology and results are explained, we will then discuss information needs and directions for model development. Thus, this exercise serves as a first pass at ranking the relative risk of nanomaterials based on the known and estimated physico-chemical properties that are used in current assessment frameworks. It will also bring out the unknowns and highlight why research in mobility, transport, aggregation, and toxicity (such as the work presented in this book) is so important in predicting the behavior and effects of nanomaterials.

First Steps: Risk Assessment from an Insurance Industry Perspective

Insurance companies routinely assess the risks associated with industrial processing. In the chemical industry, the relative risks of substances and processes involved in fabricating a product are assessed by insurance companies as a basis for calculating the appropriate premiums to charge their customers. Nanotechnology companies, like all others, will have to carry insurance on manufacturing processes. They will be subjected to actuarial protocols that assess the risk posed by a given process and determine how much a company will have to pay to an insurance provider in exchange for liability coverage. The interface between developing nanoscience and existing risk assessment frameworks presents itself as a natural starting point for translating this emerging class of novel technologies into the language of the risk community. This is already an active area of interest and development among insurance companies worldwide.

XL Insurance is a Swiss insurance company that employs a numerical rating tool to combine risk factors and calculate premiums for the processes it underwrites. With a focus on environmental pollution and health risks, the aim of the algorithm, understandably, is to quantify the relative risk of manufacturing processes with respect to financial liability of the insurer. As we have discussed, a full life-cycle assessment would build out from the impact of the manufacturing process itself to evaluate the effects of the product in question from cradle to grave. As a starting point, though, XL Insurance allowed the transparent use of their insurance protocol to rank the fabrication processes of nanomaterials.

General data requirements of risk assessment methodology

To evaluate environmental risk, a function of hazard and exposure, we have established that the effects of a substance must be combined with the likelihood that it will come in contact with the environment. The definition of risk makes clear that the level of hazard posed by a nanomaterial is a characteristic independent of the chance that it will ever realize its potential damage. Thus, in order to understand the impact a material will have on its surroundings, we have to know how dangerous it is as well as how it will interact when released to the environment. Where will it go, in which media will it tend to collect, with what organisms will it be likely to make contact, and to what extent will it have toxic effects on them when that contact occurs? We must also include information about the amount of the material and the conditions in which it is being handled or produced. We need to understand the probability of substance release and the quantity at stake in order to combine risk with exposure. An appropriate algorithm can then be used to combine the factors and determine the resulting relative impact of the material.

Just as in any risk assessment, choosing the appropriate parameters and the appropriate methodology for combining them is the crux of evaluating the environmental impact and insurance liability for this sample model. In the XL Insurance protocol, the hazard of each process is defined based on constituent substance characteristics such as carcinogenicity or lethal dose in rats, and on such relevant process characteristics as temperature, pressure, enthalpy, and fire hazard rating [27]. The exposure portion is quantified by incorporating persistence and mobility of constituent substances and by scoring expected emissions of the substances during the manufacturing process. These factors are combined with a focus on order of magnitude so that an idea of the risk range is what drives the premium. This focus on relative risk acknowledges the pragmatic reality of uncertainty inherent to the business of insuring a wide range of manufacturing industries, and a variety of specific processes within each industry. The fact that a level of uncertainty is built into the system by aiming at orders of magnitude may prove to narrow the knowledge gaps between traditional products and nano-products in practice. By applying an insurance database currently in commercial use, we are able to benchmark the risk of nanomaterials' fabrication processes against each other and against other non-nano materials from the perspective of an industrial insurer. This type of risk assessment, although not inclusive of all environmental impacts or life-cycle considerations, is representative of the type of risk assessment nanomaterials manufacturers will first encounter as insurers grapple with qualifying the relative risk of these new processes.

The following sample risk assessment ranks the fabrication processes for five different nanomaterials by their relative risk. Nanomaterials that exhibited foreseeable potential for production beyond laboratory scale were chosen for manufacturing evaluation: single-walled carbon nanotubes, buckyballs (C_{60}), quantum dots composed of zinc selenide, alumoxane nanoparticles, and nano-titanium dioxide. For each nanomaterial, a process flow chart was created to capture the published synthesis method that had the most potential for being scaled up for commercialization at the time of the study. An inventory of input materials, output materials less the final product, and waste streams for each step of fabrication was then created from the process flows. This method facilitated the collection of insurance protocol data requirements for all materials involved in the fabrication processes. Physico-chemical properties and quantities of the inventoried materials were used to qualitatively assess relative risk based on factors such as toxicity, flammability, and persistence in the environment. The substance and process data were then quantitatively combined according to the actuarial protocol in the risk database to arrive at a final numerical score for the process. This relative risk score could then be compared with the risk scores of other commonplace production processes. It is relevant to mention that in this way the risk assessment was no different from that conducted in any other new chemical process. The exercise was carried out as a first step toward bridging the gap between novel materials and the established production industry into which they are already being incorporated. The process of listing all input and output streams was a thorough, yet straightforward exercise, so it is not detailed here. Instead we begin with the input parameters to the database.

System data requirements

Appreciation of either the qualitative or quantitative conclusions of this model requires recognition of the two main sets of choices involved in setting up a risk assessment tool: which characteristics are chosen

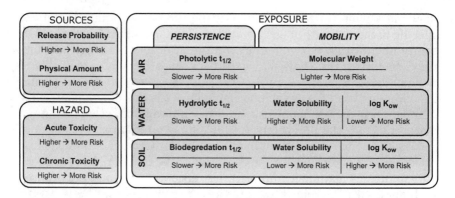

Figure 13.6   Pictorial representation of XL database required risk parameters.

for inclusion in the tool, and how those parameters are then combined. In the case of environmental risk posed by an industrial process, the hazard component of risk is represented by characteristics of the materials involved as well as some characteristics of the process itself. The exposure risk is evaluated via three pathways: air, water, and soil. Information about the likelihood of the process to result in release of the substance is combined with its expected physical quantity and with characteristics that predict the fate and transport in various media to represent the risk of exposure to a given substance. Figure 13.6 is a conceptual diagram of the parameters required by the database, offering a first pass indication of how parameter values will affect the ultimate risk score. This simplified explanation of the selected parameters shows that they were included in the database because they capture pertinent material and process characteristics in a way that will allow resulting risk to be calculated. It makes sense, for instance, that if a substance degrades slower in the air it will persist there longer and have more chance for exposure, resulting in higher risk in that medium. Readers are encouraged to review each parameter to convince themselves of the directional effects these values would have on overall risk.

Collecting, weighting, combining data parameters

With the chosen characteristics defined by measurable parameters in the system, some quantitative decisions must be made regarding how to mathematically combine the values to arrive at a final risk score. Finding real values for these parameters presents many challenges that will be expounded upon as the detailed data requirements are explained. For now, keep in mind that adapting our available information to the requirements of the system was an iterative process, resulting in the expression of each chemical and process according to the chemical and physical properties required by the insurance company's algorithm in determining its relative risk.

Required substance characteristics

The substance characteristics utilized in the database, as well as the relationships used to combine the parameters, are detailed in Figure 13.7. It should be noted that the numerical values for each of the substance data fields are mapped to relative risk classes. These risk classes group levels of risk posed by orders of magnitude. For instance, the characteristic of toxicity is represented by the data parameter LC_{50} in the

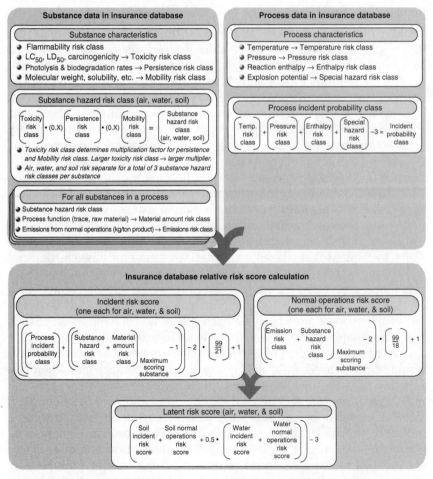

Figure 13.7 Schematic of XL Insurance database and formulation of risk scores, adapted from Robichaud, et al., 2005.

database. The whole spectrum of possible values for the LC_{50} parameter of a given substance is represented in a scale of 1 through 4. For instances where a substance characteristic value was not known, which was often the case for persistence factors such as photolytic half-life, a relative risk class was directly assigned to at least approximate the parameter's order of magnitude. This assignment was often made via comparison with similar compounds. Risk scores corresponding to toxicity, persistence, and mobility data for each constituent substance were then combined to calculate the risk associated with that constituent's interaction with air, water, and soil; this cumulative risk score for each compound was denoted as its substance hazard risk class.

Required process characteristics

Each manufacturing process is defined within the database by entering its constituent substances and conditions such as temperature, pressure, and enthalpy. Figure 13.7 depicts the process characteristics utilized in the insurance algorithm. Maximum temperature, pressure, and enthalpy values for the process are entered to represent the highest risk step in each fabrication process, and the corresponding risk classes for those condition parameters were combined to calculate the probability of an accident occurring during the process. This value is referred to as the process incident probability class. Each substance involved in the fabrication process, with the exception of the final nanomaterial product, was defined individually in terms of its risk based on hazard and exposure variables. To quantify its contribution to risk within the particular fabrication process, the amount present, its role in the process, its physical phase at the temperature and pressure of the process, and any emissions were identified. Based on the relative amount of each material used in the process, an amount category was assigned so that a primary substance present in higher amounts represents greater risk than a sparsely used substance. By specifying substance role, the model determines an exposure probability for each substance based on whether such exposure would only be expected due to an in-process accident, whether normal operations would result in emissions, or whether both pathways would be possible. The physical phase of each substance if or when it is released must be included to apply the appropriate persistence and mobility parameters for the media of air, soil, and water. An emission factor is included for any substance emitted during normal production, indicating the amount of the substance released to the environment in kg per ton of product. Like the other physical values, these emission factors are mapped to a unitless scale representing orders of magnitude. Any normal releases of a material from a level of 0.00001 kg/ton to 1000 kg/ton are represented by a set of Emissions Risk Classes ranging from 1 to 10.

The emissions were particularly challenging parameters to collect due to uncertainty of projecting industry scaled versions of published processes. Even on a smaller production scale, insufficient data regarding detailed mass balances of waste streams made accurate estimation of emissions difficult. In cases where an emission risk class could not be sufficiently determined, one was approximated based on either a similar process that already existed within the commercial database or on a stoichiometric calculation of mass balance.

Comparison of risk

At this point the assembled risk class information can be combined to arrive at some conclusions about relative risk, both qualitatively using some of the collected information, and quantitatively via the insurance protocol. Because we were starting with a populated insurance database that included many known fabrication processes, the nanomaterial manufacturing risks could be compared with each other and against other common processes previously defined in the system. For both our qualitative and quantitative results, nanomaterials' risk scores are presented alongside the scores for six other familiar processes: silicon wafer (semiconductor) production, wine production, high density plastic (polyolefin) production, automotive lead-acid battery production, petroleum refining, and aspirin production. Each of these processes and their supporting substance data were taken from the insurance database; as they are established and already in commercial use, their data and underlying assumptions were not addressed. The comparison processes were chosen to offer a benchmark of the risk posed by widely accepted or familiar operations. Silicon wafer production, an integral part of computer manufacturing, and automotive lead-acid battery production are both manufacturing processes found in or near many communities. Petrochemical complexes in industrial cities are characterized by polyolefin production and petroleum refining facilities. The well established yet constantly evolving pharmaceutical industry is represented by aspirin production. Lastly, wine production was chosen as an interesting benchmark comparison since it would be considered by most a relatively benign, and even desirable, process to be in close proximity to a community. When including this comparison data, the only change made was to remove final products from the process as we did in the case of nanomaterial production, in order to consistently compare only the manufacturing processes' contribution to insurance risk.

Qualitative risk assessment

The nanomaterial fabrication data, once collected, were first organized qualitatively to characterize and compare the relative risks. This

500   Potential Impacts of Nanomaterials

TABLE 13.1   Qualitative Risk Rankings for Single-Walled Nanotube (SWNT) Production

SWNT	Toxicity	Water solubility	Log K_{ow} (Bio-accumulation)	Flammability	Emissions
Carbon monoxide	■	■	□	■■	□-■■
Sodium hydroxide	■■	■■	□	□	■
Iron pentacarbonyl	□	□	□	□	□-■■
Carbon dioxide	□	■■	□	□	□-■■
Water	□		□	□	■

representation of the data focused on the relative risk posed by five key properties for each constituent material of each nanomaterial fabrication process. We collected data regarding each substance's toxicity (LC_{50} and LD_{50}), water solubility, log K_{ow}, flammability, and expected emissions. The risk classes in each of these five categories were represented visually in the form of squares, where filled-in squares represent higher risk and empty squares represent lower risk. The purpose of the resulting risk tables (Tables 13.1 through 13.11) is to give a subjective sense of the risk posed by manufacturing these materials, based solely on the order of magnitude of selected parameters but without mathematically combining them. Tables 13.1 to 13.11 are adapted from Robichaud et al., 2005 [26].

These qualitative graphs organize the information we know about manufacturing the chosen materials so as to allow some general observations. We can see that the nanomaterial fabrication processes appear overall to have fewer toxic ingredients and fewer total constituent materials, but they are also predicted to have relatively higher emissions compared to the non-nano processes. A couple of caveats about the differences

TABLE 13.2   Qualitative Risk Rankings for C_{60} Production

C_{60}	Toxicity	Water solubility	Log K_{ow} (Bio-accumulation)	Flammability	Emissions
Benzene	■■■	■■	□	■■	
Toluene	■■	■	■	■■	□-■
Argon	□	■	□	□	□
Nitrogen	□	■	□	□	□
Oxygen	□	■	□	■	□-■■
Soot	■■	□	■■	□	■■■
Activated carbon	□	□	■■	■	
Carbon dioxide	□	■■	□	□	■■
Water	□			□	□

TABLE 13.3   Qualitative Risk Rankings for ZnSe Quantum Dot Production

ZnSe Q-Dots	Toxicity	Water solubility	Log K_{ow} (Bio-accumulation)	Flammability	Emissions
Nitrogen	□	■	□	□	□
Formamide	■■	■■	□	□	
Heptane	■	□	■■	■■	■-■■
Poloxalene	■	■■	□	□	■-■■
Diethyl zinc	■	□	□	■■	
Hydrogen selenide	■■■	■■	□	■■	■-■■
Carbon dioxide	□	■■	□	□	■■
Carbon monoxide	■	■	□	■■	■■

in maturity between the manufacturing methods should be mentioned, as they may account for some of the observed qualitative differences.

The higher projected emissions in nano production could partially be explained by the uncertainty inherent in imagining scaled-up industrial volume versions of the published manufacturing processes. The comparison non-nano processes may have so few emissions due to being established and streamlined, resulting in improved recapturing and recycling of the multiple hazardous materials involved in their production. It would follow that fully mature versions of the nanomaterial production processes may be expected to have lower emission levels than are represented here. More materials may be involved in the established processes precisely because they have been fully developed at the industrial scale. Nanomaterial production processes might eventually require a similar number of materials as they develop further. Recycling, washing, and recapturing steps may be added to the processes, adding more chemicals to the process.

It is important to acknowledge those caveats, but we can still glean some insight by looking at this qualitative snapshot of current manufacturing methods. Of the nanomaterial production processes, alumoxane and single-walled nanotube (SWNT) production appear to

TABLE 13.4   Qualitative Risk Rankings for Alumoxane Production

Alumoxane	Toxicity	Water solubility	Log K_{ow} (Bio-accumulation)	Flammability	Emissions
Acetic acid	■■	■■	□	■	□-■
Aluminum oxide	□	□	■■	□	□-■■■
Water	□			□	

TABLE 13.5   Qualitative Risk Rankings for Nano-Titanium Dioxide Production

Nano-TiO$_2$	Toxicity	Water solubility	Log K$_{ow}$ (Bio-accumulation)	Flammability	Emissions
Methane	□	■	□	■■	
Hydrochloric acid	■■	■■	□	□	□-■■
Phosphoric acid	■■■	■■	□	□	□-■■
Titanium tetrachloride	■■■	□	□	□	□-■■
Carbon dioxide	□	■■	□	□	■

present lower risk levels. Compared with those two, ZnSe quantum dots, C$_{60}$, and nano-titanium dioxide appear to be associated with relatively more risk. Overall, the qualitative graphs show all of the nanomaterials production processes as posing relatively lower risk than either petroleum refining or polyolefin production.

TABLE 13.6   Qualitative Risk Rankings for Silicon Wafer Production

Silicon wafers	Toxicity	Water solubility	Log K$_{ow}$ (Bio-accumulation)	Flammability	Emissions
Sodium hydroxide	■■	■■	□	□	
Hydrochloric acid	■■	■■	□	□	
Phosphoric acid	■■■	■■	□	□	
Hydrogen fluoride	■■	■■	□	□	□
Sulfuric acid	■■■	■■	□	■	
Methyl-2-pyrrolidone, N-	■	■■	□	■	□
Acetone	□	■■	□	■■	□
Ethanol	□	■■	□	■■	□
Nitrogen	□	■	□	□	
Anhydrous ammonia	■■	■■	□	■■	■■
Chlorine	■■■	■■	□	■	
Hexafluoroethane	■	□	■■	□	■■
Phosphine	■■■	■	■	■■	□
Boron trifluoride	■■■	□	□	□	
Hydrogen bromide	■■	■■	□	□	
Silicon	□	□	■■	□	
Diborane	■■■	□	□	■■	
Germanium	■	□	■■	□	
Arsine	■■■	■■	□	■■	■
Oxygen	□	■	□	■	

TABLE 13.7   Qualitative Risk Rankings for Wine Production

Wine	Toxicity	Water solubility	Log K_{ow} (Bio-accumulation)	Flammability	Emissions
Zineb	■■■	■	■	□	
Maneb	■■■	□	■■	□	
Copper oxychloride	■■	□	■■	□	
Water	□			□	
Glucose	■	■■	□	□	
Sulfur	■	□	■■	□	
Sulfur dioxide	■■	■■	□	□	□

Quantitative risk assessment

Three relative risk scores are calculated by the insurance protocol for each fabrication process: Incident Risk, Normal Operations Risk, and Latent Contamination Risk. The Incident Risk score represents the risk due to an in-process accident that leads to accidental exposure. Risk posed by substances that are expected to be emitted during the fabrication process are referred to as Normal Operations Risk. To account for the potential long-term contamination of the operations site, a Latent Risk score is calculated by combining factors from the Normal Operations Risk and Incident Risk scores.

The reader will recall that risk scores corresponding to conditions such as temperature and pressure were combined to calculate the probability of an accident occurring during the process, expressed as a process incident probability class. Risk scores corresponding to toxicity, persistence, and mobility data for each constituent substance were then combined to calculate the risk associated with that constituent's interaction with air, water, and soil; this cumulative risk score for each compound was denoted as its substance hazard risk class. An amount hazard risk class was assigned for each substance based on the relative amount used in the process, wherein more of a substance enhanced the level of risk posed. The process incident probability class, amount hazard risk class, and substance hazard risk class were used to compute final risk ratings for air, water, and soil pathways due to sudden release and due to normal emissions from a given process.

Incident Risk was determined for each exposure pathway by combining the process hazard rating, the amount hazard ratings for each substance, the substance hazard risk classes, and an actuarial adjustment coefficient that forced the final risk score into a 1–100 distribution. Figure 13.7 provides a schematic of calculations, and the actuarial coefficient will be explained in the next section. Normal Operations Risk was calculated for each pathway by combining the constituent substances

TABLE 13.8   Qualitative Risk Rankings for High Density Plastic (Polyolefin) Production

High density plastics (Polyolefins)	Toxicity	Water solubility	Log K$_{ow}$ (Bio-accumulation)	Flammability	Emissions
Ethylene	□	■	□	■■	■■
Butylene	■	■	□	■■	■■
Hexane, n-	■	■	■	■■	■■
Propylene	□	■■	□	■■	■■
Hydrogen	□	□	■■	■■	
Hydrochloric acid	■■	■■	□	□	
Vinyl acetate	■■	■■	□	■■	
Polyethylene	□	□	■■	□	■■
Styrene	■■■	■	■	■	
Titanium tetrachloride	■■■	□	□	□	■■
Alumina trihydrate	■	□	■■	□	■■
Magnesium hydroxide	■	□	■■	□	
Magnesium chloride	□	■■	□	□	
Aluminum chloride	■	■■	□	□	
Cyclohexane	■	■	■	■■	
Triethyl aluminum	■■	□	□	■■	
Polypropylene	□	□	■■	□	■■
Acrylic acid	■■	■■	□	■	
Methacrylic acid	■	■■	□	■	
Methyl acrylate	■■	■■	□	■■	
Methyl methacrylate	■	■■	□	■■	
Polybutylene	□	□	■■	□	■■
Isobutane	□	■	□	■■	
Diethylaluminum chloride	■	□	□	■■	
Diethylaluminum hydride	■■	□	■■	■■	
Titanium trichloride	■	□	□	■	■■
Vanadium trichloride	■■	■	■■	■	□
Magnesium ethylate	□	□	□	■	
Methylphenol, 2, 6-di-tert-butyl-4-	■	□	■■	□	
Ethyl benzoate	■	■	■	■	
Butyl alcohol	■	■■	□	■	■■
Silicon dioxide	□	□	■■	□	□

TABLE 13.9   Qualitative Risk Rankings for Automotive Lead-Acid Battery Production

Automotive lead-acid battery	Toxicity	Water solubility	Log K_{ow} (Bio-accumulation)	Flammability	Emissions
Sulfuric acid	■■■	■■	□	■	■
Calcium sulfate	□	■■	□	□	■■■
Antimony	■■	□	■■	□	
Arsenic	■■■	□	■■	□	
Tin	■	□	■■	□	
Soot	■■	□	■■	□	
Lead	■■	□	■■	□	
Lead oxide	■■	□	■■	□	
Lead monoxide	■■	■	■■	□	■
Lead dioxide	■■	■	■■	■	
Sodium perchlorate	■	■■	□	■	
Barium sulfate	■■	□	■■	□	
Polyvinyl chloride	□	□	■■	□	
Hydrochloric acid	■■	■■	□	□	

TABLE 13.10   Qualitative Risk Rankings for Refined Petroleum Production

Refined petroleum	Toxicity	Water solubility	Log K_{ow} (Bio-accumulation)	Flammability	Emissions
Thiophene	■■■	■■	□	■■	
Benzene	■■■	■■	□	■■	■■
Ethylenediamine	■■	■■	□	■	
Xylene	■■	■	■	■	
Toluene	■■	■	■	■■	
Paraffin oil	□	□	■■	■	
Silica, crystalline	■■	□	■■	□	
Methane	□	■	□	■■	■■■
Ethylene	□	■	□	■■	
Sulfur	■	□	■■	□	
Butane	□	■	■	■■	
Butadiene, 1,3-	■■■	■	□	■■	
Sulfur dioxide	■■	■■	□	□	■■
Hydrogen sulfide	■■	■■	□	■■	
Soot	■■	□	■■	□	
Aluminum oxide	□	□	■■	□	
Vanadium pentoxide	■■■	■■	■■	□	
Chromic oxide	■■	■■	■■	□	
Bitumen	■	□	■■	□	
Anhydrous ammonia	■■	■■	□	■■	
Carbon monoxide	■	■	□	■■	■
Nitrogen dioxide	■■■	□	□	■	■■
Phenol	■■	■■	□	■	■

TABLE 13.11   Qualitative Risk Rankings for Aspirin Production

Aspirin	Toxicity	Water solubility	Log K_{ow} (Bio-accumulation)	Flammability	Emissions
Sodium phenolate	■■	■■	□	■	
Phenol	■■	■■	□	■	□
Toluene	■■	■	■	■■	
Acetic acid	■■	■■	□	■	
Salicylic acid	■	■■	□	□	□
Sodium salicylate	■	■■	□	□	
Acetic anhydride	■■	■■	□	■	
Sulfuric acid	■■■	■■	□	■	
Sodium sulfate	□	■■	□	□	□
Carbon	□	□	■■	■	
Carbon dioxide	□	■■	□	□	

emission coefficient risk categories, their substance hazard risk classes, and another actuarial adjustment coefficient. The final Incident and Normal Operations Risk ratings calculated for soil and water were combined to arrive at a Latent Risk score. Air ratings are excluded from this calculation because by the time long-term latent contamination is of concern, airborne contaminants would either have already come in contact with an organism or settled into the water or soil.

Explaining the actuarial adjustment coefficient.   Mathematically combining the system parameters is the step that assimilates the collected information and imposes a weighting system to arrive at a relative score for each process. The actuarial coefficients chosen to combine the variables in the XL Insurance database are based on order of magnitude. Essentially, the rating is set up in such a way that the increase from one risk class to the next corresponds to an increase in risk by an order of magnitude. Theoretically the incident risk score is built from 21 risk classes. Fourteen classes are reserved for the characterization of the process risk exposure, which is comprised of temperature, pressure, enthalpy, and fire risk. Three classes (1–3) are for the amount categorization (main, secondary, trace). Ten classes (1–10) are for the substance property characterization, accounting for toxicity, persistence, and mobility. Adding up all these classes yields a range of a minimum of 6 to a maximum of 27 classes, so the risk range covers 21 "orders of magnitude." In essence, by adding up the classes we may say mathematically that we multiplied by adding up exponents. The company recognizes that 21 orders of magnitude in risk exposure is a wide range, and that this may not be a completely realistic view of the risk picture. However, since perfect data are almost never available with regard to

real systems and the environment, organizing by order of magnitude provides a good way to arrive at useful conclusions given the available
information. The choice was made to adjust the scale so that relative risk
scores would range from 1 to 100. Since the risk classes are combined
additively, the number of classes were intended to be set up so as to give
equal weight to both the measure for severity of risk and the frequency
or likelihood of its occurrence (Risk = hazard * exposure). XL Insurance
considers the 11 process related classes to represent exposure likelihood and the 12 substance related classes to represent the hazard posed
by the processes. This choice in particular highlights how influential the
decisions made in the "weighting filter" can be on the final relative risk.
Other possibilities exist for combining the classes, of course. For example, some might argue that the substance characteristic of photolytic
half-life really determines more about exposure than hazard because it
describes how long the substance will remain in the air before chemically degrading.

The numerical factors added to the end of the combination expression
are there to make sure that the lowest risk score in the scheme comes
out as 1 and the highest as 100:

Case 1 (minimum risk score):

$$\left[\Big((P[=1]+T[=1]+H[=1]+G[=1])\right.$$
$$\left.+ ((S_{path}(=1)+M_{path}(=1))_{max}-1)-2\Big)\right]$$
$$*\left(\frac{99}{21}\right)+1=\left(0*\left(\frac{99}{21}\right)\right)+1=1$$

Case 2 (maximum risk score):

$$\left[\Big((P[=3]+T[=3]+H[=3]+G[=5]) - 3\Big)\right]$$
$$+ ((S_i,L+(=3)+M_i(=10))_{max}-1)-2)]$$
$$*\left(\frac{99}{21}\right)+1=\left(21 *\left(\frac{99}{21}\right)\right)+1=100$$

This calculation is done for every substance and for each of the three
pathways considered: soil, water, and air. The maximum resulting value
is taken as the overall risk score for the production process.

In the XL database, a final score in the Incident Risk and Normal Operations Risk categories corresponds to the highest of the three pathway-specific scores out of air risk, water risk, and soil risk. For each of these three transport media-based risks, the score is based on the highest risk substance in that exposure pathway. This method of taking the highest risk medium based on its highest risk substance is another of the important choices involved with combining the system parameters. It is appropriate for insurance premiums, for as we have established, the overall goal in this system is to approximate the relative order of magnitude of the risks. The practice of choosing only the highest scoring constituent material instead of accounting for multiple substances admittedly precludes differentiating processes that involve multiple hazardous substances. The same score could be calculated for both a process that includes benzene and toluene as main ingredients as well as a process that uses only benzene, since benzene corresponds to the highest substance hazard risk class. Adding toluene to a process does add to the consequent risk, but it does not change it by a full order of magnitude from an insurance perspective.

Running the XL insurance database protocol

The three types of relative risk scores were generated for each of the five chosen nanomaterial production processes. Due to the uncertain nature of predicting the final scaled-up manufacturing process volumes, expected emissions from normal operations were more difficult to estimate, and as such were only able to be approximated with the system. Three separate hypothetical cases are defined for each fabrication process, resulting in three sets of scores for Normal Operations and for Latent Risk, both of which incorporate the risk due to predicted emissions. The three cases are defined representing a range of manufacturing scenarios intended to estimate boundary conditions of "most materials emitted" and "least materials emitted." In processes for which the authors had mostly published academic work available, such as ZnSe quantum dots, the emission risk class was made to vary by a few orders of magnitude between the low risk and high risk cases. Much of the work for those nanomaterials has been carried out in the private sector, and as such less public information is available. In other processes for which more data were accessible, such as SWNT and alumoxane production, closer estimates could be made based on known practices or on reaction stoichiometry. For detailed assumptions related to all the nanomaterials fabrication processes' scores and the different cases considered, the reader is invited to refer to the 2005 *Environmental Science & Technology* article and its online supporting information [27].

Quantitative insurance-based risk results

For each nano and non-nano manufacturing process, graphs were prepared comparing Incident Risk, Normal Operations Risk, and Latent Risk as calculated by the insurance database. The Incident Risk scores in Figure 13.8 show that, by and large, the nanomaterials production processes score comparably with or lower than the traditional comparison processes. The manufacturing methods for alumoxane particles and SWNTs are low in relative risk compared with all the non-nano products, scoring close to or lower than wine production. These results are due to the lack of extreme temperature and pressure in the production processes of alumoxane and SWNTs, along with the low hazard ratings of their constituent substances. Incident risks for ZnSe quantum dots and Nano-TiO_2 are mid-range, similar to silicon wafer production, automotive lead-acid battery production, petroleum refining, and aspirin production. The highest risk nanomaterial database incident ratings were from C_{60}, which scored near the risk presented by polyolefin production. This rating can be explained by one very hazardous culprit material (benzene), which is required by our chosen method for producing C_{60}. Because the Incident Risk category is based on the identities of all constituent materials and on process conditions, a relatively confident characterization can be reached from the published nanomaterials' synthesis methods. Therefore, the order of magnitude of the fabrication incident risk is considered reliable enough so that Figure 13.8

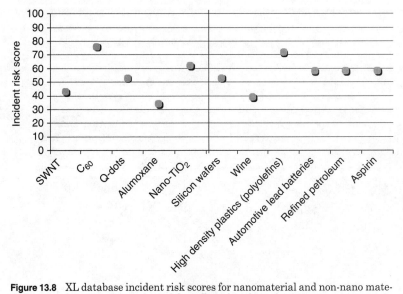

Figure 13.8 XL database incident risk scores for nanomaterial and non-nano material production processes, adapted from Robichaud, et al. 2005.

does not include ranges of risk scores. We will see why this is important to mention when we look at the Normal Operations and Latent Risk scores.

The Normal Operations scores, unlike the Incident Risk calculations, incorporate risks posed by projected emissions of any materials during the fabrication process. The uncertainty in these emissions was accounted for by including a range of probable manufacturing scenarios, scoring three different cases defined for each nanomaterial. The assumptions upon which each of the three cases is built beget different risk scores for the nanomaterials' production processes. The Normal Operations Risk scores are shown in Figure 13.9 as a range of values, with the mid-risk case represented as the main data point. A single point denotes the score of each of the non-nano production methods because they were already established in the database by XL Insurance as fully scaled-up manufacturing processes. The emissions data will not run the risk of varying by more than an order of magnitude as they would when projecting the scale-up of nanomaterial production processes, thus the variability already allowed by the protocol will suffice. For ZnSe quantum dots, changing the assumptions did not cause the final score to differ by orders of magnitude, so the range is very narrow. Alumoxane and SWNT production earned low Normal Operations Risk scores, again proving to be on the order of the lowest scoring non-nano material processes, wine and aspirin. Scores in this

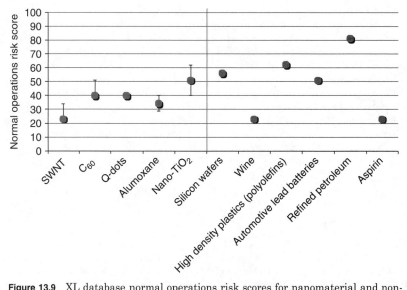

Figure 13.9 XL database normal operations risk scores for nanomaterial and non-nano material production processes, adapted from Robichaud et al., 2005.

category for C_{60}, ZnSe quantum dots, and nano-TiO_2 placed these three nanomaterials in a similar range as silicon wafers and automotive lead-acid batteries. The highest Normal Operations Risk scores were held by polyolefin production and petroleum refining, which were calculated as higher risk than all the nanomaterial manufacturing processes. When predicting industrial scale nanomaterial manufacturing practices, the emissions values were estimated conservatively by erring on the side of more emitted constituent materials and higher emissions levels for those substances. Therefore, it makes sense to predict that further clarification and streamlining of the processes would contribute to both reduced score variability and lower overall risk scores. For instance, with both SWNTs and alumoxane particles, more production information was accessible than for the other nanomaterials. The cases considered differed from each other by less, and the Normal Operations Risk scores for each of these prove to have both narrower margins of error and lower overall values. Mature processes with high Incident Risk scores such as polyethylene production and silicon wafer production offer another demonstration of the proposed trend toward lower emissions for more developed processes. These two fabrication processes employ many materials, including several very hazardous ones, yet they have very low expected emissions and thus lower Normal Operations Risk scores.

The Latent Risk scores are shown in Figure 13.10, with the 11 compared processes ranking in an order similar to what we saw in the other

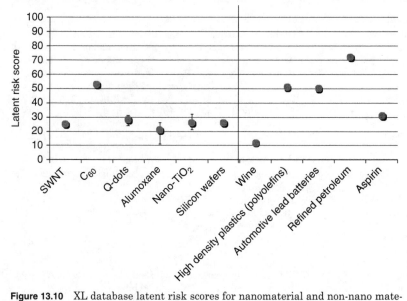

Figure 13.10   XL database latent risk scores for nanomaterial and non-nano material production processes, adapted from Robichaud et al., 2005.

two risk categories. This should be anticipated since Latent Risk is a function of the Incident and Normal Operations Risks. Except for C_{60}, all the nanomaterial fabrication processes' Latent Risk scores compare closely with silicon wafer, wine, and aspirin production, and are lower than polyolefin, automotive lead-acid battery, and refined petroleum production. While the production of C_{60} has a higher score than the other nanomaterials, it is still markedly lower than that of petroleum refining and could be considered comparable to the latent contamination score for polyolefin and automotive lead-acid battery production.

These graphs show the insurance database risk scores visually compared with one another. For a further appreciation of the exact numerical score in each of the three risk categories, refer to Table 13.12. Here the processes are listed with not only the three scores, but the culprit material(s) whose hazard score drove the Incident Risk or Normal Operations Risk score. This is included to highlight the influence of the mathematical algorithm chosen to combine the parameters. For the XL Insurance database, the properties and quantities of these few culprit materials determine the risk of the entire manufacturing processes.

TABLE 13.12   XL Insurance Database Risk Scores with Culprit Materials Responsible for the Risk Score, Adapted from Robichaud et al., 2005

Production Process	Incident Risk		Normal Operations Risk		Latent Risk
	Rank	Highest Risk Material(s)	Rank	Highest Risk Material(s)	Rank
Single-walled nanotubes	43	Carbon monoxide, Iron pentacarbonyl	23–34	Sodium hydroxide, Carbon monoxide	25
C_{60}	76	Benzene	40	Soot, toluene	52–54
Q-dots	58	Hydrogen selenide	40	Hydrogen selenide, carbon monoxide, surfactant	24–31
Alumoxane	34	Acetic acid	29–40	Acetic acid, aluminum oxide	11–26
Nano-TiO_2	62	Titanium tetrachloride	40–62	Titanium tetrachloride	21–32
Silicon wafers	53	Sulfuric acid	56	Arsine	19
Wine	39	Dithiocarbamate pesticides (zineb)	23	Sulfur dioxide	12
High density plastic (polyolefin)	72	Titanium tetrachloride, vinyl acetate	62	Titanium tetrachloride	51
Automotive lead-acid batteries	58	Lead dioxide	51	Sulfuric acid, lead monoxide	50
Refined petroleum	76	Benzene, toluene	67	Benzene, toluene, xylenes	42
Aspirin	58	Phenol, toluene	23	Phenol	31

What do we learn from this example?

This example illustrates how a slice of the nanomaterial risk landscape can be understood utilizing an existing methodology. From the point of view of insurance stakeholders, we conclude that there do not appear to be any abnormal pollution risks associated with the manufacturing portion of these five nanomaterials. Both the qualitative and quantitative insurance-based approaches suggest that differences in handling during operations affect the final risk scores as much as the identities of the constituent substances in a process. Such a conclusion offers some practical direction for manufacturers by pointing out that recycling and successful recapture of materials are viable ways to lower Normal Operations Risk scores without altering the chemicals in the process.

This exercise highlights several challenges and limitations of developing a ranking of relative risk for nanomaterials with scarce information. One key challenge is the lack of real data regarding the mass balances and yield rates needed to estimate emissions values for the various substances in nanomaterial manufacturing processes. Especially in the case of new materials such as the five nanomaterials in our example, it becomes necessary to project from the lab scale to the industrial scale volumes handled in their manufacture. In the quantitative risk results, we dealt with this data deficiency by including a range of potential scores from a variety of manufacturing scenarios. As experience with larger scale fabrication yields better information on mass balances, the risk will be more accurately characterized.

The very manner in which the risk is calculated may be tailored for the end-use. In the case of an insurance industry, it may be reasonable to evaluate risk from the standpoint of the largest single component, where from another perspective interactions between sources of risk must be accounted for. For example, the XL Insurance protocol applied here scores process risks by orders of magnitude, so it only accounts for the highest scoring material rather than accounting for multiple chemicals. This is a mathematical choice that results in acceptable accuracy for the company (e.g., if the plant blows up, do you need to worry about toxicity?). However, a model for toxicity that is focused on long-term health effects for workers may need to consider toxic interactions between chemicals.

Other life-cycle stages

The manufacture of nanomaterials considered in the above analysis involves raw materials commonly used in the production of conventional materials, and the assessment concerned only the manufacturing portion of the life cycle without the materials themselves. The actuarial risk assessment protocol used for the manufacture of nanomaterials may also be extended to other stages of the life cycle. This is especially

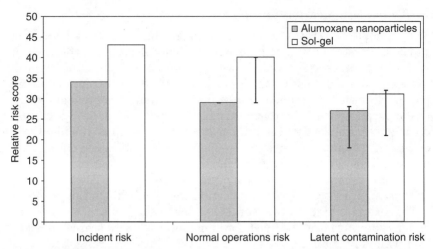

Figure 13.11   Relative risks in the manufacturing of ceramic membranes using alumox-ane nanoparticles and sol-gel technologies, including risks in the production of the precursors, adapted from Beaver et al. 2005.

straightforward when the produced nanomaterials are in turn used in other manufacturing processes. Alumoxane nanoparticles, for example, are used in the production of ceramic membranes. Wiesner et al. pointed out that, compared to the more traditional sol-gel process, the use of alumox-ane nanoparticles reduces the time and energy required and eliminates the use of organic solvents in the fabrication of ceramic membranes [28]. The risk assessment performed for alumoxane nanoparticles in the previous section can be extended to include their use in the manufacture of ceramic membranes. Such a comparison study was carried out to compare the risks associated with the production of ceramic membranes using alu-moxane nanoparticles against the more traditional sol-gel process [29]. Again, the nanomaterials themselves could not be characterized, but the effects of their use on process parameters such as temperature and pressure, as well as the on the selection of other substances required for the manufacture of the membranes, could be captured. Although this work is not detailed here, Figure 13.11 is included to illustrate the concept of carrying a relative risk assessment through to additional life-cycle stages, showing the relative risk ranking on the XL scale of 1 to 100.

Knowledge Gaps in the Life-Cycle Assessment of Nanomaterials Risks

Our ultimate goal is to incorporate information on the nanomaterials themselves into the risk assessment of the nanomaterials life cycle. The characteristics frequently used to estimate substance risks are often easily found in the literature for bulk substances, such as those used in producing the nanomaterials, but what about for the nanomaterials themselves? There is some temptation to use the information known about a bulk material that is considered closely related in composition. For instance, the MSDS for carbon black has been applied for C_{60} since they both consist of only carbon atoms. However, such descriptions are inappropriate because the size, shape, surface functionality, and structural uniformity of nanomaterials lend them unique attributes. If these engineered nanomaterials have truly unique properties that make them more desirable than their bulk counterparts for specific applications, then it is reasonable to assume that engineered nanomaterials may also have different physical chemical properties that may affect their risk profile. These distinctions may cause them to affect organisms in ways that differ from those of compositionally similar larger materials.

Although many of these engineered materials are new, exposure to nanoscale particles is not a new phenomenon for humans or the environment. For example, nanoscale particles are unintentional products of natural and anthropogenic combustion, and considerable research has been done on these ultrafine particles and their effects [30]. The molecule C_{60} may be one of many combustion products making up atmospheric ultrafine particles in urban areas. Also, it would be advantageous to consider nanomaterials in classes that represent materials with similar properties, thereby avoiding a case-by-case evaluation of every new nanomaterial. However, to assess the risks of classes of nanomaterials, they must first be classified. Rigorous nomenclature and pertinent criteria that might set one category of material apart from another do not yet exist. SWNTs exist in a variety of different lengths and chiralities; they may be multiwalled or single-walled; fullerenes can be coated, uncoated, functionalized, aggregated, or free; quantum dots can be comprised of a wide array of different base metals. The criteria for determining which of these attributes are important for classification based on potential for exposure and hazard do not yet exist.

Data needs

To incorporate the risks of nanomaterials themselves using a model such as the XL Insurance database, information on material properties analogous to that entered for the reagents must be available for the nanomaterial product. The general parameter requirements of the system we have just considered in our manufacturing assessment are shown again in Figure 13.12, a forward-looking reprise of Figure 13.6, which depicted current system requirements. For each parameter required in the database, some future studies needed to generate data sufficient for understanding these behaviors are listed. We will then discuss

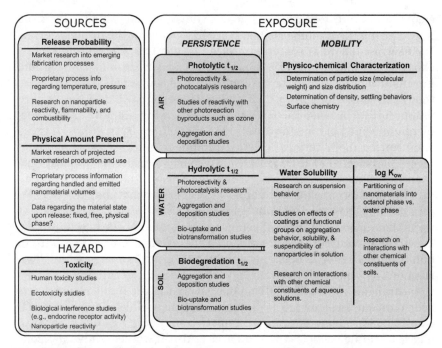

Figure 13.12   Research gaps for which data need to be collected to fit nanoparticle characteristics and behaviors into the source, hazard, mobility, and persistence parameters included in the XL protocol.

the parameters in more detail, using fullerenes as a sample nanomaterial to show what is known thus far as well as to flesh out the details of what cannot yet be predicted. This table is useful in expressing the requirements of the insurance model, and in suggesting how properties of nanomaterials might be chosen to fit into these preset parameters; however, it may not be inclusive of all the relevant parameters needed to best capture nanoparticle-specific behaviors. Potential addition of more appropriate nano-parameters will be discussed later in the chapter.

Exposure

The potential for exposure to nanomaterials begins with the production of these materials. However, one consequence of the nascent quality of the nanomaterials industry is considerable uncertainty regarding the quantity of materials to be produced, the manner in which they will be handled in day-to-day operations, and the size of potential markets.

Release probability. To understand the likelihood of a material being released, processing information must be available that reflects variations in emissions and sources of incident risk arising from variations in day-to-day operations such as extreme temperatures and pressures. These emissions will likely be the most predictable due to both the expense of nanomaterials and to tight manufacturing controls. Release probability due to an incident will also hinge on data regarding the fire and explosion dangers and catalytic reactivity levels of the nanomaterials [31]. In the case of fullerenes, the processes used to fabricate them do involve some extreme temperatures and high pressures, but thus far these process steps are carried out under extreme control with a great focus on collecting the nanomaterials yielded. It has also been suggested that the large specific surface area of some particles, especially easily oxidizable metal nanoparticles, could pose a risk of spontaneous flammability in air [32]. More work needs to be done to offer conclusive results on the degree and causes of ignition potential, but to date carbon-based fullerenes are not expected to be highly flammable, combustible, or likely to spontaneously catalyze chain reactions in air.

It could be argued that incidents such as explosions are hazards unto themselves that should be considered as true hazard endpoints, and indeed from a worker safety standpoint this is one of the primary hazards being investigated. From the perspective of this environmentally focused model, however, the potential for incident is considered instead as a possible input of these materials to the air, water, and soil.

Source estimates.   To project the physical amount present, which will contribute to the magnitude of the hazard posed by a material, some market data must be available to tell us the volume we can expect to be produced. The lion's share of this information is proprietary at the present time, as companies rush to be among the early providers of reliable, industrial volumes of these much anticipated materials. Companies such as Cientifica and Lux Research conduct detailed market research that can provide some idea of the volumes we can expect to see produced in the near term. A recent market study found that more than $30 billion in goods manufactured in 2005 incorporated emerging nanotechnology, and further projected that by 2014 a full 15 percent of total global manufacturing output would incorporate nanotechnology [33]. While these numbers do not directly translate into physical measurements describing the mass of nanomaterials in question, they offer an appreciation for the magnitude of exposure to be expected. As more companies develop methods for large-scale industrial production, further reviews of their patents will also offer quantitative insights. Information about the emissions of nanomaterials in the course of production will be necessary in evaluating this parameter as well, including amounts and states of the materials released.

The quantity of materials produced will also be related to the format in which they are produced. In some cases, "raw" nanomaterials such as C_{60}

may be produced as feedstock to downstream processes that then incorporate these materials in composite materials, fluids, and devices. Incorporation of SWNTs into a plastic composite to improve the wear characteristics of the product will likely alter the availability of the nanomaterial (physically and biologically) while introducing a new step in processing with inherent risk of handling.

Persistence and mobility in air and water.   In our sample insurance model, the persistence of a material in air is measured based on photolytic half-life, or time until absorption of light will result in half of the mass of a substance being altered via cleavage of one or more covalent bonds within the material's molecular entity. Often in practice, the photolytic half-life is not known directly for a substance, but the concept of how quickly a substance will be destabilized by exposure to sunlight will be estimated by order of magnitude. For example, a chemical known to have a half-life of five days in air would be very low risk for persisting and correspond to a risk class of 1, whereas a half-life of 90 or more days corresponds to a risk class of 3. Research on photoreactivity and photocatalysis, as well as studies of reactivity with other photoreaction by-products such as ozone, is being carried out that may deliver values for this parameter. To follow the thread of our fullerene example, most of the research into photoreactivity has focused on the effects of excited nanomaterials on other constituent chemicals in the air or solution rather than seeking the true photolytic half-life. To date, however, there does not seem to be evidence that these carbon-based nanomaterials would be rapidly altered by light absorption. Another factor likely to affect whether a material persists in the air will be aggregation; if, for example, fullerenes tend to form aggregates in the air, they will then potentially exhibit behaviors distinct from those of the individual particles, reacting differently with the environment and potentially even settling out of the media altogether.

With typical non-nanoscale substances the molecular weight is often of great importance for describing its mobility in air. For nanomaterials, this same parameter could potentially be represented instead by a group of descriptors such as particle size, density, and surface chemistry, which would allow determination of which forces dominate transport.

It is important to note that the same parameters affecting persistence and mobility of nanoparticles in air, including photoreactivity, aggregation state, and size will apply to their partitioning from air to aqueous solutions, as well as to their behavior within aqueous solutions.

Aqueous and soil persistence and mobility: Aggregation state and rate.   The molecular weight of many compounds has been related to their mobility, hydrophobicity, and propensity to bioaccumulate. Here again, particle characteristics like size, density and surface chemistry will play a key role. In addition to these important factors, the degree of aggregation (aggregation state) and the rate at which materials tend to aggregate is an important determinant of a material's persistence in the environment. Beyond determining whether the substance remains in a particular medium, the importance of aggregation with regard to reactivity and bioavailability was underscored in the earlier discussion of how the aggregation state may affect ROS generation by nanomaterials. These parameters will be linked with the functionality of the nanomaterial. Aggregation and deposition on surfaces will affect not only the net exposure and persistence of these materials but perhaps their dose as well, as in the case of nanoparticle uptake by inhalation.

Chapter 7 talked in detail about processes governing nanoparticle stability and transport in the environment. We saw here that factors such as pH of the solution and nanomaterial functionalization drive the material's behavior in water. As we saw in the sample risk assessment, the mobility of materials in both water and soil are measured in terms of water solubility and octanol water partitioning coefficient. Here again, studies on the effects of functionalized groups, coatings, and aggregation behavior promise to provide crucial information in predicting the behaviors of nanoparticles in solution. We know that buckyballs are virtually insoluble in water in their pristine form, with their true thermodynamic solubility estimated at -18 mol/liter [40], but that they can be suspended with the addition of functionalized hydroxyl groups as they are in the case of fullerol. We know that the individual molecules have a log K_{ow} value near 12.6 [40], but the question remains as to whether we can really expect to find individual molecules free in the environment.

Aqueous and soil persistence and mobility: Biological interactions. Biodegradation rates determine the persistence of a material in soils within the sample risk assessment database. Most of the research to date has concentrated on establishing the potential for uptake, but whether the materials are metabolized or degraded has yet to be established. For example, one recent study shows the uptake of nanoiron into central nervous system microglia cells, proving that these particles are assimilated into the cell [41]. Some other biological interactions are being studied with a focus on reduction of metal ions, with one body of work establishing actual nanoparticle synthesis within alfalfa plants [42]. While this is not biodegradation, it has interesting implications on our understanding of metabolic pathways in relation to metal nanoparticles. Some of the work presented in Chapter 12 will lead to a future understanding of nanomaterials' pathways in microbial environments. The same information that tells us about nanomaterial mobility in solutions, water solubility, and partitioning behaviors will also help us understand behavior in soils and propensity for bio-accumulation.

Toxicity.   Hazards such as toxicity and ecotoxicity are discussed in Chapters 11 and 12. Despite the attention that early studies have given to this element of hazard, general principles relating the characteristics of a given nanomaterial and the hazards it may pose remain elusive. Toxicity can be measured with respect to multiple endpoints, including cytotoxicity, genotoxicity, toxicity to specific organs, or the onset of acute effects or chronic effects (Chapter 6). Interference with biological processes, such as endocrine system disruption, is another potential category of hazard that should be studied and considered with respect to assessing risk. It may well be that research on an indicator parameter such as nanoparticle reactivity could assist in predicting the hazardous effects it may have on its various surroundings. Very little is known regarding fullerene metabolism or systemic elimination. Several studies have claimed various toxic effects may be induced by C_{60} fullerenes, while others have indicated there may even be therapeutic affects attributable to these materials [34–39]. In addition to the more empirical toxicity studies, priority is being given to understanding the toxokinetics within the human body. The NanoSafe II initiative in Europe includes research on modeling transport within the body to various targets, including adsorption, distribution, metabolism, and excretion [30]. The intermediate step of modeling dose based on a given level of exposure requires similar information on the physical-chemical behaviors of these materials as they are transported and transformed by organisms. In addition to data needs in modeling dose for individual organisms, interactions between organisms must be detailed to arrive at a complete ecotoxicological assessment. By beginning with cases at the base of the food web, as was presented in Chapter 12, an initial assessment is greatly simplified.

Ultimately, an understanding of mechanisms that underpin any nanomaterial toxicity will be required if we aim beyond an empirical level toward predictive modeling. Work on fundamental characteristics and behavior of nanomaterials is beginning to tie into our understanding of toxicity to shed light on some of these mechanisms.

Searching for the true risk-determining characteristics of nanomaterials

The required characteristics presented in Table 13.13 have proven useful in facilitating a discussion of the state of risk assessment knowledge, but this is by no means an exhaustive list of factors that may well be useful in describing the true risk and impacts of nanomaterials. We

TABLE 13.13   Nanomaterial Characteristics, Behaviors, and Effects That May Affect Their Relative Risk

Primary descriptors: Physical characteristics	Secondary descriptors: Exposure	Tertiary descriptors: Hazard
■ Added molecular groups ■ Chemical composition ■ Number of particles ■ Free or bound in matrix ■ Size distribution ■ Shape ■ Solubility ■ Surface area ■ Surface charge ■ Surface coating ■ Surface reactivity ■ Thermal conductivity ■ Electrical conductivity ■ Tensile strength ■ Proportion of total number of atoms at the surface of a structure ■ Molecular weight ■ Boiling point ■ Optical behaviors ■ Magnetic behaviors	■ Adsorption tendency ■ Ability to cross blood-brain barrier ■ Ability to cross placenta ■ Degree of aggregation ■ Ability to travel to deep lung ■ Ability to travel to upper lung ■ Dispersability ■ Interactions with naturally occurring chemical species in aqueous phase ■ Transformation to other compounds ■ Interactions with naturally occurring chemical species in soils ■ Bio-accumulation potential ■ Ability to travel to central nervous system	■ ROS generation ■ Oxidative stress ■ Mitochondrial perturbation ■ Inflammation ■ Protein denaturation, degradation ■ Breakdown in immune tolerance ■ Allergenicity ■ Cellular/subcellular changes ■ Damage to eyes ■ Damage to lung ■ Damage to GI tract ■ Excreted/cleared from body ■ Irritation effect ■ Damage to central nervous system

observed in several cases that the parameter values depended greatly on modifications and speciation of the underlying material. The relevant characteristics of the nanomaterial under consideration must account for these variations. Should we provide the solubility of a coated or uncoated nanoparticle? Should we adhere strictly to the pure nanomaterial or score the more likely to occur aggregate as the base substance in a risk database? We must understand in which life-cycle phases the nanomaterials have the potential to be released to the environment and in which they may be safely bound; their mere existence does not ensure that they will pose a problem to human health or the environment.

The Risk Assessment Unit of the European Commission and the United States EPA have both listed this distinction as key to enabling a practical assessment of nanomaterial risks [43, 44]. Such clarifications have to be made for non-nano materials already. For example, the LED screens in computer monitors and televisions are composed of hazardous heavy metals such as cadmium and selenium, and yet we live and work in very close proximity to the materials in their fixed form every day with no worry of their toxicity. There is a growing consensus concerning the priorities for efforts in evaluating the impacts on health and environment from nanomaterials that includes standardizing materials, developing or improving the ability to measure these materials, and agreement on what is important to measure [30, 44, 45]. An "ideal" list of parameters to be evaluated to enable an assessment of nanomaterial impacts will not necessarily coincide with the list of parameters that can actually be measured physically or feasibly. Table 13.13 summarizes suggestions from six different studies [16, 23, 24, 31, 44, 46].

As we know from some of the earlier chapters in this book, many of these descriptors, especially the physical characteristics, are already known. In Chapter 4, *Methods for Structural and Chemical Characterization of Nanomaterials*, methods to measure many physical-chemical properties were discussed, including solubility, density, melting point, dielectric constants, size, and surface exchange capacity. The second and third categories of descriptors, which we refer to as "exposure factors" and "hazard factors," are similar in that they depend on additional variables beyond the physical-chemical characteristics of the nanoparticles. They are presented as two distinct groups to set apart the different contributions to risk. Exposure factors are primarily fate and transport variables, while the hazard factors are related to toxicity. While the toxicity factors here are presented in a single list, it is likely that for research and modeling purposes a more hierarchical approach will be taken to defining and testing for toxicity. For example, different enzymes may be present based on different levels of oxidative stress, so an assessment of the extent of damage may involve testing for the presence of a specific enzyme [46]. The overall questions for the risk community are how to choose the most appropriate indicators, the most realistically measurable indicators, and how to combine these once chosen.

As has been shown from many angles throughout the book, we are now enjoying the excitement of nanotechnology's youth. There is no question about whether nanomaterials will be manufactured, widely used, and in some way interact with the environment. So much potential exists to harness the novel abilities of these new materials, and it is our hope that this chapter has shed some light on ways that nanotechnology companies, researchers, and regulators can meet this challenge in an environmentally responsible and beneficial manner.

References

1. *International Standard 14040: Environmental Management—Life Cycle Assessment—Principles and Framework*. 1997, International Organisation for Standardisation (ISO).
2. Guinee, J.B., *Handbook on Life Cycle Assessment: Operational Guide to ISO Standards*. 2002: Dordrecht: Kluwer Academic Publishers.
3. Udo de Haes, H.A., et al., *Life-Cycle Impact Assessment: Striving Towards Best Practices*. 2002, Pensacola, FL: SETAC Press.

4. Goedkoop, M., and R. Spriensma, *The Eco-indicator 99: A Damage Oriented Method for Life Cycle Impact Assessment*. 1999, Amersfoort: PRÈ Consultants.
5. Bare, J.C., et al., TRACI: The Tool for the Reduction and Assessment of Other Environmental Impacts. *Journal of Industrial Ecology*, 2003. 6(3–4): pp. 49–78.
6. Saling, P., et al., Eco-Efficiency Analysis by BASF: The Method. *International Journal of Life-Cycle Assessment*, 2002. 7(4): pp. 203–218.
7. Tanzil, D., and B. Beloff, *Assessing Impacts: Indicators and Metrics*, in *Transforming Sustainability Strategy into Action: The Chemical Industry*, B. Beloff, M. Lines, and D. Tanzil, editors. 2005, Hoboken, NJ: John Wiley & Sons, Inc..
8. Tanzil, D., and B. Beloff, Assessing Impacts: Overview on Sustainability Indicators and Metrics. *Environmental Quality Management*, 2006. 15(4): pp. 41–56.
9. Schmidt, I., et al., Managing Sustainability of Products and Processes with the Socio-Eco-Efficiency Analysis by BASF, in *Greener Management International*. 2004. pp. 79–94.
10. Graedel, T.E., *Streamlined Life-Cycle Assessment*. 1998, Upper Saddle River, NJ: Prentice Hall.
11. Andersson, K., et al., The Feasibility of Including Sustainability in LCA for Product Development. *Journal of Cleaner Production*, 1998. 6(3–4): pp. 289–298.
12. Bennett, R., et al., Environmental and Human Health Impacts of Growing Genetically Modified Herbicide-Tolerant Sugar Beet: A Life-Cycle Assessment. *Plant Biotechnology Journal*, 2004. 2(4): pp. 273–278.
13. Lloyd, S.M., and L.B. Lave, Life Cycle Economic and Environmental Implications of Using Nanocomposites in Automobiles. *Environmental Science and Technology*, 2003. 37: pp. 3458–3466.
14. Lloyd, S.M., L.B. Lave, and H.S. Matthews, Life Cycle Benefits of Using Nanotechnology to Stabilize Platinum-Group Metal Particles in Automotive Catalysts. *Environmental Science and Technology*, 2005. 39: pp. 1384–1392.
15. Opinion on the Appropriateness of Existing Methodologies to Assess the Potential Risks Associated with Engineered and Adventitious Products of Nanotechnologies. 2005, *European Commission, Scientific Committee on Emerging and Newly Identified Health Risks*, pp. 41–58.
16. Tran, C.L., et al., *A Scoping Study to Identify Hazard Data Needs for Addressing the Risks Presented by Nanoparticles and Nanotubes*. 2005, Institute of Occupational Medicine, pp. 1–48.
17. Michalsen, A., Risk Assessment and Perception. *Injury Control and Safety Promotion*, 2003. 10(4): pp. 201–204.
18. Biocca, M., Risk Communication and the Precautionary Principle. *Human and Ecological Risk Assessment*, 2005. 1: pp. 261–266.
19. Power, M., and L.S. McCarty, Perspective: Environmental Risk Management Decision-Making in a Societal Context. *Human and Ecological Risk Assessment*, 2006. 12: pp. 18–27.
20. Langsner, H., *Nanotechnology: Non-Traditional Methods for Valuation of Nanotechnology Producers*. 2005, New York City: Innovest Strategic Value Advisors. pp. 13–95.
21. Slovic, P., Perception of Risk. *Science*, 1987. 236(4799): pp. 280–285.
22. Renn, O., *Towards an Integrative Approach*. 2005, Geneva, Switzerland: International Risk Governance Council.
23. Owen, R., and M. Depledge, Editorial: Nanotechnology and the Environment: Risks and Rewards, in *Marine Pollution Bulletin*. 2005, pp. 609–612.
24. Morgan, K., Development of a Preliminary Framework for Informing the Risk Analysis and Risk Management of Nanoparticles. *Risk Analysis*, 2005. 25(6): pp. 1621–1635.
25. Guzman, K.A.D., M.R. Taylor, and J.F. Banfield, Environmental Risks of Nanotechnology: National Nanotechnology Initiative Funding, 2000–2004. *Environmental Science and Technology*, 2006. 40: pp. 1401–1407.
26. Robichaud, C.O., et al., Relative Risk Analysis of Several Manufactured Nanomaterials: An Insurance Industry Context. *Environmental Science and Technology*, 2005. 39(22): pp. 8985–8994.
27. Robichaud, C.O., et al., Relative Risk Analysis of Several Manufactured Nanomaterials: An Insurance Industry Context: Supporting Information. *Environmental Science and Technology*, 2005. 39(22): pp. 8985–8994.

28. DeFriend, K.A., M.R. Wiesner, and A.R. Barron, Alumina and Aluminate Ultrafiltration Membranes Derived from Alumina Nanoparticles. *Journal of Membrane Science*, 2003. 224(1–2): pp. 11–28.
29. Beaver, E., et al., *Implications of Nanomaterials Manufacture & Use*. 2005, Bridges to Sustainability.
30. *Characterising the Potential Risks Posed by Engineered Nanoparticles: A First UK Government Research Report*. 2005, Nanosafe II.
31. Howard, J., *Approaches to Safe Nanotechnology*, N.I.f.O.S.a. Health, Editor. 2005, Centers for Disease Control and Prevention.
32. *Explosion Hazards Associated with Nanopowders*. 2005, United Kingdom Health and Safety Executive.
33. Holman, M.W., et al., *The Nanotech Report, 4th Edition*. 2006, Lux Research.
34. Kasermann, F., and C. Kempf, Buckminsterfullerene and Photodynamic Inactivation of Viruses. *Reviews in Medical Virology*, 1998. 8(3): pp. 143–151.
35. Nakamura, E., et al., Biological Activity of Water-Soluble Fullerenes. Structural Dependence of DNA Cleavage, Cytotoxicity, and Enzyme Inhibitory Activities Including HIV-Protease Inhibition. *Bulletin of the Chemical Society of Japan*, 1996. 69(8): pp. 2143–2151.
36. Tokuyama, H., S. Yamago, and E. Nakamura, Photoinduced Biochemical Activity of Fullerene Carboxylic Acid. *Journal of the American Chemical Society*, 1993. 115: pp. 7918–7919.
37. Tabata, Y., Y. Murakami, and Y. Ikada, Antitumor Effect of Poly(Ethylene Glycol) Modified Fullerene. *Fullerene Science and Technology*, 1997. 5(5): pp. 989–1007.
38. Tabata, Y., Y. Murakami, and Y. Ikada, Photodynamic Effect of Polyethylene Glycol-Modified Fullerene on Tumor. *Japanese Journal of Cancer Research*, 1997. 88: pp. 1108–1116.
39. Tsao, N., et al., Inhibition of Escherichia Coli-Induced Meningitis by Carboxyfullerence. *Antimicrob Agents Chemother*, 1999. 43: pp. 2273–2277.
40. Abraham, M.H., C.E. Green, and J.W.E. Acree, Correlation and Prediction of the Solubility of Buckminsterfullerene in Organic Solvents; Estimation of Some Physicochemical Properties. *The Royal Society of Chemistry*, 2000(2): pp. 281–286.
41. Wiesner, M.R., et al., Assessing the Risks of Manufactured Nanomaterials. *Environmental Science and Technology*, 2006. 40(14): pp. 4336–4345.
42. Gardea-Torresday, J.L., et al., Formation and Growth of Au Nanoparticles Inside Live Alfalfa Plants. *Nano Letters*, 2002. 2(4): pp. 397–401.
43. Nanotechnologies: A Preliminary Risk Analysis on the Basis of a Workshop March 2004, in *Nanotechnologies: A Preliminary Risk Analysis*. 2004. Brussels: European Commission: Community Health and Consumer Protection Directorate General of the European Commission.
44. *Nanotechnology White Paper*, J. Morris and J. Willis, editors. 2005, Washington, DC: United States Environmental Protection Agency.
45. Roco, M.C., and B. Karn, Environmentally Responsible Development of Nanotechnology. *Environmental Science and Technology, A Pages*, 2005. 39(5): pp. 106A–112A.
46. Nel, A., et al., Toxic Potential of Materials at the Nanolevel. *Science*, 2006. 311: pp. 622–627.

Index